Toxicity Characterization, Detection and Remediation of Contaminants in Soils and Groundwater

Toxicity Characterization, Detection and Remediation of Contaminants in Soils and Groundwater

Editors

Junhao Qin
Peidong Su
Feng Zhu
Lin Ding

MDPI

Basel • Beijing • Wuhan • Barcelona • Belgrade • Novi Sad • Cluj • Manchester

Editors

Junhao Qin
Natural Resources
and Environment
South China Agricultural
University
Guangzhou
China

Peidong Su
Chemical and Environmental
Engineering
China University of Mining
and Technology-Beijing
Beijing
China

Feng Zhu
School of Metallurgy
and Environment
Central South University
Changsha
China

Lin Ding
College of Environmental and
Chemical Engineering
Nanchang Hangkong University
Nanchang
China

Editorial Office
MDPI
St. Alban-Anlage 66
4052 Basel, Switzerland

This is a reprint of articles from the Special Issue published online in the open access journal *Toxics* (ISSN 2305-6304) (available at: www.mdpi.com/journal/toxics/special_issues/VNN0TS7S9Y).

For citation purposes, cite each article independently as indicated on the article page online and as indicated below:

Lastname, A.A.; Lastname, B.B. Article Title. *Journal Name* **Year**, *Volume Number*, Page Range.

ISBN 978-3-7258-1398-8 (Hbk)
ISBN 978-3-7258-1397-1 (PDF)
doi.org/10.3390/books978-3-7258-1397-1

Contents

About the Editors

Junhao Qin

Junhao Qin works at the Department of Ecology, College of Natural Resources and Environment, South China Agricultural University, Tianhe District, Guangzhou.

His interests are as follows: the effects of the atmospheric oxidation deposition of hydrogen peroxide on the biogeochemical cycle of environmental pollutants (such as arsenic, nitrogen, herbicides, antibiotics, and microplastics). The remediation technology of heavy metal pollution uses agricultural biodiversity utilization.

Peidong Su

Peidong Su graduated from China University of Mining and Technology (Beijing) and earned his bachelor degree in 2013, and then he received his master's degree at the same university in 2016. He then finished his Ph.D in Jackson State University in the United States in 2019. He won the Overseas High-Caliber Personnel (Level C) and received key support from the China Postdoctoral Science Foundation (China Postdoctoral Dispatch Program—Chinese University of Hong Kong) (120 people nationwide). He has published more than 45 SCI papers, including 18 SCI papers completed and published as the first/corresponding author in the past three years, and 1 of them was selected as ACS Editors' paper of choice. He served as an Early-Career Board Member of Reviews of Environmental Contamination and an Toxicology (IF=7.9) and Eco-Environment and Health, and he also served as the Guest Editor of *Toxics* (IF=4.6) and Editor of *Modern Chemical Research* (ISSN 1672-8114). He has also served as a reviewer of the *Journal of Hazardous Materials, Journal of Cleaner Production, Environmental International*, and other mainstream environmental journals.

Feng Zhu

Zhu Feng is an associate professor and doctoral supervisor at the School of Metallurgy and Environment, Central South University. He has long been engaged in research on the large-scale consumption of bulk solid waste and the ecological restoration of mining and metallurgical sites. He has presided over national key research and development projects, two National Natural Science Foundation projects, the Hunan Provincial Outstanding Youth Fund, etc. He has also taken on the role of a first/corresponding author. He has published more than 30 SCI-indexed papers; 6 ESI 1% highly cited papers; and 2 popular papers in the *Chemical Engineering Journal*, the *Journal Of Hazardous Materials*, and other authoritative ecological and environmental journals. He has also overseen 10 authorized national invention patents, and he has published one of the 100 most influential domestic academic papers in China, winning the Hunan Province Outstanding Doctoral Thesis. He serves as a member of the Youth Working Committee of the Chinese Soil Society, a member of the Metallurgical Slag Academic Committee of the Chinese Ceramic Society, and a youth editorial board member of Transactions of Nonferrous Metals Society of China and Carbon Research. Funded by national science and technology projects, the phenomenon of soil occurrence in red mud dumps was confirmed for the first time, and "physical–chemical–biological combination" ecological restoration technology for red mud dumps was innovatively developed; furthermore, a calcium/iron/silicon-based fitness control system for degraded soil was developed. Materials, creatively proposed as an ecological restoration technology, were used for non-ferrous smelting sites based on microbial process control. He has also carried out ecological restoration project demonstrations of more than 100,000 m2 in lead, zinc, antimony, aluminum, and other smelting sites, promoting the ecological environmental protection and sustainable development of the non-ferrous metal industry.

Lin Ding

Ding Lin is a female PhD holder and a master's graduate advisor. She is an expert in ecological and environmental damage assessment and evaluation in Jiangxi Province. Her research focuses on the design, preparation, and application of environmental functional materials for the control and resource utilization of heavy metal pollution. Currently, she leads seven national and provincial natural science foundation projects, as well as science and technology projects funded by the provincial department of education, involving both vertical and horizontal topics. She has published over 30 SCI papers in internationally renowned journals, such as *Applied Catalysis B: Environmental,* the *Chemical Engineering* journal, and *ACS Sustainable Chemistry & Engineering.* Additionally, she has co-authored an English academic monograph published by Elsevier and holds eight national invention patents.

toxics

Article

Spatial Differentiation Characteristics and Evaluation of Cu and Cd in Paddy Soil around a Copper Smelter

Yuan Ding [1,2,*], Li Xi [1,2], Yujing Wu [1,2], Yihong Chen [1,2], Xiaoping Guo [1,2], Hong Shi [1,3,*] and Shuo Cai [1,3]

1 National-Local Joint Engineering Research Center of Heavy Metals Pollutants Control and Resource Utilization, Nanchang Hangkong University, Nanchang 330063, China
2 College of Environment and Chemical Engineering, Nanchang Hangkong University, Nanchang 330063, China
3 Jiangxi Key Laboratory of Agricultural Efficient Water-Saving and Non-Point Source Pollution Preventing, Jiangxi Central Station of Irrigation Experiment, Nanchang 330063, China
* Correspondence: 39011@nchu.edu.cn (Y.D.); 8659979@163.com (H.S.)

Abstract: To accurately evaluate the distribution and bioavailability of potentially toxic elements (PTEs) such as Cu and Cd in farmlands near a copper smelter, we determined the total concentrations (Cu-T and Cd-T), various speciation concentrations of Cu and Cd and physicochemical properties of 18 paddy soil (or colloid) samples in Guixi town, Jiangxi province, China. The results showed that the concentrations of Cu-T and Cd-T in the soil around the smelter far exceeded the standard limits. Specifically, Cu ranged from 97.47 to 1294.63 mg·kg^{-1}, with a coefficient of variation (CV) of 0.95; Cd ranged from 0.14 to 9.06 mg·kg^{-1}, and the CV was 1.68. Furthermore, the pollution of PTEs continued to accumulate, posing a significant risk to the environment and human health. The findings from the analysis of soil and colloid indicated that the distribution characteristics of Cu and Cd speciations did not align with the total concentrations. The highest pollution points were found to be shifted to the residual fraction of Cu, organic fraction, and crystalline iron oxide fraction of Cd in soil. The dominant fraction of Cu in colloid was the amorphous iron oxide fraction, whereas Cd was the crystalline iron oxide fraction. The assessment of Cu and Cd migration (M_R) revealed that Cd posed a greater ecological risk. Further examination of the properties of iron oxides in soil and colloid revealed that they played a crucial role in the migration and transformation of soil PTEs.

Keywords: copper smelter; potentially toxic elements (PTEs); migration rate (M_R); speciation; soil colloid; iron oxides

check for updates

Citation: Ding, Y.; Xi, L.; Wu, Y.; Chen, Y.; Guo, X.; Shi, H.; Cai, S. Spatial Differentiation Characteristics and Evaluation of Cu and Cd in Paddy Soil around a Copper Smelter. *Toxics* **2023**, *11*, 647. https://doi.org/10.3390/toxics11080647

Academic Editor: Rafael Clemente

Received: 20 June 2023
Revised: 21 July 2023
Accepted: 22 July 2023
Published: 26 July 2023

1. Introduction

Copper smelting is an industry notorious for its heavy pollution, characterized by the emission of atmospheric particulates and other waste materials containing heavy metals such as Cu, Cd, Pb, and Zn [1]. Research findings indicated that Cu and Cd were two of the primary pollutants in the soil surrounding the copper smelter [2]. These two elements are classified as potential toxic elements (PTEs), known for their highly toxic and non-degradable properties [3]. Moreover, they can be transferred to the human body through the food chain, and their toxicity can result in bioaccumulation or biomagnification, which can lead to many diseases, and are poisonous, mutagenic, and carcinogenic to humans [4,5].

The assessment of contaminated soil is typically based on total concentrations of potentially toxic elements (PTEs) because of the ease of monitoring and lower cost compared to other methods [6]. In response to the multi-element co-contamination issues surrounding copper enterprises, evaluation indexes are often used to comprehensively assess the pollution situation of multiple elements. Evaluation indexes, such as EF (enrichment factor) [7], IPI (integrated pollution index) [8], and I_{geo} (geo-accumulation index) [9], provide effective methods for evaluating the pollution degree of PTEs in soil. By comparing the observed data of PTEs with the background values, I_{geo} produces a concise score to reflect the pollution condition [10]. This approach has been widely adopted, as it enhances our comprehension of the extent of associated PTEs' pollution [11].

While most evaluation index methods based on total concentrations can indicate soil pollution to a certain extent, they don't account for the bioavailability and potential risk of PTEs [12]. The bioavailability, migration, and transformation of PTEs in soil are mainly affected by the physicochemical properties of soil, such as pH, redox potential, iron oxides, and organic matter, except for the total concentrations of PTEs [10]. Research by Liu showed that iron oxides in soil significantly affected the speciation of PTEs [13]. PTEs were transformed into more stable speciations during the process of iron species transformation and iron redox mediated by iron-reducing bacteria [14]. Chemical extraction methods, such as the Tessier extraction method, have been widely accepted by researchers to express the distribution characteristics and bioavailability of Cu and Cd in soil [15]. To explore the effects of crystalline and amorphous iron oxides on the activities of Cu and Cd, the modified Tessier extraction method was adopted in this paper. The speciation of Cu and Cd in soil was divided into the exchangeable and the carbonate fraction (F1), readily reducible iron and manganese oxide fraction (F2), organic fraction (F3), crystalline iron oxide fraction (F4), and residual fraction (F5) [16], with decreasing bioavailability of Cu and Cd in soil. To accurately depict the intricate dynamics of heavy metal redistribution and fixation in soil, Han and Banin [17] devised a novel evaluation method, the reduced partition index (I_R). This approach was based on a sequential selective dissolution procedure, enabling the assessment of heavy metal fixation in soil. I_R represented the weighted sum of each heavy metal fraction, with the weight of each fraction gradually increasing as its activity decreased. In the pursuit of tracing the heavy metal fractions loosely bound to the soil, Burachevskaya [18] employed Miller's method as a sequential selective extraction technique combined with I_R, and effectively gauged the levels of Cu, Pb, and Zn pollution surrounding a coal-fired power plant. It provided a comprehensive understanding of the redistribution and fixation characteristics of heavy metals in soil. Conversely, to conveniently compare the activity and migration degree of Cu and Cd in the soil at different sites and highlight the influence of the exchangeable and carbonate fractions, readily reducible iron and manganese oxide fractions, the modified weight calculation method based on I_R was used in this paper to quantitatively calculate the migration of Cu and Cd in soil—migration rate M_R. However, the weights of each fraction in this method decrease gradually with the decrease in activity.

Guixi town in Jiangxi province, China, is home to the most advanced and largest flash copper smelter in Asia. Early research results showed that the total concentrations of Cu and Cd in the paddy soil under the comprehensive slag field dam of the smelter ranged from 102.31 to 415.54 $mg{\cdot}kg^{-1}$ and from 0.33 to 6.87 $mg{\cdot}kg^{-1}$, respectively (MEEP, 2018b Cu < 50 $mg{\cdot}kg^{-1}$, Cd < 0.3 $mg{\cdot}kg^{-1}$) [19,20]. Moreover, survey results on paddy soil in the village of Shuiquan and Zhushan showed that concentrations of Cu and Cd were from 6.17 to 11.7 and from 3.02 to 3.41 times as much as the background, respectively. The tested villages were located in the southwest direction of the copper smelters, which was the annual downwind direction [21]. This copper smelter has both non-point and point source pollution characteristics, resulting in soil quality that is worse than the farmland standard. While non-point source pollution caused by the slag field dam has been effectively addressed [22], the long-term and positioned emission of soot from the smelter would continue to increase the concentrations of Cu and Cd in the soil as point source pollutions [23]. Therefore, further investigation and analysis of PTEs' pollution around the smelter are necessary.

As mentioned above, the bioavailability and migration of Cu and Cd in soil not only depends on the total concentrations of PTEs but are also affected by the physicochemical properties of the soil. This paper focuses on the paddy soils surrounding a typical copper smelter in southern China and aims to analyze: (1) the current situation of soil Cu and Cd pollution around the smelter on a small scale; (2) the distribution characteristics of various Cu and Cd species in southern China polluted by non-point sources and atmospheric point sources; and (3) the spatial heterogeneity and bioavailable evaluation of Cu and Cd distribution in this region.

2. Materials and Methods

2.1. Study Site

The copper smelter was situated in the northeastern region of Jiangxi Province, located within the intermediary zone between the Wuyi Mountain and the Poyang Lake Plain. The climatic conditions of the area are characterized by a subtropical monsoon climate, which is marked by high temperatures, ample sunshine, and copious rainfall. The principal wind direction in the region is easterly, with northeasterly winds being the secondary dominant direction throughout the year.

The copper smelter under study has adopted the world's most advanced oxygen-enriched flash smelting technology and double contact double adsorption acid production technology. The primary pollutants discharged from the plant are PTEs, such as Cu and Cd.

2.2. Soil Sampling

In November 2020, we collected soil samples from 18 sites in the farmland surrounding the copper smelter, from the surface layer (0–20 cm). The selection of sampling sites adhered to the fan-type distribution principle of the atmospheric point source, as prescribed by "the Technical Specification for Soil Environmental Monitoring (HJT 166–2004)" published by the Ministry of Ecology and Environment of the People's Republic of China (MEEP 2004) [24]. To account for the pollution characteristics in the smelter area, we intensified the sampling frequency in the non-point source region, as shown in Figure 1.

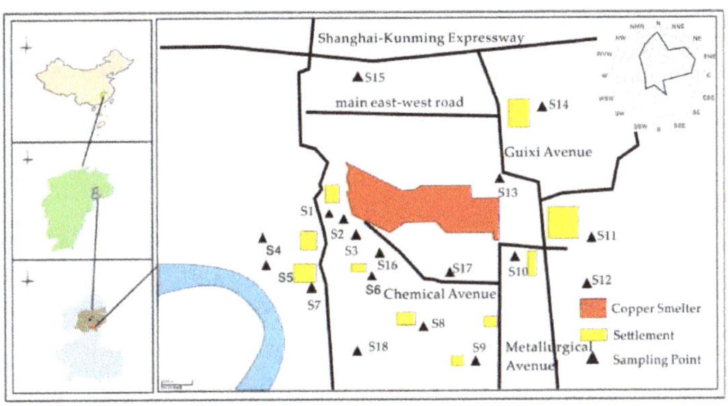

Figure 1. Map of sampling points.

Three sampling points were selected from each site, with a minimum distance of 1 m between them. About 500–750 g of samples were collected from each point and mixed to get a composite sample. Before sample collection, extraneous materials such as small gravel, waste plastics, and dead branches were removed from the sampling area. The samples were air-dried in the laboratory, ground, and passed through a nylon sieve with a 2 mm aperture before being sealed for later use.

2.3. Sample Analysis

The physicochemical properties of the soil were analyzed according to the method of Bao [25]. The soil pH was measured using a pH meter (PHS-3C, Shanghai) at a water-to-soil ratio of 2.5:1 (V/m). Soil organic matter (SOM) was determined using the $K_2Cr_2O_7$ oxidation method. Available phosphorus (AP) was extracted by 0.5 mol·L^{-1} $NaHCO_3$ (soil-to-water ratio of 20:1) and measured using the molybdenum–antimony colorimetric method. Free iron (Fe_d) in the soil, which includes amorphous ferric oxide (Fe_o) and crystalline ferric oxide (Fe_c), was extracted according to the method of Zhang [26]. The total concentrations of Cu and Cd (Cu-T and Cd-T) in the soil samples were determined using ICP-MS after digestion with $HCl–HNO_3–HClO_4$.

To investigate the effects of soil iron oxides on the migration and transformation of PTEs, different fractions of Cu and Cd in the soil samples were sequentially extracted using a modified Tessier method to determine the distribution of Cu and Cd speciations around the smelter (Table 1) [16]. This method allowed for the identification of Cu and Cd speciation in the exchangeable and carbonate fraction (F1), readily reducible iron and manganese oxide fraction (F2), organic fraction (F3), crystalline iron oxide fraction (F4), and residual fraction (F5).

Table 1. Outline of the sequential extraction procedure for Cu and Cd [16].

Fraction	Extractants	Extraction Procedures
exchangeable and carbonate fraction (F1)	16 mL NaAc–HAc (1 M)	25 °C, 6 h shaking
readily reducible iron and manganese oxide fraction (F2)	20 mL $NH_2OH–HCl$ (0.04 M)	96 °C, 6 h shaking in the dark
organic fraction (F3)	3 mL HNO_3 (0.02 M) + 5 mL H_2O_2 (30%), pH = 2; 5 mL 3.2 M CH_3COONH_4	5 h in a water basin at 85 °C; cool, add CH_3COONH_4 and shake for 30 min, 25 °C
crystalline iron oxide fraction (F4)	10 mL DCB + $Na_2S_2O_4$	15 min in a water basin at 80 °C, repeat twice
residual fraction (F5)	2 mL HCl/6 mL HNO_3/2 mL HF	digestion

Liu [27] mentioned the preparation of soil colloid, where the heavy metal fractions present in the soil colloid can be classified into three distinct species, namely the amorphous iron oxide fraction, the crystalline iron oxide fraction, and the residual fraction [16].

2.4. *Methods of Pollution Assessment*

2.4.1. Geo-Accumulation Index

The index of geo-accumulation (I_{geo}) is used to assess soil contamination. I_{geo} is calculated by the following equation [28]:

$$I_{geo} = \log_2[C_i/1.5B_n] \quad (1)$$

where C_i is the concentration of the examined metal in the soil or sediment, B_n is the geochemical background value of the metal in the equation in the shale, and factor 1.5 accounts for the possible variations in the background values.

The I_{geo} scale consists of seven grades, as shown in Table 2.

Table 2. Classification of the geo-accumulation index [28].

Class	Index	Pollution Level
0	$I_{geo} \leq 0$	Unpolluted
1	$0 < I_{geo} < 1$	Slightly polluted
2	$1 < I_{geo} < 2$	Moderately polluted
3	$2 < I_{geo} < 3$	Moderately to heavily polluted
4	$3 < I_{geo} < 4$	Heavily polluted
5	$4 < I_{geo} < 5$	Heavily to extremely polluted
6	$I_{geo} > 5$	Extremely polluted

2.4.2. Migration Rate (M_R)

We used M_R to quantify the distribution and migration characteristics of Cu and Cd, and M_R is calculated by the following equation [18]:

$$M_R = \frac{\sum_{i=1}^{k}(F_i \times (k-i+1)^n)}{k^n} \quad (2)$$

where, F is the percentage content of each fraction of PTEs; i is the extraction step number; k is the number of total extraction steps and equals 5 in this paper; n is set to 2 in this paper.

2.5. Data Analysis

The statistical analysis, including normal distribution testing and correlation analysis, was performed using IBM SPSS Statistics 26.0 software. The figures were generated by Origin 2021. The spatial distribution characteristics maps of PTEs' pollution were generated by ArcMap 10.5.

3. Results

3.1. Characteristics of the Soil around a Copper Smelter

The physicochemical properties of the soil around the copper smelter are displayed in Table 3. The soil pH ranged from 4.04 to 5.16, with a coefficient of variation (CV) of 0.06, which was typical of soil acidity in the middle and lower branches of the Yangtze River, China. The concentrations of SOM in the study area showed an approximately normal distribution, with 66.67% of samples meeting the third-level standard, while the others were lower than the middle level, according to the classification standard of soil nutrients in the Second National Soil Survey of China [29]. The concentrations of AP showed an approximately log-normal distribution, with 83% of samples meeting level 3 or greater than the standard. Furthermore, the total iron (Fe-T) concentrations in the soil met level 2 or greater than the standard. Therefore, there was no nutrient stress in this region, and the impact of nutrient stress on soil PTEs would not be considered below.

Table 3. Soil characteristics and PTE concentrations around the copper smelter.

Items	Maximum	Minimum	Mean	SD	Skewness	Kurt.	CV	K–S Test
pH	5.16	4.04	4.78	0.29	−1.53	2.24	0.06	Non-normal
SOM (%)	4.43	1.09	2.27	0.67	1.84	6.48	0.30	Approximately normal
AP (mg·kg^{-1})	123.29	0.75	31.17	30.71	2.18	4.66	0.99	Approximately log-normal
Fe-T (g·kg^{-1})	64.34	10.37	23.74	15.37	1.82	2.33	0.65	Non-normal
Fe_d (g·kg^{-1})	31.47	4.87	9.95	7.01	2.08	4.43	0.70	Non-normal
Fe_o (g·kg^{-1})	22.64	0.79	4.73	5.43	2.52	6.79	1.15	Non-normal
Cu (mg·kg^{-1})	1294.63	97.47	317.70	301.13	2.39	6.29	0.95	Approximately log-normal
Cd (mg·kg^{-1})	9.06	0.14	1.14	2.04	4.09	17.09	1.79	Approximately log-normal

Note: CV represents the coefficient of variation; SD represents standard deviation; K–S test represents the Kolmogorov–Smirnov test.

The average concentrations of Cu and Cd in the study area were higher than the local background levels by 15.65 and 10.56 times, respectively [21], and by 6.35 and 3.8 times the soil heavy metal pollution risk screening value (Table S1) [20], suggesting that Cu and Cd were severely polluting the study area. The distribution characteristics of Cu and Cd demonstrated that the CVs were 0.95 and 1.68, respectively, indicating that the heterogeneity of soil heavy metals in the study area was significant, which was consistent with industrial areas such as smelters.

3.2. Distribution and Evaluation of the Total PTEs

The concentrations of Cu and Cd in the soil in the study area ranged from 97.47–1294.63 mg·kg^{-1} and 0.14–9.06 mg·kg^{-1}, respectively. Wang [30] previously investigated the region and reported that the concentrations of Cu and Cd in the vicinity of the smelter were 38.14–586.21 mg·kg^{-1} and 0.48–3.79 mg·kg^{-1}, respectively. These results indicate that the emission of the smelter has led to a continued accumulation of PTEs' pollution in the area, necessitating further investigation.

The distributions of PTEs in the study area were assessed using Cu-T and Cd-T (Figure 2a,b). The maximum concentrations of both Cu and Cd were found to occur in the downwind region (west) of the local smelter, with the highest pollution points being S16 and S6, respectively. These findings were consistent with previous investigations into the spatial distributions of Cu and Cd in the soil of the region [31] and confirmed that the heterogeneity of Cu and Cd distribution was caused by the complex pollution sources and the different migration characteristics of Cu and Cd [32].

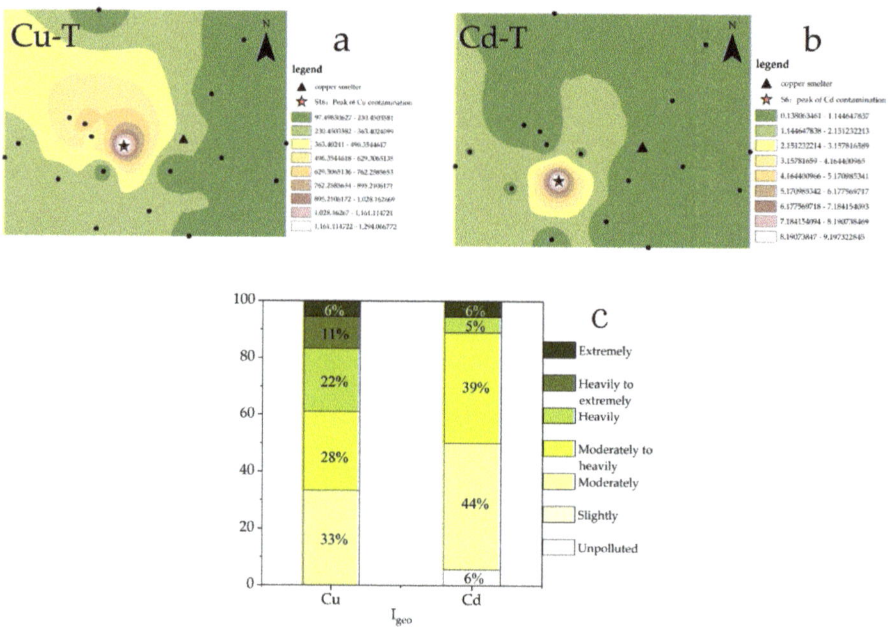

Figure 2. (**a**). Spatial distribution of Cu-T; (**b**). Spatial distribution of Cd-T; (**c**). Evaluation of Cu and Cd.

To visually evaluate the accumulation of Cu and Cd in the study area, I_{geo} was utilized (Figure 2c). The results indicated that all sampling points in the study area were above moderately polluted for Cu, with approximately 39% of points classified as heavily polluted or worse. In contrast, while some sampling points for Cd were unpolluted, most (89%) were classified as moderately to heavily polluted. In summary, the pollution of Cu in the study area was found to be more severe than Cd.

3.3. Spatial Distribution of PTEs' Speciations

We have refined the Tessier method (2.3 Materials and Method) to comprehensively consider the effects of F2 and F4 on the bioavailability of Cu and Cd and have analyzed the spatial distribution characteristics of various PTE speciations (Figures 3 and 4).

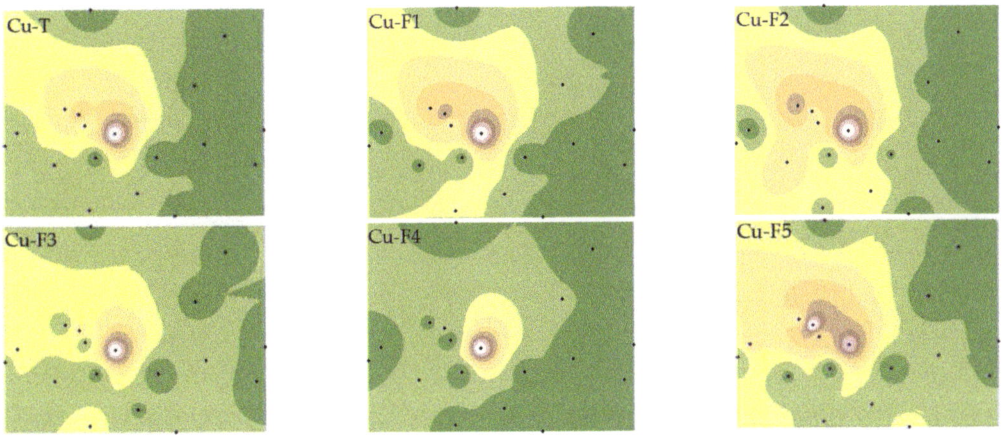

Figure 3. Spatial distributions of Cu total and fraction concentrations. Note: Exchangeable and carbonate fractions (F1), readily reducible iron and manganese oxide fractions (F2), organic fractions (F3), crystalline iron oxide fractions (F4), residual fractions (F5).

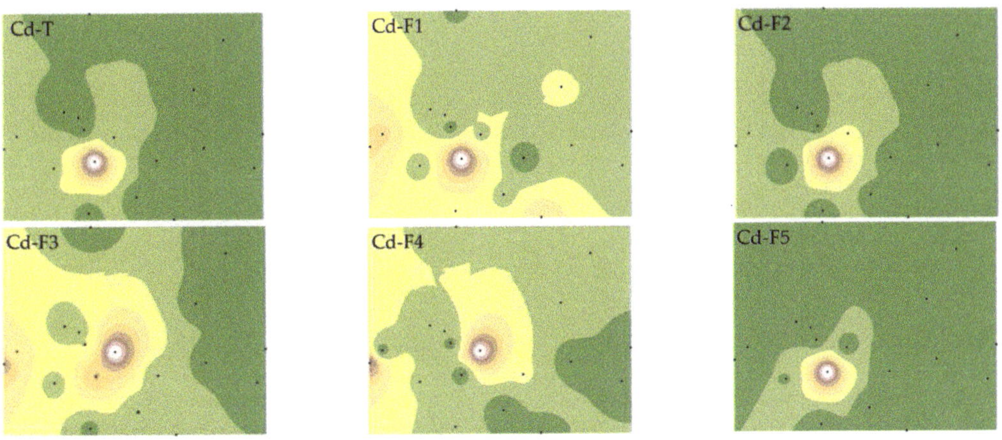

Figure 4. Spatial distributions of Cd total and fraction concentrations. Note: Exchangeable and carbonate fractions (F1), readily reducible iron and manganese oxide fractions (F2), organic fractions (F3), crystalline iron oxide fractions (F4), residual fractions (F5).

As depicted in Figure 3, the western region of the smelter was most severely affected by Cu pollution, with pollution severity increasing closer to the smelter. The distribution characteristics of Cu speciations were distinct from Cu-T, with the highest pollution point of Cu-F5 varying, while the highest pollution point and highest secondary pollution point of Cu-F1, Cu-F2, and Cu-F5 coexisted. This phenomenon may be attributed to the differences in iron oxides and pH in the study area, which significantly impact the spatial distribution characteristics of Cu [33,34].

The spatial distributions of Cd-T and various fractions differed from Cu. Except for Cd-F2 and Cd-F5, which showed high similarity with Cd-T, the others were quite distinct, particularly the significant variation of Cd-F3 and Cd-F4 [35]. This suggests that Cu and Cd in soil may have different sources, or distinct migration characteristics due to the reactions between PTEs and other materials (such as iron oxides, SOM, AP, etc.) in the soil [36].

Further statistical analysis of soil Cu and Cd speciations (Table S2) revealed that the CVs of Cu and Cd speciations ranged from 0.95 to 1.44 and from 0.86 to 2.65, respectively, indicating that the spatial heterogeneity of Cd in the study area was higher and greatly influenced by human factors. Cu was primarily present in non-residual fractions, with the content of Cu-F4 being relatively low (2.63–14.30%) (Table S1). The primary fractions of Cd were F1 (18–74%) and F5 (11–73%) (Table S1), confirming that the non-active components of the two elements existed differently, and that the total concentration could not fully express the bioavailability of Cu and Cd.

3.4. Distribution of PTEs in Soil Colloid

A conspicuous difference was observed between the distributions of Cu and Cd speciations in the investigated area, which was influenced by soil physicochemical properties such as pH, AP, SOM, and iron oxides, among others [37,38]. As is well-known, soil colloid boasts a vast specific surface area and carries negative charges, thereby playing an important role in the distribution of PTE speciations and determining the concentration ratio and composition of amorphous and crystalline iron oxide fractions of PTEs in soil [39]. Consequently, this paper focused on the distribution of PTEs speciations in soil colloid, and the results are presented in Figure 5.

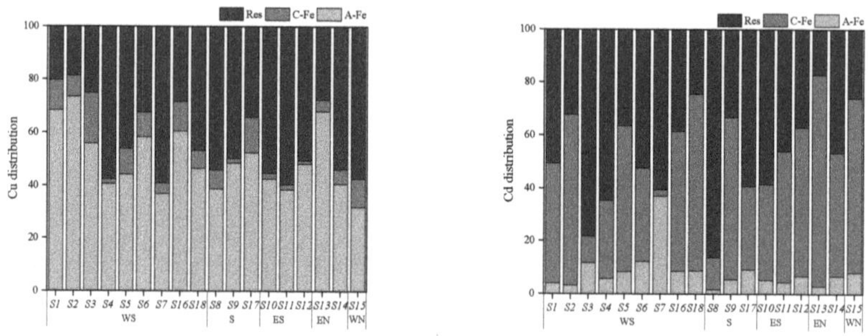

Figure 5. Distribution of each form of Cu and Cd in soil colloids. Note: A-Fe: amorphous iron oxide bound fraction; C-Fe: crystalline iron oxide bound fraction; Res: residual fraction.

The concentrations of Cu and Cd in soil colloid ranged from 185.45 to 1727.15 mg·kg^{-1} and from 0.25 to 9.95 mg·kg^{-1}, respectively, which were significantly higher than those in soil, underscoring the fact that soil colloid served as the main carrier of PTEs in soil. Cu in colloid was dominated by the amorphous iron oxide fraction (32–73%) and residual fraction (20–59%), whereas the main occurrence fractions of Cd were the residual fraction (17–86%) and crystalline iron oxide fraction (10–80%). This could be one of the principal reasons for the spatial heterogeneity of Cu and Cd distribution.

4. Discussion

4.1. Factors Affecting the Spatial Heterogeneity of Soil PTEs

The present study utilized correlation analysis to investigate the impact of soil physicochemical properties on the distribution of Cu and Cd speciations in the farmlands surrounding the smelter. The results, as depicted in Figure 6, demonstrate that soil physicochemical properties, such as SOM, AP, and pH, did not exhibit any significant correlation with Cu and Cd. This finding was consistent with the characteristics of the area with continuous external pollution [40]. However, iron oxides (Fe-T, Fe_d, and Fe_o) showed a positive correlation with the non-residual Cu, Cd-F3, and Cd-F4, indicating that soil iron oxides were the main factors regulating the migration and transformation of PTEs [41]. Previous studies have established that iron oxides were the primary carriers of PTEs, which influenced the bioavailability of PTEs [42]. Iron oxides mainly retain PTEs through surface complexation, surface precipitation, and structural incorporation [43]. In combination with Figure 5, it was observed that Cu had a high affinity with Feo, and unlike Cu, Cd has a high affinity with Fec. This finding was a testament to the fact that the species of iron oxide significantly affected the binding, distribution, and bioavailability of PTEs [44].

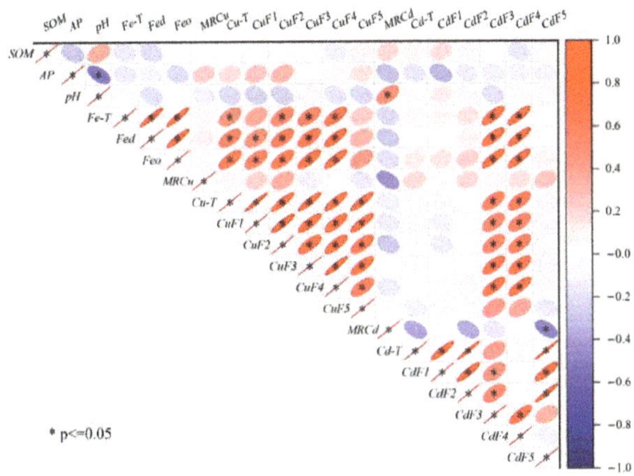

Figure 6. Correlation among Cu-T, Cd-T, fraction concentrations, and soil basic properties.

Cu-T exhibited a positive correlation with the speciations, with Cu-F1 displaying the highest correlation, which was consistent with the spatial distribution of Cu-T. On the other hand, a significant correlation was observed between Cd-T and Cd-F5, whereas the correlation between Cd-T and Cd-F3 (or Cd-F5) was not significant, similar to their spatial distribution. This observation further confirms that total concentrations cannot entirely describe the migration and transformation of PTEs. The properties of PTEs were also one of the main factors affecting their distribution [45]. Cu and Cd were significantly correlated only with F3 and F4, and also with iron oxides, which proved that the interaction between iron oxides and SOM had an important influence on the regulation of the migration and transformation of PTEs [46].

4.2. Assessment of PTE Pollution around the Smelter

This study assessed the pollution characteristics of the study area, with a focus on two main pollutants, namely Cu and Cd, and their respective spatial distribution characteristics. Based on the analysis of total concentrations, it was observed that the pollution of Cu around the smelter was more severe than Cd (Figure 2c). However, research indicates that the ecological risk and toxicity of Cd are higher than Cu, owing to its high bioavailability and mobility [47]. To assess the risk of PTE pollution, conventional evaluation methods typically involve weighting the toxicity coefficient of elements based on their total concentrations, for example, the Ecological Risk Index (RI) [48].

Based on speciation analysis, this paper employed a weighted summation method to calculate the migration (M_R) of PTEs, to emphasize the role of F1, F2, F3, and other active fractions in the migration. The calculation results are presented in Figure 7. The higher the M_R value, the higher the potential risk. The M_R value served to reflect the availability of PTEs to a certain extent. For instance, if PTEs in soil exist solely in the fraction of F1, the M_R value is maximum, which is 1. Conversely, if PTEs exist only in the fraction of F5, the M_R value is minimum, which is 0.04. The average single-element M_R value of PTEs around the smelter was higher for Cd (0.59) than for Cu (0.49), indicating that the potential risk of Cd was greater. This approach enabled an accurate and digital assessment of the ecological risk under the continuous external pollution to the surrounding area, while reflecting the geochemical behavior of PTEs in the soil.

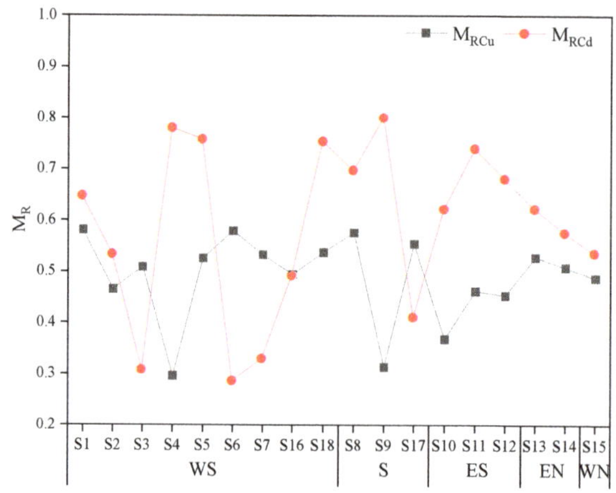

Figure 7. M_R of Cu and Cd in the study area.

5. Conclusions

(1) The concentrations of Cu and Cd in the soil surrounding the copper smelter far exceeded the risk screening values of the soil environmental quality standard (GB15618–2018). The concentration of Cu ranged from 97.47 to 1294.63 mg·kg^{-1} with a CV of 0.95, indicating moderate variation, while Cd ranged from 0.14 to 9.06 mg·kg^{-1} with a CV of 1.68, indicating strong variation. Moreover, concentrations of Cu and Cd were found to be continuously accumulating.

(2) The spatial distribution characteristics of all fractions of Cu and Cd were not entirely consistent with the total concentration. For Cu, the highest points of pollution with different orientations appeared in fractions F5. For Cd, the spatial distributions of fractions F1, F3, and F4 were quite distinct from that of Cd-T. Correlation analysis results indicated that soil physicochemical properties, particularly iron oxides, played a crucial role in the migration and transformation of Cu and Cd.

(3) The heterogeneity distribution of Cu and Cd pollution was significant, with Cu accumulation being more severe in the study area and mainly resulting from the production activities of the smelter. Nonetheless, the results of M_R analysis emphasized the influence of active fractions of Cu and Cd, with Cd exhibiting a higher value of M_R and posing a greater potential risk to the environment.

Supplementary Materials: The following supporting information can be downloaded at: https://www.mdpi.com/article/10.3390/toxics11080647/s1, Table S1: The background value of soil elements and the screening value of soil pollution risk of agricultural land locally; Table S2: The distribution statistics of each form of Cu and Cd.

Author Contributions: Experiments conception and design by Y.D. Performance of the experiments, data analysis, figures and/or tables preparation, article authored or drafts reviewed, and final draft approved by Y.D., L.X. and Y.C. Data analysis, figures and/or tables preparation, and final draft approved by H.S., Y.W. and X.G. Data analysis, article authored or drafts reviewed, and final draft approved by S.C. All authors have read and agreed to the published version of the manuscript.

Funding: This work was supported by the Program for National Natural Science Foundation of China (No. 41967021), the National Key Research and Development Program of China (No. 2022YFD1700800, 2022YFD1700805), and the Program for open foundation of National-Local Joint Engineering Research Center of Heavy Metals Pollutants Control and Resource Utilization (No. EL202280080).

Institutional Review Board Statement: Not applicable.

Informed Consent Statement: Not applicable.

Data Availability Statement: The authors declare that data supporting the findings of this study are available within the article.

Conflicts of Interest: The authors declare no conflict of interest.

References

1. Hashem, S.; Azam, J.; Vahidreza, J.; Rainer, S. Spatial distribution of copper and other elements in the soils around the Sarcheshmeh copper smelter in southeastern Iran. *Atmos. Pollut. Res.* **2020**, *11*, 1681–1691.
2. Liu, L.; Wu, L.; Luo, Y.; Zhang, C.; Jiang, Y.; Qiu, X. The impact of a copper smelter on adjacent soil zinc and cadmium fractions and soil organic carbon. *J. Soil Sediment.* **2010**, *10*, 808–817. [CrossRef]
3. Zhang, Q.; Wang, C. Natural and Human Factors Affect the Distribution of soil heavy metal pollution: A review. *Water Air Soil Pollut.* **2020**, *231*, 350. [CrossRef]
4. Dasharathy, S.; Arjunan, S.; Maliyur, B.A.; Murugasen, V.; Ramachandran, S.; Keshav, R.; Murugan, R. Mutagenic, carcinogenic, and teratogenic effect of heavy metals. *Evid.-Based Complement. Altern. Med.* **2022**, *2022*, 8011953. [CrossRef] [PubMed]
5. Aminur, R.; Kazuhiro, Y.; Monirul, M.I.; Genta, K. Investigation of efficient adsorption of toxic heavy metals (chromium, lead, cadmium) from aquatic environment using orange peel cellulose as adsorbent. *Sustainability* **2023**, *15*, 4470.
6. Zhang, X.; Zhong, T.; Liu, L.; Ouyang, X. Impact of soil heavy metal pollution on food safety in China. *PLoS ONE* **2017**, *10*, e0135182. [CrossRef]
7. Li, Y.; Zhou, H.; Gao, B.; Xu, D. Improved enrichment factor model for correcting and predicting the evaluation of heavy metals in sediments. *Sci. Total Environ.* **2021**, *755 Pt 1*, 142437. [CrossRef]
8. Salih, M.-A.; Jawdat, A.-J.-M.; Zaki, A.-H. Urban geochemistry assessment using pollution indices: A case study of urban soil in Kirkuk, Iraq. *Environ. Earth Sci.* **2019**, *78*, 587.
9. Ju, T.; Lei, M. Geo-accumulation index method to optimize the evaluation method of polymetallic environment quality: Taking developed agricultural areas as an example. *Chin. J. Environ. Sci.* **2022**, *43*, 957–964.
10. Yan, F.; Liu, C.; Wei, B. Evaluation of heavy metal pollution in the sediment of Poyang Lake based on stochastic geo-accumulation model (SGM). *Sci. Total Environ.* **2019**, *659*, 1–6.
11. Hazzeman, H.; Juen, L.L.; Zaharin, A.A.; Farhanna, M.N.; Ayunie, N.A.; Md, Y.F.; Bakar, S.A.; Mangala, P.S. Geo-accumulation index and contamination factors of heavy metals (Zn and Pb) in urban river sediment. *Environ. Geochem. Health* **2017**, *39*, 1259–1271.
12. Mahanta, M.J.; Bhattacharyya, K.G. Total concentrations, fractionation and mobility of heavy metals in soils of urban area of Guwahati, India. *Environ. Monit. Assess.* **2011**, *173*, 221–240. [CrossRef]
13. Qu, C.; Chen, J.; Mortimer, M.; Wu, Y.; Cai, P.; Huang, Q. Humic acids restrict the transformation and the stabilization of Cd by iron (hydr)oxides. *J. Hazard. Mater.* **2022**, *430*, 128365. [CrossRef]
14. Liu, Q.; Chen, Z.; Chen, Z.; Pan, X.; Luo, J.; Huang, F.; Zhang, X.; Lin, Q. Microbial community characteristics of cadmium speciation transformation in soil after iron-based materials application. *Appl. Soil Ecol.* **2023**, *183*, 104745. [CrossRef]
15. Burachevskaya, M.; Minkina, T.; Mandzhieva, S.; Bauer, T.; Chaplygin, V.; Sushkova, S.; Tsitsuashvili, V.; Popileshko, Y. Chemical partitioning of Zn in soil: Application of two sequential extraction procedures. *Geochem. Explor. Environ. Anal.* **2019**, *19*, 93–100. [CrossRef]
16. Wang, J.J. The Speciations of Heavy Metals in Soils Around a Mining Area in Southern China and the Transformation of Heavy Metal under Reductive Conditions. Master's Thesis, Huazhong Agricultural University, Wuhan, China, 2018. (In Chinese).
17. Han, F.X.; Banin, A. Long-term transformations and redistribution of potentially toxic heavy metals in arid-zone soils incubated: I. Under saturated conditions. *Water Air Soil Pollut.* **1997**, *95*, 399–423.
18. Burachevskaya, M.; Minkina, T.; Mandzhieva, S.; Bauer, T.; Chaplygin, V.; Zamulina, I.; Sushkova, S.; Fedorenko, A.; Ghazaryan, K.; Movsesyan, H.; et al. Study of copper, lead, and zinc speciation in the Haplic Chernozem surrounding coal-fired power plant. *Appl. Geochem.* **2019**, *104*, 102–108. [CrossRef]
19. Liang, J.; Zhou, J.; Ma, Y.; Cui, J.; Xu, L. Status and assessment of heavy metals pollution in paddy soil around the spoil area of smeltery. *J. Agro-Environ. Sci.* **2009**, *28*, 877–882. (In Chinese)
20. *GB 15618–2018*; Soil Environmental Quality-Risk Control Standard for Soil Contamination of Agricultural Land. Ministry of Ecology and Environment of the People's Republic of China: Beijing, China, 2018.
21. Long, A.-H.; Liu, J.-J.; Ni, C.-Y.; Huang, G.-F.; Tang, H.-Y. Assessment on the characteristic of heavy metals contaminated farmland soil around Guixi Smeltery Jiangxi Province. *Chin. J. Soil Sci.* **2006**, *37*, 1212–1217. (In Chinese)
22. Zhou, J.; Cui, H.-B. Engineering and prospect for remediating large-scale arable land contaminated by heavy metals with "demonstration project of soil remediation on the periphery of Guixi smelter" as example. *Bull. Chin. Acad. Sci.* **2014**, *29*, 272, 336–343. (In Chinese)
23. Zhang, J.; Sun, H.; Xie, L.; Hou, D. Changes in the ecological risk of heavy metals after soil remediation in a typical Brownfield: A case study of Guixi smelter in Jiangxi Province. *Acta Ecol. Sin.* **2017**, *37*, 6128–6137.
24. *HJ/T 166–2004*; The Technical Specification for Soil Environmental Monitoring. Ministry of Ecology and Environment of the People's Republic of China: Beijing, China, 2004.

25. Bao, S.D. *Soil and Agricultural Chemistry Analysis*, 3rd ed.; China Agriculture Press: Beijing, China, 2005. (In Chinese)
26. Zhang, Z.-W.; Zhu, Z.-X.; Fu, W.-L.; Wen, Z.-L. Morphology of soil iron oxides and its correlation with soil-forming process and forming conditions in a karst mountain. *Environ. Sci.* **2012**, *33*, 2013–2020.
27. Liu, Z.; Zhou, M.; Liao, W.; Liu, J.; Luo, C.; Lu, C.; Chen, Z.; Zhu, H. Fertilizer-holding performance of graphene on soil colloids based on double electric layer theory. *Materials* **2023**, *16*, 2578. [CrossRef]
28. Li, X.; Li, L.; Zhou, Z.; Li, T.; An, J.; Zhang, S.; Xu, X.; Pu, Y.; Wang, G.; Jia, Y.; et al. Soil potentially toxic element pollution at different urbanization intensities: Quantitative source apportionment and source-oriented health risk assessment. *Environ. Earth Sci.* **2023**, *251*, 114550. [CrossRef]
29. National Soil Survey Office. *Soil Species of China*; China Agriculture Press: Beijing, China, 1993; Volume 1. (In Chinese)
30. Wang, X.; Li, L.; Guo, N.; Xin, Z.; Li, X.; Sun, X.; Li, Y. A comprehensive exploration on pollution characteristics and ecological risks of heavy metals in surface paddy soils around a large copper smelter, southeast China. *Sustainability* **2021**, *13*, 13359. [CrossRef]
31. Zhao, Z.; Shi, Y.; Mao, X.; Li, H. Characteristics assessment and source analysis of heavy metal pollution in soil surrounding a smelter. *Chin. J. Inorg. Anal. Chem.* **2020**, *10*, 22–27. (In Chinese)
32. Sun, B.; Cao, Y. Spatial variation and affecting factors of Cu and Cd pollution in paddy soil in hilly region. *J. Agro-Environ. Sci.* **2006**, *25*, 922–928. (In Chinese)
33. Xiao, H.-Y.; Jiang, S.-Y.; Wu, D.-S.; Zhou, W.-B. Risk element (As, Cd, Cu, Pb, and Zn) contamination of soils and edible vegetables in the vicinity of Guixi smelter, south China. *Soil Sediment Contam.* **2011**, *20*, 592–604. [CrossRef]
34. Chowdhury, M.A.R.; Singer, D.M. Trace metal enrichment in the colloidal fraction in soils developing on abandoned mine spoils. *Minerals* **2022**, *12*, 1290. [CrossRef]
35. Rahman, M.S.; Ahmed, Z.; Seefat, S.M.; Alam, R.; Islam, A.; Choudhury, T.R.; Begum, B.A.; Idris, A.M. Assessment of heavy metal contamination in sediment at the newly established tannery industrial Estate in Bangladesh: A case study. *Environ. Chem. Ecotoxicol.* **2022**, *4*, 1–12. [CrossRef]
36. Tang, X.; Wu, Y.; Han, L.; Lan, Z.; Rong, X. Characteristics of heavy metal migration in farmland. *Environ. Earth Sci.* **2022**, *81*, 338. [CrossRef]
37. Hu, B.; Guo, P.; Wu, Y.; Deng, J.; Su, H.; Li, Y.; Nan, Y. Study of soil physicochemical properties and heavy metals of a mangrove restoration wetland. *J. Clean. Prod.* **2021**, *291*, 125965. [CrossRef]
38. Chen, J.; Zhang, H.; Wei, Q.; Farooq, U.; Zhang, Q.; Lu, T.; Wang, X.; Chen, W.; Qi, Z. Mobility of water-soluble aerosol organic matters (WSAOMs) and their effects on soil colloid-mediated transport of heavy metal ions in saturated porous media. *J. Hazard. Mater.* **2022**, *440*, 129733. [CrossRef]
39. Zhao, W.; Gu, C.; Ying, H.; Feng, X.; Zhu, M.; Wang, M.; Tan, W.; Wang, X. Fraction distribution of heavy metals and its relationship with iron in polluted farmland soils around distinct mining area. *Appl. Geochem.* **2021**, *130*, 104969. [CrossRef]
40. Oral, R.; Pagano, G.; Siciliano, A.; Toscanesi, M.; Gravina, M.; Nunzio, A.D.; Palumbo, A.; Thomas, P.J.; Tommasi, F.; Burić, P.; et al. Soil pollution and toxicity in an area affected by emissions from a bauxite processing plant and a power plant in Gardanne (southern France). *Ecotoxicol. Environ. Saf.* **2019**, *170*, 55–61. [CrossRef]
41. Vasarevičius, S.; Danila, V.; Januševičius, T. Immobilisation of cadmium, copper, lead, and nickel in soil using nano zerovalent iron particles: Ageing effect on heavy metal retention. *Water Air Soil Pollut.* **2020**, *231*, 496. [CrossRef]
42. Yang, J.; Liu, J.; Dynes, J.J.; Peak, D.; Regier, T.; Wang, J.; Zhu, S.; Shi, J.; Tse, J.S. Speciation and distribution of copper in a mining soil using multiple synchrotron-based bulk and microscopic techniques. *Environ. Sci. Pollut. Res.* **2013**, *21*, 2943–2954. [CrossRef]
43. Shi, M.; Min, X.; Ke, Y.; Lin, Z.; Yang, Z.; Wang, S.; Peng, N.; Yan, X.; Luo, S.; Wu, J.; et al. Recent progress in understanding the mechanism of heavy metals retention by iron (oxyhydr)oxides. *Sci. Total Environ.* **2021**, *752*, 141930. [CrossRef]
44. Shen, Q.; Demisie, W.; Zhang, S.; Zhang, M. The association of heavy metals with iron oxides in the aggregates of naturally enriched soil. *Bull. Environ. Contam. Toxicol.* **2020**, *104*, 144–148. [CrossRef]
45. Zhang, X.; Yang, H.; Cui, Z. Migration and speciation of heavy metal in salinized mine tailings affected by iron mining. *Water Sci. Techenol.* **2017**, *76*, 1867–1874. [CrossRef]
46. Maya, E.; Juan, S.-L.-P.; Vincent, N.; Kristin, B.; Scott, F. Organic compounds alter the preference and rates of heavy metal adsorption on ferrihydrite. *Sci. Total Environ.* **2021**, *750*, 141485.
47. Ye, Y.; Li, Y.; Cao, Z.; Liu, S.; Zhao, Y. Experimental and numerical study on Cu and Cd migration in different functional-area soils under simulated rainfall conditions. *Environ. Res.* **2021**, *208*, 112239. [CrossRef]
48. Cupara, N.; Nikolić, I.; Đurović, D.; Milašević, I.; Medin, D.; Krivokapić, S. Heavy metal assessment in agricultural soils and vegetables in the vicinity of industrial pollutants in the Pljevlja municipality (Montenegro): Ecological and health risk approach. *Environ. Monit. Assess.* **2022**, *194*, 819. [CrossRef]

toxics MDPI

Article

The Migration Mechanism of BTEX in Single- and Double-Lithology Soil Columns under Groundwater Table Fluctuation

Jingwei Zheng [1], Yang Yang [2,*], Juan Li [2,*], Hao Zhang [2] and Yan Ma [1,*]

1 School of Chemical and Environmental Engineering, China University of Mining and Technology-Beijing, Beijing 100083, China; zjwljwater@foxmail.com
2 Technical Centre for Soil, Agriculture and Rural Ecology and Environment, Ministry of Ecology and Environment, Beijing 100012, China; zhanghao@tcare-mee.cn
* Correspondence: yangyang@tcare-mee.cn (Y.Y.); lijuan@craes.org.cn (J.L.); mayan2202@163.com (Y.M.)

Abstract: The migration of light non-aqueous phase liquids (LNAPLs) trapped in porous media is a complex phenomenon. Groundwater table fluctuation can not only affect contaminant migration but also redox conditions, bacterial communities, and contaminant degradation. Understanding LNAPLs' (e.g., benzene, toluene, ethylbenzene, and xylene (BTEX)) behavior within porous media is critical for the high efficiency of most in situ remediation systems. A laboratory study of single- and double-lithology soil column investigation of the groundwater table fluctuation effect on BTEX transport, using benzene and toluene as typical compounds, in a typical representative model of aquifers subjected to water table fluctuation was undertaken in this study. The results show that benzene and toluene migration in single-lithology soil columns packed with sand was mainly affected by flushing due to the hydraulic force induced by water table fluctuations and that the double-lithology soil column packed with sand and silt was significantly affected by retention due to the higher adsorption induced by 10 cm of silt. The dissolution mainly correlated with the BTEX migration in saturated zones, and the contaminant concentration increased when the water table fell and decreased when the water table rose. For a contaminated site with a single-lithology structure consisting of sand, more attention should be paid to organic contaminant removal within the groundwater, and a double-lithology structure containing silt is more suited to the removal of organic contaminants from the silt layer. The difference in biodegradation kinetics between the groundwater table fluctuation (GTF) zone and the saturated zone should be better understood for the remediation of BTEX compounds.

Keywords: groundwater table fluctuation; benzene and toluene; migration mechanism; single- and double-lithology soil columns

check for updates

Citation: Zheng, J.; Yang, Y.; Li, J.; Zhang, H.; Ma, Y. The Migration Mechanism of BTEX in Single- and Double-Lithology Soil Columns under Groundwater Table Fluctuation. *Toxics* **2023**, *11*, 630. https://doi.org/10.3390/toxics11070630

Academic Editor: Junhao Qin

Received: 14 June 2023
Revised: 9 July 2023
Accepted: 13 July 2023
Published: 20 July 2023

1. Introduction

Subsurface contamination by non-aqueous phase liquids (NAPLs) is a widespread problem. NAPLs cause concern because of their persistence in the subsurface and their ability to contaminate large volumes of soil and groundwater. NAPLs that are lighter than water, such as gasoline and BTEX, are referred to as light non-aqueous phase liquids (LNAPLs) [1]. LNAPLs present extensive contamination risks due to their ability to partition into both air and water phases, creating pathways for exposure to human health and the environment [2]. Volatile petroleum hydrocarbons (PHCs) released in the form of vapor from LNAPLs, known as petroleum vapor intrusion (PVI), represent the most likely pathway for human health risks [3]. In recent years, significant efforts have been made to optimize the development of effective methods for the remediation of contaminants in soils and groundwater, such as air sparging and bioventing [4]. However, these techniques can be expensive and, in some cases, ineffective [4]. Significant efforts have been made in recent years to develop effective methods to remove pollutants from contaminated soil and groundwater. Not only have chemical and biological interactions been researched [5] but studies have also focused on the diffusion and migration of pollutants in groundwater [6,7]. Pollutant behavior within porous media is critical for the high efficiency of most in situ remediation systems [8].

The descriptions of water flow and phase distribution are the basis for understanding the transport of dissolved organic pollutants from the soil surface to groundwater. When a small amount of an LNAPL is released into the subsurface, it migrates through the unsaturated zone until it distributes itself as a residual saturation. If the release is large enough, it reaches the capillary fringe, where it spreads laterally above the saturated zone [9]. The outcome for organic chemicals not only depends on their interactions with the solid phase but also on the water flow velocity in the soil [10,11]. Water flow largely depends on the amounts and intensities of precipitation and evapotranspiration, as well as the movement of the free water table in the subsoil or aquifer. Drying and wetting processes driven by water potential differences may occur simultaneously at different depths under field conditions [12,13]. The groundwater table fluctuation (GTF) zone, where the groundwater table rises and falls due to the removal or recharge of groundwater, tidal patterns, or industrial activities has a significant effect on contaminant migration. However, little is known about the impact of GTF zones on contaminants [14,15]. The GTF zone causes drying and wetting processes, which, linked to the redox of hydrochemistry [16], can lead to the dissolution and subsequently more complicated transport of petroleum (e.g., benzene and toluene) in contaminated subsurface environments. The rise and fall of the groundwater table may promote contaminant migration through the groundwater level, where the contaminants pass through the water boundary and contaminate the lower groundwater regions, resulting in pollution plume dispersion [17,18].

With repeated rises and falls due to GTF, LNAPLs that float on the surface layer are particularly susceptible to adsorption into the porous region surrounding the water table, known as the smear zone [19,20]. Previous studies have shown that water level fluctuation has a significant influence on the migration of organic pollutants throughout the water table [21], suggesting that the dissolution and biodegradation of LNAPL components is likely to increase [22]. Due to the potentially widespread and dispersed nature of petroleum hydrocarbon plumes, the remediation process can be lengthy, requiring cost-effective and practical methods to ensure the long-term control of contaminated groundwater [23]. Compared to other remediation techniques, natural attenuation (NA) has been recognized as one of the remediation options that can meet the aforementioned requirements [24]. NA is a passive remediation measure that relies on natural mechanisms to degrade and remove contaminants from the soil and groundwater [25]. Due to its cost-effectiveness and minimal environmental impact, NA is considered a potential approach for the remediation of sites contaminated with petroleum hydrocarbons [26]. Within intrinsic biotic and abiotic mechanisms, intrinsic bioremediation (including natural aerobic and anaerobic biodegradation processes) plays key roles in plume control and pollutant removal [26]. The biodegradation of petroleum hydrocarbons occurs naturally with indigenous microbes and is an inexpensive and sustainable option [27]. During biodegradation, petroleum hydrocarbons break down into less hazardous substances. The whole mass loss of LNAPLs due to partitioning and physical and biological degradation is termed natural source zone depletion (NSZD) [28]. Many studies have investigated the microbial communities and physical conditions that may lead to the natural source zone depletion (NSZD) of petroleum hydrocarbons [29]. These studies have considered the roles of different redox conditions in biodegradation [30] and how groundwater fluctuations can cause changes in the redox environment, thereby promoting microbial degradation [31]. Therefore, it is essential to better understand the behavior and transformation mechanisms of organic pollutants in water table fluctuation zones in order to improve the effectiveness of remediation methods for groundwater pollutants.

GTF can not only affect contaminant migration but also redox conditions, bacterial communities, and contaminant degradation. Water table cycle fluctuations can accelerate the attenuation of pollutants in groundwater [32]; for example, the removal rate of NH_4-N has been found to be increased by higher rates of water level fluctuation [1], and diesel oil has also been found to be increased by GTF [33]. The redox conditions for a fluctuating water table were not in equilibrium but were subjected to spatial and temporal variations. The microbial processes in the column that was well-defined in terms of water flow, gas flow, and initial mineral phase resulted in heterogeneity that had consequences for organic pollutants [34].

The toxic aromatic hydrocarbons benzene, toluene, ethylbenzene, and m-xylene (BTEX) are common petroleum pollutants in the subsurface environment [35]. Due to the low density and solubility of benzene and toluene, they behave as LNAPLs and are absorbed into the surface layer of the water table. The migration of benzene and toluene to the subsurface environment, e.g., advection, dispersion, adsorption, and degradation [36], is mainly affected by the flow patterns and porous media characteristics. Field investigation studies have shown that cyclical fluctuations have a significant influence on the migration and distribution of BTEX in groundwater [37,38]; however, the migration mechanisms of BTEX associated with the hydraulic force and different soil structures of GTF have rarely been studied.

Therefore, this study explored the BTEX migration mechanism in a laboratory soil column GTF system. Benzene and toluene was chosen as BTEX compound. Two typical porous media were used in the experiment, consisting of single- and double-lithology soil columns. This simulated groundwater table system allows investigation of the effects of hydraulic force and allows investigation of different porous media characteristics controls on a representative subunit of a GTF system to further understand benzene and toluene migration from petroleum-contaminated shallow groundwater and the effects of geological media, as well as to identify benzene and toluene abundance zone and assist strategies for the remediation of benzene and toluene compounds.

2. Methodology

2.1. Conceptual Model for Pollutants in Water Table Fluctuation Zone

Generally, once organic pollutants are released into the subsurface, most migrate downwards through the unsaturated vadose zone [21,22] to the capillary fringe and groundwater table [22]. Since shallow subsurface environments can be saturated or unsaturated, the migration and transformation of pollutants via different soil layers are affected by a large number of different environmental factors, particularly in GTF zones with frequently alternating cycles of saturation and unsaturation. The transport and fate of benzene and toluene in the subsurface were subject to four main fundamental mechanisms as follows: the infiltration and migration of benzene and toluene in the subsurface under the influence of gravitational and capillary forces, the dissolution and consequent advection–dispersion in the downward-flowing water phase, the transport of benzene and toluene vapor as the soil pore gas, and the influence of chemical and biological functions on benzene and toluene. As shown in Figure 1, when the groundwater table rises, the porous media becomes saturated and contaminants can be dispersed or caused to migrate with the water flow (Figure 1A); when the groundwater table falls, the porous media becomes unsaturated and the residual contaminants can be suspended in a three-phase gas–liquid–solid state (Figure 1B).

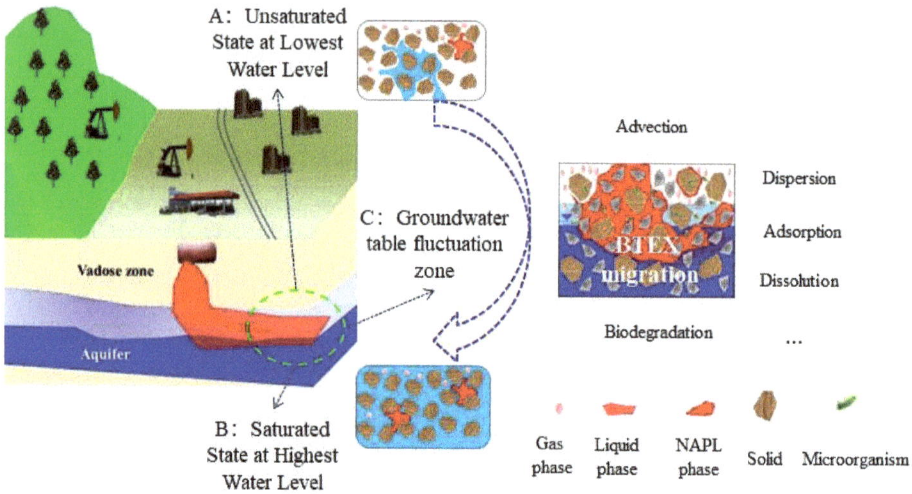

Figure 1. Conceptual model of petroleum contamination in groundwater fluctuation zones. Saturated state at highest water level (**A**); unsaturated state at lowest water level (**B**); groundwater table fluctuation zone (**C**). Red represents the pollutant and blue represents groundwater.

2.2. Experimental Setup

A schematic diagram of the experimental setup is shown in Figure 2, and it is used to simulate the GTF zone. The migration mechanisms of benzene and toluene were assessed under varying hydraulic conditions, including single- and double-lithology soil and groundwater table fluctuation conditions. The columns were made of organic glass tubes (height of 100 cm, from top 0 cm to bottom −100 cm; diameter of 10 cm). Pretreated sand and silt were uniformly packed into the columns. Column I was a single-lithology structure packed with sand only. Column II was a double-lithology structure packed with sand and silt, and the silt was located in the middle of the column. A total of 600 mL of pure benzene (300 mL) and toluene (300 mL) liquid was evenly loaded into each column, and the top of the column was sealed immediately to prevent loss of benzene and toluene by evaporation [39].

To simulate the hydraulic dynamic conditions experienced in the GTF zones, the water tank location was adjusted to represent the change in the groundwater table. The amplitude of the fluctuation was 20 cm, and the whole duration of the experiment was 54 days. At the beginning of the experiment, the groundwater table was −45 cm from 0 to 6 days, −35 cm from 7 to 12 days, −25 cm from 13 to 18 days, −35 cm from 19 days to 24 days, −45 cm from 25 days to 30 days, −55 cm from 31 days to 36 days, −65 cm from 37 days to 42 days, −55 cm from 43 days to 48 days, and −45 cm from 49 days to 54 days. Each water sample was collected at the end of each groundwater level using a soil solution collector (Rhizon, Holland). The locations of the collected water samples were at −50 cm (Column I 1–1 and Column II 2–1), −60 cm (Column I 1–2 and Column II 2–2), and −70 cm (Column I 1–3 and Column II 2–3). The number of water samples collected by column I and column II was 9 and 9, respectively. Four soil samples from each column were collected to assess the groundwater bacterial community characteristics, and the locations of the soil samples were at −30 cm, −40 cm, −50 cm, and −60 cm, respectively.

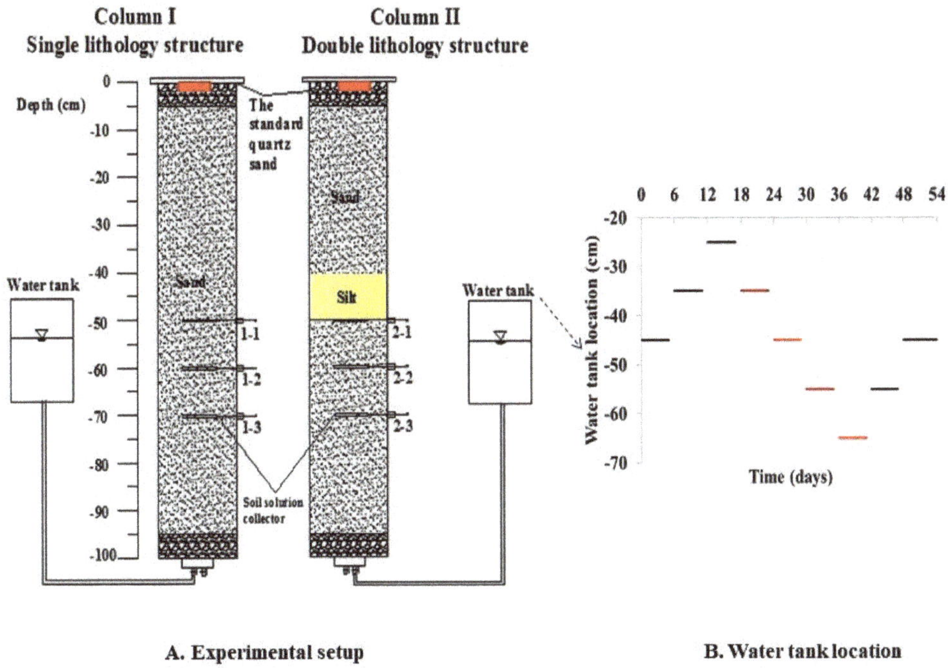

Figure 2. In the schematic of the experimental apparatus (**A**), red is the leakage point of pollutants, and in the water table location (**B**), red represents the stage of water level decline.

2.3. Simulation Materials

Two types of porous media were prepared. Column I was packed with sand only, and column II was packed with sand and silt. Table 1 summarizes the basic characteristics of the porous media. The simulated groundwater was collected from a shallow aquifer without benzene and toluene contamination (below the detection limit). And the physical and chemical characteristics of benzene and toluene are listed in Table 2.

Table 1. The basic physical and chemical properties of media in experiment columns [40].

Porous Media Type	Sand	Silt
Media Size (mm)	0.5–1.25	0.075–0.1
Soil Bulk Density (g cm^{-3})	1.51	1.38
Initial Porosity	0.30	0.28
Initial Water Content (%)	20.5	24.8
Permeability Coefficient (cm s^{-1})	1.2×10^{-3}	7.9×10^{-4}
pH	8.51	8.45
Organic Matter Content (%)	0.78	3.33

Table 2. The physical and chemical properties of benzene and toluene [41].

Pollutant	Molecular Formula	Molecular Weight	Density (g cm^{-3}, 20 °C)	Solubility in Water (mg L^{-1}, 25 °C)	Log K_{ow}	K_{oc}
Benzene	C_6H_6	78.1	0.8765	1800	1.6~2.4	10~12.9
Toluene	$C_6H_5CH_3$	92.1	0.8669	347~707	2.1~3.0	11.9~14.8

2.4. Analytical Methods

A total of 18 water samples were collected from the columns during the whole experiment process. Water samples from the columns were analyzed for benzene and toluene concentrations. Benzene and toluene concentrations were analyzed using gas chromatography–mass spectrometry (GC-MS 2010QP-Ultra, SHIMADZU, Japan). The temperature of the chromatographic column was 40 °C. Groundwater bacteria were analyzed using high-throughput sequencing, and the genomic DNA of all tiny organisms in the environmental sample was tested using a shotgun [42,43].

3. Results

3.1. Benzene and Toluene Change with Depth

Benzene and toluene concentration variation along with the depth of the column are compared in Figure 3 for column I and column II. The change range in benzene concentration in column I was 1.26 to 65.16 mg/L, 0.0064 to 5.73 mg/L, and 0.0011 to 1.97 mg/L at 1–1, 1–2, and 1–3, respectively; in column II, it was was 0.25 to 28.98 mg/L, 0.00027 to 1.52 mg/L, and 0.00044 to 0.42 mg/L at 2–1, 2–2, and 2–3, respectively. The change range in toluene concentration in column I was 0.99 to 60.68 mg/L, 0.0025 to 14.44 mg/L, and 0.0007 to 1.84 mg/L at 1–1, 1–2, and 1–3, respectively; in column II, it was was 0.0074 to 4.99 mg/L, 0.00025 to 0.88 mg/L, and 0.00038 to 0.24 mg/L at 2–1, 2–2, and 2–3, respectively. The amplitude of benzene and toluene concentration variation for column I was almost twice or more than that shown in column II with the addition of 10 cm silt, and the concentration at the same depth for the column was higher than that in column II, showing the additional contaminant retention provided by silt in the porous media in comparison to sand media alone. Therefore, the concentration amplitude for sand only was also faster than that for the porous media containing silt, highlighting the negative impact on BTEX migration in column II due to the inclusion of silt in the porous media. When comparing the BTEX migration mechanism between porous media types, the contaminant concentration amplitude was higher in the media composed of sand only than in the porous media containing silt.

$$logK_{OC} = 0.623logK_{OW} + 0.873 \quad (1)$$

Koc represents the soil/sediment sorption coefficient. K_{OW} is the water/octanol partition coefficient [44].

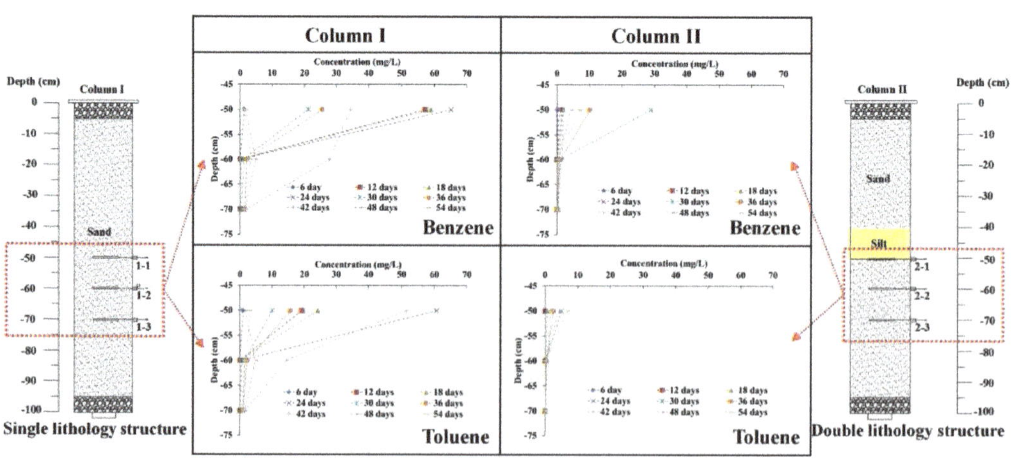

Figure 3. The concentration of benzene and toluene changes with soil column depth.

$$C_{soil} = C_{water} \times K_{oc} \quad (2)$$

C_{soil} represents the solute concentration in soil, C_{water} represents the solute concentration in water, and K_{oc} represents the organic carbon–water partition coefficient in soil/sediment [45].

Using Equation (1), the *Koc* value can be calculated. Using Equation (2), the concentrations of benzene and toluene in the soil can be calculated. Figure 4 shows the predicted concentrations of benzene and toluene in the soil. The highest predicted concentration of benzene in the soil was 746 mg/kg for column 1 and 331 mg/kg for column 2. The highest predicted concentration of toluene was 686 mg/kg for column 1 and 93 mg/kg for column 2. As the water level fluctuated, the pollutant concentrations gradually decreased. This may be associated with microbial activity [46].

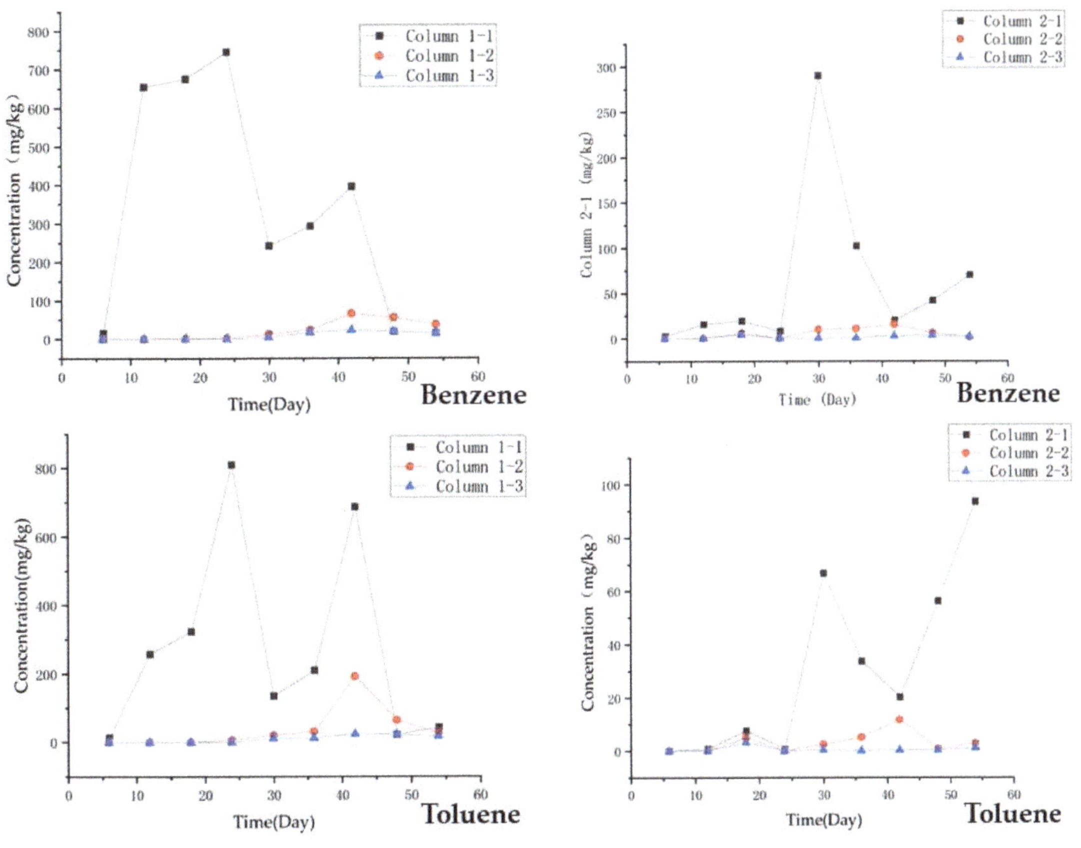

Figure 4. The concentration changes of benzene and toluene in the soil.

3.2. Benzene and Toluene Changes with GTF and Bacterial Community Structure

Water samples from different depths (−50 cm, −60 cm, −70 cm) in the two columns were analyzed for BTEX change trends under water table fluctuations, as shown in Figure 5 (left). In general, benzene and toluene concentrations at the same depth showed the same change trends in columns I and II with water table fluctuation, showing a higher concentration amplitude in column I. The benzene and toluene concentrations always changed with the water table fluctuation at −50 cm in the whole period, showing a lower concentration amplitude in the initial period followed by a higher amplitude at −60 and −70 cm after several water table up-down cycles. Notable differences were observed between column I and column II. First, the BTEX concentration showed a higher trend in column I than that in column II, especially at −50 cm, despite the same concentration and volume of benzene and toluene being applied to each column. The highest benzene and toluene concentrations were 65.16 mg/L and 60.68 mg/L in column I, 28.98 mg/L and 4.99 mg/L in column II, and almost all the concentrations of the samples in column II were lower than that in column I. Thereafter, there were different changes in benzene and toluene concentrations between column I and column II, especially in the whole period at −50 cm, and in the last period at −60 cm and −70 cm.

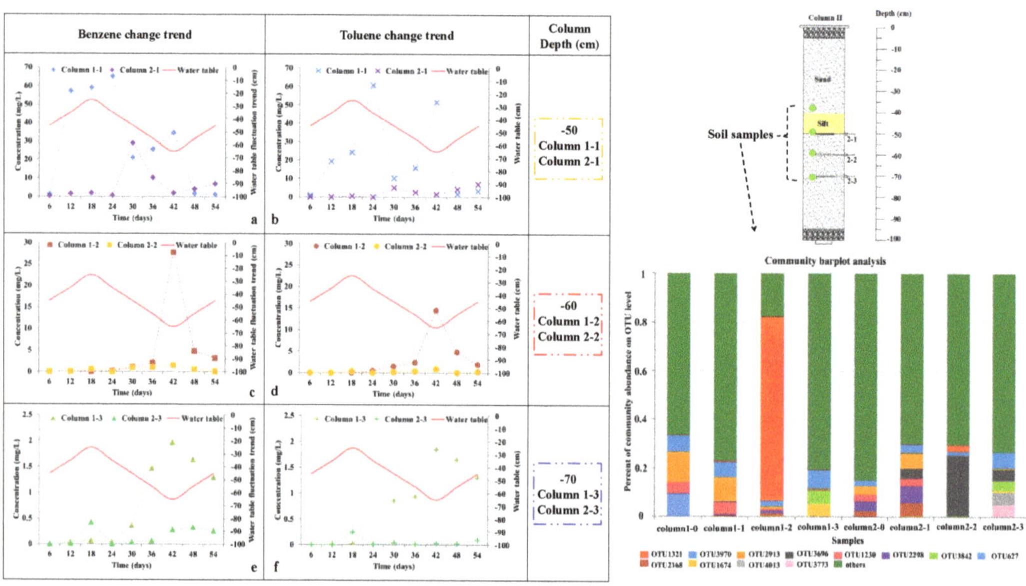

Figure 5. The benzene and toluene concentration changes under groundwater table fluctuation, and the bacterial community with depth after the experiment.

The benzene concentration change trends at −50 cm, −60 cm, and −70 cm are shown in Figure 5a,c,e, respectively. As shown in Figure 5a, when the groundwater table rose by 20 cm from 6 days to 18 days, the benzene concentration noticeably increased in column 1–1 and slightly in column 2–1, and the increased amplitude was 57.73 mg/L and 1.67 mg/L. When the groundwater table declined by 40 cm from 18 days to 42 days, the benzene concentration initially decreased, followed by an increase in column 1–1; however, it initially increased followed by a decrease in column 2–1. When the groundwater table rose by 20 cm from 42 days to 54 days, the benzene concentration decreased in column 1–1 and increased in column 2–1, showing the opposite trend. As shown in Figure 5c,e, the benzene concentration was stabilized from 6 days to 24 days. Thereafter, the benzene concentration increased as the groundwater table declined from 24 days to 42 days and decreased as the groundwater table rose from 42 days to 56 days.

The toluene concentration change trends at −50 cm, −60 cm, and −70 cm were shown in Figure 4b,e,f, respectively. The toluene concentration change trend was similar to that of benzene at −50 cm, −60 cm, and −70 cm. The main reason for the benzene and toluene concentration change trends may be the 10 cm silt in column II, which changed the soil column structure, including hydrodynamic force, permeability coefficient, adsorption, and biodegradation conditions. And the sample points were at different depths; columns

1–1, 1–2, 2–1, and 2–2 were in the groundwater table fluctuation zone with "unsaturation and saturation cycles", and columns 1–3 and 2–3 were in the saturation zone. Therefore, different points underwent different water conservation conditions.

Additionally, soil samples were also collected from different depths in two columns for bacterial community analysis after the experiment, as shown in Figure 5 (right). The compositions and abundance of the bacterial communities at the OTU level are shown as −40 cm (columns 1–0 and 2–0), −50 cm (columns 1–1 and 2–1), −60 cm (columns 1–2 and 2–2), and −70 cm (columns 1–3 and 2–3). Three samples were collected at each depth for the repeatability test. The classified sequences at each depth were affiliated with 13 bacterial OTUs (>5 %). For all 12 OTUs except others (green color in Figure 5), OTU3970 belonged to Actinobacteria, OTU2168 belonged to Cyanobacteria, OTU3696 was unclassified, and the others in 12 OTUs belonged to Proteobacteria at the phyla level. For column I and column II, OTU3970 existed along with depth, and at each depth, the approximate relative abundances were larger than 1%. OTU2913 and OTU1230 also existed in the two columns; however, the relative abundances were obviously higher at the top of the GTF zone (−40 and −50 cm) than that at the bottom of the GTF zone (−60 and −70 cm). Additionally, OTU3824 relative abundances were obviously higher at −70 cm.

For column 1–0 and column 2–0, the differences were OTU627 (9.0%) in column 1–0, and OTU2298 (3.5%) and OTU2168 (2.6%) in column 2–0. For column 1–1 and column 2–1, the differences were OTU3696 (3.8%) in column 1-1, and OTU2298 (6.9%) and OTU2168 (5.3%) in column 2–1. For column 1–2 and column 2–2, the differences were OTU1321 (76%) in column 1–2, and OTU1321 (2.9%) and OTU3696 (25%) in column 2-2. For column 1–3 and column 2–3, the differences were OTU1674 (5.3%) in column 1–3, and OTU3696 (4.3%), OTU4013 (5.3%), and OTU3773 (5%) in column 2–3. Significant variations were detected between column I and column II.

4. Discussion

4.1. Single- and Double-Lithology Structure Effect on BTEX Migration Mechanisms

Column I was a single-lithology structure with sand only, and column II was a double-lithology structure with sand and 10 cm silt. The benzene and toluene concentrations in column 2–1 were significantly lower than those in column 1–1 during the whole GTF period, showing a higher level of benzene and toluene retention within the porous media of silt. Therefore, a contaminated site with a double-lithology structure containing silt was more contaminated with BTEX, and a contaminated site with a single-lithology structure consisting of sand was more suited for the removal of organic contaminants from within the groundwater.

The benzene and toluene retention within the silt has a significant effect on BTEX migration, mainly because of BTEX adsorption and dissolution from the porous media. Adsorption and dissolution are influenced by hydro-chemical and geological characteristics [39,47,48]. Parameters such as pH, initial porosity, specific surface area, organic matter content, and media size were subjected to factor analysis. The basic physical and chemical properties of porous media are shown in Figure 3. Compared to sand, silt has a smaller media size and initial porosity, which indicates that silt has a higher specific surface area. Additionally, silt has higher organic matter content. Notably, the differences in these parameters show the different effects of sand and silt on benzene and toluene migration. When assessing the correlation of the parameters of media size with benzene and toluene adsorption, there was a negative correlation. Smaller media sizes will have a higher specific surface area, which leads to easier adsorption of benzene and toluene on silt than on sand. Therefore, the media size and specific surface area were significantly correlated with the benzene and toluene adsorption ratio but showed an inverse correlation with benzene and toluene dissolution. However, the initial porosity of the porous media was positively correlated with benzene and toluene dissolution. The initial porosity of porous media was higher, and benzene and toluene were more easily desorbed from the porous media. Therefore, due to the silt in column II, benzene and toluene concentrations were significantly lower than those in column 1–1 during the whole GTF period. This confirms that the dissolution and transport actions of the contaminants within the water could be reduced with higher levels of adsorption of contaminants into porous media [49].

4.2. Water Table Fluctuation Effect on BTEX Migration Mechanisms

The schematic water table fluctuation in the soil column and the porous media state are shown in Figure 6. The migration mechanism and distribution characteristics of BTEX in the soil column are discussed via image analysis. During the water table fluctuations, the contaminated porous media experienced periods of saturation and unsaturation, so contaminants such as benzene and toluene were in a continuous state of transformation between saturated and unsaturated environments. In theory, in this scenario, benzene and toluene exist in three different states: free, dissolved, and residual [8]. According to benzene and toluene solubility in 1800 mg/L and 347~707 mg/L of water at 25 °C in Table 2, the benzene and toluene concentrations of all the water samples did not exceed the criterion. The highest concentrations of benzene and toluene measured in column I were 65.16 mg/L and 60.68 mg/L, respectively, whereas in column II, they were 28.98 mg/L and 4.99 mg/L, respectively. These results are related to the adsorption of benzene and toluene by the soil and soil organic matter. Research by Jing Sun et al. showed that soil with an average particle size of 0.08 mm has a maximum adsorption capacity of 376.39 mg/kg for benzene [50]. The sorption behavior of soils also depends on the chemical conformation and physical structural properties of the soil organic matter (SOM) fraction [51]. Among the various factors influencing the sorption behavior, SOM is considered a key factor controlling the adsorption process as the distribution phase for NAPL uptake [52]. For example, Lion et al. suggested that the organic carbon (f_{oc}) was a dominant factor controlling the sorption capacity of American soils even if $f_{oc} < 0.1\%$ [53]. Therefore, the dissolved benzene and toluene would migrate with the water flow, where the dissolution of contaminants has a significant impact on pollutant migration. In addition, benzene and toluene are light non-aqueous phase liquids; if the volume of contaminants is enough, they would form a lens above the groundwater table, and some contaminants would migrate with the water flow in free form. Therefore, groundwater table fluctuation would generate a hydrodynamic force, and free and dissolved toluene would be transported with the water flow. The residual state would adsorb onto the porous media and could not be transported with the water flow. However, the saturated and unsaturated environments of the porous media would cause benzene and toluene to migrate among free, dissolved, and residual states, leading to benzene and toluene concentrations in water fluctuation.

Soil water collectors at different depths experienced diverse saturation and unsaturation conditions. As shown in Figure 6, columns 1–1, 2–1, 1–2, and 2–2 experienced saturation from the water table (WT) to the highest water table (WTmax), unsaturation from WTmax to the lowest water table (WTmin), and then to saturation from WTmin to WT. Columns 1–3 and 2–3 always experienced saturation from WT to WTmax, and from WTmin to WT. Therefore, the benzene and toluene concentration fluctuations were different, as shown in Figure 5 (left).

For column I, the BTEX formed a lens above the water table and existed as free contaminants, and benzene and toluene dissolved in water. On day 6, the concentration was very low at −40, −50, and −60 cm. As the water table rose from WT to WTmax, the contaminant concentration in column 1–1 increased. Because the residual contaminants in the unsaturated zone were only occasionally in contact with water, these contaminants were less depleted in water, so more contaminants were transferred from the free and residual state to dissolve into the water and, consequently, the contaminant concentration increased as the water table increased. However, the contaminant concentrations in columns 1–2 and 1–3 remained stabilized because the water table increased, and free and residual state contaminants could not contact these depths. As the water table fell from WTmax to WTmin, the contaminant concentrations in column 1–1 decreased initially and thereafter increased. Because water initially promoted the dilution of dissolved state contaminants, as the water table fell, free and dissolved contaminants were transported with the water flow, and contaminant concentrations increased. Also, in columns 1–2 and 1–3, as the water table fell, the contaminant concentrations increased at the end of the water table fall period. This is because more dissolved state contaminants were transported to the depth with the water flow. As the water table rose from WTmin to WTmax again, the contaminant concentrations in columns 1–1, 1–2, and 1–3 decreased. As the water table rose, it promoted the dilution of the dissolved state contaminants.

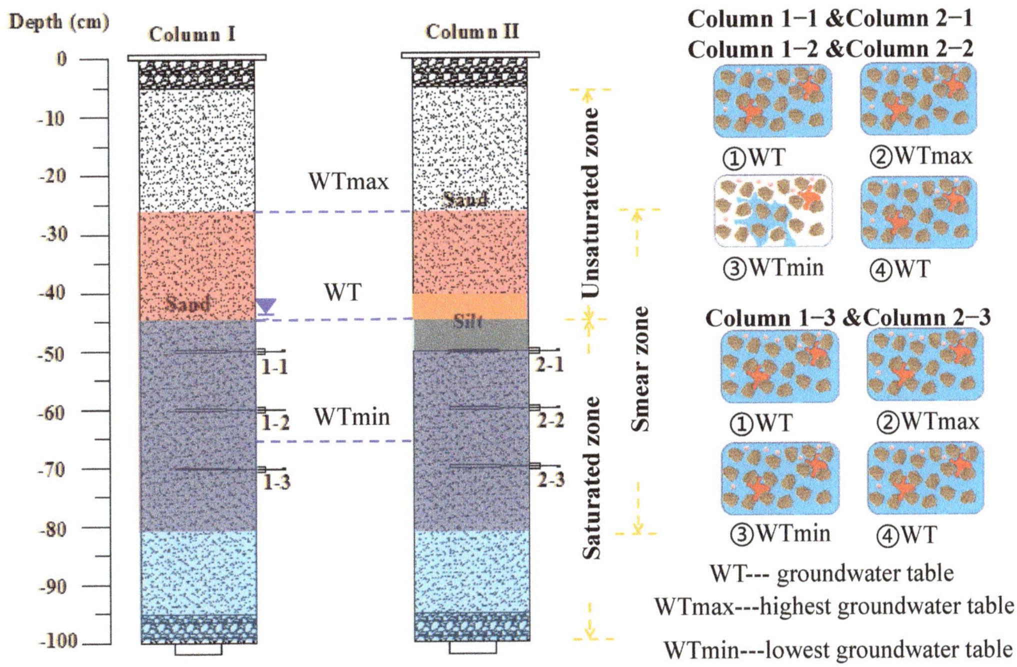

Figure 6. Schematic water table fluctuation with respect to soil column and porous media state. The blue arrow indicates the groundwater level, and the yellow arrow indicates the saturated and unsaturated zone of the groundwater.

For column II, the contaminant concentrations in columns 2–2 and 2–3 presented similar fluctuation trends to columns 1–2 and 1–3, and the concentration was smaller than that in column I. Although the contaminant concentration in column 2–1 was smaller than that in column 1–1, it presented different fluctuation trends, especially in the water table fall period from WTmax to WTmin. The 10 cm silt in column II, which had a higher retention of contaminants, led to this difference. At the beginning of the water table fall period, some dissolved and residual state contaminants were transported with the water flow to column 2–1, and the contaminant concentration increased. However, later in the water table fall period, more free-state contaminants were adsorbed onto the porous media, and fewer contaminants were transported with the water flow; as the water promoted the dilution of dissolved state contaminants, the contaminant concentration decreased.

Therefore, the benzene and toluene migration was mainly affected by flushing due to the hydraulic force induced by the water fluctuations, causing dissolution and adsorption onto porous media [7]. Because silt had a higher retention of benzene and toluene, which affected the benzene and toluene migration mechanism, more free and dissolved benzene and toluene adsorbed onto the porous media and formed residual benzene and toluene, and the effect of GTF on benzene and toluene migration in column I was higher than that in column II for the 10 cm silt. For a contaminated site with a single-lithology structure consisting of sand, GTF led to the downward migration of benzene and toluene, and the contaminant concentration in the saturated zone increased, which is why more attention should be paid to organic contaminant removal within the groundwater. A contaminated site with a double-lithology structure containing silt was more suited to the removal of organic contaminants from the silt layer.

4.3. Bacterial Community Structure Effect on BTEX Migration Mechanisms

As shown in Figure 5, the bacterial OTU numbers and abundance at each depth were significantly different. However, most of them belonged to Proteobacteria, which was also detected in many BTEX-contaminated sites. Proteobacteria are one of the main BTEX-degrading bacteria [54]. Biodegradation is an important process that can affect the dissolved state of the BTEX concentration, which affects BTEX migration. Biodegradation kinetic changes in response to water table fluctuations were demonstrated in several studies, e.g., Rainwater et al. (1993), Sinke et al. (1998), Dobson et al. (2007), and Rezanezhad et al. (2014) [33,34,55,56]. These studies suggest that biodegradation enhancement during episodes of a rising water table is attributed to aquifer oxygenation [38], and biodegradation mainly occurs under aerobic conditions. But some studies show that under anaerobic conditions, BTEX can also be degraded [57]. Because the GTF and soil lithology structures affect the aerobic and anaerobic conditions, the bacterial community presented obvious ifferences.

Therefore, it is speculated that oxygen is enough in the unsaturated zone and that aerobic biodegradation and hydrodynamic forces affect BTEX migration. In the GTF zone, the aerobic and anaerobic conditions always change, and biodegradation kinetics may be lower than the hydrodynamic force, which would mainly affect BTEX migration. In the saturated zone, anaerobic biodegradation and hydrodynamic forces affect BTEX migration. Additionally, it is speculated that under some conditions during the GTF period, biodegradation kinetics become higher than the mass transfer coefficient caused by the hydrodynamic force, which results in a strong reduction in the BTEX concentration.

Mass transfer continually removes the BTEX mass from LNAPL entrapped in the pores, with the aqueous BTEX mineralized by native bacterial metabolism. The quantification of BTEX mass removal is a key issue in natural source zone depletion (NSZD) assessment (ITRC, 2009); however, in many cases, the available monitoring data are scarce and cannot identify a clear decreasing concentration tendency, inducing incorrect interpretations [58,59]. Further research on bacterial community structure research can provide important data for BTEX biodegradation. Therefore, biodegradation kinetics should be considered in long-term prediction models of groundwater table fluctuations.

5. Conclusions

This study provides important insights into the migration behavior of light non-aqueous phase liquids (LNAPLs) in porous media, aiding the design and implementation of more efficient in situ remediation strategies and optimizing the repair of BTEX compounds. Experiments on the degradation of organic pollutants under groundwater table fluctuations indicate the need for a better understanding of the differences in biodegradation kinetics between the groundwater table fluctuation zone and the saturation zone to effectively remediate benzene and toluene pollution in groundwater. The groundwater table simulation system investigates the hydraulic intervention on BTEX migration, as well as the control of representative subunits of the groundwater table fluctuation system according to different porous media characteristics, further enhancing the understanding of BTEX migration and the influence of geological media in shallow petroleum-contaminated groundwater. Overall, this study highlights the significant influence of GTF on benzene and toluene behavior in single- and double-lithology soil column environments. The main conclusions can be summarized as follows:

(1) As demonstrated, groundwater table fluctuations can promote considerable LNAPL dissolution and, consequently, significant changes in the BTEX concentration over time. The migration of benzene and toluene in a single-lithology soil column packed with sand is mainly affected by flushing due to the hydraulic force induced by water table fluctuations, causing dissolution and adsorption onto the porous media. However, benzene and toluene migration in a double-lithology soil column packed with sand and silt is significantly affected by retention due to the higher adsorption induced by the 10 cm silt. The effect of GTF is lower in the single-lithology soil column than in the double-lithology soil column.

(2) The migration of benzene and toluene in the GTF zone is complex, and there is no clear relationship between the contaminant concentration change trend and the water table fluctuation. However, the contaminant concentration near and in the saturated zone increases as the water table falls, and decreases as the water table rises, especially when the water table near the sample collection point and dissolution mainly correlate with the BTEX migration in the saturated zone.

(3) For a contaminated site with a single-lithology structure consisting of sand, GTF promotes downward BTEX migration, and the contaminant concentration in the saturated zone increases, which is why more attention should be paid to organic contaminant removal within the groundwater. A contaminated site with a double-lithology structure containing silt has more retention contamination with BTEX and is more suited to the removal of organic contaminants from the silt layer. The bacterial community structures in the two columns are very different, especially in the GTF zone, and further research can provide important data for BTEX biodegradation. Biodegradation kinetics should be considered in long-term prediction models of groundwater table fluctuations and should be better understood for the remediation of BTEX compounds.

Author Contributions: All authors contributed to the writing of this paper. Conceptualization, Y.Y. and J.Z.; methodology, Y.Y.; software, J.Z.; validation, J.L., Y.Y. and H.Z.; formal analysis, J.Z.; investigation, Y.Y.; resources, Y.M.; data curation, Y.Y.; writing—original draft preparation, Y.Y.; writing—review and editing, J.Z.; visualization, J.L.; supervision, Y.Y.; project administration, H.Z.; funding acquisition, Y.M. All authors have read and agreed to the published version of the manuscript.

Funding: This study was funded by the National Natural Science Foundation of China (No. 42077169 and 41703114) and the National Key R&D Program of China (2020YFC1808304-001 and 2020YFC1806301-02).

Institutional Review Board Statement: Not applicable.

Informed Consent Statement: Not applicable.

Data Availability Statement: The data that support the findings of this study are available from the corresponding author (Yang Y.) upon reasonable request.

Acknowledgments: The authors are grateful for the financial support provided by the National Natural Science Foundation of China (No. 42077169 and 41703114) and the National Key R&D Program of China (2020YFC1808304-001 and 2020YFC1806301-02). The authors would like to thank the reviewers for their constructive suggestions.

Conflicts of Interest: The authors declare no conflict of interest.

References

1. Li, Y.; Bai, G.; Zou, X.; Qu, J.; Wang, L. Nitrogen Migration and Transformation Mechanism in the Groundwater Level Fluctuation Zone of Typical Medium. *Water* **2021**, *13*, 3626. [CrossRef]
2. Technical Guide For Addressing Petroleum Vapor Intrusion At Leaking Underground Storage. U.S. Environmental Protection Agence, Office of Underground Storage Tanks, Washington, DC, USA. 2015. Available online: https://www.epa.gov/vaporintrusion/technical-guide-addressing-petroleum-vapor-intrusion-leaking-underground-storage (accessed on 10 June 2023).

3. Murphy, C.W.; Davis, G.B.; Rayner, J.L.; Walsh, T.; Bastow, T.P.; Butler, A.P.; Puzon, G.J.; Morgan, M.J. The role of predicted chemotactic and hydrocarbon degrading taxa in natural source zone depletion at a legacy petroleum hydrocarbon site. *J. Hazard. Mater.* **2022**, *430*, 128482. [CrossRef] [PubMed]
4. Lari, K.S.; Rayner, J.L.; Davis, G.B.; Johnston, C.D. LNAPL Recovery Endpoints: Lessons Learnt Through Modeling, Experiments, and Field Trials. *Groundw. Monit. Remediat.* **2020**, *40*, 21–29. [CrossRef]
5. Teng, M.; Zhou, Y.; Song, M.; Dong, K.; Chen, X.; Wang, C.; Bi, S.; Zhu, W. Chronic Toxic Effects of Flutolanil on the Liver of Zebrafish (*Danio rerio*). *Chem. Res. Toxicol.* **2019**, *32*, 995–1001. [CrossRef]
6. Yang, Y.; Li, J.; Lv, N.; Wang, H.; Zhang, H. Multiphase migration and transformation of BTEX on groundwater table fluctuation in riparian petrochemical sites. *Environ. Sci. Pollut. Res.* **2023**, *30*, 55756–55767. [CrossRef]
7. Yang, Y.; Li, P.; Zhang, X.; Li, M.; Lu, Y.; Xu, B.; Yu, T. Lab-based investigation of enhanced BTEX attenuation driven by groundwater table fluctuation. *Chemosphere* **2017**, *169*, 678–684. [CrossRef]
8. Cavelan, A.; Golfier, F.; Colombano, S.; Davarzani, H.; Deparis, J.; Faure, P. A critical review of the influence of groundwater level fluctuations and temperature on LNAPL contaminations in the context of climate change. *Sci. Total Environ.* **2021**, *806*, 150412. [CrossRef]
9. Kamon, M.; Li, Y.; Flores, G.; Inui, T.; Katsumi, T. Experimental and Numerical Study on Migration of LNAPL under the Influence of Fluctuating Water Table in Subsurface. *Annu. Disas Prev. Res. Inst.* **2006**, *49*, 383–392.
10. Boutt, D.F. Assessing hydrogeologic controls on dynamic groundwater storage using long-term instrumental records of water table levels. *Hydrol. Process.* **2017**, *31*, 1479. [CrossRef]
11. Jeong, J.; Jeong, J.; Park, E.; Lee, B.S.; Song, S.-H.; Han, W.S.; Chung, S. Development of an efficient data-driven method to estimate the hydraulic properties of aquifers from groundwater level fluctuation pattern features. *J. Hydrol.* **2020**, *590*, 125453. [CrossRef]
12. Hillel, D.J.N. *Fundamentals of Soil Physics*; Academic Press: San Diego, CA, USA, 1980.
13. Lehmann, P.; Stauffer, F.; Hinz, C.; Dury, O.; Flühler, H. Effect of hysteresis on water flow in a sand column with a fluctuating capillary fringe. *J. Contam. Hydrol.* **1998**, *33*, 81–100. [CrossRef]
14. Cui, M.; Li, Y.; Xu, D.; Lu, J.; Gao, B. Geochemical characteristics and ecotoxicological risk of arsenic in water-level-fluctuation zone soils of the Three Gorges Reservoir, China. *Sci. Total Environ.* **2023**, *881*, 163495. [CrossRef] [PubMed]
15. Ye, C.; Cheng, X.; Liu, W.; Zhang, Q. Revegetation impacts soil nitrogen dynamics in the water level fluctuation zone of the Three Gorges Reservoir, China. *Sci. Total Environ.* **2015**, *517*, 76–85. [CrossRef] [PubMed]
16. Nanditha, H.S.; Reshmidevi, T.V.; Simha, L.U.; Kunhikrishnan, P. Statistical analysis of rainfall and groundwater interaction in Bhadra catchment. *Environ. Dev. Sustain.* **2023**, 1–21. [CrossRef]
17. Sun, Y.; Yue, G.; Ma, J. Transport and natural attenuation of benzene vapor from a point source in the vadose zone. *Chemosphere* **2023**, *323*, 138222. [CrossRef]
18. Cao, Z.; Yang, M.; Tan, T.; Song, X. Vertical Transportation Diversity of Petroleum Pollutants under Groundwater Fluctuations and the Instructions for Remediation Strategy. *Sustainability* **2023**, *15*, 6514. [CrossRef]
19. Gupta, P.K.; Yadav, B.; Yadav, B.K.; Sushkova, S.; Basu, S. Engineered Bioremediation of NAPL Polluted Sites: Experimental and Simulation-Optimization Approach under Heterogeneous Moisture and Temperature Conditions. *J. Environ. Eng.* **2021**, *147*, 04021023. [CrossRef]
20. Rani, P.; Piegari, E.; Di Maio, R.; Vitagliano, E.; Soupios, P.; Milano, L. Monitoring time evolution of self-potential anomaly sources by a new global optimization approach. Application to organic contaminant transport. *J. Hydrol.* **2019**, *575*, 955–964. [CrossRef]
21. Mao, B.; Liu, Z.; Liu, S.; Zhang, M.; Lu, T. Investigation of relative permeability, saturation and capillary pressure relations of NAPL-contaminated sands. *J. Soils Sediments* **2019**, *20*, 1609–1620. [CrossRef]
22. Liu, Y.; Sun, Y.; Yu, J.; Xia, X.; Ding, A.; Zhang, D. Impacts of groundwater level fluctuation on soil microbial community, alkane degradation efficiency and alkane-degrading gene diversity in the critical zone: Evidence from an accelerated water table fluctuation simulation. *Environ. Sci. Pollut. Res.* **2022**, *29*, 83060–83070. [CrossRef] [PubMed]
23. DeBofsky, A.; Xie, Y.; Jardine, T.D.; Hill, J.E.; Jones, P.D.; Giesy, J.P. Effects of the husky oil spill on gut microbiota of native fishes in the North Saskatchewan River, Canada. *Aquat. Toxicol.* **2020**, *229*, 105658. [CrossRef] [PubMed]
24. Schriks, M.; Heringa, M.B.; van der Kooi, M.M.E.; de Voogt, P.; van Wezel, A.P. Toxicological Relevance of Emerging Contaminants for Drinking Water Quality. *Water Res.* **2010**, *44*, 461–476. [CrossRef] [PubMed]
25. Sun, C.; Zhang, J.; Ma, Q.; Chen, Y. Human Health and Ecological Risk Assessment of 16 Polycyclic Aromatic Hydrocarbons in Drinking Source Water from a Large Mixed-Use Reservoir. *Int. J. Environ. Res. Public Health* **2015**, *12*, 13956–13969. [CrossRef] [PubMed]
26. Da Silva, M.L.B.; Corseuil, H.X. Groundwater Microbial Analysis to Assess Enhanced BTEX Biodegradation by Nitrate Injection at a Gasohol-Contaminated Site. *Int. Biodeterior. Biodegrad.* **2012**, *67*, 21–27. [CrossRef]
27. Gupta, P.K.; Yadav, B.K. Remediation and Management of Petrochemical-Polluted Sites Under Climate Change Conditions. *Environ. Contam. Ecol. Implic. Manag.* **2019**, *14*, 25–47. [CrossRef]
28. Lari, K.S.; Davis, G.B.; Rayner, J.L.; Bastow, T.P.; Puzon, G.J. Natural source zone depletion of LNAPL: A critical review supporting modelling approaches. *Water Res.* **2019**, *157*, 630–646. [CrossRef]

29. Blanco-Enríquez, E.G.; de la Serna, F.J.Z.-D.; Peralta-Pérez, M.D.R.; Ballinas-Casarrubias, L.; Salmerón, I.; Rubio-Arias, H.; Rocha-Gutiérrez, B.A. Characterization of a Microbial Consortium for the Bioremoval of Polycyclic Aromatic Hydrocarbons (PAHs) in Water. *Int. J. Environ. Res. Public Health* **2018**, *15*, 975. [CrossRef]
30. Vigneron, A.; Bishop, A.; Alsop, E.B.; Hull, K.; Rhodes, I.; Hendricks, R.; Head, I.; Tsesmetzis, N. Microbial and Isotopic Evidence for Methane Cycling in Hydrocarbon-Containing Groundwater from the Pennsylvania Region. *Front. Microbiol.* **2017**, *8*, 593. [CrossRef]
31. Zhou, A.-X.; Zhang, Y.-L.; Dong, T.-Z.; Yu-Ling, Z.; Su, X.-S. Response of the microbial community to seasonal groundwater level fluctuations in petroleum hydrocarbon-contaminated groundwater. *Environ. Sci. Pollut. Res.* **2015**, *22*, 10094–10106. [CrossRef]
32. Ishida, C.K.; Kelly, J.J.; Gray, K.A. Effects of variable hydroperiods and water level fluctuations on denitrification capacity, nitrate removal, and benthic-microbial community structure in constructed wetlands. *Ecol. Eng.* **2006**, *28*, 363–373. [CrossRef]
33. Rainwater, K.; Mayfield, M.P.; Heintz, C.; Claborn, B.J. Enhanced *in situ* biodegradation of diesei fuel by cyclic vertical water table movement: Preliminary studies. *Water Environ. Res.* **1993**, *65*, 717–725. [CrossRef]
34. Sinke, A.; Dury, O.; Zobrist, J. Effects of a fluctuating water table: Column study on redox dynamics and fate of some organic pollutants. *J. Contam. Hydrol.* **1998**, *33*, 231–246. [CrossRef]
35. Kalantar, M.; Zamir, S.M.; Ferdowsi, M.; Shojaosadati, S.A. Removal of toluene in a biotrickling filter in the presence of methanol vapors: Experimental study, mathematical modeling, and kinetic parameters optimization. *J. Environ. Chem. Eng.* **2020**, *9*, 104617. [CrossRef]
36. Aminnaji, M.; Rabbani, A.; Niasar, V.J.; Babaei, M. Effects of Pore-Scale Heterogeneity on Macroscopic NAPL Dissolution Efficiency: A Two-Scale Numerical Simulation Study. *Water Resour. Res.* **2019**, *55*, 8779–8799. [CrossRef]
37. Lee, C.-H.; Lee, J.-Y.; Cheon, J.-Y.; Lee, K.-K. Attenuation of Petroleum Hydrocarbons in Smear Zones: A Case Study. *J. Environ. Eng.* **2001**, *127*, 639–647. [CrossRef]
38. Teramoto, E.H.; Chang, H.K. Field data and numerical simulation of btex concentration trends under water table fluctuations: Example of a jet fuel-contaminated site in Brazil. *J. Contam. Hydrol.* **2017**, *198*, 37–47. [CrossRef]
39. Xianyin, Z.; Lianying, A.; Chunge, L.; Li, C. The Dissolution Experiment of Polyhalite at Different Temperature and Pressure. *Acta Geol. Sin. (Engl. Ed.)* **2014**, *88*, 412–413.
40. Ministry of Housing and Urban-Rural Development of the People's Republic of China. *Code for Investigation of Geotechnical Engineering GB50021-2001*; China Planning Press: Beijing, China, 2002.
41. Olivella, L.; Figueras, M.; Fraile, J.; Vilanova, M.; Ginebreda, A.; Barcelo, D. Fate of MTBE and DCPD Compounds Relative to BTEX in Gasoline Contaminated Aquifers. *J. Sci. World J.* **2002**, *2*, 1108–1114. [CrossRef]
42. Handelsman, J.; Rondon, M.R.; Brady, S.F.; Clardy, J.; Goodman, R.M. Molecular biological access to the chemistry of unknown soil microbes: A new frontier for natural products. *Chem. Biol.* **1998**, *5*, R245–R249. [CrossRef]
43. Knight, R.; Jansson, J.; Field, D.; Fierer, N.; Desai, N.; A Fuhrman, J.; Hugenholtz, P.; van der Lelie, D.; Meyer, F.; Stevens, R.; et al. Unlocking the potential of metagenomics through replicated experimental design. *Nat. Biotechnol.* **2012**, *30*, 513–520. [CrossRef]
44. Piao, H.; Tao, S.; Hu, H.; Lu, X. Estimation of sorption coefficient so for organic compounds with K OW. *Environ. Sci. Technol.* **1999**, *4*, 8–13.
45. Gao, C.; Govind, R.; Tabak, H.H. Predicting Soil Sorption Coefficients of Organic Chemicals Using a Neural Network Model. *Environ. Toxicol. Chem.* **1996**, *15*, 1089–1096. [CrossRef]
46. Padhi, S.K.; Gokhale, S. Benzene biodegradation by indigenous mixed microbial culture: Kinetic modeling and process optimization. *Int. Biodeterior. Biodegrad.* **2017**, *119*, 511–519. [CrossRef]
47. Muller, F.L.L.; Tankéré-Muller, S.P.C. Seasonal Variations in Surface Water Chemistry at Disturbed and Pristine Peatland Sites in the Flow Country of Northern Scotland. *Sci. Total Environ.* **2012**, *435–436*, 351–362. [CrossRef]
48. Dehnavi, S.M.; Ebrahimipour, G. Biostimulation of Petroleum-Contaminated Soils with Synthetic and Natural Sources of NPK Fertilizer. *Soil Sediment Contam. Int. J.* **2023**, 1–14. [CrossRef]
49. Resende, J.F.; Paulino, I.M.R.; Bergamasco, R.; Vieira, M.F.; Vieira, A.M.S. Hydrogels produced from natural polymers: A review on its use and employment in water treatment. *Braz. J. Chem. Eng.* **2022**, *40*, 23–38. [CrossRef]
50. Sun, J.; Lin, G.; Henghua, Z.; Tang, X.; Zhang, L. Study on Adsorption-Desorption of Benzene in Soil. *IOP Conf. Series Earth Environ. Sci.* **2020**, *546*, 042041. [CrossRef]
51. Han, C.; Zhang, H.; Gu, Q.; Guo, G.; Li, Y.; Li, F. Toluene sorption behavior on soil organic matter and its composition using three typical soils in China. *Environ. Earth Sci.* **2012**, *68*, 741–747. [CrossRef]
52. Liu, X.; Luo, Y.; Zhang, H.; Wu, J.; Zhu, R.; Wang, H. Spatial heterogeneity of particulate organic matter for the sorption of ciprofloxacin at the microstructure scale. *Sci. Total Environ.* **2022**, *847*, 157326. [CrossRef]
53. Lion, L.W.; Stauffef, T.B.; MacIntyre, W.G. Sorption of hydrophobic compounds on aquifer materials: Analysis methods and the effect of organic carbon. *J. Contam. Hydrol.* **1990**, *5*, 215–234. [CrossRef]
54. Viggi, C.C.; Tucci, M.; Resitano, M.; Palushi, V.; Crognale, S.; Matturro, B.; Papini, M.P.; Rossetti, S.; Aulenta, F. Enhancing the Anaerobic Biodegradation of Petroleum Hydrocarbons in Soils with Electrically Conductive Materials. *Bioengineering* **2023**, *10*, 441. [CrossRef] [PubMed]
55. Dobson, R.; Schroth, M.H.; Zeyer, J. Effect of water-table fluctuation on dissolution and biodegradation of a multi-component, light nonaqueous-phase liquid. *J. Contam. Hydrol.* **2007**, *94*, 235–248. [CrossRef] [PubMed]

56. Rezanezhad, F.; Couture, R.-M.; Kovac, R.; O'connell, D.; Van Cappellen, P. Water table fluctuations and soil biogeochemistry: An experimental approach using an automated soil column system. *J. Hydrol.* **2013**, *509*, 245–256. [CrossRef]
57. Abed, R.M.M.; Safi, N.M.D.; Köster, J.; de Beer, D.; El-Nahhal, Y.; Rullkötter, J.; Garcia-Pichel, F. Microbial Diversity of a Heavily Polluted Microbial Mat and Its Community Changes following Degradation of Petroleum Compounds. *Appl. Environ. Microbiol.* **2002**, *68*, 1674–1683. [CrossRef]
58. Jiang, J.; Liu, Z.; Li, B.; Yuan, S.; Lin, R.; Yu, X.; Liu, X.; Zhang, X.; Li, K.; Xiao, D.; et al. Ecotoxicological risk assessment of 14 pesticides and corresponding metabolites to groundwater and soil organisms using China-PEARL model and RQ approach. *Environ. Geochem. Health* **2022**, *45*, 3653–3667. [CrossRef]
59. Teramoto, E.H.; Pede, M.A.Z.; Chang, H.K. Impact of water table fluctuations on the seasonal effectiveness of the pump-and-treat remediation in wet–dry tropical regions. *Environ. Earth Sci.* **2020**, *79*, 1–15. [CrossRef]

toxics

MDPI

Review

Toxicity of Heavy Metals and Recent Advances in Their Removal: A Review

Manar K. Abd Elnabi [1,2], Nehal E. Elkaliny [1], Maha M. Elyazied [1], Shimaa H. Azab [1], Shawky A. Elkhalifa [1], Sohaila Elmasry [3], Moustafa S. Mouhamed [1], Ebrahim M. Shalamesh [1], Naira A. Alhorieny [1], Abeer E. Abd Elaty [1], Ibrahim M. Elgendy [1], Alaa E. Etman [1], Kholod E. Saad [1], Konstantina Tsigkou [4], Sameh S. Ali [1,5,*], Michael Kornaros [4,*] and Yehia A.-G. Mahmoud [1]

1 Botany Department, Faculty of Science, Tanta University, Tanta 31527, Egypt; manar.abdelnabi@ejust.edu.eg (M.K.A.E.); nehal31012429@science.tanta.edu.eg (N.E.E.); maha31012879@science.tanta.edu.eg (M.M.E.); shaimaa31012609@science.tanta.edu.eg (S.H.A.); shawqe31012093@science.tanta.edu.eg (S.A.E.); mostafa30974505@science.tanta.edu.eg (M.S.M.); ibrahem231058915@science.tanta.edu.eg (E.M.S.); nira231058550@science.tanta.edu.eg (N.A.A.); abeer31012615@science.tanta.edu.eg (A.E.A.E.); ibrahim31012107@science.tanta.edu.eg (I.M.E.); alaa231058671@science.tanta.edu.eg (A.E.E.); khaloud31012947@science.tanta.edu.eg (K.E.S.); yehiamahmoud@science.tanta.edu.eg (Y.A.-G.M.)

2 Biotechnology Program, Institute of Basic and Applied Science (BAS), Egypt-Japan University of Science and Technology, New Borg El-Arab City 21934, Egypt

3 Microbiology Department, Faculty of science, Damanhour University, Behaira 22514, Egypt; s.abdelazeez00804@sci.dmu.edu.eg

4 Department of Chemical Engineering, University of Patras, 1 Karatheodori str, 26504 Patras, Greece; ktsigkou@chemeng.upatras.gr

5 Biofuels Institute, School of the Environment and Safety Engineering, Jiangsu University, Zhenjiang 212013, China

* Correspondence: samh@ujs.edu.cn or samh_samir@science.tanta.edu.eg (S.S.A.); kornaros@chemeng.upatras.gr (M.K.)

Citation: Abd Elnabi, M.K.; Elkaliny, N.E.; Elyazied, M.M.; Azab, S.H.; Elkhalifa, S.A.; Elmasry, S.; Mouhamed, M.S.; Shalamesh, E.M.; Alhorieny, N.A.; Abd Elaty, A.E.; et al. Toxicity of Heavy Metals and Recent Advances in Their Removal: A Review. *Toxics* **2023**, *11*, 580. https://doi.org/10.3390/toxics11070580

Academic Editors: Junhao Qin, Peidong Su, Feng Zhu and Lin Ding

Received: 28 May 2023
Revised: 14 June 2023
Accepted: 24 June 2023
Published: 3 July 2023

Abstract: Natural and anthropogenic sources of metals in the ecosystem are perpetually increasing; consequently, heavy metal (HM) accumulation has become a major environmental concern. Human exposure to HMs has increased dramatically due to the industrial activities of the 20th century. Mercury, arsenic lead, chrome, and cadmium have been the most prevalent HMs that have caused human toxicity. Poisonings can be acute or chronic following exposure via water, air, or food. The bioaccumulation of these HMs results in a variety of toxic effects on various tissues and organs. Comparing the mechanisms of action reveals that these metals induce toxicity via similar pathways, including the production of reactive oxygen species, the inactivation of enzymes, and oxidative stress. The conventional techniques employed for the elimination of HMs are deemed inadequate when the HM concentration is less than 100 mg/L. In addition, these methods exhibit certain limitations, including the production of secondary pollutants, a high demand for energy and chemicals, and reduced cost-effectiveness. As a result, the employment of microbial bioremediation for the purpose of HM detoxification has emerged as a viable solution, given that microorganisms, including fungi and bacteria, exhibit superior biosorption and bio-accumulation capabilities. This review deals with HM uptake and toxicity mechanisms associated with HMs, and will increase our knowledge on their toxic effects on the body organs, leading to better management of metal poisoning. This review aims to enhance comprehension and offer sources for the judicious selection of microbial remediation technology for the detoxification of HMs. Microbial-based solutions that are sustainable could potentially offer crucial and cost-effective methods for reducing the toxicity of HMs.

Keywords: heavy metals; bioremediation; heavy metals toxicity; biosorption; biomineralization; biotransformation

1. Introduction

Heavy metals (HMs) are defined as metallic elements with a greater density than water [1]. HMs are naturally occurring elements with a large atomic weight and at least five times the density of water [2]. Assuming that heaviness and toxicity are intertwined, HMs also include metalloids, such as arsenic (As), that can induce toxicity at low exposure levels [3]. Various heavy metals, such as chromium (Cr), cadmium (Cd), nickel (Ni), copper (Cu), zinc (Zn), lead (Pb), mercury (Hg), and As, are recognized as biologically dispensable and detrimental to the aquatic ecosystem [4]. Anthropogenic sources, including but not limited to the nonferrous metallurgical industry, mining, mineral processing, electroplating, leather tanning, and chemical industries, are commonly attributed to the discharge of hazardous metals such as HMs [5]. The level of toxicity exhibited by these hazardous materials is contingent upon the amount and length of time that living organisms are exposed to them. Figure 1 depicts the origins and impacts of HMs on humans through the food chain. Prolonged exposure to HMs through dermal contact, ingestion of contaminated food, and inhalation can result in the development of diverse disorders in both humans and other organisms [6]. In addition, the issue of soils contaminated with HM has become an increasingly significant environmental concern worldwide [7].

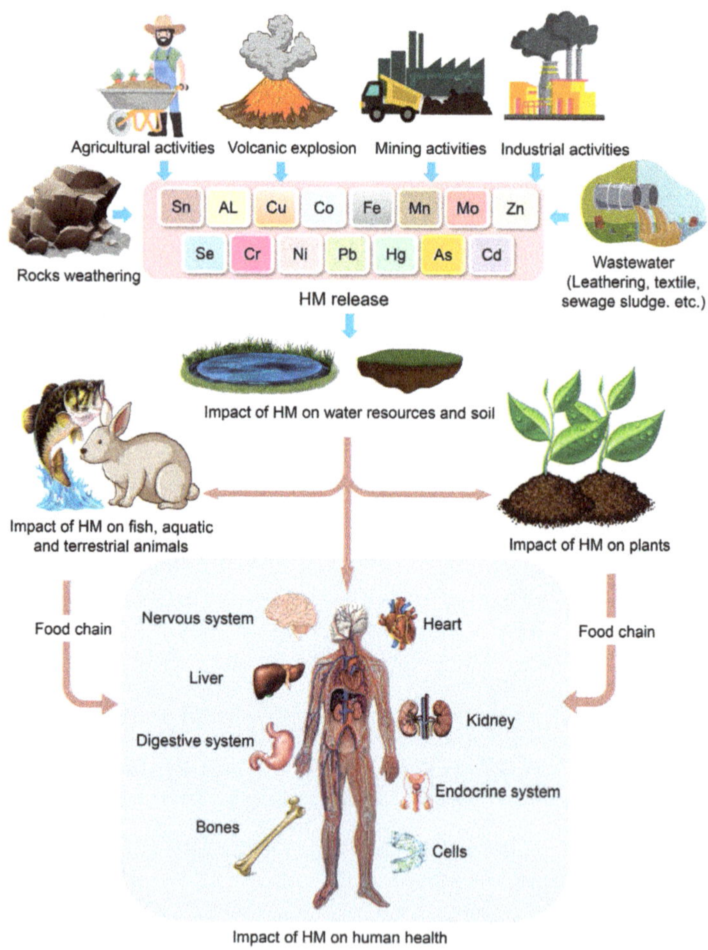

Figure 1. Origins and impacts of HMs on humans through the food chain.

The presence of HMs in composting materials poses a considerable obstacle to their utilization. Consequently, there has been a significant focus on mitigating the toxicity of HMs during the composting process. The composting of HMs primarily impacts microorganisms, and the toxicity levels of different HMs exhibit variability [8]. Various techniques, including chemical, physical, or a combination of both, can be employed to degrade HMs. A considerable number of these techniques are neither ecologically sustainable nor financially viable. Microbial remediation is a viable alternative technique due to the efficient adaptability of microorganisms to their environment [9–14]. The remediation of HMs through microbial bioremediation involves various mechanisms, including bioaugmentation, biosorption, biomineralization, and biotransformation [15]. The implementation of various techniques such as remediation, microbial fuel cells, biofilm, nanomaterials, cell immobilization, and genetic engineering have shown great potential in the removal of HMs. The aim of this review is to provide updated knowledge regarding the abundance of HMs and their potential toxicity on living organisms and plants, as well as their negative impact on human health. It additionally offers mechanistic insights and highlights research uncertainties pertaining to HM remediation via biological and green remediation approaches.

2. Impact of HMs on Human Health

The exposure of humans to HMs is a result of industrial activities and can manifest through various means, such as ingestion, inhalation, and dermal absorption [16]. HMs are categorized based on their function in biological systems as either essential or non-essential HMs. The presence of excessive amounts of essential or non-essential metals can lead to physiological or morphological disorders or genetic mutations [17]. Table 1 categorizes HMs according to their adverse impacts on the human body.

Table 1. Classification of HMs based on their impact on the human body.

HM Compounds	Impact HM on Human
Lead (Pb), mercury (Hg), arsenic (As), and cadmium (Cd)	High degree of toxicity, carcinogenic, and induce damage to multiple organs
Tin (Sn) and aluminum (Al)	Less toxic to human body
Barium (Ba) and lithium (Li)	Non-essential for physiological and biochemical functions
Copper (Cu), zinc (Zn), nickel (Ni), manganese (Mn), molybdenum (Mo), cobalt (Co), selenium (Se), chromium (Cr), and iron (Fe)	Essential for physiological and biochemical functions

The presence of HMs in the ecosystem is attributed to anthropogenic activities, which are the primary sources of human exposure to these substances [18]. HMs are found in the Earth's crust in the form of ores, which are extracted as minerals during mining operations. The majority of ores contain HMs, including but not limited to As, Fe, Pb, Zn, gold (Au), Ni, silver (Ag), and Co, in the form of sulfides. Conversely, some ores contain metals such as Mn, Al, and Se in the form of oxides. In the course of mining operations, the extraction process results in the discharge of HMs from the ore, which are subsequently dispersed into the surrounding environment. These metals remain in the soil and may be transported to other locations via air and water currents. Moreover, the utilization of HMs in industrial applications results in the discharge of certain elements into the atmosphere through combustion or into soil and water systems as effluents. Additionally, industrial commodities such as paints, cosmetics, pesticides, and herbicides can also function as origins of heavy metals (Figure 1).

HMs have the potential to be conveyed to various locations on soil and water bodies via erosion, run-off, or acid rain. The diagram presented in Figure 2 illustrates the trajectory of HM origins and the subsequent impact on human exposure. The toxicity of HMs to biological systems is caused by their ability to form bonds with sulfhydryl groups and generate reactive oxygen species (ROS). The aforementioned phenomenon results in the deactivation of crucial macromolecules, leading to the manifestation of oxidative stress and depletion of glutathione. Upon exposure to toxic metals and subsequent entry into the body, a multitude of processes occur, including the potential interaction or inhibition of certain metabolic pathways [19]. Consequently, a plethora of deleterious impacts on both human and animal populations are observed. The aforementioned conditions encompass a range of medical issues such as organ dysfunction, metabolic irregularities, hormonal imbalances, congenital anomalies, compromised immune system function, and malignancy [20,21]. Consequently, numerous international organizations establish regulations pertaining to the existence of metals in the environment, foods, and drinking water. The subsequent section delineates the origin and toxicity of distinct HMs on human health.

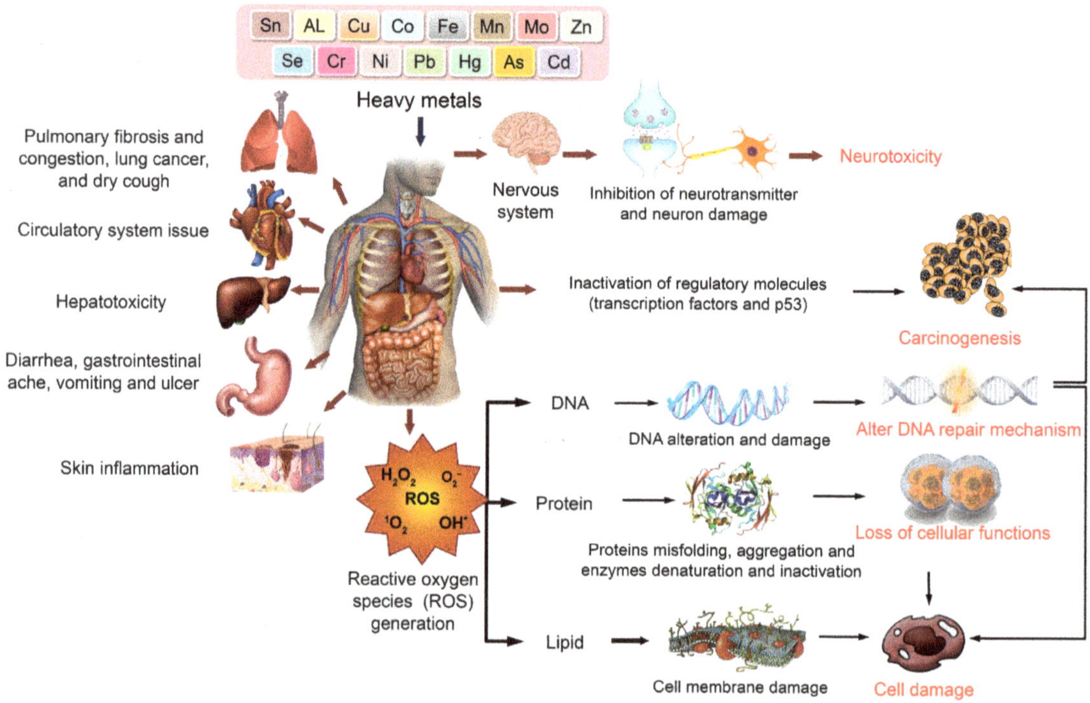

Figure 2. Oxidative stress and human organ toxicity following exposure to HMs.

2.1. *Toxicity in Humans following Exposure to HMs*

2.1.1. Effect of Arsenic

Arsenic (As) is a hazardous metallic element that can be found in various geological formations, as well as in water and air. As exhibits carcinogenic properties in humans and is associated with various detrimental health outcomes in both the short and prolonged periods [22]. The process of As biotransformation within the human body involves the methylation of various As compounds, ultimately resulting in the production of detrimental end metabolites such as mono-methylmalonic acid (MMA) and di-methyl arsenic acid (DMA). The thiol group of cells is impacted, leading to impaired functionality in cell enzymes, respiration, and mitotic division [8]. A correlation has been observed between exposure to As and the incidence of cardiovascular disease, including hypertension. The impact of exposure to As and its metabolites, namely MMA and DMA, on hypertension during pregnancy has been reported [23]. The findings revealed that pregnant women with low concentrations of DMA experienced an increase in systolic, diastolic, and main arterial pressure, which could potentially result in adverse cardiovascular health outcomes for both the mother and the child [23].

2.1.2. Effect of Mercury

Mercury (Hg) is a hazardous HM that is ubiquitous in the environment. It has the potential to undergo methylation, leading to the formation of methylmercury (MeHg), which can accumulate in the food web. The consumption of seafood has been linked to human exposure to Hg [24]. Lipoperoxidation and nitrite concentration were found to be elevated while total antioxidant capacity was reduced as a result of the presence of Hg in the forms of HgII and MeHg [25]. Multiple studies have demonstrated that exposure to Hg can have significant adverse effects on children. Specifically, prenatal exposure to Hg has been found to negatively impact child growth, which may be attributed to a decrease in parasympathetic modulation of cardiac autonomic function in children [26]. A positive correlation has been reported between exposure to Hg and blood pressure measures during childhood [27]. There is also a positive association between high levels of Hg and dyslipidemia in adults [28]. Several studies on the effect of Hg on liver function found that liver enzymes increased significantly with Hg exposure [29].

2.1.3. Effect of Lead

Lead (Pb) is a toxic HM that has the tendency to accumulate in various tissues including blood, bones, and most body organs [30]. Exposure to Pb has been associated with various neurological disorders such as Alzheimer's disease, Parkinson's disease, amyotrophic lateral sclerosis, and attention-deficit/hyperactivity disorder [31]. There is an observed association between Pd and cardiovascular diseases, and it is regarded as a risk factor for vascular complications in individuals with diabetes [32]. The impact of Pb exposure on liver and kidney functions, as well as on white blood cell count, serum urea, creatinine, aspartate transaminase, alanine transaminase, hemoglobin, and hematocrit concentrations, was observed to be more pronounced in the high blood lead level group as compared to the low blood lead level group [33]. Numerous studies have extensively examined the impact of Pb on children. The findings of these studies have demonstrated that exposure to Pb has adverse effects on the physical growth of children, particularly boys [34]. Exposure to Pd has the potential to alter the concentration of sex hormones, thereby impacting the functioning of the reproductive system [35].

2.1.4. Effect of Chromium

The accumulation of chromium (Cr) in body organs has been found to have potential implications for human health [36]. Cr has a deleterious impact on bronchial epithelial cells, potentially through the aberrant modulation of apoptosis-related proteins, cytoskeletal proteins, and energy metabolism-associated proteins [37]. The impact of Cr on fetal growth during pregnancy has been investigated [38]. The findings of the study suggest that Cr may have a toxic effect on fetal growth. Cr is regarded as a carcinogen that has been linked to the development of lung cancer [39]. The manifestation of hyperpigmented skin is a consequence of exposure to elevated levels of Cr [40].

2.1.5. Effect of Copper

Copper (Cu) is a vital micronutrient for humans. However, excessive levels of Cu can lead to toxic and harmful effects [41]. The accumulation of Cu has been observed to induce mutations and aggregation of mitochondrial protein, leading to a decrease in the activity of the primary antioxidant enzyme and an increase in the production of toxic ROS [42]. A correlation between the elevation of Cu and fibrosis in renal tissues has been established [43]. The presence of Cu has a detrimental impact on male reproductive function, leading to a decrease in sperm count and mobility [44]. There exists a correlation between exposure to copper and obesity [45]. The presence of Cu has an impact on the maintenance of homeostasis of crucial elements such as Ca, Fe, and Mn, which are linked to the induction of oxidative stress. This, in turn, can result in the development of neurodegenerative disorders [46].

2.1.6. Effect of Nickel

Nickel (Ni) is ubiquitously distributed in various environmental compartments such as air, water, and soil [47]. The adverse impacts of Ni on humans have been explored, particularly in the context of pregnancy. The findings revealed a positive association between Ni exposure in pregnant women and preterm delivery [48]. Individuals diagnosed with diabetes exhibited elevated concentrations of Ni in their urine [49]. Offspring may experience congenital heart defects that are associated with maternal exposure to Ni [50]. Ni has been found to have a correlation with immunological disorders, as well as the occurrence of Type I hypersensitivity and Type IV immune reactions in individuals experiencing chronic systemic symptoms associated with Ni [51].

2.1.7. Effect of Uranium

Recent ecological research has examined the potential correlation between prolonged consumption of uranium (U) through drinking water and an elevated risk of leukemia as well as kidney, lung, and colorectal cancer in both genders [52]. It has been evidenced that the absorption of U leads to an increase in ROS, DNA strand breaks, and alterations in gene expression, all of which have adverse clinical consequences [53]. The hexavalent uranyl ion (UO_2^{2+}) has been observed to accumulate in kidney and bone tissues. This accumulation has been found to cause acute and chronic damage to the kidneys, while also increasing the likelihood of osteosarcoma and ontogenesis [54]. Exposure to U either accidentally or through contamination of food or water can lead to a decline in hematopoietic function or bone marrow illness, which can subsequently result in a range of systemic effects [55].

2.1.8. Effect of Cadmium

Cadmium (Cd) is a toxic HM that has detrimental effects on human health [47]. The impact of Cd on the vascular endothelium has been observed to stimulate the release of antithrombolytic substances, leading to the production of various inflammatory indicators [56]. Vijayakumar et al. [57] have conducted a study to ascertain the biological effects of Cd on the proliferation and metastasis of prostate cancer and basal breast tumors. Breast tumor cells may experience a reduction in their anti-oxidative defenses, leading to the initiation of ROS formation upon exposure to Cd [58]. Furthermore, the proximal tubule is the site of Cd accumulation and nephron destruction [59].

2.1.9. Effect of Iron

Iron (Fe) is an essential element for various biological processes, including but not limited to DNA replication, mitochondrial respiration, and oxygen transport, that are crucial for the survival of almost all living organisms. Fe is a redox-active element that plays a role in the generation of ROS, which can cause damage to cellular membranes, DNA, and proteins [60]. Fe is an indispensable constituent of human physiology and participates in numerous cellular metabolic pathways, such as the conveyance of oxygen [61]. Fe deficiency anemia is the most commonly occurring form of anemia worldwide [62]. Fe deficiency can have adverse effects on the development and functioning of the immune system [63]. Fe deficiency during pregnancy can pose risks to both the mother and fetus [64]. In contrast, an overabundance of iron has been associated with heightened susceptibility to cardiovascular disease, gestational diabetes, and fetal complications, as well as the generation of oxidative stress and cellular harm [65].

2.1.10. Effect of Vanadium

Vanadium (V) is present in various environmental compartments such as soil, water, and air. Various tissues and organs, such as the kidney, central nervous system, lung, lymphoid organs, and immune system, undergo histological and physiological changes as a result of this HM [66]. Exposure to V has been associated with various adverse health effects, including respiratory dysfunction, kidney toxicity, biochemical and hematologic changes, toxicity for both reproduction and development, mutagenicity, neurotoxicity, and immunotoxicity [67]. V has an impact on the digestive, respiratory, and cardiovascular systems [68]. The condition known as olfactory dysfunction is the result of exposure to low concentrations of V through the intranasal route. This condition is marked by a decrease in the volume of the olfactory bulb and a decline in dopaminergic neurotransmission to the olfactory bulb [69].

2.1.11. Effect of Cobalt

Cobalt (Co) is a mineral that occurs naturally and can be detected in over a hundred organic and inorganic compounds [70]. The exposure to Co has been associated with inflammations of the upper respiratory tract, including rhinitis and bronchitis, as well as respiratory ailments affecting the lower respiratory system. Simultaneous exposure to certain substances can result in the development of fibrotic alterations in pulmonary tissue, which can subsequently lead to the onset of asthma [71]. Exposure to CO may result in various adverse effects, including lung fibrosis, hepatotoxicity, and carcinogenesis [72]. Exposure to Co can result in various neural system disorders such as memory loss, neuropathies, optic atrophy, and bilateral nerve deafness [73]. The impact of Co on the heart can result in instances of single cardiomyopathy, hypertension, and reversible electrocardiographic alterations [74].

2.1.12. Effect of Thallium

Thallium (TI) exhibits greater toxicity than Hg, Cd, and Pb [75]. Chronic TI poisoning, characterized by symptoms such as anorexia and headaches, can result from prolonged exposure to low levels of TI [76]. Respiratory muscle paralysis may lead to a state of coma in severe instances [77]. One of the characteristic indications of TI poisoning is the occurrence of hair loss subsequent to the contraction of hair follicles. Additional symptoms encompass issues related to digestion, discomfort, mental health, and cardiovascular system complications [78]. The incidence of fetal mortality, preterm birth or low birth weight has been found to be associated with Tl toxicity during pregnancy [79].

2.2. Mechanism of Intoxication following Exposure to HMs

Upon ingestion via food or water, HMs undergo acidification within the stomach's acidic environment. Under acidic conditions, the aforementioned elements undergo oxidation and attain their respective oxidative states (Zn^{2+}, As^{3+}, Cd^{2+}, As^{2+}, and Pb^{2+}). These states can effectively interact with biological molecules, including proteins and enzymes, to form robust and enduring bonds [80]. The aggregation of proteins may be induced by HM, as evidenced by the observed As-induced protein aggregation, which was found to be dependent on concentration. Furthermore, the aggregates exhibited a diverse array of proteins that were notably enriched in functionalities pertaining to metabolic processes, protein folding, protein synthesis, and stabilization [81]. The ability of these agents to induce protein aggregation in vivo is likely contingent upon their cellular uptake/export efficiency and their unique biological mechanisms. Figure 3 depicts the diverse mechanisms involved in HM intoxication.

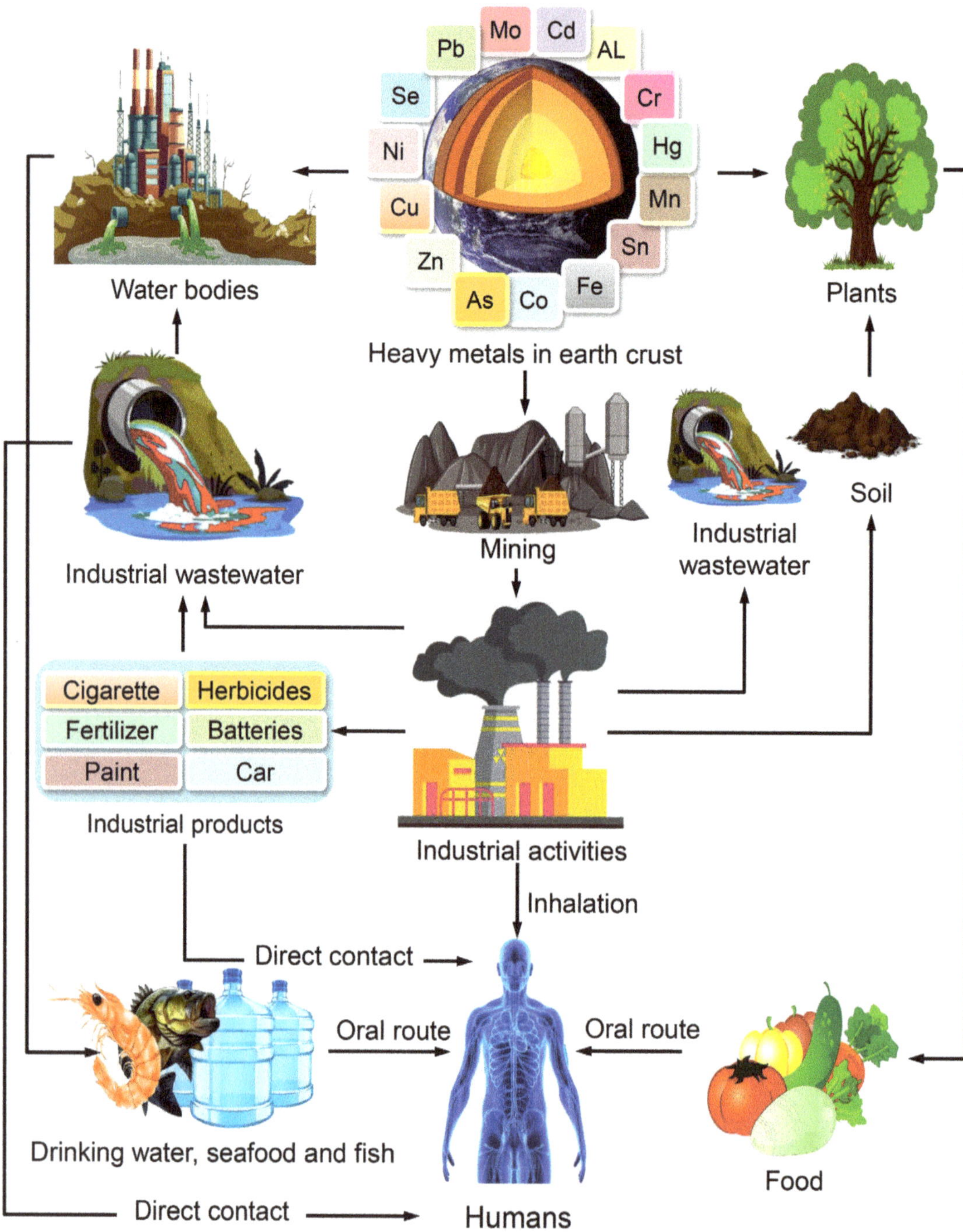

Figure 3. Intoxication mechanisms in humans following exposure to HMs.

3. Impact of HMs on Aquatic Animals

HMs are naturally occurring elements that are also generated through human activities, such as industrialization and agriculture. These HMs can enter aquatic ecosystems through direct and indirect means. The transportation of HMs occurs in tandem with the water cycle, whereby they are conveyed from upstream to downstream along river courses, ultimately depositing in vast water bodies such as large lakes, rivers, or oceans [82]. Whilst HMs have the capability to enter water directly, a considerable quantity of HMs are assimilated into the ground and sediment, which function as a conveyance or prospective origin of HMs in aquatic environments [83]. It has been observed that sediment-bound HMs have the potential to migrate into water, and the rate of their release is influenced by various factors such as the type of HMs, as well as the physico-chemical properties of both the water and sediment, including pH, salinity, specific surface area, and cation exchange capacity [83]. Aquatic invertebrates can ingest and accumulate HMs in lakes, rivers, and marine environments through either water or food pathways. HMs have the potential to undergo transfer from aquatic invertebrates to higher trophic levels, ultimately leading to biomagnification. This process can result in harmful consequences for all organisms, including humans. The ingestion of various HMs by aquatic invertebrates has the potential to impact aquatic ecosystems, as depicted in Figure 4.

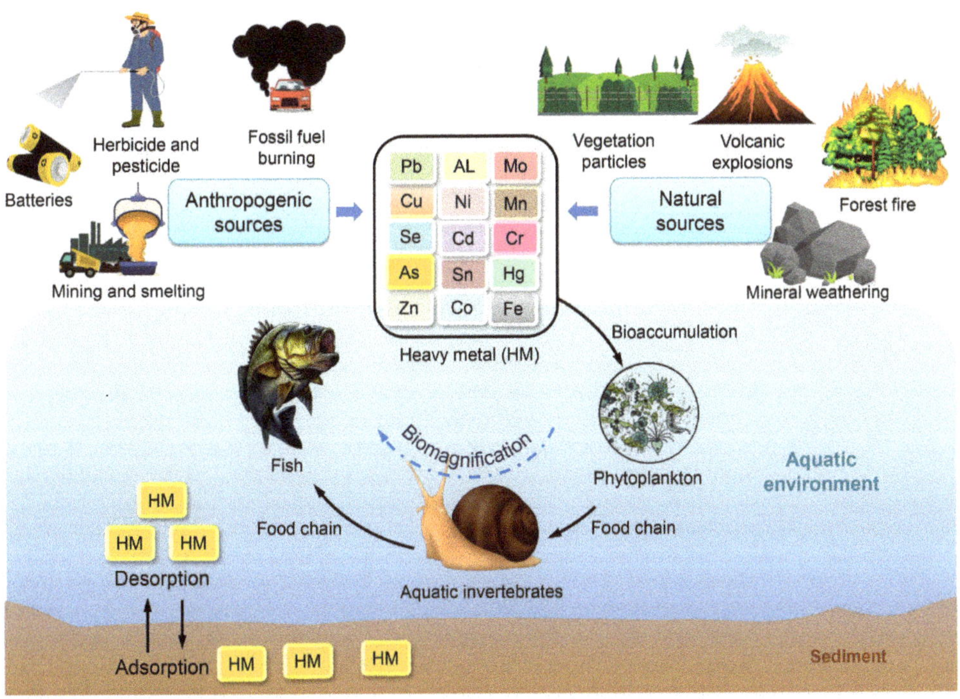

Figure 4. Distribution of HMs in aquatic environments.

The biological properties of pure elements and their compounds may exhibit differences in toxicity when considering metals, owing to variations in solubility, oxidation state, and bioavailability. The presence of Cr has been found to impede the growth and reproductive capabilities of aquatic invertebrates, induce oxidative stress and DNA damage, and potentially result in teratogenic and carcinogenic effects [84]. Cu has been found to have various detrimental impacts on aquatic organisms, including but not limited to inducing oxidative stress, modifying enzyme activity, impeding growth or reproduction, interfering with the endocrine system, and diminishing energy acquisition [85]. The exposure of aquatic invertebrates to Zn has the potential to induce oxidative stress and alter various physiological processes, including immune responses, metabolism, and detoxification [86]. As has the potential to cause various physiological and biochemical toxic effects, including but not limited to growth inhibition, oxidative stress, fecundity reduction, apoptosis, and immunotoxicity [87]. Even at low concentrations, Cd demonstrates deleterious impacts on aquatic invertebrates. The impact of Cd resulted in a reduction in the growth rate and reproductive capacity in freshwater fleas. Additionally, the substance caused neuronal toxic effects in the oyster [88]. The presence of Pb in aquatic ecosystems can result in toxicity, which may manifest as the inhibition of crucial enzymatic activity and the occurrence of oxidative stress-induced harm to cell membranes [89]. Hg exhibits a notable attraction towards the sulfur (-SH) group and can readily attach to the cysteine component found in proteins. This phenomenon leads to functional impairment and toxicity in aquatic invertebrates [90].

4. Impact of HMs on Soil and Plants

The stress induced by HMs is a significant factor that adversely impacts the agricultural yield of plants. HMs are widely acknowledged as a constituent of soil. However, their concentration in soil and plants can have detrimental effects on the environment [91]. HMs have the ability to transfer through the food chain after being absorbed by plants, ultimately accumulating in the bodies of animals and humans and posing potential health risks [92]. HMs have the potential to infiltrate groundwater or surface waters and subsequently undergo absorption by plants or turn into emissions as gases that are sent into the atmosphere. Alternatively, they may form semi-permanent associations with soil constituents such as clay or organic matter, ultimately impacting human health at a later time. It is imperative to keep soil pollution levels below the prescribed toxicity thresholds to ensure sustainability [93]. The utilization of chemical fertilizers and pesticides has been found to elevate the likelihood of soil contamination by HMs, which can subsequently accumulate in the tissue of crops cultivated in such contaminated soil [94]. The mobility of HMs in soil is significantly influenced by the presence of total organic matter and pH levels. Upon introduction of HMs to soil, their downward movement is limited by the soil's capacity to retain the metals or by the interaction of the metals with the adjacent waste matrix, which may enhance their mobility. High concentrations of HMs in soil can have adverse effects on the development, growth, and nitrogen fixation of legumes [95]. Microbial remediation refers to the biological process by which microorganisms adsorb or transform heavy metals into less toxic products. Microbial remediation is a process that primarily involves various mechanisms such as biosorption, bioaccumulation, bioleaching, biovolatilization, and biomineralization [96], as illustrated in Figure 5, for the purpose of addressing HM contamination in soil.

There was a decrease in the concentration of HMs in the soil as the depth increased [97]. The transport of HMs ions is influenced by both their properties and the texture of the soil. The duration required for the infiltration of HMs in sandy and compound soil is comparatively shorter than that in loess. The migration resistance of loess to all metals is significantly higher when compared to sandy and compound soil. HMs are immobilized in the soil through the formation of complexes with ligands, which can be either organic or inorganic in nature. When subjected to leaching conditions, HMs present in soil have the potential to contaminate surface water and groundwater, thereby exacerbating water pollution and adversely impacting aquatic organisms [98].

The detection of Hg in soil solutions poses a challenge owing to its robust affinity with soil particles and organic substances. Inorganic Hg can cause toxicity to humans and disrupt soil biota across all trophic levels [99]. The conversion of Hg^{2+} in soil into methylmercury, a highly toxic substance that accumulates in plants and is subsequently enriched in the food chain, poses a significant risk to human health through chronic poisoning. This process occurs through various mechanisms, including absorption, migration, and transformation [100]. The presence of Hg in soil has a notable effect on both the nitrogen cycle of the soil and the prevalence of microorganisms inhabiting the soil [99].

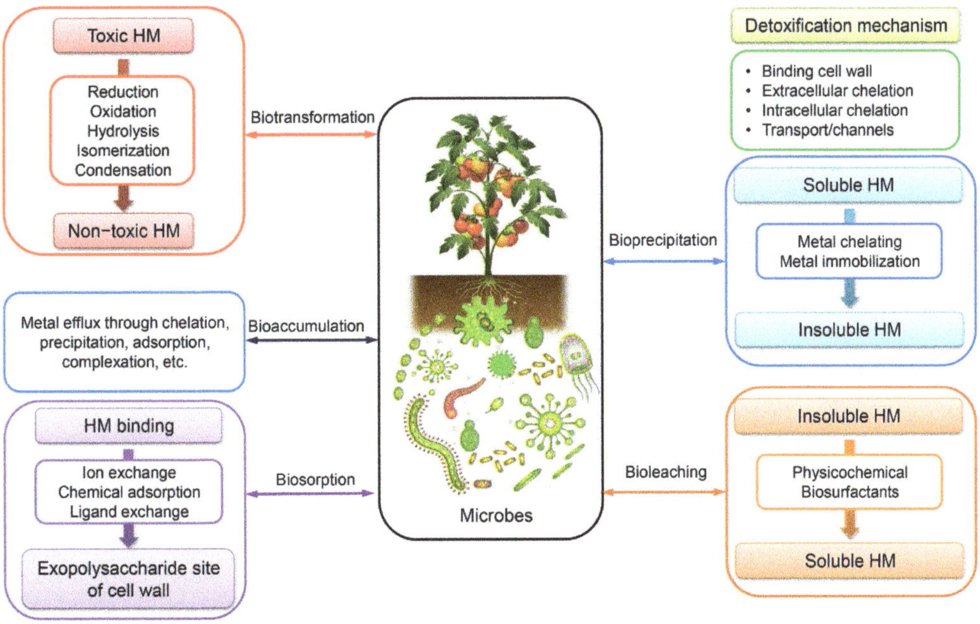

Figure 5. Microbial remediation mechanisms of soils contaminated with HMs.

Elevated concentrations of Cu have deleterious effects on various metabolic pathways in plants, including but not limited to photosynthesis, respiration, and enzymatic activity. Cu has been found to enhance the production of ROS in plants and upregulate the expression of genes associated with oxidative stress response [101]. Elevated concentrations of Cd have a significant impact on the development of soil microorganisms and the activities of enzymes. In contrast to the majority of other metals, it has been observed that Cd exhibits a higher degree of plant absorption and can translocate into the edible portions prior to the onset of phytotoxicity [91].

Cr is a noxious element for plants. Cr is a naturally occurring element that is ubiquitously distributed in the environment, encompassing the soil, air, and water compartments [102]. Cr is present in two oxidation states, namely trivalent and hexavalent, in soil. However, hexavalent Cr is comparatively more toxic and mobile than its trivalent counterpart [103]. The presence of Cr in soil has a negative impact on both soil fertility and microbial activity, ultimately resulting in decreased plant yield [104]. The presence of Cr compounds can have detrimental effects on the growth and development of plants. This can manifest in altered germination processes and hindered development of roots, stems, and leaves, ultimately leading to reduced yield and dry matter production [105]. The elevated concentration of soil Pb has been associated with certain chemical fertilizer production enterprises. Pb has a tendency to accumulate in the roots and lower above-ground portions of plants. While plant roots are not highly receptive to Pb absorption, the presence of this element can negatively impact various metabolic processes in plants, including root development, photosynthesis, and water uptake [93]. The enzymatic activity of soil biota is hindered by Pb, leading to the buildup of partially digested organic matter [106].

Phytoremediation techniques represent an environmentally friendly technology that facilitates the detoxification of polluted metal-lands, water, and groundwater through the implementation of various processes such as phytoextraction, phytostabilization, phytovolatilization, rhizofiltration, and phytofiltration [107]. The utilization of plants for the remediation of contaminated sites, known as phytoremediation, is a sustainable approach that has been extensively investigated in both field and laboratory settings. HMs exhibit resistance to remediation through both microbial and physiological mechanisms, and they persist in soil for extended periods, thereby posing a potential threat to the ecosystem [108]. It is imperative to implement remedial measures to mitigate the consumption of metals by the atmospheric, terrestrial, and aquatic ecosystems, and to curtail the contamination of groundwater [109]. Numerous phytoremediation techniques entail the utilization of rhizodegradation and phytodegradation mechanisms to facilitate the decomposition of both organic and inorganic pollutants. The diagram presented in Figure 6 illustrates the implementation of phytoremediation as a means of addressing the issue of HMs in plants.

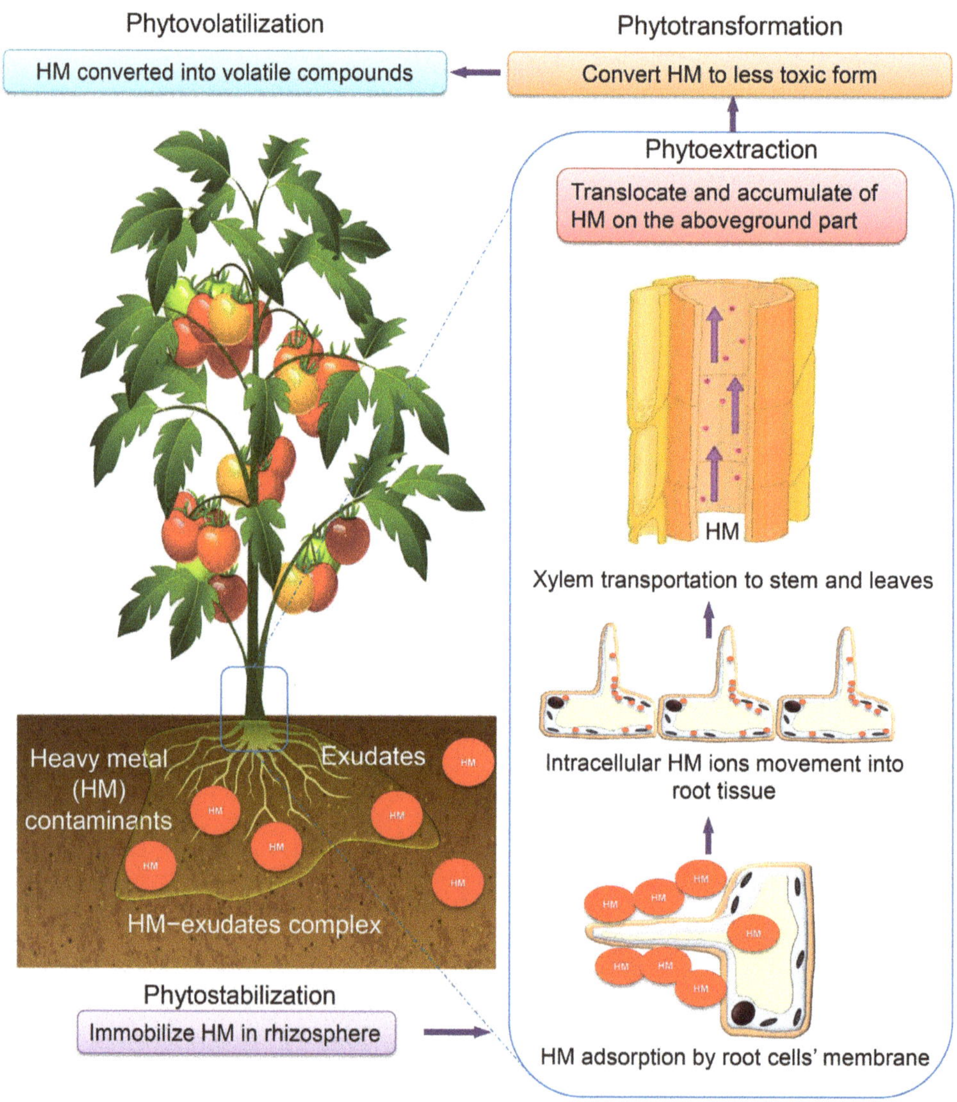

Figure 6. Transfer route and phytoremediation mechanism of HMs into plant cells and tissues.

5. Toxic Effect of HMs on Microbial Cells

The toxicity of HMs is attributed to multiple mechanisms, which include degradation of enzymatic functions leading to lethality, functioning as redox catalysts in the generation of ROS, disruption of ion regulation, and direct effects on the composition of DNA and proteins. Cu has the ability to catalyze the production of ROS, thereby serving as soluble electron carriers. The aforementioned phenomenon has the potential to induce significant harm to cytoplasmic molecules, DNA, lipids, and other proteins [110]. HMs have an impact on the rate of enzyme activity, which is influenced by various soil factors such as pH, organic content, and chemical composition. The negative effects of toxic metal-contaminated soil can be attributed to the decrease in urease, catalase, and lipolytic activity. There exists an inverse correlation between the concentration of HMs and soil enzyme activity. A decrease in bioavailability of Cd, Cu, and Pb in treated soil results in an increase in soil enzyme activity [111]. Figure 7 illustrates the adverse impacts of HMs on microbial cells. Cr has the potential to induce oxidative harm and denaturation of microorganisms, thereby impeding their bioremediation efficacy. The electric interaction between the negatively charged phosphate groups of DNA and intracellular cationic Cr complexes could potentially affect transcription and replication processes, leading to mutagenesis [110]. Both Cd and Pb exhibit deleterious impacts on microorganisms, causing harm to their cell membranes and disrupting the integrity of their DNA structure. The aforementioned impairment arises due to the dislodgment of metallic elements from their initial binding locations or bonding interactions [112]. The manipulation of DNA structure can have an impact on microbial morphology, metabolism, and growth by inducing functional disruption, cell membrane disruption, inhibition of enzyme activity, and oxidative phosphorylation [113].

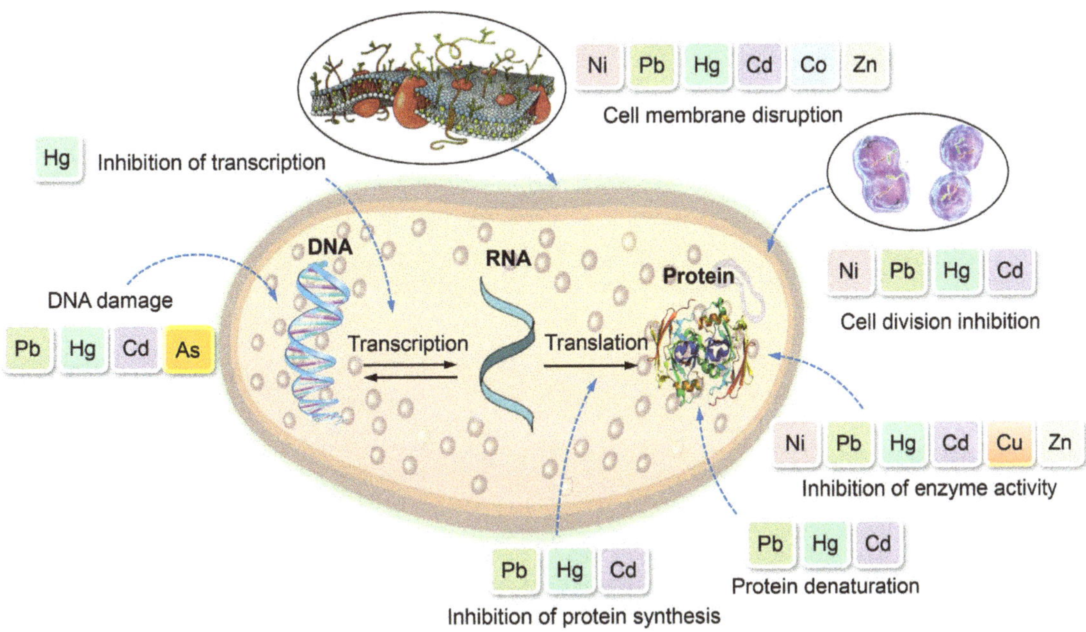

Figure 7. Toxicity of HMs on microorganisms.

6. Methods for the Remediation of HMs

Removal of HMs is typically accomplished by physical remediation, chemical remediation, and bioremediation [114–119]. The advantages and disadvantages of remediation technologies are summarized in Table 2.

Table 2. Remediation approaches of HMs.

Remediation Type	Approach	HM	Advantages	Disadvantages	References
Physical	Replacement	Ni, Cd, Cr, Zn, Cu, Pb	Appropriate for smaller contaminated sites	Expensive	[114]
	Thermal desorption	Hg	• Effective • Safe • Lack of secondary pollutants • Short pretreatment period	• Expensive devices • Long desorption time	[115]
Chemical	Oxidation/reduction	Cu, As, Sb, Cr, Pb, Se	Suitable for soils that are heavily polluted with HMs	• Expensive • Need large quantity of chemicals	[116]
	Leaching	Cr, Zn, Cu, Pb, Ni, Cd	Suitable for soils that are heavily polluted with HMs	Formation of secondary pollutants	[117]
	Electrokinetic remediation-permeable reactive barrier	Cd, Pb, Cr, Ni	Suitable for soils that are heavily polluted with HMs	Time-consuming	[118]
	Stabilization	Pb, Cd, Cu, Mn, Zn, As, Fe, Ni	• Effective • Low cost • Convenient	Pollutants cannot be removed thoroughly	[118]
Bioremediation	Microbial remediation	Cr, Pb, Ni, Zn, As,	• Effective • Low cost • Lack of secondary pollutants	• Concentration limit • Time-consuming	[119]
	Phytoremediation	Se, Cd, Cu, Pb, Zn	• Convenient • Eco-friendly • Low cost	• Time-consuming • Non-specific • Low efficiency	[119]

7. Bioremediation: Interaction of Microorganisms with HMs

Bioremediation is an innovative technology that employs a variety of biological agents, including bacteria, fungi, algae, yeasts, moulds, and plants, to effectively eliminate, detoxify, transform, or neutralise the adverse effects of HMs [120]. In contrast to several physico-chemical techniques commonly employed for mitigating HM pollution, bioremediation presents a variety of economically viable advantages owing to its notable efficacy in HM removal, cost-efficiency, ease of management, and widespread availability in both contaminated soil and water [121]. Microorganisms are a significant factor in the bioremediation of HMs among biological agents. These organisms are not only proficient in the dissolution of HMs, but also actively participate in the oxidation and reduction of transition HMs. The utilisation of microorganisms' metabolic abilities for the eradication of HM pollution is a form of green technology. Overall, most HMs have been classified as toxic. However, various biological organisms have developed distinct resistance mechanisms and intricate intracellular pathways to facilitate the utilization, interaction, adaptation, and detoxification of HMs for the purpose of cellular regeneration [122]. Figure 8 illustrates the various mechanisms employed by microorganisms for the elimination of HMs, including detoxification, biosorption, degradation, mineralization, and transformation from highly toxic to low toxic forms [122,123]. Furthermore, it has been suggested that the utilization of fungi and bacteria may have a potential role in the bioremediation of HM, as depicted in Table 3 [124–140].

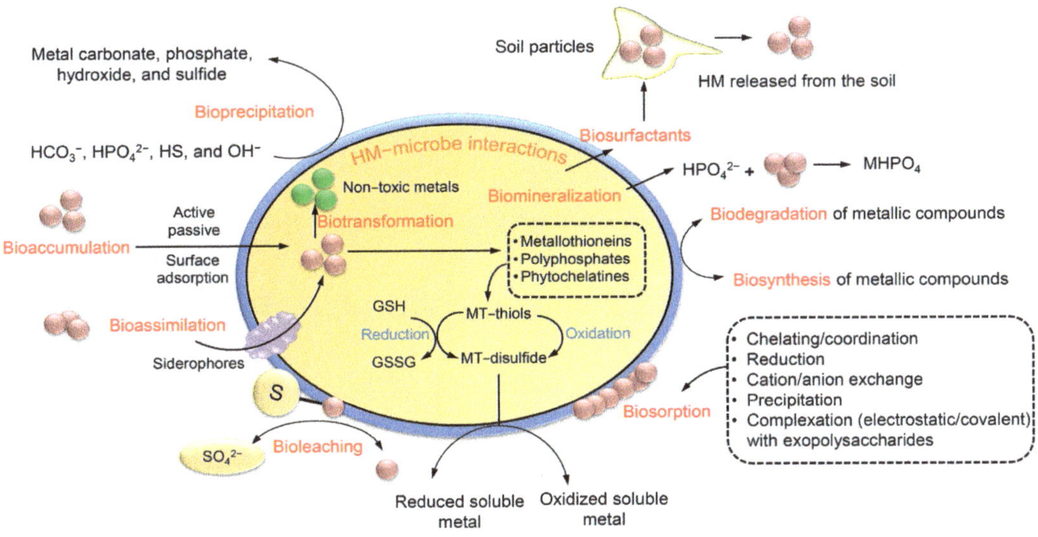

Figure 8. Mechanisms of microbial bioremediation of heavy metals. GSH, glutathione; GSSG, glutathione disulfide; MT, Metallothionein.

Table 3. Efficient bacterial and fungal species used in HM removal.

Microorganisms	HM	Efficiency of Removal (%)	References
Penicillium spp. XK10	Cd	32.2	[124]
Cladosporium sp. XM01	Mn	91.5	[125]
Penicillium janthinillum GXCR	Pb, Fe, Cu	85–99	[126]
Mucor rouxii	Pb, Zn,	11–35	[127]
Aspergilus niger	Cd, Cr	43–98	[128]
Aspergillus flavus	As, Pb, Cr, Ni	46–97	[129]
Trichoderma brevicompactum	Cu, Cr, Cd, Zn	4–64	[130]
Aspergillus fumigatus	Cd, Cr	69–79	[131]
Penicillium simplicissimum	Cr, Pb, Cu, Cd, Zn	28–88	[131]
Trichoderma virens	Cu, Cr	63–70	[132]
Pseudomonas putida	Cu	25.4	[133]
Bacillus subtilis	Cu	37.8	[133]
Pseudomonas aeruginosa	Hg	60	[134]
Sporosarcina pastaurii	Cd, Zn, Pb	94–98	[135]
Variovorax boronicumulans	Cd, Zn, Pb	73–95	[135]
Stenotrophomonas rhizophila	Cd, Zn, Pb	63–96	[135]
Citrobacter freundii	Cd, Cu, Fe, Mn, Zn	40–80	[136]
Pseudomonas aeruginosa	Cu, Zn	18–65	[137]
Acinetobacter sp. B9	Ni	69	[138]
Bacillus thuringiensis	Hg	62	[139]
Streptomyces sp.	Pb	83	[140]

7.1. Mechanisms of HM Removal by Fungi

The most common mechanisms of HM mycoremediation are bioaugmentation, bioaccumulation, biosorption, biomineralization, and biotransformation [41] (Figure 9). The implementation of fungi in bioaugmentation exhibits the most superior performance in terms of both metal elimination and enzymatic activity [141]. Bioaccumulation is a biological phenomenon wherein microorganisms synthesize proteins to recycle HMs for utilization in various cellular processes such as enzyme catalysis, signaling, and stabilization of charges in biomolecules [142]. Fungi have the ability to mitigate the harmful impact of HMs by generating inorganic acids and organic acid complexes that facilitate the leaching of HMs. This characteristic makes fungi a promising candidate for the sustainable management of wastewater. Certain types of fungi have the ability to synthesize metal transport proteins that facilitate the efflux of heavy metals from the cell or their sequestration into vacuoles [143]. The elevated levels of proteins resulting from the heightened concentration of iron facilitate the synthesis of crucial proteins that participate in the detoxification processes of *Aspergillus flavus*, thereby mitigating the deleterious effects of iron [144].

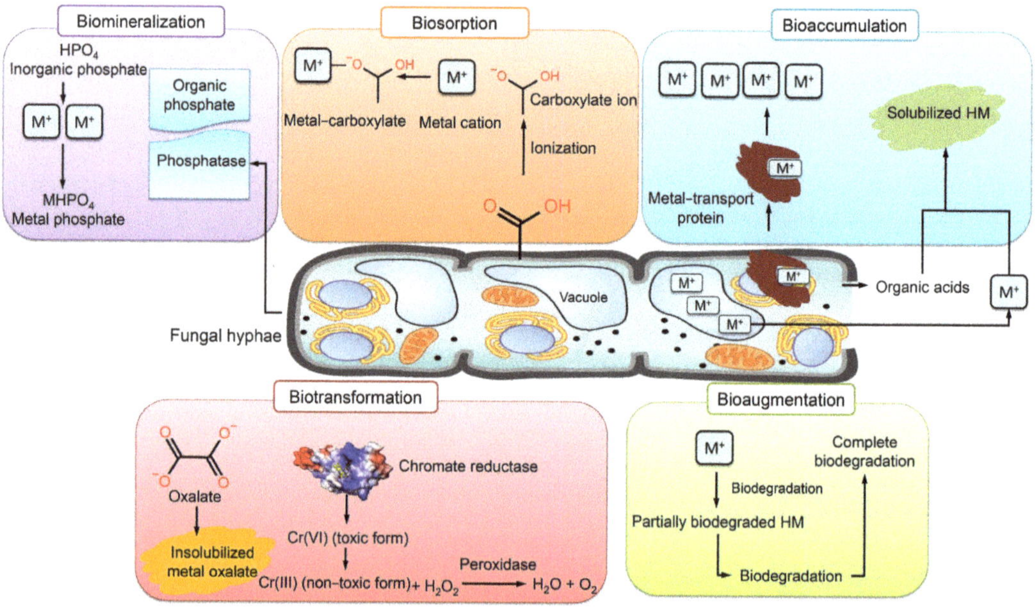

Figure 9. Mechanisms of HM removal by fungi.

Biosorption has emerged as a viable alternative for HM remediation. Fungi are known to primarily employ this technique through the ion exchange mechanism that occurs between their sorption materials or various polymeric substances, such as EPS [145]. The present approach involves the utilization of adsorbents as biological agents that facilitate the binding of HMs to non-living biomass of microorganisms, particularly through the functional groups of cell wall sorption components. The binding of metal ions is contingent upon the pH level, which influences the electrostatic interaction between metal ions and a functional group. At low pH levels, two distinct phenomena take place: firstly, the functional groups carried by the mycelium undergo ionization, resulting in the generation of a negative charge. Secondly, the occurrence of protonation in a solution can be attributed to the presence of free hydrogen ions resulting from an acidic pH. As a result of these two mechanisms, a significant quantity of unbound protons present in the solution adhere to the negatively charged sites on the cellular wall of the fungus [145]. Under optimal pH conditions, the fungus exhibits an elevated binding affinity and establishes electrostatic interactions with positively charged metal ions. The influx of metal complexes into the fungal cell results in a decrease in the quantity of metal ions absorbed from the solution, thereby affecting the fungus's retention capacity and ultimately leading to a reduction in its biosorption potential [146]. The augmentation of the incubation period results in an elevation of both biomass production and functional groups, leading to a corresponding increase in the quantity of biosorbent metal attached to said groups.

The metal that is biosorbed in the fungal cell wall is bound by metal-chelating proteins, and subsequently either restored in vacuoles or precipitated in the fungal inner cell surface through the formation of intracellular organic acid [41]. The process of metal binding to cell surface binding sites is contingent upon a variety of factors, including the intricate structure of microorganisms, extracellular precipitation, intracellular accumulation, oxidation and reduction, and methylation/demethylation [147]. Fungi possess inner walls with multiple layers on their adsorbent surface, which are accountable for the sorption of metals. As a result of metal adsorption, the surface of fungi appears rough. The fungal surface is comprised of a capsule layer that is characterized by the presence of carboxyl groups. These groups are instrumental in facilitating the mechanism of metal biosorption [148]. Metal cations can effectively bind to the fungal surface with the aid of anionic ligands, including carboxyl and amino groups. Metal ions exhibit an affinity for functional groups containing atoms with electron-donating properties. The carboxyl group of galacturonic acid, which is the primary constituent of fungal capsules, facilitates cation binding by attracting charges. The presence of oxygen in carboxylate ions facilitates anionic binding and enhances the attraction toward metal cations [148].

The process of biomineralization pertains to the capacity of living organisms to form minerals as a means of mitigating the harmful effects of metal toxicity. The biomineralization of metal phosphate is initiated by the production of acid phosphatase by fungi in a medium containing phosphate, leading to the formation of phosphate binders [149]. Certain fungi exhibit uneven diffusion of metals such as Pb^{+2} in their mycelia, resulting in the manifestation of green fluorescence with granular distribution. This phenomenon is attributed to the biomineralization of metals [150]. The concentration of precipitated calcium carbonate increases concomitantly with the rise in calcium oxide levels within the fungal culture inhabiting the contaminated concrete cube. The aforementioned process induces ion activity, thereby facilitating the mineralization of the solution, leading to the substantial precipitation of calcium carbonate. The entrapment of Cr by *Candida orthopsilosis* resulting in the formation of calcium carbonate precipitate has been observed to enhance the removal of HM and facilitate the mineralization of the surfaces of concrete cubes [151].

Biotransformation involves the conversion of toxic metal forms into less toxic forms through oxidation–reduction or mineralization–demineralization reactions [152]. The capacity of fungi to generate oxalate, which combines with HMs to form coordination compounds that function as chelators, enables them to convert soluble metals into insoluble particles, specifically insoluble metal oxalates [145]. A range of fungi have the ability to undergo biotransformation of chromium, arsenic, and other HMs into non-toxic forms. Additionally, these fungi are capable of accumulating the transformed metals in their hyphae, thereby preventing the entry of toxic metals into the roots of plants [153].

7.2. Mechanisms of HM Removal by Bacteria

Biosorption is a biological physicochemical process that is employed for the removal of recalcitrant compounds, including metal ions [112]. The utilization of this approach is characterized by its cost-effectiveness, high efficiency in eliminating toxicity, absence of the requirement for supplementary nutrients, and minimal production of chemical and/or biological sludge [112]. The analysis of the supernatant derived from suspension cultures of bacteria revealed that the primary factor responsible for metal sequestration was the soluble exopolysaccharides (EPSs). EPS is composed of carbohydrates and proteins, which jointly facilitate the biosorption of Zn and Pb [154]. The secretion of EPS has been observed in various bacterial strains, including *Paenibacillus jamilae*, *Bacillus firmus*, *Bacillus licheniformis* KX657843, *Herbaspirillium* sp., and *Paenibacillus peoriae* TS7. The biosorption capacity of the substance in question exhibited notable efficacy in the elimination of various metals, including Zn^{2+}, Pb^{2+}, Ni^{2+}, Cu^{2+}, Cd^{2+}, Co^{2+}, and Hg. This can be attributed to the electrostatic interactions that occur between the negatively charged functional groups of EPS and the positively charged metal ion (Figure 10A) [155]. The biosorption process is primarily influenced by the quantity of biosorbent utilized. Bacterial immobilization is a biological process that relies on the utilization of bacteria's inherent characteristics to eliminate various pollutants. The fundamental principle of the mechanism involves the interplay between bacteria and HMs, facilitated by the immobilization of the biocatalyst onto diverse materials that serve as carriers, providing safeguarding against external factors. This immobilization enables the microbial cells to be readily available for degradation purposes [156]. Two mechanisms, namely the attachment method and the encapsulation method, can be utilized to elucidate bacterial immobilization mechanisms, as depicted in Figure 10B.

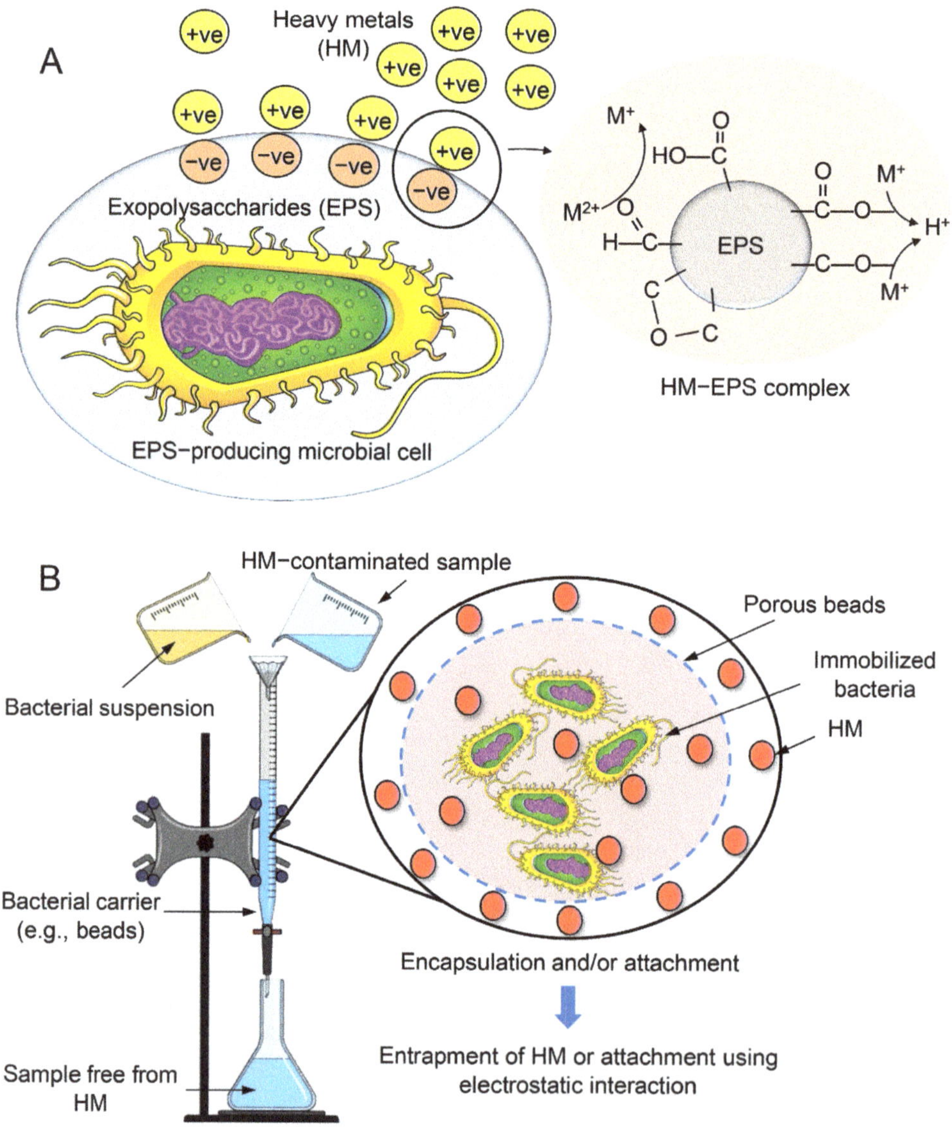

Figure 10. Mechanisms of bacterial exopolysaccharide (**A**) and encapsulation (**B**) for HM removal.

Bioleaching is a microbial-mediated process employed for the purpose of solubilizing minerals from waste solids. The process of bioleaching has been found to be both cost-effective and efficient in the removal of a diverse range of HMs, including but not limited to Cd, Cu, Cr, Fe, Ni, Pb, and Zn. The involvement of mesophilic bacteria in metal leaching is attributed to their ability to generate acids such as sulphuric acid, which facilitate the mobilization of metals. The reduction in pH and increase in oxidation–reduction potential (ORP) can effectively create an optimal environment for the removal of heavy metals (HMs) from sludge [157]. Sulfur-oxidizing and iron-oxidizing bacteria are the predominant bacterial strains employed in bioleaching [8]. The process of bioleaching can occur either through the direct metabolic activity of leaching bacteria or indirectly through the by-products of bacterial metabolism [8]. The direct mechanism involves the binding of bacteria with HMs, leading to the oxidation of the latter to their soluble metal forms. Bacteria have been observed to convert sulphur to sulfuric acid through an indirect mechanism. This process results in a reduction in the pH of the medium and an increase in the solubility of HMs, as demonstrated in Figure 11 [157]. Biomineralization refers to the process by which various solid minerals, such as carbonates, phosphates, silicates, and sulphates are formed from metal ions that are subsequently precipitated by microorganisms that are actively involved in the process. Microbial-induced precipitation is the governing mechanism of biomineralization, which encompasses microbial-induced carbonate perception that is contingent upon urea hydrolysis, pH, and temperature [135]. The process of biomineralization involves the participation of various microorganisms, including photosynthetic microorganisms and sulfate-reducing bacteria, which contribute significantly to mineral precipitation through autotrophic and heterotrophic pathways [158].

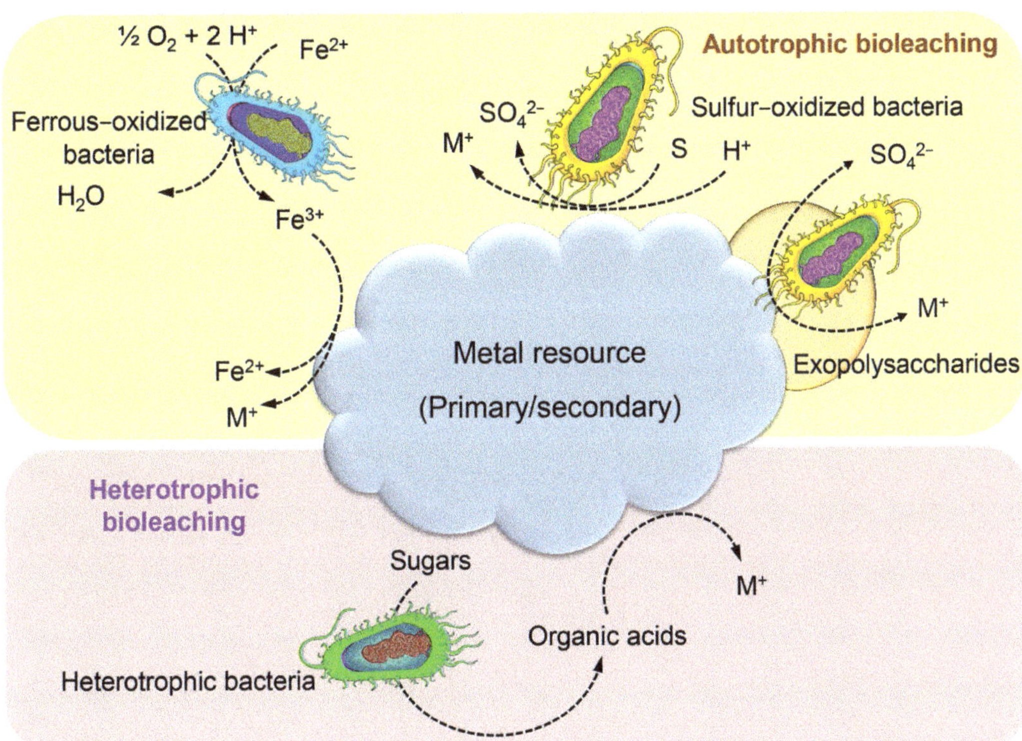

Figure 11. Bacterial bioleaching mechanism for HM removal.

7.3. Advances in Microbial Technologies for HM Removal

Extensive endeavors have been undertaken over a prolonged period to tackle the protracted persistence of HMs in the environment. Notwithstanding, the issue persists as a result of the increasing utilization of HM and the absence of feasible eradication remedies [159]. In order to mitigate the potential ecological ramifications of HM pollution, it is imperative to develop innovative and robust ecological technologies that can effectively extract and reclaim HMs from contaminated environments. The utilization of microbial bioremediation has been identified as a cost-effective approach to address the issue of HM contamination in environmental sites. Several techniques that hold promise for the removal of HMs from soil, including microbial fuel cells, biofilm, nanotechnology, and genetic engineering have been identified [160].

7.3.1. Microbial Fuel Cells

Microbial fuel cells (MFCs) are bio-electrochemical devices that facilitate the decomposition of organic waste into smaller molecules, thereby liberating protons (H^+) and electrons (e^-), which subsequently culminates in the production of energy. The diagram presented in Figure 12A illustrates the general structure and functioning of the MFC. Microorganisms generate electrons and protons at the anode through the process of oxidizing organic materials. Additionally, these microorganisms assimilate metal ions into their biomass. The transportation of electrons occurs through an external circuit, while the diffusion of protons toward the cathode takes place through the solution [161]. In principle, it is possible for any chemical possessing a high redox potential to function as the cathodic electron acceptor within the context of an MFC. Certain heme molecules exhibit significantly elevated redox potentials. Hence, they can serve as electron acceptors, wherein metallic ions amalgamate with protons and undergo reduction, thereby facilitating their elimination from the HMs and subsequent retrieval [162].

MFCs have been identified as a feasible approach to mitigate environmental contamination. The proposed approach presents a viable remedy for the treatment of wastewater and soil, along with the removal and recovery of HMs [163]. MFCs have the capability to convert chemical energy into electrical energy via oxidation mechanisms facilitated by microorganisms or enzymatic catalysis. MFCs provide a number of advantages in comparison to conventional fuel cells. The production of fuel can be achieved through the utilization of diverse organic or inorganic materials, including but not limited to soil sediments and organic waste [164]. The utilization of this technology has the potential to offer significant benefits in regions lacking access to electrical power and to improve soil characteristics. Notwithstanding its potential benefits, the technology presents a number of obstacles, including suboptimal rates of electrical generation. Consequently, various research endeavors have been undertaken to enhance electrical efficiency [165]. The removal of HMs from wastewater and soil has been the subject of extensive research utilizing MFC-based techniques [164–166]. Given the limited power output of MFCs, it is advisable to employ them primarily as sensors for detecting the toxicity of HMs [167].

7.3.2. Biofilm Technology

The microbial communities comprising bacteria, fungi, microalgae, and yeast are crucial in mitigating the deleterious impact of inorganic metallic salts, HMs, and organic waste on soil and water quality [168]. The replacement of physicochemical procedures with microbial metabolism for efficient utilization is a more environmentally friendly, secure, and effective strategy [169]. The microbial population generates EPSs, which enhance the efficacy of HM elimination. EPS substances are a class of biogenic macromolecules composed of proteins, uronic acid, and polysaccharides. These compounds are synthesized by microorganisms as a mechanism of self-preservation in response to various environmental stressors, such as extreme temperatures, acidic or alkaline conditions, and other physiological challenges. The aforementioned stressors enhance the efficacy of metal sequestration in the bioremediation of heavy metals [169]. The formation of biofilms is a consequence of the association of one or more species, and the composition and configuration of EPS can be influenced by environmental factors [170]. EPS contains a significant amount of anionic charge. The sequestration of metallic ions occurs in substantial quantities [171]. The utilization of microbial biofilm has demonstrated a high degree of efficacy in the removal of HMs from both wastewater and soil.

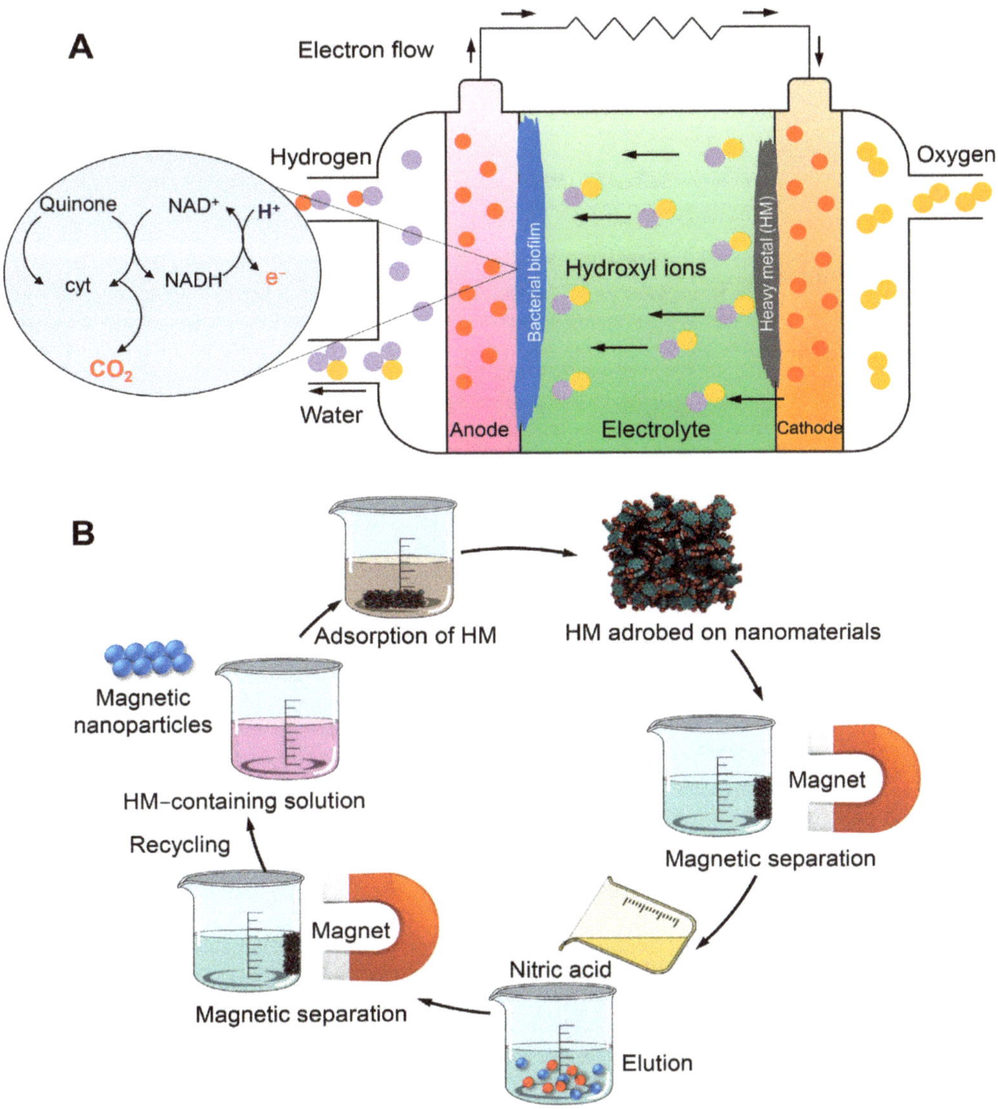

Figure 12. Mechanisms of HM removal via microbial fuel cells (**A**) and adsorption of metal ions across mesoporous magnetite nanoparticles (**B**).

7.3.3. Nanomaterial for HM Removal

Carbon nanotubes (CNTs) have demonstrated exceptional adsorption capabilities toward various HMs such as Cu, Pb, and Cr, owing to their potential adsorption active sites. The utilization of CNTs for the purpose of removing HMs from wastewater is impeded by the considerable expenses associated with their production, as well as the difficulty of isolating CNTs from the wastewater stream [172]. The utilization of silica-based nanomaterials has been investigated as a means of extracting HMs from polluted sites, owing to their exceptional surface properties [173]. The promotion of HM adsorption can be achieved through the utilization of nano-silica as the foundation for nanocomposites or by modifying its surface with functional groups such as -NH2 and -SH. Silica-based nanomaterials possess remarkable surface properties and are non-toxic, which makes them advantageous [174]. The employment of nanotechnology exhibits considerable potential in the realm of environmental remediation, particularly in the removal of HMs. The manipulation of material properties has been utilized for the purpose of removing HMs from the environment. The material exhibits exceptional surface properties and quantum effects due to its reduction to the nanoscale [172]. The principal modality by which nanomaterials eliminate HMs is through their elevated adsorption capacity and discernment for HMs (Figure 12B).

7.3.4. Biotechnological Tools

The application of contemporary biotechnological techniques such as genetic engineering and recombinant DNA technology has expedited the process of biodegradation and remediation of HMs [173]. Metagenomics and metabolic investigations offer insights into microbial diversity, population, and functional composition with respect to metal resistance genes. These findings can be leveraged to improve the efficacy of microbial strains in the removal or degradation of heavy metals. The field of genetic engineering has enabled the transfer of advantageous traits from one species to another, resulting in the development of specific strains for the purpose of bioremediating soil, sludge, or polluted water [175].

8. Conclusions

The issue of toxicity caused by HMs has emerged as a significant environmental challenge, with detrimental impacts on agricultural productivity and food security. The toxic effects of HMs result in their accumulation in the environment, which has adverse effects on human health, plant growth, and marine ecosystems. HMs can gain entry into the human body through various routes, such as ingestion via drinking water or food, inhalation through air, or through dermal exposure. After being absorbed, HMs are retained and accumulate within the human body. The accumulation of noxious metals in biological systems results in a range of deleterious consequences on diverse body tissues and organs. Various physical and chemical approaches have been suggested as potential remedies for the pollution caused by HMs. Nonetheless, the effectiveness of these techniques has been limited, and progress in this area has been slow. Therefore, alternative methodologies utilizing bioremediation have been explored. The employment of microbial bioremediation has emerged as an effective, reliable, and eco-friendly alternative for HM detoxification. However, HMs can interfere with the cellular activities of microorganisms depending on the existing concentration. The metal-tolerant microbes can enhance HM bioremediation. Microbes exhibit varying cellular structures, and their mechanisms for dealing with HM toxicity are typically dependent on the affinity of cellular biomolecules for metal ions and the stability of the specific metal in question. Microorganisms employ enzymatic systems or develop resistance through various mechanisms as a means of self-protection against HMs. However, the development of large-scale applications of combined remediation technology is recommended. In the context of practical application, one can acquire knowledge from both domestic and international experiences and enhance the judicious utilization of cutting-edge technologies. The mitigation of harmful effects caused by HMs necessitates the implementation of both scientific and community-based remediation technologies. Furthermore, forthcoming studies will derive advantages from assessing novel targets as protective procedures against organ toxicity triggered by HMs.

Author Contributions: Writing—original draft, M.K.A.E., N.E.E., M.M.E., S.H.A., S.A.E., S.E., M.S.M., E.M.S., N.A.A., A.E.A.E., I.M.E., A.E.E., and K.E.S., K.T.; Software, S.S.A.; Writing—review and editing, Validation, Visualization, S.S.A., M.K., and Y.A.-G.M. All authors have read and agreed to the published version of the manuscript.

Funding: This research received no external funding.

Institutional Review Board Statement: Not applicable.

Informed Consent Statement: Not applicable.

Data Availability Statement: Data will be given upon request.

Conflicts of Interest: The authors declare that they have no known competing financial interests or personal relationships that could have appeared to influence the work reported in this paper.

References

1. Fergusson, J.E. *The Heavy Elements: Chemistry, Environmental Impact and Health Effects*; Pergamon Press: Oxford, UK, 1990; Volume 614.
2. Tchounwou, P.B.; Yedjou, C.G.; Patlolla, A.K.; Sutton, D.J. Heavy Metal Toxicity and the Environment. *Exp. Suppl.* **2012**, *101*, 133–164.
3. Duffus, J.H. "Heavy Metals" a Meaningless Term? (IUPAC Technical Report). *Pure Appl. Chem.* **2002**, *74*, 793–807. [CrossRef]
4. Jacob, J.M.; Karthik, C.; Saratale, R.G.; Kumar, S.S.; Prabakar, D.; Kadirvelu, K.; Pugazhendhi, A. Biological Approaches to Tackle Heavy Metal Pollution: A Survey of Literature. *J. Environ. Manag.* **2018**, *217*, 56–70. [CrossRef]
5. Sarker, A.; Al Masud, M.A.; Deepo, D.M.; Das, K.; Nandi, R.; Ansary, M.W.R.; Islam, A.R.M.T.; Islam, T. Biological and Green Remediation of Heavy Metal Contaminated Water and Soils: A State-of-the-Art Review. *Chemosphere* **2023**, *332*, 138861. [CrossRef] [PubMed]
6. Anyanwu, B.; Ezejiofor, A.; Igweze, Z.; Orisakwe, O. Heavy Metal Mixture Exposure and Effects in Developing Nations: An Update. *Toxics* **2018**, *6*, 65. [CrossRef] [PubMed]
7. Wu, C.; Li, F.; Yi, S.; Ge, F. Genetically Engineered Microbial Remediation of Soils Co-Contaminated by Heavy Metals and Polycyclic Aromatic Hydrocarbons: Advances and Ecological Risk Assessment. *J. Environ. Manag.* **2021**, *296*, 113185. [CrossRef]
8. Chen, S.-Y.; Wu, J.-Q.; Sung, S. Effects of Sulfur Dosage on Continuous Bioleaching of Heavy Metals from Contaminated Sediment. *J. Hazard. Mater.* **2022**, *424*, 127257. [CrossRef] [PubMed]
9. Ali, S.S.; Abdelkarim, E.A.; Elsamahy, T.; Al-Tohamy, R.; Li, F.; Kornaros, M.; Zuorro, A.; Zhu, D.; Sun, J. Bioplastic Production in Terms of Life Cycle Assessment: A State-of-the-Art Review. *Environ. Sci. Ecotechnol.* **2023**, *15*, 100254. [CrossRef] [PubMed]
10. Koutra, E.; Mastropetros, S.G.; Ali, S.S.; Tsigkou, K.; Kornaros, M. Assessing the Potential of Chlorella Vulgaris for Valorization of Liquid Digestates from Agro-Industrial and Municipal Organic Wastes in a Biorefinery Approach. *J. Clean. Prod.* **2021**, *280*, 124352. [CrossRef]
11. Ali, S.S.; Elsamahy, T.; Abdelkarim, E.A.; Al-Tohamy, R.; Kornaros, M.; Ruiz, H.A.; Zhao, T.; Li, F.; Sun, J. Biowastes for Biodegradable Bioplastics Production and End-of-Life Scenarios in Circular Bioeconomy and Biorefinery Concept. *Bioresour. Technol.* **2022**, *363*, 127869. [CrossRef]
12. Abdelfattah, A.; Ali, S.S.; Ramadan, H.; El-Aswar, E.I.; Eltawab, R.; Ho, S.-H.; Elsamahy, T.; Li, S.; El-Sheekh, M.M.; Schagerl, M.; et al. Microalgae-Based Wastewater Treatment: Mechanisms, Challenges, Recent Advances, and Future Prospects. *Environ. Sci. Ecotechnol.* **2022**, *13*, 100205. [CrossRef] [PubMed]
13. Ali, S.S.; El-Sheekh, M.; Manni, A.; Ruiz, H.A.; Elsamahy, T.; Sun, J.; Schagerl, M. Microalgae-Mediated Wastewater Treatment for Biofuels Production: A Comprehensive Review. *Microbiol. Res.* **2022**, *265*, 127187. [CrossRef] [PubMed]
14. Al-Tohamy, R.; Ali, S.S.; Li, F.; Okasha, K.M.; Mahmoud, Y.A.-G.; Elsamahy, T.; Jiao, H.; Fu, Y.; Sun, J. A Critical Review on the Treatment of Dye-Containing Wastewater: Ecotoxicological and Health Concerns of Textile Dyes and Possible Remediation Approaches for Environmental Safety. *Ecotoxicol. Environ. Saf.* **2022**, *231*, 113160. [CrossRef] [PubMed]
15. Kumar, L.; Bharadvaja, N. Microbial Remediation of Heavy Metals. In *Microbial Bioremediation & Biodegradation*; Springer: Singapore, 2020.
16. Ali, H.; Khan, E.; Ilahi, I. Environmental Chemistry and Ecotoxicology of Hazardous Heavy Metals: Environmental Persistence, Toxicity, and Bioaccumulation. *J. Chem.* **2019**, *2019*, 1–14. [CrossRef]
17. Mitra, S.; Chakraborty, A.J.; Tareq, A.M.; Emran, T.B.; Nainu, F.; Khusro, A.; Idris, A.M.; Khandaker, M.U.; Osman, H.; Alhumaydhi, F.A.; et al. Impact of Heavy Metals on the Environment and Human Health: Novel Therapeutic Insights to Counter the Toxicity. *J. King Saud Univ.-Sci.* **2022**, *34*, 101865. [CrossRef]
18. Verma, S.; Bhatt, P.; Verma, A.; Mudila, H.; Prasher, P.; Rene, E.R. Microbial Technologies for Heavy Metal Remediation: Effect of Process Conditions and Current Practices. *Clean Technol. Environ. Policy* 2021. [CrossRef]
19. Rehman, A.U.; Nazir, S.; Irshad, R.; Tahir, K.; ur Rehman, K.; Islam, R.U.; Wahab, Z. Toxicity of Heavy Metals in Plants and Animals and Their Uptake by Magnetic Iron Oxide Nanoparticles. *J. Mol. Liq.* **2021**, *321*, 114455. [CrossRef]

20. Li, R.; Wu, H.; Ding, J.; Fu, W.; Gan, L.; Li, Y. Mercury Pollution in Vegetables, Grains and Soils from Areas Surrounding Coal-Fired Power Plants. *Sci. Rep.* **2017**, *7*, 46545. [CrossRef]
21. Deng, Y.; Wang, M.; Tian, T.; Lin, S.; Xu, P.; Zhou, L.; Dai, C.; Hao, Q.; Wu, Y.; Zhai, Z.; et al. The Effect of Hexavalent Chromium on the Incidence and Mortality of Human Cancers: A Meta-Analysis Based on Published Epidemiological Cohort Studies. *Front. Oncol.* **2019**, *9*, 24. [CrossRef]
22. Negahdari, S.; Sabaghan, M.; Pirhadi, M.; Alikord, M.; Sadighara, P.; Darvishi, M.; Nazer, M. Potential Harmful Effects of Heavy Metals as a Toxic and Carcinogenic Agent in Marine Food-An Overview. *Egypt. J. Vet. Sci.* **2021**, *52*, 379–385. [CrossRef]
23. Wang, X.; Wu, Y.; Sun, X.; Guo, Q.; Xia, W.; Wu, Y.; Li, J.; Xu, S.; Li, Y. Arsenic Exposure and Metabolism in Relation to Blood Pressure Changes in Pregnant Women. *Ecotoxicol. Environ. Saf.* **2021**, *222*, 112527. [CrossRef] [PubMed]
24. Sundseth, K.; Pacyna, J.; Pacyna, E.; Pirrone, N.; Thorne, R. Global Sources and Pathways of Mercury in the Context of Human Health. *Int. J. Environ. Res. Public Health* **2017**, *14*, 105. [CrossRef] [PubMed]
25. Teixeira, F.B.; de Oliveira, A.C.A.; Leão, L.K.R.; Fagundes, N.C.F.; Fernandes, R.M.; Fernandes, L.M.P.; da Silva, M.C.F.; Amado, L.L.; Sagica, F.E.S.; de Oliveira, E.H.C.; et al. Exposure to Inorganic Mercury Causes Oxidative Stress, Cell Death, and Functional Deficits in the Motor Cortex. *Front. Mol. Neurosci.* **2018**, *11*, 125. [CrossRef] [PubMed]
26. Papadopoulou, E.; Botton, J.; Caspersen, I.H.; Alexander, J.; Eggesbø, M.; Haugen, M.; Iszatt, N.; Jacobsson, B.; Knutsen, H.K.; Meltzer, H.M.; et al. Maternal Seafood Intake during Pregnancy, Prenatal Mercury Exposure and Child Body Mass Index Trajectories up to 8 Years. *Int. J. Epidemiol.* **2021**, *50*, 1134–1146. [CrossRef]
27. Farzan, S.F.; Howe, C.G.; Chen, Y.; Gilbert-Diamond, D.; Korrick, S.; Jackson, B.P.; Weinstein, A.R.; Karagas, M.R. Prenatal and Postnatal Mercury Exposure and Blood Pressure in Childhood. *Environ. Int.* **2021**, *146*, 106201. [CrossRef]
28. Kang, P.; Shin, H.Y.; Kim, K.Y. Association between Dyslipidemia and Mercury Exposure in Adults. *Int. J. Environ. Res. Public Health* **2021**, *18*, 775. [CrossRef]
29. Lee, S.; Cho, S.-R.; Jeong, I.; Park, J.B.; Shin, M.-Y.; Kim, S.; Kim, J.H. Mercury Exposure and Associations with Hyperlipidemia and Elevated Liver Enzymes: A Nationwide Cross-Sectional Survey. *Toxics* **2020**, *8*, 47. [CrossRef]
30. Charkiewicz, A.E.; Backstrand, J.R. Lead Toxicity and Pollution in Poland. *Int. J. Environ. Res. Public Health* **2020**, *17*, 4385. [CrossRef]
31. Wang, T.; Zhang, J.; Xu, Y. Epigenetic Basis of Lead-Induced Neurological Disorders. *Int. J. Environ. Res. Public Health* **2020**, *17*, 4878. [CrossRef]
32. Obeng-Gyasi, E.; Ferguson, A.C.; Stamatakis, K.A.; Province, M.A. Combined Effect of Lead Exposure and Allostatic Load on Cardiovascular Disease Mortality—A Preliminary Study. *Int. J. Environ. Res. Public Health* **2021**, *18*, 6879. [CrossRef]
33. Nakhaee, S.; Amirabadizadeh, A.; Brent, J.; Mehrpour, O. Impact of Chronic Lead Exposure on Liver and Kidney Function and Haematologic Parameters. *Basic Clin. Pharmacol. Toxicol.* **2018**, *124*, 621–628. [CrossRef]
34. Zhou, C.-C.; He, Y.-Q.; Gao, Z.-Y.; Wu, M.-Q.; Yan, C.-H. Sex Differences in the Effects of Lead Exposure on Growth and Development in Young Children. *Chemosphere* **2020**, *250*, 126294. [CrossRef] [PubMed]
35. Dehghan, S.F.; Mehrifar, Y.; Ardalan, A. The Relationship between Exposure to Lead-Containing Welding Fumes and the Levels of Reproductive Hormones. *Ann. Glob. Health* **2019**, *85*, 125. [CrossRef] [PubMed]
36. Chakraborty, R.; Renu, K.; Eladl, M.A.; El-Sherbiny, M.; Elsherbini, D.M.A.; Mirza, A.K.; Vellingiri, B.; Iyer, M.; Dey, A.; Valsala Gopalakrishnan, A. Mechanism of Chromium-Induced Toxicity in Lungs, Liver, and Kidney and Their Ameliorative Agents. *Biomed. Pharmacother.* **2022**, *151*, 113119. [CrossRef] [PubMed]
37. Xia, B.; Yuan, J.; Pang, L.; He, K. Chromium [Cr(VI)] Exposure Causes Cytotoxicity of Human Bronchial Epithelial Cells (16-HBE) and Proteomic Alterations. *Int. J. Toxicol.* **2022**, *41*, 225–233. [CrossRef]
38. Peng, Y.; Hu, J.; Li, Y.; Zhang, B.; Liu, W.; Li, H.; Zhang, H.; Hu, C.; Chen, X.; Xia, W.; et al. Exposure to Chromium during Pregnancy and Longitudinally Assessed Fetal Growth: Findings from a Prospective Cohort. *Environ. Int.* **2018**, *121*, 375–382. [CrossRef]
39. Baszuk, P.; Janasik, B.; Pietrzak, S.; Marciniak, W.; Reszka, E.; Białkowska, K.; Jabłońska, E.; Muszyńska, M.; Lesicka, M.; Derkacz, R.; et al. Lung Cancer Occurrence-Correlation with Serum Chromium Levels and Genotypes. *Biol. Trace Elem. Res.* **2021**, *199*, 1228–1236. [CrossRef]
40. Al Hossain, M.M.A.; Yajima, I.; Tazaki, A.; Xu, H.; Saheduzzaman, M.; Ohgami, N.; Ahsan, N.; Akhand, A.A.; Kato, M. Chromium-Mediated Hyperpigmentation of Skin in Male Tannery Workers in Bangladesh. *Chemosphere* **2019**, *229*, 611–617. [CrossRef]
41. Kumar, V.; Dwivedi, S.K. Bioremediation Mechanism and Potential of Copper by Actively Growing Fungus Trichoderma Lixii CR700 Isolated from Electroplating Wastewater. *J. Environ. Manag.* **2021**, *277*, 111370. [CrossRef]
42. Kahlson, M.A.; Dixon, S.J. Copper-Induced Cell Death. *Science* **2022**, *375*, 1231–1232. [CrossRef]
43. Niu, Y.; Zhang, Y.; Zhu, Z.; Zhang, X.; Liu, X.; Zhu, S.; Song, Y.; Jin, X.; Lindholm, B.; Yu, C. Elevated Intracellular Copper Contributes a Unique Role to Kidney Fibrosis by Lysyl Oxidase Mediated Matrix Crosslinking. *Cell Death Dis.* **2020**, *11*, 211. [CrossRef] [PubMed]
44. Tvrda, E.; Peer, R.; Sikka, S.C.; Agarwal, A. Iron and Copper in Male Reproduction: A Double-Edged Sword. *J. Assist. Reprod. Genet.* **2015**, *32*, 3–16. [CrossRef] [PubMed]
45. Yang, H.; Liu, C.-N.; Wolf, R.M.; Ralle, M.; Dev, S.; Pierson, H.; Askin, F.; Steele, K.E.; Magnuson, T.H.; Schweitzer, M.A.; et al. Obesity Is Associated with Copper Elevation in Serum and Tissues. *Metallomics* **2019**, *11*, 1363–1371. [CrossRef]

46. Witt, B.; Stiboller, M.; Raschke, S.; Friese, S.; Ebert, F.; Schwerdtle, T. Characterizing Effects of Excess Copper Levels in a Human Astrocytic Cell Line with Focus on Oxidative Stress Markers. *J. Trace Elem. Med. Biol.* **2021**, *65*, 126711. [CrossRef] [PubMed]
47. Genchi, G.; Carocci, A.; Lauria, G.; Sinicropi, M.S.; Catalano, A. Nickel: Human Health and Environmental Toxicology. *Int. J. Environ. Res. Public Health* **2020**, *17*, 679. [CrossRef] [PubMed]
48. Chen, X.; Li, Y.; Zhang, B.; Zhou, A.; Zheng, T.; Huang, Z.; Pan, X.; Liu, W.; Liu, H.; Jiang, Y.; et al. Maternal Exposure to Nickel in Relation to Preterm Delivery. *Chemosphere* **2018**, *193*, 1157–1163. [CrossRef] [PubMed]
49. Titcomb, T.J.; Liu, B.; Lehmler, H.; Snetselaar, L.G.; Bao, W. Environmental Nickel Exposure and Diabetes in a Nationally Representative Sample of US Adults. *Expo. Health* **2021**, *13*, 697–704. [CrossRef]
50. Yu, M.; Zhang, J. Serum and Hair Nickel Levels and Breast Cancer: Systematic Review and Meta-Analysis. *Biol. Trace Elem. Res.* **2017**, *179*, 32–37. [CrossRef]
51. Büyüköztürk, S.; Gelincik, A.; Ünal, D.; Demirtürk, M.; Çelik, D.D.; Erden, S.; Çolakoğlu, B.; Erdem Kuruca, S. Oral Nickel Exposure May Induce Type I Hypersensitivity Reaction in Nickel-Sensitized Subjects. *Int. Immunopharmacol.* **2015**, *26*, 92–96. [CrossRef]
52. Redvers, N.; Chischilly, A.M.; Warne, D.; Pino, M.; Lyon-Colbert, A. Uranium Exposure in American Indian Communities: Health, Policy, and the Way Forward. *Environ. Health Perspect.* **2021**, 129. [CrossRef]
53. Surdyk, S.; Itani, M.; Al-Lobaidy, M.; Kahale, L.A.; Farha, A.; Dewachi, O.; Akl, E.A.; Habib, R.R. Weaponised Uranium and Adverse Health Outcomes in Iraq: A Systematic Review. *BMJ Glob. Health* **2021**, *6*, e004166. [CrossRef] [PubMed]
54. Wang, X.; Dai, X.; Shi, C.; Wan, J.; Silver, M.A.; Zhang, L.; Chen, L.; Yi, X.; Chen, B.; Zhang, D.; et al. A 3,2-Hydroxypyridinone-Based Decorporation Agent That Removes Uranium from Bones In Vivo. *Nat. Commun.* **2019**, *10*, 2570. [CrossRef] [PubMed]
55. Joseph, S.J.; Arunachalam, K.D.; Murthy, P.B.; Ramalingam, R.; Musthafa, M.S. Uranium Induces Genomic Instability and Slows Cell Cycle Progression in Human Lymphocytes in Acute Toxicity Study. *Toxicol. Vitr.* **2021**, *73*, 105149. [CrossRef] [PubMed]
56. Sears, C.G.; Poulsen, A.H.; Eliot, M.; Howe, C.J.; James, K.A.; Harrington, J.M.; Roswall, N.; Overvad, K.; Tjønneland, A.; Raaschou-Nielsen, O.; et al. Urine Cadmium and Acute Myocardial Infarction among Never Smokers in the Danish Diet, Cancer and Health Cohort. *Environ. Int.* **2021**, *150*, 106428. [CrossRef] [PubMed]
57. Vijayakumar, V.; Abern, M.R.; Jagai, J.S.; Kajdacsy-Balla, A. Observational Study of the Association between Air Cadmium Exposure and Prostate Cancer Aggressiveness at Diagnosis among a Nationwide Retrospective Cohort of 230,540 Patients in the United States. *Int. J. Environ. Res. Public Health* **2021**, *18*, 8333. [CrossRef] [PubMed]
58. Amadou, A.; Praud, D.; Coudon, T.; Danjou, A.M.N.; Faure, E.; Deygas, F.; Grassot, L.; Leffondré, K.; Severi, G.; Salizzoni, P.; et al. Exposure to Airborne Cadmium and Breast Cancer Stage, Grade and Histology at Diagnosis: Findings from the E3N Cohort Study. *Sci. Rep.* **2021**, *11*, 23088. [CrossRef]
59. Yan, L.-J.; Allen, D.C. Cadmium-Induced Kidney Injury: Oxidative Damage as a Unifying Mechanism. *Biomolecules* **2021**, *11*, 1575. [CrossRef]
60. Rajan, M.; Anderson, C.P.; Rindler, P.M.; Romney, S.J.; Ferreira dos Santos, M.C.; Gertz, J.; Leibold, E.A. NHR-14 Loss of Function Couples Intestinal Iron Uptake with Innate Immunity in C. Elegans through PQM-1 Signaling. *Elife* **2019**, *8*. [CrossRef]
61. Verna, G.; Sila, A.; Liso, M.; Mastronardi, M.; Chieppa, M.; Cena, H.; Campiglia, P. Iron-Enriched Nutritional Supplements for the 2030 Pharmacy Shelves. *Nutrients* **2021**, *13*, 378. [CrossRef]
62. Killip, S.; Bennett, J.M.; Chambers, M.D. Iron Deficiency Anemia. *Am. Fam. Physician* **2007**, *75*, 671–678.
63. Weyh, C.; Krüger, K.; Peeling, P.; Castell, L. The Role of Minerals in the Optimal Functioning of the Immune System. *Nutrients* **2022**, *14*, 644. [CrossRef]
64. Means, R.T. Iron Deficiency and Iron Deficiency Anemia: Implications and Impact in Pregnancy, Fetal Development, and Early Childhood Parameters. *Nutrients* **2020**, *12*, 447. [CrossRef]
65. Quezada-Pinedo, H.G.; Cassel, F.; Duijts, L.; Muckenthaler, M.U.; Gassmann, M.; Jaddoe, V.W.; Reiss, I.K.; Vermeulen, M.J. Maternal Iron Status in Pregnancy and Child Health Outcomes after Birth: A Systematic Review and Meta-Analysis. *Nutrients* **2021**, *13*, 2221. [CrossRef]
66. Rojas-Lemus, M.; López-Valdez, N.; Bizarro-Nevares, P.; González-Villalva, A.; Ustarroz-Cano, M.; Zepeda-Rodríguez, A.; Pasos-Nájera, F.; García-Peláez, I.; Rivera-Fernández, N.; Fortoul, T.I. Toxic Effects of Inhaled Vanadium Attached to Particulate Matter: A Literature Review. *Int. J. Environ. Res. Public Health* **2021**, *18*, 8457. [CrossRef]
67. Imtiaz, M.; Rizwan, M.S.; Xiong, S.; Li, H.; Ashraf, M.; Shahzad, S.M.; Shahzad, M.; Rizwan, M.; Tu, S. Vanadium, Recent Advancements and Research Prospects: A Review. *Environ. Int.* **2015**, *80*, 79–88. [CrossRef] [PubMed]
68. Filler, G.; Kobrzynski, M.; Sidhu, H.K.; Belostotsky, V.; Huang, S.-H.S.; McIntyre, C.; Yang, L. A Cross-Sectional Study Measuring Vanadium and Chromium Levels in Paediatric Patients with CKD. *BMJ Open* **2017**, *7*, e014821. [CrossRef] [PubMed]
69. Ngwa, H.A.; Kanthasamy, A.; Jin, H.; Anantharam, V.; Kanthasamy, A.G. Vanadium Exposure Induces Olfactory Dysfunction in an Animal Model of Metal Neurotoxicity. *Neurotoxicology* **2014**, *43*, 73–81. [CrossRef]
70. Mahey, S.; Kumar, R.; Sharma, M.; Kumar, V.; Bhardwaj, R. A Critical Review on Toxicity of Cobalt and Its Bioremediation Strategies. *SN Appl. Sci.* **2020**, *2*, 1–12. [CrossRef]
71. Wahlqvist, F.; Bryngelsson, I.-L.; Westberg, H.; Vihlborg, P.; Andersson, L. Dermal and Inhalable Cobalt Exposure—Uptake of Cobalt for Workers at Swedish Hard Metal Plants. *PLoS ONE* **2020**, *15*, e0237100. [CrossRef] [PubMed]
72. Magaye, R.; Zhao, J.; Bowman, L.; Ding, M. Genotoxicity and Carcinogenicity of Cobalt-, Nickel- and Copper-Based Nanoparticles. *Exp. Ther. Med.* **2012**, *4*, 551–561. [CrossRef] [PubMed]

73. Zhang, N.; Yang, S.; Yang, J.; Deng, Y.; Li, S.; Li, N.; Chen, X.; Yu, P.; Liu, Z.; Zhu, J. Association between Metal Cobalt Exposure and the Risk of Congenital Heart Defect Occurrence in Offspring: A Multi-Hospital Case-Control Study. *Environ. Health Prev. Med.* **2020**, *25*, 38. [CrossRef] [PubMed]
74. Linna, A.; Uitti, J.; Oksa, P.; Toivio, P.; Virtanen, V.; Lindholm, H.; Halkosaari, M.; Sauni, R. Effects of Occupational Cobalt Exposure on the Heart in the Production of Cobalt and Cobalt Compounds: A 6-Year Follow-Up. *Int. Arch. Occup. Environ. Health* **2020**, *93*, 365–374. [CrossRef] [PubMed]
75. Karbowska, B. Presence of Thallium in the Environment: Sources of Contaminations, Distribution and Monitoring Methods. *Environ. Monit. Assess.* **2016**, *188*, 1–19. [CrossRef]
76. Kremer, D.; Riemersma, N.L.; Groothof, D.; Sotomayor, C.G.; Eisenga, M.F.; Post, A.; Knobbe, T.J.; Touw, D.J.; Bakker, S.J.L. Plasma Thallium Concentration, Kidney Function, Nephrotoxicity and Graft Failure in Kidney Transplant Recipients. *J. Clin. Med.* **2022**, *11*, 1970. [CrossRef]
77. Lin, G.; Yuan, L.; Bai, L.; Liu, Y.; Wang, Y.; Qiu, Z. Successful Treatment of a Patient with Severe Thallium Poisoning in a Coma Using Prussian Blue and Plasma Exchange: A Case Report. *Medicine* **2019**, *98*, e14629–e14629. [CrossRef] [PubMed]
78. Ahmed, M.J.; Mia, M.L. A New Simple, Highly Sensitive and Selective Spectrofluorimetric Method for the Speciation of Thallium at Pico-Trace Levels in Various Complex Matrices Using N-(Pyridin-2-Yl)-Quinoline-2-Carbothioamide. *RSC Adv.* **2021**, *11*, 32312–32328. [CrossRef]
79. Xia, W.; Du, X.; Zheng, T.; Zhang, B.; Li, Y.; Bassig, B.A.; Zhou, A.; Wang, Y.; Xiong, C.; Li, Z.; et al. A Case–Control Study of Prenatal Thallium Exposure and Low Birth Weight in China. *Environ. Health Perspect.* **2016**, *124*, 164–169. [CrossRef]
80. Zwolak, I. Epigallocatechin Gallate for Management of Heavy Metal-Induced Oxidative Stress: Mechanisms of Action, Efficacy, and Concerns. *Int. J. Mol. Sci.* **2021**, *22*, 4027. [CrossRef]
81. Tamás, M.J.; Sharma, S.K.; Ibstedt, S.; Jacobson, T.; Christen, P. Heavy Metals and Metalloids as a Cause for Protein Misfolding and Aggregation. *Biomolecules* **2014**, *4*, 252–267. [CrossRef]
82. Patel, P.; Raju, N.J.; Reddy, B.C.S.R.; Suresh, U.; Sankar, D.B.; Reddy, T.V.K. Heavy Metal Contamination in River Water and Sediments of the Swarnamukhi River Basin, India: Risk Assessment and Environmental Implications. *Environ. Geochem. Health* **2017**, *40*, 609–623. [CrossRef]
83. Miranda, L.S.; Wijesiri, B.; Ayoko, G.A.; Egodawatta, P.; Goonetilleke, A. Water-Sediment Interactions and Mobility of Heavy Metals in Aquatic Environments. *Water Res.* **2021**, *202*, 117386. [CrossRef]
84. Kim, B.-M.; Kim, B.; Nam, S.-E.; Eom, H.-J.; Lee, S.; Kim, K.; Rhee, J.-S. Reductive Transformation of Hexavalent Chromium in Ice Decreases Chromium Toxicity in Aquatic Animals. *Environ. Sci. Technol.* **2022**, *56*, 3503–3513. [CrossRef] [PubMed]
85. Morroni, L.; Gaion, A.; Broccoli, A.; Ferrari, S.; Pellegrini, D.; Sartori, D. Influence of Salinity on Copper Toxicity in Paracentrotus Lividus and Arbacia Lixula Embryos. *Water* **2022**, *15*, 65. [CrossRef]
86. Wang, N.; Kunz, J.L.; Cleveland, D.M.; Steevens, J.A.; Hammer, E.J.; Van Genderen, E.; Ryan, A.C.; Schlekat, C.E. Evaluation of Acute and Chronic Toxicity of Nickel and Zinc to 2 Sensitive Freshwater Benthic Invertebrates Using Refined Testing Methods. *Environ. Toxicol. Chem.* **2020**, *39*, 2256–2268. [CrossRef] [PubMed]
87. Byeon, E.; Kang, H.-M.; Yoon, C.; Lee, J.-S. Toxicity Mechanisms of Arsenic Compounds in Aquatic Organisms. *Aquat. Toxicol.* **2021**, *237*, 105901. [CrossRef] [PubMed]
88. Perić, L.; Stinga Perusco, V.; Nerlović, V. Differential Response of Biomarkers in the Native European Flat Oyster Ostrea Edulis and the Non-Indigenous Pacific Oyster Crassostrea Gigas Co-Exposed to Cadmium and Copper. *J. Exp. Mar. Bio. Ecol.* **2020**, *523*, 151271. [CrossRef]
89. Meng, J.; Wang, W.-X.; Li, L.; Zhang, G. Tissue-Specific Molecular and Cellular Toxicity of Pb in the Oyster (Crassostrea Gigas): MRNA Expression and Physiological Studies. *Aquat. Toxicol.* **2018**, *198*, 257–268. [CrossRef]
90. Xu, X.; Shi, L.; Wang, M. Comparative Quantitative Proteomics Unveils Putative Mechanisms Involved into Mercury Toxicity and Tolerance in Tigriopus Japonicus under Multigenerational Exposure Scenario. *Environ. Pollut.* **2016**, *218*, 1287–1297. [CrossRef]
91. Alengebawy, A.; Abdelkhalek, S.T.; Qureshi, S.R.; Wang, M.-Q. Heavy Metals and Pesticides Toxicity in Agricultural Soil and Plants: Ecological Risks and Human Health Implications. *Toxics* **2021**, *9*, 42. [CrossRef]
92. Chen, D.; Liu, X.; Bian, R.; Cheng, K.; Zhang, X.; Zheng, J.; Joseph, S.; Crowley, D.; Pan, G.; Li, L. Effects of Biochar on Availability and Plant Uptake of Heavy Metals – A Meta-Analysis. *J. Environ. Manag.* **2018**, *222*, 76–85. [CrossRef]
93. Sabra, M.; Aboulnasr, A.; Franken, P.; Perreca, E.; Wright, L.P.; Camehl, I. Beneficial Root Endophytic Fungi Increase Growth and Quality Parameters of Sweet Basil in Heavy Metal Contaminated Soil. *Front. Plant Sci.* **2018**, *9*, 1726. [CrossRef] [PubMed]
94. Kladsomboon, S.; Jaiyen, C.; Choprathumma, C.; Tusai, T.; Apilux, A. Heavy Metals Contamination in Soil, Surface Water, Crops, and Resident Blood in Uthai District, Phra Nakhon Si Ayutthaya, Thailand. *Environ. Geochem. Health* **2019**, *42*, 545–561. [CrossRef] [PubMed]
95. Mondal, N.K.; Das, C.; Datta, J.K. Effect of Mercury on Seedling Growth, Nodulation and Ultrastructural Deformation of *Vigna radiata* (L) Wilczek. *Environ. Monit. Assess.* **2015**, *187*, 241. [CrossRef]
96. Verma, S.; Kuila, A. Bioremediation of Heavy Metals by Microbial Process. *Environ. Technol. Innov.* **2019**, *14*, 100369. [CrossRef]
97. Wu, Q.; Leung, J.Y.S.; Geng, X.; Chen, S.; Huang, X.; Li, H.; Huang, Z.; Zhu, L.; Chen, J.; Lu, Y. Heavy Metal Contamination of Soil and Water in the Vicinity of an Abandoned E-Waste Recycling Site: Implications for Dissemination of Heavy Metals. *Sci. Total Environ.* **2015**, *506*, 217–225.

98. Wang, Z.; Lei, G. Study on Penetration Effect of Heavy Metal Migration in Different Soil Types. *IOP Conf. Ser. Mater. Sci. Eng.* **2018**, *394*, 52033. [CrossRef]
99. Mahbub, K.R.; King, W.L.; Siboni, N.; Nguyen, V.K.; Rahman, M.M.; Megharaj, M.; Seymour, J.R.; Franks, A.E.; Labbate, M. Long-Lasting Effect of Mercury Contamination on the Soil Microbiota and Its Co-Selection of Antibiotic Resistance. *Environ. Pollut.* **2020**, *265*, 115057. [CrossRef]
100. Xu, J.; Zhang, J.; Lv, Y.; Xu, K.; Lu, S.; Liu, X.; Yang, Y. Effect of Soil Mercury Pollution on Ginger (Zingiber Officinale Roscoe): Growth, Product Quality, Health Risks and Silicon Mitigation. *Ecotoxicol. Environ. Saf.* **2020**, *195*, 110472. [CrossRef]
101. Htwe, T.; Onthong, J.; Duangpan, S.; Techato, K.; Chotikarn, P.; Sinutok, S. Effect of Copper Contamination on Plant Growth and Metal Contents in Rice Plant (*Oryza sativa* L.). *Commun. Soil Sci. Plant Anal.* **2020**, *51*, 2349–2360. [CrossRef]
102. Sharma, A.; Kapoor, D.; Wang, J.; Shahzad, B.; Kumar, V.; Bali, A.S.; Jasrotia, S.; Zheng, B.; Yuan, H.; Yan, D. Chromium Bioaccumulation and Its Impacts on Plants: An Overview. *Plants* **2020**, *9*, 100. [CrossRef]
103. Nasiri, A.; Jamshidi-Zanjani, A.; Khodadadi Darban, A. Application of Enhanced Electrokinetic Approach to Remediate Cr-Contaminated Soil: Effect of Chelating Agents and Permeable Reactive Barrier. *Environ. Pollut.* **2020**, *266*, 115197. [CrossRef]
104. Abbas, A.; Azeem, M.; Naveed, M.; Latif, A.; Bashir, S.; Ali, A.; Bilal, M.; Ali, L. Synergistic Use of Biochar and Acidified Manure for Improving Growth of Maize in Chromium Contaminated Soil. *Int. J. Phytoremediation* **2019**, *22*, 52–61. [CrossRef] [PubMed]
105. Singh, D.; Sharma, N.L.; Singh, C.K.; Yerramilli, V.; Narayan, R.; Sarkar, S.K.; Singh, I. Chromium (VI)-Induced Alterations in Physio-Chemical Parameters, Yield, and Yield Characteristics in Two Cultivars of Mungbean (*Vigna radiata* L.). *Front. Plant Sci.* **2021**, *12*, 735129. [CrossRef] [PubMed]
106. Liu, J.; Liu, F.; Ding, C.; Ma, F.; Yu, H.; Shi, Y.; Zhang, X. Response of Trametes Hirsuta to Hexavalent Chromium Promotes Laccase-Mediated Decolorization of Reactive Black 5. *Ecotoxicol. Environ. Saf.* **2020**, *205*, 111134. [CrossRef] [PubMed]
107. Wang, L.; Hou, D.; Shen, Z.; Zhu, J.; Jia, X.; Ok, Y.S.; Tack, F.M.G.; Rinklebe, J. Field Trials of Phytomining and Phytoremediation: A Critical Review of Influencing Factors and Effects of Additives. *Crit. Rev. Environ. Sci. Technol.* **2019**, *50*, 2724–2774. [CrossRef]
108. Suman, J.; Uhlik, O.; Viktorova, J.; Macek, T. Phytoextraction of Heavy Metals: A Promising Tool for Clean-Up of Polluted Environment? *Front. Plant Sci.* **2018**, *9*, 1476. [CrossRef]
109. Yan, A.; Wang, Y.; Tan, S.N.; Mohd Yusof, M.L.; Ghosh, S.; Chen, Z. Phytoremediation: A Promising Approach for Revegetation of Heavy Metal-Polluted Land. *Front. Plant Sci.* **2020**, *11*, 359. [CrossRef]
110. Igiri, B.E.; Okoduwa, S.I.R.; Idoko, G.O.; Akabuogu, E.P.; Adeyi, A.O.; Ejiogu, I.K. Toxicity and Bioremediation of Heavy Metals Contaminated Ecosystem from Tannery Wastewater: A Review. *J. Toxicol.* **2018**, *2018*, 1–16. [CrossRef]
111. Matinian, N.N.; Gusareva, A.L.; Bakhmatova, K.A.; Sheshukova, A.A. Microbiological Indicators and Heavy Metal Concentration in Ecological Assessment of Urban Soils of Saint Petersburg, Russia. *Geogr. Environ. Sustain.* **2020**, *13*, 214–223. [CrossRef]
112. Priyadarshanee, M.; Das, S. Biosorption and Removal of Toxic Heavy Metals by Metal Tolerating Bacteria for Bioremediation of Metal Contamination: A Comprehensive Review. *J. Environ. Chem. Eng.* **2021**, *9*, 104686. [CrossRef]
113. Ayangbenro, A.; Babalola, O. A New Strategy for Heavy Metal Polluted Environments: A Review of Microbial Biosorbents. *Int. J. Environ. Res. Public Health* **2017**, *14*, 94. [CrossRef]
114. Gong, Y.; Zhao, D.; Wang, Q. An Overview of Field-Scale Studies on Remediation of Soil Contaminated with Heavy Metals and Metalloids: Technical Progress over the Last Decade. *Water Res.* **2018**, *147*, 440–460. [CrossRef] [PubMed]
115. Zhao, C.; Dong, Y.; Feng, Y.; Li, Y.; Dong, Y. Thermal Desorption for Remediation of Contaminated Soil: A Review. *Chemosphere* **2019**, *221*, 841–855. [CrossRef] [PubMed]
116. Liang, W.; Wang, G.; Peng, C.; Tan, J.; Wan, J.; Sun, P.; Li, Q.; Ji, X.; Zhang, Q.; Wu, Y.; et al. Recent Advances of Carbon-Based Nano Zero Valent Iron for Heavy Metals Remediation in Soil and Water: A Critical Review. *J. Hazard. Mater.* **2022**, *426*, 127993. [CrossRef] [PubMed]
117. Hu, W.; Niu, Y.; Zhu, H.; Dong, K.; Wang, D.; Liu, F. Remediation of Zinc-Contaminated Soils by Using the Two-Step Washing with Citric Acid and Water-Soluble Chitosan. *Chemosphere* **2021**, *282*, 131092. [CrossRef]
118. Wang, D.; Li, G.; Qin, S.; Tao, W.; Gong, S.; Wang, J. Remediation of Cr(VI)-Contaminated Soil Using Combined Chemical Leaching and Reduction Techniques Based on Hexavalent Chromium Speciation. *Ecotoxicol. Environ. Saf.* **2021**, *208*, 111734. [CrossRef]
119. Wang, J.; Hou, L.; Yao, Z.; Jiang, Y.; Xi, B.; Ni, S.; Zhang, L. Aminated Electrospun Nanofiber Membrane as Permeable Reactive Barrier Material for Effective In-Situ Cr(VI) Contaminated Soil Remediation. *Chem. Eng. J.* **2021**, *406*, 126822. [CrossRef]
120. Jeyakumar, P.; Debnath, C.; Vijayaraghavan, R.; Muthuraj, M. Trends in Bioremediation of Heavy Metal Contaminations. *Environ. Eng. Res.* **2022**, *28*, 220630–220631. [CrossRef]
121. Medfu Tarekegn, M.; Zewdu Salilih, F.; Ishetu, A.I. Microbes Used as a Tool for Bioremediation of Heavy Metal from the Environment. *Cogent Food Agric.* **2020**, *6*, 1783174. [CrossRef]
122. Sreedevi, P.R.; Suresh, K.; Jiang, G. Bacterial Bioremediation of Heavy Metals in Wastewater: A Review of Processes and Applications. *J. Water Process Eng.* **2022**, *48*, 102884. [CrossRef]
123. Jasrotia, S.; Kansal, A.; Mehra, A. Performance of Aquatic Plant Species for Phytoremediation of Arsenic-Contaminated Water. *Appl. Water Sci.* **2015**, *7*, 889–896. [CrossRef]
124. He, Y.; Li, C.; Sun, Z.; Zhang, W.; He, J.; Zhao, Y.; Xu, Z.; Zhao, W. Penicillium Spp. XK10, Fungi with Potential to Repair Cadmium and Antimony Pollution. *Appl. Sci.* **2023**, *13*, 1228. [CrossRef]

125. Wang, M.; Xu, Z.; Huang, Y.; Dong, B. Static Magnetic Field Enhances Cladosporium Sp. XM01 Growth and Fungal Mn(II) Oxidation. *J. Hazard. Mater.* **2022**, *437*, 129332. [CrossRef] [PubMed]
126. Wang, R.; Fan, X.-W.; Li, Y.-Z. Efficient Removal of a Low Concentration of Pb(II), Fe(III) and Cu(II) from Simulated Drinking Water by Co-Immobilization between Low-Dosages of Metal-Resistant/Adapted Fungus Penicillium Janthinillum and Graphene Oxide and Activated Carbon. *Chemosphere* **2022**, *286*, 131591. [CrossRef]
127. Rusu, L.; Grigoraș, C.-G.; Simion, A.-I.; Suceveanu, E.-M.; Istrate, B.; Harja, M. Biosorption Potential of Microbial and Residual Biomass of Saccharomyces Pastorianus Immobilized in Calcium Alginate Matrix for Pharmaceuticals Removal from Aqueous Solutions. *Polymers* **2022**, *14*, 2855. [CrossRef] [PubMed]
128. Khan, I.; Aftab, M.; Shakir, S.; Ali, M.; Qayyum, S.; Rehman, M.U.; Haleem, K.S.; Touseef, I. Mycoremediation of Heavy Metal (Cd and Cr)–Polluted Soil through Indigenous Metallotolerant Fungal Isolates. *Environ. Monit. Assess.* **2019**, 191. [CrossRef]
129. Kumar, V.; Pandita, S.; Singh Sidhu, G.P.; Sharma, A.; Khanna, K.; Kaur, P.; Bali, A.S.; Setia, R. Copper Bioavailability, Uptake, Toxicity and Tolerance in Plants: A Comprehensive Review. *Chemosphere* **2021**, *262*, 127810. [CrossRef] [PubMed]
130. Zhang, D.; Yin, C.; Abbas, N.; Mao, Z.; Zhang, Y. Multiple Heavy Metal Tolerance and Removal by an Earthworm Gut Fungus Trichoderma Brevicompactum QYCD-6. *Sci. Rep.* **2020**, *10*, 6940. [CrossRef]
131. Chen, S.H.; Cheow, Y.L.; Ng, S.L.; Ting, A.S.Y. Mechanisms for Metal Removal Established via Electron Microscopy and Spectroscopy: A Case Study on Metal Tolerant Fungi Penicillium Simplicissimum. *J. Hazard. Mater.* **2019**, *362*, 394–402. [CrossRef]
132. Tansengco, M.; Tejano, J.; Coronado, F.; Gacho, C.; Barcelo, J. Heavy Metal Tolerance and Removal Capacity of Trichoderma Species Isolated from Mine Tailings in Itogon, Benguet. *Environ. Nat. Resour. J.* **2018**, *16*, 39–57.
133. Melo, A.; Quintelas, C.; Ferreira, E.C.; Mesquita, D.P. The Role of Extracellular Polymeric Substances in Micropollutant Removal. *Front. Chem. Eng.* **2022**, *4*, 778469. [CrossRef]
134. Imron, M.F.; Kurniawan, S.B.; Abdullah, S.R.S. Resistance of Bacteria Isolated from Leachate to Heavy Metals and the Removal of Hg by Pseudomonas Aeruginosa Strain FZ-2 at Different Salinity Levels in a Batch Biosorption System. *Sustain. Environ. Res.* **2021**, *31*, 14. [CrossRef]
135. Jalilvand, N.; Akhgar, A.; Alikhani, H.A.; Rahmani, H.A.; Rejali, F. Removal of Heavy Metals Zinc, Lead, and Cadmium by Biomineralization of Urease-Producing Bacteria Isolated from Iranian Mine Calcareous Soils. *J. Soil Sci. Plant Nutr.* **2019**, *20*, 206–219. [CrossRef]
136. Gomaa, E.Z.; El-Meihy, R.M. Bacterial Biosurfactant from Citrobacter Freundii MG812314.1 as a Bioremoval Tool of Heavy Metals from Wastewater. *Bull. Natl. Res. Cent.* **2019**, *43*, 69.
137. Mulligan, C.N.; Yong, R.N.; Gibbs, B.F. Heavy Metal Removal from Sediments by Biosurfactants. *J. Hazard. Mater.* **2001**, *85*, 111–125. [CrossRef]
138. Bhattacharya, A.; Gupta, A. Evaluation of Acinetobacter Sp. B9 for Cr (VI) Resistance and Detoxification with Potential Application in Bioremediation of Heavy-Metals-Rich Industrial Wastewater. *Environ. Sci. Pollut. Res.* **2013**, *20*, 6628–6637. [CrossRef]
139. Saranya, K.; Sundaramanickam, A.; Shekhar, S.; Swaminathan, S.; Balasubramanian, T. Bioremediation of Mercury by Vibrio Fluvialis Screened from Industrial Effluents. *Biomed Res. Int.* **2017**, *2017*, 6509648. [CrossRef]
140. Hamdan, A.M.; Abd-El-Mageed, H.; Ghanem, N. Biological Treatment of Hazardous Heavy Metals by Streptomyces Rochei ANH for Sustainable Water Management in Agriculture. *Sci. Rep.* **2021**, *11*, 9314. [CrossRef]
141. Hassan, A.; Pariatamby, A.; Ossai, I.C.; Hamid, F.S. Bioaugmentation Assisted Mycoremediation of Heavy Metal and/Metalloid Landfill Contaminated Soil Using Consortia of Filamentous Fungi. *Biochem. Eng. J.* **2020**, *157*, 107550. [CrossRef]
142. Diep, P.; Mahadevan, R.; Yakunin, A.F. Heavy Metal Removal by Bioaccumulation Using Genetically Engineered Microorganisms. *Front. Bioeng. Biotechnol.* **2018**, *6*, 157. [CrossRef]
143. Khilji, S.A.; Aqeel, M.; Maqsood, M.F.; Khalid, N.; Tufail, A.; Sajid, Z.A.; Al-Surhanee, A.A.; Hashem, M.; Alamri, S.; Al-Mutairi, K.A.; et al. Hemarthria Compressa—Aspergillus Niger—Trichoderma Pseudokoningii Mediated Trilateral Perspective for Bioremediation and Detoxification of Industrial Paper Sludge. *Sustainability* **2021**, *13*, 12266. [CrossRef]
144. El-Sayed, M.T.; Ezzat, S.M.; Taha, A.S.; Ismaiel, A.A. Iron Stress Response and Bioaccumulation Potential of Three Fungal Strains Isolated from Sewage-Irrigated Soil. *J. Appl. Microbiol.* **2022**, *132*, 1936–1953. [CrossRef]
145. Geetha, N.; Bhavya, G.; Abhijith, P.; Shekhar, R.; Dayananda, K.; Jogaiah, S. Insights into Nanomycoremediation: Secretomics and Mycogenic Biopolymer Nanocomposites for Heavy Metal Detoxification. *J. Hazard. Mater.* **2021**, *409*, 124541. [CrossRef] [PubMed]
146. Sharma, R.; Jasrotia, T.; Sharma, S.; Sharma, M.; Kumar, R.; Vats, R.; Kumar, R.; Umar, A.; Akhtar, M.S. Sustainable Removal of Ni(II) from Waste Water by Freshly Isolated Fungal Strains. *Chemosphere* **2021**, *282*, 130871. [CrossRef] [PubMed]
147. Kumar, V.; Dwivedi, S.K. Mycoremediation of Heavy Metals: Processes, Mechanisms, and Affecting Factors. *Environ. Sci. Pollut. Res.* **2021**, *28*, 10375–10412. [CrossRef]
148. Sundararaju, S.; Manjula, A.; Kumaravel, V.; Muneeswaran, T.; Vennila, T. Biosorption of Nickel Ions Using Fungal Biomass Penicillium Sp. MRF1 for the Treatment of Nickel Electroplating Industrial Effluent. *Biomass Convers. Biorefinery* **2020**, *12*, 1059–1068. [CrossRef]
149. Coelho, E.; Reis, T.A.; Cotrim, M.; Mullan, T.K.; Renshaw, J.; Rizzutto, M.; Corrêa, B. Talaromyces Amestolkiae Uses Organic Phosphate Sources for the Treatment of Uranium-Contaminated Water. *BioMetals* **2022**, *35*, 335–348. [CrossRef] [PubMed]
150. Luo, D.; Qiang, S.; Geng, R.; Shi, L.; Song, J.; Fan, Q. Mechanistic Study for Mutual Interactions of Pb2+ and Trichoderma Viride. *Ecotoxicol. Environ. Saf.* **2022**, *233*, 113310. [CrossRef]

151. Schaefer, S.; Steudtner, R.; Hübner, R.; Krawczyk-Bärsch, E.; Merroun, M.L. Effect of Temperature and Cell Viability on Uranium Biomineralization by the Uranium Mine Isolate Penicillium Simplicissimum. *Front. Microbiol.* **2021**, *12*, 802926. [CrossRef]
152. Saeid, A.; Cepoi, L.; Jastrzębska, M.; Nomngongo, P.N. Bioremediation of Heavy Metal Ions Contaminated Soil. *Soil Bioremediation* **2021**, 87–114.
153. Husna; Hussain, A.; Shah, M.; Hamayun, M.; Qadir, M.; Iqbal, A. Heavy Metal Tolerant Endophytic Fungi Aspergillus Welwitschiae Improves Growth, Ceasing Metal Uptake and Strengthening Antioxidant System in *Glycine max* L. *Environ. Sci. Pollut. Res.* **2021**, *29*, 15501–15515. [CrossRef]
154. Pagliaccia, B.; Carretti, E.; Severi, M.; Berti, D.; Lubello, C.; Lotti, T. Heavy Metal Biosorption by Extracellular Polymeric Substances (EPS) Recovered from Anammox Granular Sludge. *J. Hazard. Mater.* **2022**, *424*, 126661. [CrossRef] [PubMed]
155. Sivashankar, R.; Sathya, A.B.; Kanimozhi, J.; Deepanraj, B. Characterization of the Biosorption Process. In *Biosorption for Wastewater Contaminants*; John Wiley & Sons: Hoboken, NJ, USA, 2021; pp. 102–116.
156. Bouabidi, Z.B.; El-Naas, M.H.; Zhang, Z. Immobilization of Microbial Cells for the Biotreatment of Wastewater: A Review. *Environ. Chem. Lett.* **2018**, *17*, 241–257. [CrossRef]
157. Kamizela, T.; Worwag, M. Processing of Water Treatment Sludge by Bioleaching. *Energies* **2020**, *13*, 6539. [CrossRef]
158. Maity, J.P.; Chen, G.-S.; Huang, Y.-H.; Sun, A.-C.; Chen, C.-Y. Ecofriendly Heavy Metal Stabilization: Microbial Induced Mineral Precipitation (MIMP) and Biomineralization for Heavy Metals within the Contaminated Soil by Indigenous Bacteria. *Geomicrobiol. J.* **2019**, *36*, 612–623. [CrossRef]
159. Kumbhar, P.; Savla, N.; Banerjee, S.; Mathuriya, A.S.; Sarkar, A.; Khilari, S.; Jadhav, D.A.; Pandit, S. Microbial Electrochemical Heavy Metal Removal: Fundamental to the Recent Development. *Wastewater Treat.* **2021**, 521–542. [CrossRef]
160. Tian, Y.; Zhu, Q.; Yuan, J.; Kneepkens, R.; Yue, Y.; Zhang, C. Direct Embryotoxicity of Chromium (III) Exposure during Preimplantation Development. *J. Reprod. Dev.* **2021**, *67*, 283–290. [CrossRef]
161. Lim, S.S.; Fontmorin, J.-M.; Pham, H.T.; Milner, E.; Abdul, P.M.; Scott, K.; Head, I.; Yu, E.H. Zinc Removal and Recovery from Industrial Wastewater with a Microbial Fuel Cell: Experimental Investigation and Theoretical Prediction. *Sci. Total Environ.* **2021**, *776*, 145934. [CrossRef]
162. Ezziat, L.; Elabed, A.; Ibnsouda, S.; El Abed, S. Challenges of Microbial Fuel Cell Architecture on Heavy Metal Recovery and Removal From Wastewater. *Front. Energy Res.* **2019**, *7*, 1. [CrossRef]
163. Gustave, W.; Yuan, Z.-F.; Li, X.; Ren, Y.-X.; Feng, W.-J.; Shen, H.; Chen, Z. Mitigation Effects of the Microbial Fuel Cells on Heavy Metal Accumulation in Rice (*Oryza sativa* L.). *Environ. Pollut.* **2020**, *260*, 113989. [CrossRef]
164. Wang, L.; Xu, D.; Zhang, Q.; Liu, T.; Tao, Z. Simultaneous Removal of Heavy Metals and Bioelectricity Generation in Microbial Fuel Cell Coupled with Constructed Wetland: An Optimization Study on Substrate and Plant Types. *Environ. Sci. Pollut. Res.* **2021**, *29*, 768–778. [CrossRef]
165. Zhang, J.; Cao, X.; Wang, H.; Long, X.; Li, X. Simultaneous Enhancement of Heavy Metal Removal and Electricity Generation in Soil Microbial Fuel Cell. *Ecotoxicol. Environ. Saf.* **2020**, *192*, 110314. [CrossRef] [PubMed]
166. Kabutey, F.T.; Antwi, P.; Ding, J.; Zhao, Q.; Quashie, F.K. Enhanced Bioremediation of Heavy Metals and Bioelectricity Generation in a Macrophyte-Integrated Cathode Sediment Microbial Fuel Cell (MSMFC). *Environ. Sci. Pollut. Res.* **2019**, *26*, 26829–26843. [CrossRef] [PubMed]
167. Adekunle, A.; Gomez Vidales, A.; Woodward, L.; Tartakovsky, B. Microbial Fuel Cell Soft Sensor for Real-Time Toxicity Detection and Monitoring. *Environ. Sci. Pollut. Res.* **2020**, *28*, 12792–12802. [CrossRef] [PubMed]
168. Wei, Z.; Tang, M.; Huang, Z.; Jiao, H. Mercury Removal from Flue Gas Using Nitrate as an Electron Acceptor in a Membrane Biofilm Reactor. *Front. Environ. Sci. Eng.* **2021**, *16*, 20. [CrossRef]
169. Khosravi, A.; Javdan, M.; Yazdanpanah, G.; Malakootian, M. Removal of Heavy Metals by *Escherichia coli* (*E. coli*) Biofilm Placed on Zeolite from Aqueous Solutions (Case Study: The Wastewater of Kerman Bahonar Copper Complex). *Appl. Water Sci.* **2020**, *10*, 167. [CrossRef]
170. Henagamage, A.P.; Peries, C.M.; Seneviratne, G. Fungal-Bacterial Biofilm Mediated Heavy Metal Rhizo-Remediation. *World J. Microbiol. Biotechnol.* **2022**, *38*, 85. [CrossRef]
171. Grujić, S.; Vasić, S.; Radojević, I.; Čomić, L.; Ostojić, A. Comparison of the Rhodotorula Mucilaginosa Biofilm and Planktonic Culture on Heavy Metal Susceptibility and Removal Potential. *Water Air Soil Pollut.* **2017**, *228*, 73. [CrossRef]
172. Rikta, S.Y. Application of Nanoparticles for Disinfection and Microbial Control of Water and Wastewater. *Nanotechnol. Water Wastewater Treat.* **2019**, 159–176.
173. Shahpiri, A.; Mohammadzadeh, A. Mercury Removal by Engineered Escherichia Coli Cells Expressing Different Rice Metallothionein Isoforms. *Ann. Microbiol.* **2018**, *68*, 145–152. [CrossRef]
174. Yang, J.; Hou, B.; Wang, J.; Tian, B.; Bi, J.; Wang, N.; Li, X.; Huang, X. Nanomaterials for the Removal of Heavy Metals from Wastewater. *Nanomaterials* **2019**, *9*, 424. [CrossRef]
175. Shehata, M.; Yamazaki, K. ichi Using Recombinant *E. coli* Displaying Surface Heavy Metal Binding Proteins for Removal of Pb2+ from Contaminated Water. *J. Bioremediation Biodegrad.* **2018**, *9*. [CrossRef]

toxics MDPI

Article

The Treatment of Aquaculture Wastewater with Biological Aerated Filters: From the Treatment Process to the Microbial Mechanism

Jiafeng Ding [1,2], Yunjuan Meng [1], Shihuan Lu [2], Yiwen Peng [3], Wen Yan [1], Wenbing Li [1], Jinchun Hu [4], Ting Ye [4], Yuchi Zhong [1,*] and Hangjun Zhang [1,2]

1 School of Life and Environmental Sciences, Hangzhou Normal University, Hangzhou 311121, China; djfphd@163.com (H.Z.)
2 School of Engineering, Hangzhou Normal University, Hangzhou 311121, China
3 Zhe Jiang Sunda Public Environmental Protection Co., Ltd., Hangzhou 311000, China
4 Quzhou Aquatic Technology Extension Station, Quzhou 324000, China
* Correspondence: sll20200103@163.com; Tel.: +86-571-28865328; Fax: +86-571-28865333

Abstract: Algal cell proliferation has posed significant problems for traditional water treatment facilities; these problems are attributed to surface hydrophilicity and electrostatic repulsion. Biological aerated filters (BAFs) have been extensively used in wastewater treatment to remove pollutants such as algal cells by utilizing the adsorption and separation capabilities of the filter media. In this study, a BAF was supplemented with biological filter medium (*Marchantia polymorpha*) to assess its effectiveness of pretreating aquaculture wastewater. In terms of process performance, steady and consistent treatment was achieved by the BAF with *M. polymorpha* (BAF2) under an algal cell density as high as 1.65×10^8 cell/L, with average removal rates for NH_4^+-N and algae cells of 74.4% and 81.9%, respectively. The photosynthetic activity parameters (rETRmax, α, Fv/Fm, and I_k) of the influent and effluent were quantitatively assessed, and *M. polymorpha* was found to remove algae by disrupting the photosynthetic system of the algal cells. Furthermore, the addition of the *M. polymorpha* filter medium enhanced the community structure of the functional microbes in the BAF system. The highest microbial community richness and diversity were observed in the BAF2. Meanwhile, *M. polymorpha* promoted an increase in the abundance of denitrifying bacteria, including *Bdellovibrio* and *Pseudomonas*. Overall, this work offers a unique perspective on the aquaculture wastewater pretreatment process and BAF design.

Keywords: biological aerated filters; aquaculture wastewater; algae cells; *M. polymorpha*; ammonia nitrogen

Citation: Ding, J.; Meng, Y.; Lu, S.; Peng, Y.; Yan, W.; Li, W.; Hu, J.; Ye, T.; Zhong, Y.; Zhang, H. The Treatment of Aquaculture Wastewater with Biological Aerated Filters: From the Treatment Process to the Microbial Mechanism. *Toxics* **2023**, *11*, 478. https://doi.org/10.3390/toxics11060478

Academic Editors: Junhao Qin, Peidong Su, Feng Zhu and Lin Ding

Received: 3 April 2023
Revised: 13 May 2023
Accepted: 23 May 2023
Published: 25 May 2023

1. Introduction

The random release of a massive volume of nitrogen- and phosphorus-rich aquaculture wastewater gives rise to the frequent occurrence of water eutrophication in the natural environment [1]. Aquaculture wastewater has also become a breeding ground for harmful algae [2]. Algae cells are difficult to separate from water due to electrostatic repulsion and surface hydrophilicity, which has caused severe challenges for conventional drinking water treatment facilities [3]. Ammonia nitrogen (NH_4^+-N) is a contaminant commonly detected in aquaculture wastewater [4,5]. Excessive concentrations of NH_4^+-N (>1.5 mg/L) can cause dissolved oxygen depletion and water quality decline and pose a threat to the survival of aquatic organisms [6]. Moreover, as coexisting pollutants, NH_4^+-N and algae cells may also react with chlorine during the chlorination disinfection to generate disinfection byproducts (DBPs), which cause harm to human health [7]. However, conventional drinking water treatment processes such as photocatalysis [8], coagulation–flocculation [9], and advanced oxidation processes [10] have difficulty removing algae cells and NH_4^+-N simultaneously and efficiently. Thus, the development of a pretreatment process for aquaculture wastewater is of great significance for maintaining drinking water safety.

Compared with conventional drinking water pretreatment utilities, biological aerated filters (BAFs) have overwhelming predominance owing to their compact footprint, large treatment volume capacity, and great resistance to shock loading [11]. BAFs can combine a short hydraulic retention time with a long sludge age organic unity which not only promotes the enrichment of nitrification bacteria with long generation periods but also obtains a high NH_4^+-N removal efficiency [12,13]. Therefore, BAFs have great potential for the treatment of aquaculture wastewater. In fact, the choice of filter medium has a significant impact on the treatment efficiency of a BAF. On one hand, suspended solids and micropollutants in the influent are removed by the filtration effect of the filter medium. On the other hand, microorganisms attached to the filter medium remove micropollutants and fine particles by flocculation. In addition, the multiplicity of filter medium will provide more possibilities to treat different wastewaters using BAFs [14]. Regarding textile wastewater, BAFs with zeolite media have advantages in removing color and organic matter [15]. Dong et al. [16] designed a high-performance BAF packed with porous ceramsites for the treatment of organic wastewater. Thus, the method of selecting appropriate filter medium for the effective pretreatment of aquaculture wastewater is of great research significance.

Marchantia polymorpha could be a potential filter medium for achieving a high removal efficiency of algal cells from aquaculture wastewater in BAF systems. *M. polymorpha* is a primordial terrestrial plant that releases secondary metabolites such as ethereal terpenoids and lipophilic aromatics, which can inhibit the growth of algal cells [17]. Thus, turning *M. polymorpha* into a biological resource to inhibit the growth of algal cells from aquaculture wastewater should be considered. *M. polymorpha* grows in a broad range of environments, making the material more accessible. Therefore, in response to the treatment requirements of aquaculture wastewater in this research, a unique BAF is set up, using *M. polymorpha* as the filler medium.

A conventional BAF was applied to remove pollution such as nitrogen and organic matter from wastewater via the activity of the microorganisms attached to the filler medium. To date, very few studies have reported the use of a natural filler medium in combination with a BAF to treat aquaculture wastewater. In this work, two aerating upflow biofilters were utilized to investigate the response of *M. polymorpha* to the removal efficiency of NH_4^+-N and algal cells. The difference between the two reactors was whether *M. polymorpha* was used. In parallel, the algae removal mechanism of *M. polymorpha* in aquaculture wastewater was revealed using batch experiments. With a focus on process performance, the dynamic changes in the microbial communities of the biofilms attached to *M. polymorpha* were also thoroughly plumbed in this study. The major purpose of this research was to uncover the underlying response mechanisms of *M. polymorpha* in BAF systems, as well as the role it plays in BAF operations, which would provide guidance for aquaculture wastewater pretreatment.

2. Materials and Methods

2.1. Characteristics of Aquaculture Wastewater and Seeding Sludge

The archetypical harmful cyanobacteria, *Microcystis aeruginosa* strain FACHB-905, was obtained from the Freshwater Algae Culture Collection of the Institute of Hydrobiology (Wuhan, China). A sterile BG-11 medium was used for culturing the test strain at 25 °C with light and dark periods (12/12 h) and 2000 lux illumination. Prior to the experiment, the test strain needed to reach exponential growth. Synthetic aquaculture wastewater was prepared following the recipe described in Table 1 and was used as the influent for the reactors. The biological aerated filter (BAF) was inoculated with 500 mL of activated sludge. The activated sludge was acquired from a denitrification reactor in our laboratory.

Table 1. Characteristics of BAF system influents.

NH_4^+-N (mg/L)	NO_2^--N (mg/L)	NO_3^--N(mg/L)	Algae Density (10^8 Cell/L)
15 ± 0.58	0	5 ± 0.42	1.65 ± 0.64

2.2. BAF Setup and Operation

Two biological aerated filters named BAF1 and BAF2 were constructed in this study (Figure 1). All BAFs were composed of a transparent polyethylene plastic column and a PVC pipe. The column had a depth of 450 mm and a diameter of 150 mm, with an effective working capacity of 5 L. Both BAF1 and BAF2 contained zeolite and activated carbon as biomass carriers, and the filling height of both the zeolite and activated carbon was 50 mm. The gravel was 15~25 mm below the filter media as a support layer. The difference between BAF1 and BAF2 was the addition of a biological filler (*M. polymorpha*). BAF2 was utilized as an experimental group to investigate the influence of the addition of a biological filler on the nitrogen removal performance of the reactor, while BAF1 was a control group without the addition of a biological filler. In addition, the component of the biological filter medium was derived from the land plant *M. polymorpha*. Whole *M. polymorpha* was collected from the edge of an artificial lake at Hangzhou Normal University (N30°19′0.79″, E120°23′56.57″). Deionized water was used to remove dirt that remained on the *M. polymorpha*, which was dried to a constant weight at 60 °C.

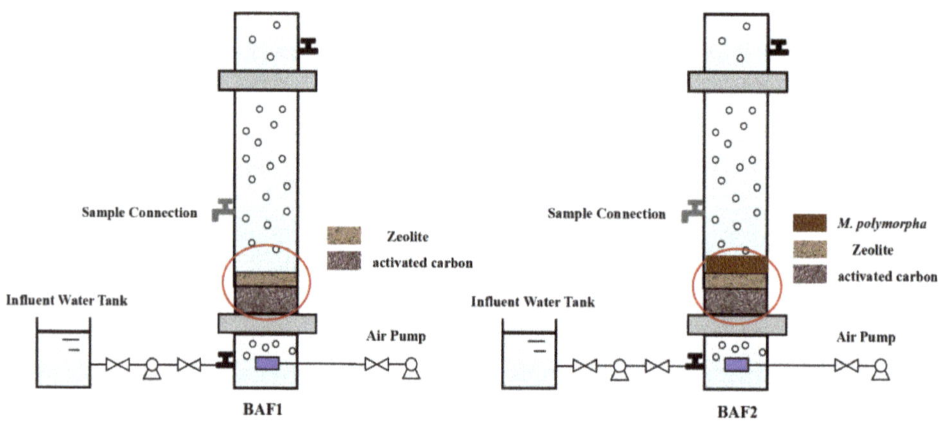

Figure 1. Schematic diagram of BAF1 and BAF2.

To explore the nitrogen and algae removal performance of the BAFs after the addition of the biological filler, the reactors were operated for 50 days without accident. Reactor operation was divided into two stages according to the hydraulic retention time (HRT) and gas-to-liquid ratio: the domestication stage (phase I: HRT = 8 h, gas-to-liquid ratio = 10:1, 1–30 days) and the stabilization stage (phase II: HRT = 4 h, gas-to-liquid ratio = 4:1, 31–50 days). The dissolved oxygen (DO) was maintained between 4.0 and 5.0 by regulating the air flow rate.

2.3. Collection and Characterization of Aquaculture Wastewater

The influents and effluents from BAFs were collected for regular testing. Water samples were tested daily for NH_4^+-N, NO_2^--N, and NO_3^--N using conventional methods [18]. Excitation–emission matrix (EEM) fluorescence spectra of the influent and effluent were obtained via a fluorescence spectrophotometer (F-4600, Hitachi Co. Ltd., Hitachi, Japan). The TOC concentration was acquired using a TOC analyzer (TOC-L, Shimadzu, Kyoto, Japan). The cell density and concentration of chlorophyll-a (Chl-a) are common indicators for expressing the growth of algal cells. The density of algal cells in aquaculture wastewater was measured via the hemocytometer method [19]. Chlorophyll a (Chl-a) content was used as an indicator of the potential photosynthetic capacity of *M. aeruginosa*, whose concentration affects the growth and photosynthesis of algal cells [20]. Chl-a was extracted with 95% ethanol at 4 °C for 24 h [21]. In brief, 10 mL of each group was collected according to the previous sampling schedule and centrifuged for 10 min at 8000 r/min to obtain a clear supernatant. Then, the absorbance of the supernatant was determined at 665 nm and 649 nm in a spectrometer. The Chl-a content (mg/L) was calculated by Equation (1):

$$\text{Chl-a} = 13.95 \times OD_{650} - 6.88 \times OD_{649} \quad (1)$$

where Chl-a denotes the concentration of chlorophyll-a, while OD_{649} and OD_{650} denote the absorbance at 649 and 650 nm, respectively.

Additionally, the maximum photochemical quantum yield (Fv/Fm), relative electron transport rate (rETRmax), and photosynthesis efficiency (initial slope, alpha) in the aquaculture wastewater were measured with a phytoplankton analyzer (WALZ, Rohrdorf, Germany), according to Wang et al. [22] All samples were exposed to a dark environment for 10 min before measurement. All physiological indicators were measured in triplicate and reported as the mean standard deviation.

2.4. Microbial Community Analysis

As previously reported, high-throughput sequencing of 16S rRNA genes was conducted for to characterize the microbial communities of the BAFs [23]. In brief, the bacterial primers 515F and 907R were used to amplify the V3-V4 region of the 16S rRNA genes [24]. Personal Biotechnology Co., Ltd. (Shanghai, China) constructed libraries and sequenced amplicons from samples. The sequencing findings were analyzed using QIIME software, which included quality filtering and taxonomy classification [25]. A value of 97% was classified as the threshold for describing the operational taxonomic units (OTUs).

3. Results and Discussion

3.1. Nitrogen Removal Performance of BAFs

The NH_4^+-N removal performances of the two BAFs are shown in Figure 2a,b. Over 50 days, BAF1 and BAF2 operated quite well, with average NH_4^+-N removal percentages of 70.2% and 74.4%, respectively. Nevertheless, the concentrations of NH_4^+-N in the BAF effluents exhibited different trends during the whole period. After the domestication stage, changing the water/gas ratio and HRT in BAF2 had no significant negative impact on the NH_4^+-N removal performance. In contrast, the NH_4^+-N removal performance of BAF2 was effectively improved in phase II, and the average NH_4^+-N removal percent was increased to 82.1%. The average NH_4^+-N removal percent of BAF1 was 71.5% in phase II. Compared to BAF1, BAF2, with the biological filler medium, demonstrated better NH_4^+-N removal performance. As displayed in Figure 2c, the effluents of the two BAFs had comparatively low NO_3^--N concentrations. The average concentrations of NO_3^--N in the BAF1 and BAF2 effluents were 6.07 and 8.17 mg/L, respectively. In addition, the average concentrations of NO_3^--N in BAF1 and BAF2 remained at low levels during the whole period at 0.41 and 0.28 mg/L, respectively. Based on these findings, the ammonium in the influent could easily be converted into nitrate throughout the BAF system, and there was no large accumulation of nitrite. This result confirmed that the addition of *M. polymorpha* did not negatively affect the denitrification performance of the BAF.

3.2. Algae Removal Efficiency of BAFs

The massive presence of ammonia nitrogen in natural water can generate harmful blooms [11]. Therefore, it is important to investigate the algae removal efficiency of the BAF with *M. polymorpha* as a filler medium. As shown in Figure 3a, the algae removal efficiencies of BAF1 and BAF2 demonstrated some fluctuations during the operation of the reactors. The average algae removal efficiencies of BAF1 and BAF2 were 49.6% and 81.9%, respectively. Previous studies have shown that a composite BAF packed with a suspended biological carrier, zeolite, and granular activated carbon had an average algae removal efficiency of 78.6% [26]. Compared to the composite BAF, the addition of *M. polymorpha* as a filler medium into the BAF was more beneficial in enhancing the removal of algal cells. Furthermore, BAF1 could reduce the effluent algal cell density, which may be attributed to the presence of the two filter media in BAF1: zeolite and activated carbon. Zeolite and activated carbon have large specific surface areas and microporous structures which can remove algae by adsorption [27]. However, the algae removal efficiency of BAF1 on days 42 and 49 was significantly lower compared to that on Day 35. Presumably, with the long-term operation of BAFs, massive algal cells adsorb on the surface of the filter media, which increases the clogging risk of the reactor and thereby decreases the algae removal capacity. In addition, the higher algae removal efficiency of BAF2 may benefit from the addition of *M. polymorpha*. On one hand, the algal cells of the influent were first intercepted by *M. polymorpha*, which reduced the algal cell density of the effluent. On the other hand, *M. polymorpha* contained a variety of sesquiterpenes, which can effectively inhibit the growth of algae cells. Chlorophyll-a (chl-a) concentration was used as an indicator of the potential photosynthetic capacity of algal cells. As shown in Figure 3b, after 50 days, the average chl-a concentrations of the effluents in BAF1 and BAF2 were 0.585 mg/L and 0.229 mg/L, respectively, which were much lower than that of the influents (0.864 mg/L). This result was consistent with the algae removal efficiencies of BAF1 and BAF2, suggesting that the addition of the *M. polymorpha* filler enhanced the algae removal ability of the BAF system.

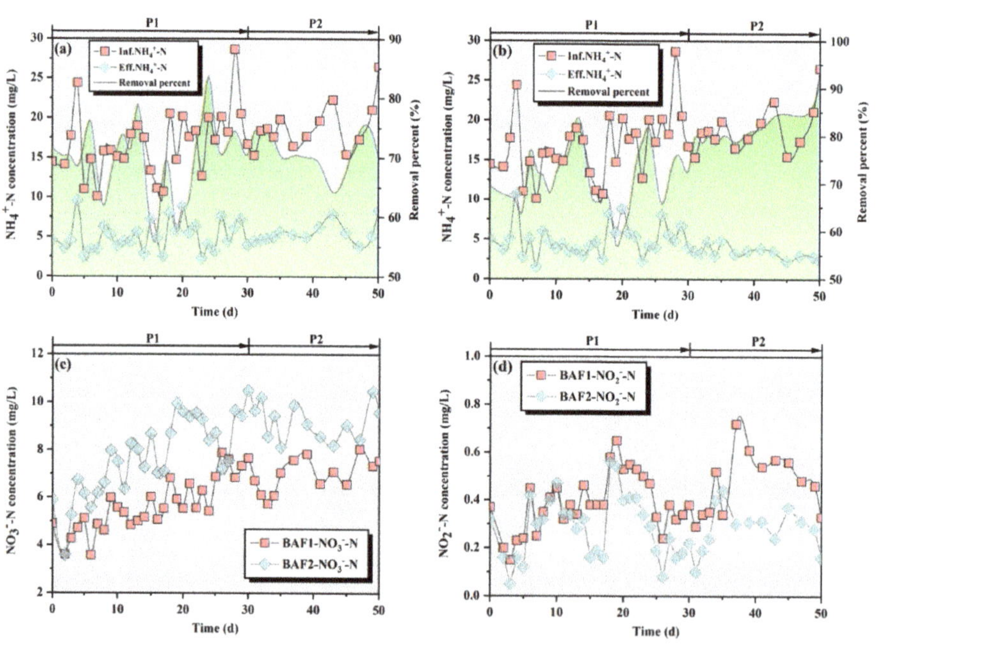

Figure 2. NH_4^+-N removal performance of BAF1 (**a**) and BAF2 (**b**) during the whole period; the effluent concentrations of NO_3^--N (**c**) and NO_2^--N (**d**) for BAF1 and BAF2.

To more comprehensively and systematically study the algae removal mechanism of the BAF system with the *M. polymorpha* filler, we further measured the photosynthetic activity parameters of the influent and effluent. As shown in Figure 4a,b, the rETRmax and α values in the effluent were noticeably lower than those in the effluent after the BAF1 treatment (without the *M. polymorpha* filler) and BAF2. The rETRmax and α values indicate the photoacclimation state of the algae, which is often used as a response to the energy transfer and capacity of the electron transport chain in algal cells [28]. Compared to BAF1, the rETRmax and α values of the BAF2 effluent decreased to a greater extent, and the average values were 22.221 and 0.049, respectively. This result suggests that the electron transport chain transmission of the algal cells was inhibited by the BAF system with the *M. polymorpha* filler. The Fv/Fm and Ik values of the effluent were also altered by the BAF system (Figure 4c,d). Fv/Fm and Ik are important indicators of the light absorbed by PSII. When the Fv/Fm value is below 0.3, this indicates poor algal health [29]. The average Fv/Fm values for the BAF1 and BAF2 effluents were 0.295 and 0.145, respectively. At the same time, the Ik value of the BAF2 effluent was significantly lower than that of the influent. These results indicate that the BAF system with the *M. polymorpha* filler caused more severe damage to the PSII of the algal cells.

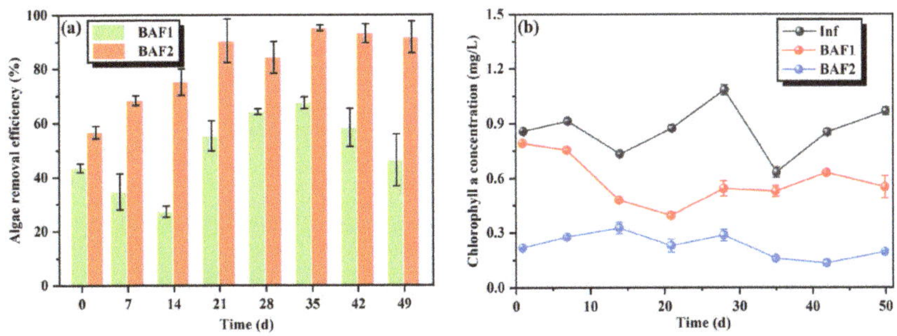

Figure 3. Algae cell removal efficiencies (**a**) and chlorophyll a concentrations (**b**) of BAF1 and BAF2 during the whole period.

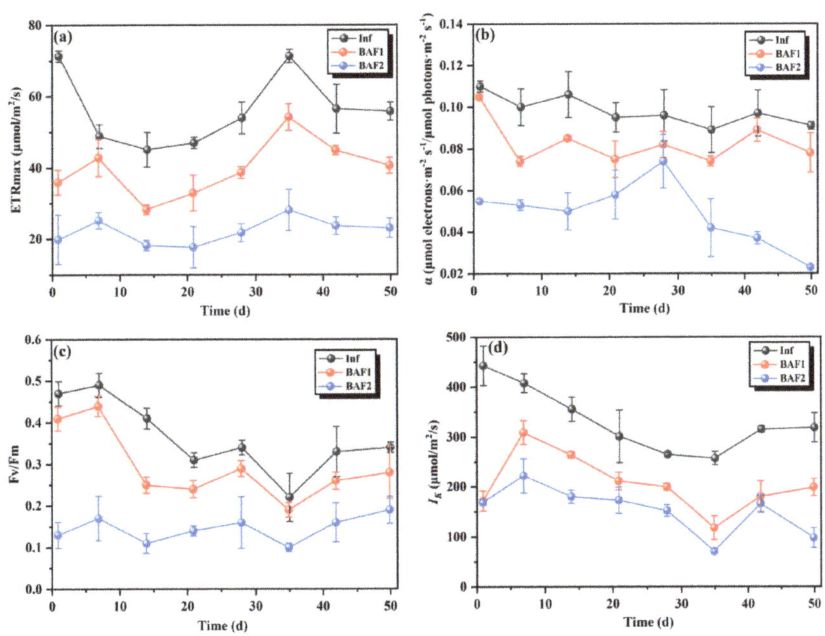

Figure 4. Performances of BAF1 and BAF2 in the aquaculture wastewater degradation experiment based on the maximal electron transport rates (rETRmax) (**a**), α (**b**), maximum efficiency of photosystem II (Fv/Fm) (**c**), and semi−light saturation point (Ik) (**d**) during the whole period.

Algae is an autotrophic species that relies on its photosynthetic system for energy conversion [30]. The *M. polymorpha* filler medium severely interfered with the capture and conversion of light energy by the algal cells in the aquaculture wastewater, leading to a complete collapse of the photosynthetic system. Therefore, biological aerated filters with *M. polymorpha* as the filler medium have potential development prospects for application in aquaculture wastewater treatment.

3.3. TOC Removal Performance of the BAFs

As shown in Figure 5, there was a significant difference in the TOC removal performance between BAF1 and BAF2 during the whole operation. The reduction in TOC may be attributed to the microbial activity in the sludge. On one hand, microorganisms consume large amounts of carbon for their own growth and metabolism [31]. On the other hand, the microorganisms in the sludge also need to obtain carbon from the wastewater for the energy to carry out denitrification [32]. In the domestication stage (P1), the average TOC removal rates in the reactors were 15.53% (BAF1) and 23.71% (BAF2). In general, it takes a long time to completely adapt the microbial community of the inoculum to the BAF system. At this stage, the microbes were unbound and readily oxidized, resulting in limited microbial activity [11]. Therefore, the TOC removal rates of BAF1 and BAF2 fluctuated greatly in P1, and the average removal rate was poor. However, in phase II, the TOC removal rate of BAF2 gradually increased, and the average removal rate of TOC in BAF2 increased to 37.49%. This result may be attributed to the microbial activity in the second stage of BAF2 being better than in the first stage, and excessive organic matter in the algae-containing wastewater being metabolized by heterotrophic microorganisms [33]. Furthermore, the average TOC removal rates in the BAFs were 16.52% (BAF1) and 28.38% (BAF2) during the whole operation, indicating that BAF2 exhibited the best performance in TOC removal. It is reasonable to assume that the addition of the *M. polymorpha* filler medium could enhance the community structure of the functional microbes in the BAF system.

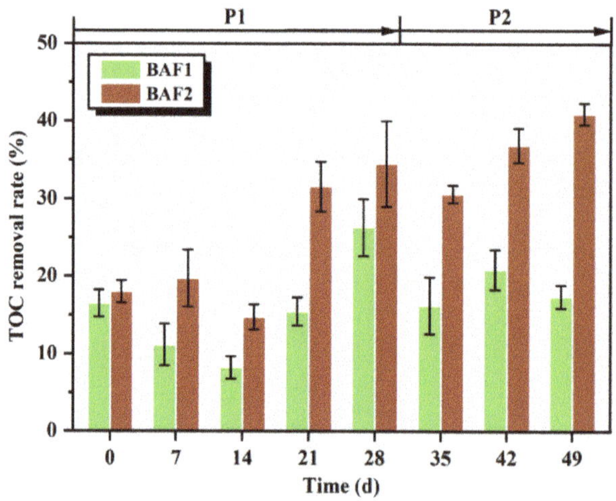

Figure 5. Performances of BAF1 and BAF2 in the aquaculture wastewater degradation experiment based on TOC concentration during the whole period.

3.4. EEM Fluorescence Spectral Analysis

In this study, changes in the type and content of organic matter in the aquaculture wastewater were detected via EEM fluorescence spectra during phase I and II. As shown in Figure 6, two peaks can be identified in the EEM spectra. The first peak represented aromatic proteins (Ex/Em: 220–250/330–360 nm) [34]. The second peak represented soluble microbial products (SMPs) (Ex/Em: 250–300/330–360 nm) [34]. Aquaculture wastewater has a massive amount of algal cells, and the algal cells release numerous extracellular compounds to the external environment; these compounds are mainly protein-like substances [35]. Therefore, the peak with the highest fluorescence intensity in the EEM fluorescence spectra of influents was identified as belonging to a protein-like substance (Figure 6a,b). The fluorescence intensity of the protein-like substance (PLS) in the effluent water decreased significantly after BAF treatment. The fluorescence intensity of the protein-like substance was higher in BAF1 than in BAF2 (Figure 6c,e). In BAF2, the *M. polymorpha* filler medium decreased the algal cell density in the influent, which limited the release of extracellular substances. Hence, the PLS fluorescence intensity in the BAF2 effluent was lower than that in the BAF1 effluent. This result was also consistent with the algae removal performance of BAF2. Additionally, in the phases I and II, the SMP peak fluorescence intensity of the BAF2 effluent was higher than that of the BAF1 effluent. In BAF2, the *M. polymorpha* filler medium provided an additional carbon source for microorganisms; thus, the SMP peak fluorescence of the BAF2 effluent was higher than that of the BAF1 effluent. Moreover, microorganisms consume carbon sources to maintain their metabolic activities and denitrification, which will also lead to an increase in the fluorescence intensity of SMPs [23]. According to the results of 3.1 and 3.3, the addition of the *M. polymorpha* filler medium improved the performance of the removal of TOC and NH_4^+-N from BAF2. Simultaneously, the microorganisms in BAF2 were adapted to the environment with the *M. polymorpha* filler medium, so a higher SMP fluorescence intensity from the BAF2 effluent was detected.

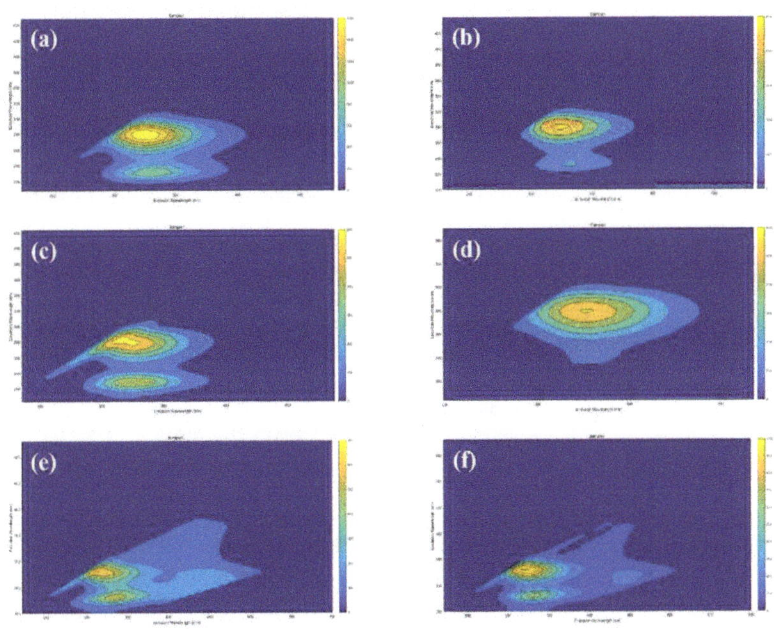

Figure 6. EEM fluorescence spectra of influents and effluents from the different BAFs during the experiments; (**a**,**b**)—EEM fluorescence spectra for influents in phases I and II; (**c**,**d**)—EEM fluorescence spectra for effluents from BAF1 and BAF2, respectively, in phase I; (**e**,**f**)—EEM fluorescence spectra for effluents from BAF1 and BAF2, respectively, in phase II.

3.5. Microbial Community Analysis

The microbial community structures of the biofilms attached to the *M. polymorpha* filler medium from BAF1 and BAF2 was measured via high-throughput sequencing. As shown in Table 2, 44,677, 45,474, and 51,531 effective sequences were attained for the microbial samples on the initial sludge (BAF0), BAF1, and BAF2, respectively. The Chao 1 and ACE indexes of BAF2 were higher than those of BAF1, which indicated that the microbial community richness for BAF2 was greater than for those attached to BAF1. Furthermore, the Simpson and Shannon indexes for BAF2 were also significantly higher than those of BAF0, suggesting that the biofilm in BAF2 had a high level of microbial diversity. Based on the above results, the addition of the *M. polymorpha* filler medium was conducive to microbial enrichment and improved the richness and diversity of the microbial community.

Table 2. The richness and evenness indices of BAF0, BAF1, and BAF2 microbial communities.

Sample	Sequence	Chao1	ACE	Simpson	Shannon
BAF0	44,677	446.0	412.9	0.9510	5.7765
BAF1	45,474	744.3	689.2	0.9841	7.4083
BAF2	51,531	835.3	753.6	0.9716	7.0655

As shown in Figure 7a, *Proteobacteria* was the most dominant phylum in all the microbial samples from BAF0, BAF1, and BAF2, accounting for 82.8%, 46.7%, and 64.6% of the total effective reads, respectively. Compared with BAF1, the abundance of *Proteobacteria* and *Bacteroidetes* was higher in BAF2. *Proteobacteria* have been widely detected in drinking water, wastewater, soils, and sludge [36]. In addition, bacteria belonging to *Proteobacteria* can remove nitrogen in anaerobic, aerobic, and anoxic environments [37]. A higher abundance of such bacteria would be conducive to improving the nitrogen removal performance of BAF. *Bacteroidetes* are essential for denitrification because they participate in the anaerobic or anoxic hydrolysis of macromolecular molecules [38]. The nitrogen removal performances of the two reactors differed owing to differences in the relative abundances of *Bacteroidetes*. Moreover, *Firmicutes* was also one of the most dominant phyla in the two reactors. The significance of *Firmicutes* in the denitrification process, utilizing solid carbon sources, has been widely mentioned [39]. *Acidobacteria* showed higher abundances in BAF1 and BAF2 than BAF0, which was ascribed to the addition of activated carbon and zeolite in the BAF system.

To obtain a better comprehension of the bacterial structure, the bacterial abundances of BAF0, BAF1, and BAF2 at the genus level were investigated. As shown in Figure 7b, at the genus level, *Actinobacteria* was the most dominant genus in BAF1 and BAF2, accounting for 16.6% and 15.5%, respectively. *Actinobacteria* is a genus of sheathed filamentous bacteria that has been discovered in several wastewater treatment systems [40]. After 50 days of reactor operation, *Bdellovibrio* and *Pseudomonas* had notably higher abundances in BAF2, increasing significantly and accounting for 7.6% and 17.8%, respectively. Previous research has shown that *Bdellovibrio* plays an essential role in NO_3^--N removal under denitrifying circumstances [41]. *Pseudomonas*, an aerobic denitrifying bacterium, has an excellent ability to convert nitrate into nitrogenous gases [42].

Overall, the above results indicate that BAF2, with *M. polymorpha* filler medium, was the most microbiologically active reactor. Meanwhile, *M. polymorpha* filler medium enriched the dominant phylum that was capable of nitrification and correspondingly played a major role in the denitrification of aquaculture wastewater. Hence, the *M. polymorpha* filler medium exhibits promise in aquaculture wastewater treatment which is worth investigating in the future.

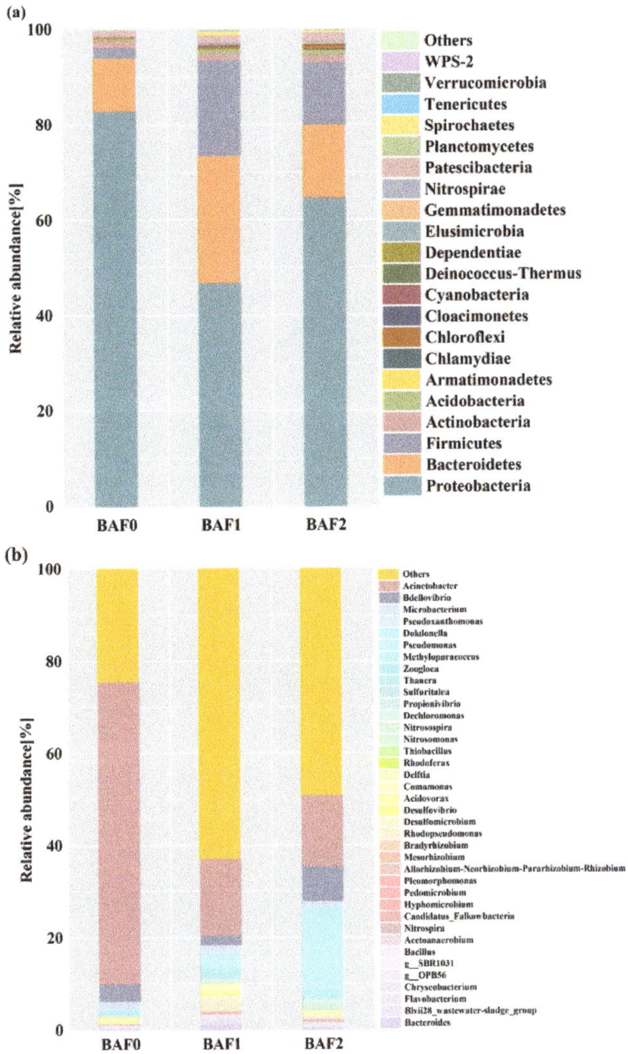

Figure 7. Microbial population dynamics of the initial sludge (BAF0), BAF1, and BAF2 microbial samples at the phylum (**a**) and genus (**b**) levels.

4. Conclusions

A biological aerated filter equipped with *M. polymorpha* filler medium was proposed for the first time in this research, which contributed to establishing an efficient and reliable pretreatment system for aquaculture wastewater. In this work, the use of the low-cost and easily available terrestrial plant, *M. polymorpha,* as a filler medium greatly enhanced the removal of ammonium ions (NH_4^+) and algal cells from aquaculture wastewater. The main contribution of *M. polymorpha* was the removal of algae cells from aquaculture wastewater by interfering with the photosynthetic system of the algae cells. On the other hand, the *M. polymorpha* filler medium could enrich denitrifying microorganisms such as *Bdellovibrio* and *Pseudomonas* and improve the denitrification performance of the BAF. This study provides vital insights into aquaculture wastewater pretreatment and a theoretical basis for the development of BAFs with *M. polymorpha* filler media in the future.

Author Contributions: Conceptualization, J.D. and Y.M.; methodology, Y.M.; software, S.L.; validation, Y.P. and S.L.; formal analysis, Y.P.; investigation, W.Y.; resources, W.L.; data curation, W.Y.; writing—original draft preparation, T.Y.; writing—review and editing, J.D. and Y.Z.; visualization, Y.Z.; supervision, J.H.; project administration, Y.Z.; funding acquisition, J.D. and H.Z. All authors have read and agreed to the published version of the manuscript.

Funding: This research was funded by the Key Research & Development project of Zhejiang Province (Grant No. 2021C02048), the National Natural Science Foundation of China (Grant No. 42107255).

Institutional Review Board Statement: Not applicable.

Informed Consent Statement: Not applicable.

Data Availability Statement: Not applicable.

Conflicts of Interest: The authors declare no conflict of interest.

References

1. Huang, J.; Xiao, Y.; Chen, B. Nutrients removal by Olivibacter jilunii immobilized on activated carbon for aquaculture wastewater treatment: ppk1 gene and bacterial community structure. *Bioresour. Technol.* **2023**, *370*, 128494. [CrossRef]
2. Barnharst, T.; Rajendran, A.; Sun, X.; Hu, B. Process optimization of aquaculture wastewater treatment using a mycoalgae biofilm. *Algal Res.* **2023**, *70*, 103020. [CrossRef]
3. Wan, Y.; Xie, P.; Wang, Z.; Ding, J.; Wang, J.; Wang, S.; Wiesner, M.R. Comparative study on the pretreatment of algae-laden water by UV/persulfate, UV/chlorine, and UV/H2O2: Variation of characteristics and alleviation of ultrafiltration membrane fouling. *Water Res.* **2019**, *158*, 213–226. [CrossRef] [PubMed]
4. Wu, H.; Wang, H.; Zhang, Y.; Antonopoulou, G.; Ntaikou, I.; Lyberatos, G.; Yan, Q. In situ biogas upgrading via cathodic biohydrogen using mitigated ammonia nitrogen during the anaerobic digestion of Taihu blue algae in an integrated bioelectrochemical system (BES). *Bioresour. Technol.* **2021**, *341*, 125902. [CrossRef]
5. Zhang, Q.; Xu, X.; Zhang, R.; Shao, B.; Fan, K.; Zhao, L.; Ji, X.; Ren, N.; Lee, D.-J.; Chen, C. The mixed/mixotrophic nitrogen removal for the effective and sustainable treatment of wastewater: From treatment process to microbial mechanism. *Water Res.* **2022**, *226*, 119269. [CrossRef] [PubMed]
6. Gogoi, M.; Bhattacharya, P.; Kumar Sen, S.; Mukherjee, I.; Bhushan, S.; Chaudhuri, S.R. Aquaculture effluent treatment with ammonia remover *Bacillus albus* (ASSF01). *J. Environ. Chem. Eng.* **2021**, *9*, 105697. [CrossRef]
7. Sheng, D.; Bu, L.; Zhu, S.; Deng, L.; Shi, Z.; Zhou, S. Transfer organic chloramines to monochloramine using two-step chlorination: A method to inhibit N-DBPs formation in algae-containing water treatment. *J. Hazard. Mater.* **2023**, *443*, 130343. [CrossRef]
8. Lu, S.; Shen, L.; Li, X.; Yu, B.; Ding, J.; Gao, P.; Zhang, H. Advances in the photocatalytic reduction functions of graphitic carbon nitride-based photocatalysts in environmental applications: A review. *J. Clean. Prod.* **2022**, *378*, 134589. [CrossRef]
9. Zhang, H.; Li, X.; Wu, D.; Yu, B.; Lu, S.; Wang, J.; Ding, J. A novel strategy for efficient capture of intact harmful algal cells using Zinc oxide modified carbon nitride composites. *Algal Res.* **2023**, *69*, 102932. [CrossRef]
10. Sun, P.; Liu, Y.; Mo, F.; Wu, M.; Xiao, Y.; Xiao, X.; Wang, W.; Dong, X. Efficient photocatalytic degradation of high-concentration moxifloxacin over dodecyl benzene sulfonate modified graphitic carbon nitride: Enhanced photogenerated charge separation and pollutant enrichment. *J. Clean. Prod.* **2023**, *393*, 136320. [CrossRef]
11. Liu, H.; Zhu, L.; Tian, X.; Yin, Y. Seasonal variation of bacterial community in biological aerated filter for ammonia removal in drinking water treatment. *Water Res.* **2017**, *123*, 668–677. [CrossRef] [PubMed]
12. Cui, B.; Yang, Q.; Zhang, Y.; Liu, X.; Wu, W.; Li, J. Improving nitrogen removal in biological aeration filter for domestic sewage treatment via adjusting microbial community structure. *Bioresour. Technol.* **2019**, *293*, 122006. [CrossRef] [PubMed]
13. Ashour, M.; Alprol, A.E.; Heneash, A.M.M.; Saleh, H.; Abualnaja, K.M.; Alhashmialameer, D.; Mansour, A.T. Ammonia Bioremediation from Aquaculture Wastewater Effluents Using Arthrospira platensis NIOF17/003: Impact of Biodiesel Residue and Potential of Ammonia-Loaded Biomass as Rotifer Feed. *Materials* **2021**, *14*, 5460. [CrossRef] [PubMed]
14. Abou-Elela, S.I.; Fawzy, M.E.; El-Gendy, A.S. Potential of using biological aerated filter as a post treatment for municipal wastewater. *Ecol. Eng.* **2015**, *84*, 53–57. [CrossRef]
15. Chang, W.-S.; Hong, S.-W.; Park, J. Effect of zeolite media for the treatment of textile wastewater in a biological aerated filter. *Process Biochem.* **2002**, *37*, 693–698. [CrossRef]
16. Dong, J.; Wang, Y.; Wang, L.; Wang, S.; Li, S.; Ding, Y. The performance of porous ceramsites in a biological aerated filter for organic wastewater treatment and simulation analysis. *J. Water Process Eng.* **2020**, *34*, 101134. [CrossRef]
17. Tanaka, M.; Esaki, T.; Kenmoku, H.; Koeduka, T.; Kiyoyama, Y.; Masujima, T.; Asakawa, Y.; Matsui, K. Direct evidence of specific localization of sesquiterpenes and marchantin A in oil body cells of *Marchantia polymorpha* L. *Phytochemistry* **2016**, *130*, 77–84. [CrossRef]
18. Rice, E.W.; Baird, R.B.; Eaton, A.D.; Clesceri, L.S. *Standard Methods for the Examination of Water and Wastewater*; American Public Health Association: Washington, DC, USA, 2012; Volume 10.

19. Lu, S.; Li, X.; Yu, B.; Ding, J.; Zhong, Y.; Zhang, H. Efficient Microcystis aeruginosa coagulation and removal by palladium clusters doped g-C3N4 with no light irradiation. *Ecotoxicol. Environ. Saf.* **2022**, *246*, 114148. [CrossRef]
20. Zhu, Y.; Cheng, S.; Wang, P.; Chen, H.; Zhang, X.; Liu, L.; Li, X.; Ding, Y. A possible environmental-friendly removal of Microcystis aeruginosa by using pyroligneous acid. *Ecotoxicol. Environ. Saf.* **2020**, *205*, 111159. [CrossRef]
21. Lichtenthaler, H.K.; Wellburn, A.R. Determinations of Total Carotenoids and Chlorophylls a and b of Leaf Extracts in Different Solvents. *Analysis.* **1983**, *11*, 591–592. [CrossRef]
22. Zhao, W.; Zheng, Z.; Zhang, J.; Roger, S.-F.; Luo, X. Allelopathically inhibitory effects of eucalyptus extracts on the growth of Microcystis aeruginosa. *Chemosphere* **2019**, *225*, 424–433. [CrossRef] [PubMed]
23. Ding, J.; Chen, B.; Zhang, Y.; Ye, X.; Li, Y.; Zhou, D.; Ding, Y.; Zhu, W.; Zhang, H. Effects of poly (1, 4-butanediol succinate) carrier on the nitrogen removal performance and microbial community of sequencing batch reactors. *J. Clean. Prod.* **2021**, *291*, 125279. [CrossRef]
24. Zhang, Q.; Wu, J.; Yu, Y.-Y.; He, Y.-J.; Huang, Y.; Fan, N.-S.; Huang, B.-C.; Jin, R.-C. Microbial and genetic responses of anammox process to the successive exposure of different antibiotics. *Chem. Eng. J.* **2021**, *420*, 127576. [CrossRef]
25. Tian, Z.; Palomo, A.; Zhang, H.; Luan, X.; Liu, R.; Awad, M.; Smets, B.F.; Zhang, Y.; Yang, M. Minimum influent concentrations of oxytetracycline, streptomycin and spiramycin in selecting antibiotic resistance in biofilm type wastewater treatment systems. *Sci. Total Environ.* **2020**, *720*, 137531. [CrossRef]
26. Xiang, S.; Han, Y.; Jiang, C.; Li, M.; Wei, L.; Fu, J.; Zhu, L. Composite biologically active filter (BAF) with zeolite, granular activated carbon, and suspended biological carrier for treating algae-laden raw water. *J. Water Process Eng.* **2021**, *42*, 102188. [CrossRef]
27. Li, H.; Zheng, F.; Wang, J.; Zhou, J.; Huang, X.; Chen, L.; Hu, P.; Gao, J.-m.; Zhen, Q.; Bashir, S.; et al. Facile preparation of zeolite-activated carbon composite from coal gangue with enhanced adsorption performance. *Chem. Eng. J.* **2020**, *390*, 124513. [CrossRef]
28. Wang, X.; Szeto, Y.T.; Jiang, C.; Wang, X.; Tao, Y.; Tu, J.; Chen, J. Effects of Dracontomelon duperreanum leaf litter on the growth and photosynthesis of Microcystis aeruginosa. *Bull. Environ. Contam. Toxicol.* **2018**, *100*, 690–694. [CrossRef]
29. Zhou, S.; Shao, Y.; Gao, N.; Deng, Y.; Qiao, J.; Ou, H.; Deng, J. Effects of different algaecides on the photosynthetic capacity, cell integrity and microcystin-LR release of Microcystis aeruginosa. *Sci. Total Environ.* **2013**, *463*, 111–119. [CrossRef]
30. Li, J.; Hu, J.; Cao, L.; Yuan, Y. Growth, physiological responses and microcystin-production/-release dynamics of Microcystis aeruginosa exposed to various luteolin doses. *Ecotoxicol. Environ. Saf.* **2020**, *196*, 110540. [CrossRef]
31. Yang, Y.; Zhang, Y.; Li, Z.; Zhao, Z.; Quan, X.; Zhao, Z. Adding granular activated carbon into anaerobic sludge digestion to promote methane production and sludge decomposition. *J. Clean. Prod.* **2017**, *149*, 1101–1108. [CrossRef]
32. Peng, H.; Zhang, Y.; Tan, D.; Zhao, Z.; Zhao, H.; Quan, X. Roles of magnetite and granular activated carbon in improvement of anaerobic sludge digestion. *Bioresour. Technol.* **2018**, *249*, 666–672. [CrossRef] [PubMed]
33. Wang, X.; Wang, J. Removal of nitrate from groundwater by heterotrophic denitrification using the solid carbon source. *Sci. China Ser. B Chem.* **2009**, *52*, 236–240. [CrossRef]
34. Chen, W.; Westerhoff, P.; Leenheer, J.A.; Booksh, K. Fluorescence excitation−emission matrix regional integration to quantify spectra for dissolved organic matter. *Environ. Sci. Technol.* **2003**, *37*, 5701–5710. [CrossRef] [PubMed]
35. Zhang, H.; Li, X.; Yu, B.; Wang, J.; Lu, S.; Zhong, Y.; Ding, J. Fabrication of amorphous carbon-based zinc oxide for efficient capture of intact Microcystis aeruginosa: Lysis in sedimentation process. *J. Environ. Chem. Eng.* **2022**, *10*, 108793. [CrossRef]
36. Vaz-Moreira, I.; Nunes, O.C.; Manaia, C.M. Ubiquitous and persistent Proteobacteria and other Gram-negative bacteria in drinking water. *Sci. Total Environ.* **2017**, *586*, 1141–1149. [CrossRef]
37. Becerra-Castro, C.; Macedo, G.; Silva, A.M.; Manaia, C.M.; Nunes, O.C. Proteobacteria become predominant during regrowth after water disinfection. *Sci. Total Environ.* **2016**, *573*, 313–323. [CrossRef]
38. Pishgar, R.; Dominic, J.A.; Sheng, Z.; Tay, J.H. Denitrification performance and microbial versatility in response to different selection pressures. *Bioresour. Technol.* **2019**, *281*, 72–83. [CrossRef]
39. Huang, Y.; Yang, H.; Li, K.; Meng, Q.; Wang, S.; Wang, Y.; Zhu, P.; Niu, Q.; Yan, H.; Li, X.; et al. Red mud conserved compost nitrogen by enhancing nitrogen fixation and inhibiting denitrification revealed via metagenomic analysis. *Bioresour. Technol.* **2022**, *346*, 126654. [CrossRef]
40. Zhang, Q.; Zhang, X.; Bai, Y.-H.; Xia, W.-J.; Ni, S.-K.; Wu, Q.-Y.; Fan, N.-S.; Huang, B.-C.; Jin, R.-C. Exogenous extracellular polymeric substances as protective agents for the preservation of anammox granules. *Sci. Total Environ.* **2020**, *747*, 141464. [CrossRef]
41. Li, Y.; Zhang, Y.; Xu, Z.; Quan, X.; Chen, S. Enhancement of sludge granulation in anaerobic acetogenesis by addition of nitrate and microbial community analysis. *Biochem. Eng. J.* **2015**, *95*, 104–111. [CrossRef]
42. Chen, Q.; Li, T.; Gui, M.; Liu, S.; Zheng, M.; Ni, J. Effects of ZnO nanoparticles on aerobic denitrification by strain Pseudomonas stutzeri PCN-1. *Bioresour. Technol.* **2017**, *239*, 21–27. [CrossRef] [PubMed]

toxics

MDPI

Article

Simultaneously Cationic and Anionic Dyes Elimination via Magnetic Hydrochar Prepared from Copper Slag and Pinewood Sawdust

Huabin Wang [1,*], Yi Wu [1], Yi Wen [1], Dingxiang Chen [1], Jiang Pu [2], Yu Ding [3], Sailian Kong [4], Shuaibing Wang [5] and Rui Xu [1,*]

[1] School of Energy and Environment Science, Yunnan Normal University, Kunming 650500, China; wuyimax@foxmail.com (Y.W.); wyaquarius@foxmail.com (Y.W.); ynnuchendx@foxmail.com (D.C.)
[2] Shiping Center for Rural Energy and Environment, Honghe 661400, China; 13867455685@163.com
[3] Baoshan City Longyang Rural Energy Workstation, Baoshan 678000, China
[4] Development Center for Rural Affairs of Jiangchuan District, Yuxi 651100, China; yunnanxt@foxmail.com
[5] College of Chemistry Biology and Environment, Yuxi Normal University, Yuxi 653100, China; wshuaibing@yxnu.edu.cn
* Correspondence: hbwang@ynnu.edu.cn (H.W.); ecowatch_xr@163.com (R.X.); Tel./Fax: +86-27-87792151 (R.X.)

Abstract: In practical wastewater, cationic and anionic dyes usually coexist, while synergistic removal of these pollutants is difficult due to their relatively opposite properties. In this work, copper slag (CS) modified hydrochar (CSHC) was designed as functional material by the one-pot method. Based on characterizations, the Fe species in CS can be converted to zero-valent iron and loaded onto a hydrochar substrate. The CSHC exhibited efficient removal rates for both cationic dyes (methylene blue, MB) and anionic dyes (methyl orange, MO), with a maximum capacity of 278.21 and 357.02 mg·g^{-1}, respectively, which was significantly higher than that of unmodified ones. The surface interactions of MB and MO between CSHC were mimicked by the Langmuir model and the pseudo-second-order model. In addition, the magnetic properties of CSHC were also observed, and the good magnetic properties enabled the adsorbent to be quickly separated from the solution with the help of magnets. The adsorption mechanisms include pore filling, complexation, precipitation, and electrostatic attraction. Moreover, the recycling experiments demonstrated the potential regenerative performance of CSHC. All these results shed light on the co-removal of cationic and anionic contaminates via these industrial by-products derived from environmental remediation materials.

Keywords: modified hydrochar; copper slag; dyes; simultaneous removal; adsorption mechanisms

Citation: Wang, H.; Wu, Y.; Wen, Y.; Chen, D.; Pu, J.; Ding, Y.; Kong, S.; Wang, S.; Xu, R. Simultaneously Cationic and Anionic Dyes Elimination via Magnetic Hydrochar Prepared from Copper Slag and Pinewood Sawdust. *Toxics* **2023**, *11*, 484. https://doi.org/10.3390/toxics11060484

Academic Editors: Junhao Qin, Peidong Su, Feng Zhu and Lin Ding

Received: 11 April 2023
Revised: 21 May 2023
Accepted: 23 May 2023
Published: 25 May 2023

1. Introduction

With the rapid development of social industry and textile scale, the immature technology of dye wastewater treatment systems leads to a large number of compounds and intermediates participating in the process of dye production and dyeing. As the wastewater is discharged into external water bodies, the treatment capacity of dye wastewater also decreases [1,2]. The ecological and environmental problems caused by printing and dyeing wastewater are inevitable and ubiquitous in daily life [3]. The printing and dyeing wastewater is a typical high-pollution wastewater, which is characterized by large discharge, high chroma, high salt concentration, strong acid, strong alkali, strong resistance to microorganisms, and so on [4,5]. Some organic compounds and toxic heavy metals pose carcinogenic, teratogenic, and mutagenic risks to humans [6]. In addition, due to the variety of dyes and complex industrial production, there are always different kinds of dye residues in the wastewater, that is, anionic dyes and cationic dyes always coexist [7]. However, the physical and chemical properties of anionic and cationic dyes are basically opposite, and it is usually difficult to remove both dyes at the same time.

Coagulation and flocculation, precipitation, microbial, electrolysis, adsorption, and membrane separation are currently commonly used to remove dyes [8,9]. However, each method has its advantages and disadvantages. For example, although the microbial method can utilize cheap renewable resources, the microbial growth and culture path is long, so it is easy to produce more by-products in the follow-up experiment, and the reaction cycle is long. The membrane separation method is simple to operate and has little pollution, but with the increase of the use time, the surface of the membrane will be polluted, which reduces the performance. Moreover, this method only separates and transfers the pollutants and cannot fundamentally achieve adsorption. According to previous studies, the adsorption method still has advantages, which are far ahead of other methods in terms of application value and actual removal effect. At the same time, it has the advantages of good removal effect, short time, low cost, simple operation, good cycling performance, and not easy to cause harm to the environment. The high-efficiency adsorbent for printing and dyeing wastewater is the core issue of the adsorption.

In recent years, zero-valent iron (ZVI) nanoparticles with abundant REDOX active sites have received a lot of focus [10]. However, ZVI tends to accumulate in solution, which reduces its reactivity. Therefore, some porous materials are used to support ZVI to improve the stability of ZVI, such as activated carbon, zeolite, biochar, and so on. Nevertheless, the synthesis of ZVI composites is high-cost and low-load, and it is necessary to optimize ZVI composites. HC is a product prepared under high pressure and low temperature, with rich functional groups and porous structure [11,12], which shows great potential in removing pollution [13,14]. Hence, HC seems to be an ideal material to load ZVI. At present, the main sources of ZVI are $K_3[Fe(C_2O_4)]\cdot 3H_2O$ [15] and $FeCl_3\cdot 6H_2O$ [16,17], reduced iron powder, and iron ores, which mainly includes limonite, siderite, and magnetite [18]. These materials have the disadvantages of complex preparation methods and high cost [19], although they have the advantages of environmental protection, economy, and efficient use [20,21]. The composite material of hydrochar and ZVI prepared with the above modifiers is at variance with the concept of sustainable development. Therefore, it is urgent to find new modified materials.

The accumulated stock of copper slag (CS) in China has reached 120 million tons [22], but the actual utilization of CS is rare. The iron in CS mainly exists in the form of weakly magnetic iron, and the effect of metal recovery by traditional methods is ineffective. The long-term and large accumulation of CS not only causes land occupation but also damages the soil and water environment [23]. Iron, copper, and other valuable metals are prevalent in CS, and it is worth noting that the iron content in CS (Fe% = 35%) is significantly greater than that recovered from iron ore (Fe% > 27%) [24]. Our research group has used CS as the modified material to prepare the relevant functional bio-adsorbent via the hydrothermal reduction method to remove heavy metal (Selenium) from the solution, and the absorption efficiency of the bio-adsorbent was great. However, whether the magnetic hydrochar modified by CS can be used in dye wastewater treatment and whether it can achieve cationic and anionic dyes elimination simultaneously remain to be studied. Moreover, the adsorption mechanism and industrial feasibility also need to be explored.

Therefore, in this research, we prepared magnetic hydrochar modified by CS and explored the removal efficiency of cationic dyes (methylene blue, MB) and anionic dyes (methyl orange, MO) in printing wastewater. The intention of this work was to obtain adsorbent magnetic hydrochar modified by CS via the hydrothermal reduction called CSHC and use CSHC for the adsorption of dyes. In this study, we believe that pine sawdust is used as the precursor to prepare hydrochar, and CS was innovatively modified with the help of high cellulose content and rich lignin in the structure. The combination of the two can not only effectively expand the specific surface area but the pore system of adsorbent materials. Moreover, the ZVI reduced in the preparation process could also be effectively attached to the hydrochar, thus providing a greater possibility for the adsorption of target pollutants. Therefore, the modified hydrochar was prepared by the one-pot method using CS as an iron source and pine sawdust as a carrier to explore its removal efficiency of cationic dyes (methylene blue, MB) and anionic dyes (methyl orange, MO) in printing and dyeing wastewater is a promising solution. Modified hydrochar (CSHC) was prepared from copper slag and pine sawdust as a high-performance adsorbent and applied to the removal of MB and MO from printing and dyeing wastewater. Various characterization methods were used to investigate the physical and chemical properties of the prepared CSHCS before and after modification. The simultaneous removal of MB and MO was tested by batch adsorption experiments. Thermodynamic analysis and kinetic model were used to illustrate the surface interaction and adsorption types, respectively. The reusability of the composite functional environmental protection material was evaluated by a recycling test. All this work could provide a promising and potential solution for CS reutilization and wastewater purification.

2. Resources and Techniques

2.1. Research Materials

The CS was obtained from Xuanwei Phoenix Steel Co, Ltd., from Qujing City, China. The pinewood sawdust was collected from Kunming City, China. The methylene blue ($C_{16}H_{18}ClN_3S$, solubility in water: 40 g·L^{-1} at 25 °C, molar mass: 373.9 g·mol^{-1}), methyl orange ($C_{14}H_{14}N_3NaO_3S$, solubility in water: 200 mg·L^{-1} at 25 °C, molar mass: 327.3 g·mol^{-1}), NaOH, and other reagents were provided from Beijing Chemical Reagent Factory. The pinewood SD was rinsed with distilled water and dried at 80 °C, and then the dried pinewood sawdust was crushed and passed through 80 mesh.

2.2. Adsorbent Preparation Methods

The preparation of CSHC is shown in Scheme 1 [25]. Firstly, weighed CS (4 g) and CS (4 g) were mixed. Secondly, the mixture was steeped in NaOH solution (200 mg·L^{-1}, 40 mL) and distilled water (60 mL) in order to reduce the crystallinity of cellulose in SD and improve the mixing degree, and then it was mixed by ultrasonic oscillation (25 °C, 0.5 h) and stirred by a magnetic stirring apparatus (25 °C, 2 h). Thirdly, the mixture had a hydrothermal reaction (120 °C, 10 h). Finally, the dried slurry was placed in a tube furnace and underwent a pyrolysis reaction (heating rate was 10 °C·min^{-1}, pyrolysis temperature was 400, 600, or 800 °C, and holding time was 2 h, N_2 protection). The three samples were washed with deionized water three times and were labeled CSHC400, CSHC600, and CSHC800, respectively. In addition, 8.0 g SD and 8.0 g CS were weighed and pyrolyzed (heating rate was 10 °C·min^{-1}, pyrolysis temperature was 600 °C, and holding time was 2 h, N_2 protection), respectively, and the products were labeled HC600 and CS600.

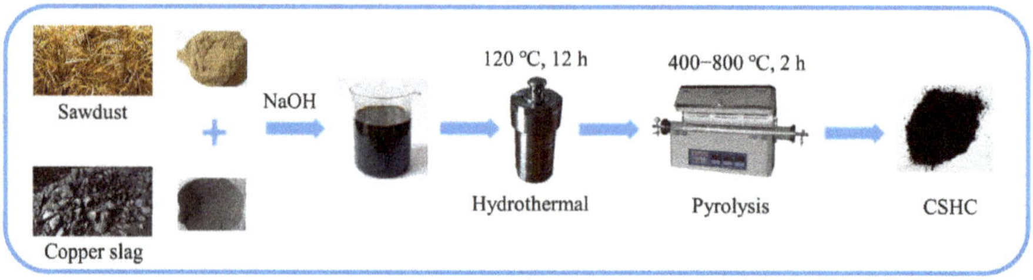

Scheme 1. Flowchart of preparation of CSHC.

2.3. Analysis Methods

The surface functional groups of the hydrochar were measured with Fourier transform infrared spectroscopy (FTIR, Spectrum Two, Perkin-Elmer, Waltham, MA, USA) using the KBr disk technique. Scanning electron microscopy (SEM, FEI Quanta 200FEG, Thermo Scientific Quanta, Hillsboro, OR, USA) was used to detect the surface morphology. The specific surface area (SSA) of biochar was detected by N_2 adsorption isotherms at 77 K using a Micropore Analyzer (ASAP 2460, Micrometrics, Norcross, GE, USA). The crystal structure was characterized by X-ray diffractometry (XRD, D8 Advance Sox-l, Bruker Co., Billerica, MA, USA) with Cu K-alpha radiation at 40 kV (λ = 0.15418 nm). An X-ray photoelectron spectrometer (XPS, ESCALAB 250Xi, Thermo Fisher Scientific, Waltham, MA, USA) was used to characterize the surface composition and chemical state of the modified bone hydrochar before and after the adsorption of the organic dyes.

2.4. Batch Experiment

All sorption experiments were performed in a 10 mL centrifuge tube with varied environmental factors at optimum operational parameters. The initial solution pH was adjusted with trace amounts of 0.001–0.1 mol·L^{-1} HCl/NaOH solutions. The adsorption kinetics were investigated by dispersing 10.0 mg of adsorbent into 10 mL of aqueous MB and MO solutions containing different initial concentrations (i.e., 50 or 500 mg·L^{-1}) at pH = 12 for MB and pH = 3 for MO, T = 318 K. The adsorption isotherm experiments were carried out by vigorously shaking 10 mL of a solution containing varied levels of organic dyes (MB and MO from 25 to 500 mg·L^{-1}) mixed with 10.0 mg of CSHC for 3 h to reach equilibrium at pH = 12.0 and pH = 3.0, respectively, T = 318 K.

Adsorption experiments were carried out in an orbital shaker (HNY-100B, Honour instrument shaker, Tianjin, China) at 120 rpm·min^{-1} for 3 h to ensure that the sorption reached equilibrium. The supernatants were then filtered with a 0.22 μm filter membrane, followed by determining the MB and MO concentrations (C_e, mg·L^{-1}) with an Ultraviolet-visible spectrophotometer (UV-722, Evolution 201 & 220, Thermo Fisher Scientific, Waltham, MA, USA) to measure the quantitative absorbance of these organic dyes under the influence of an adsorbent. Then, the MB and MO adsorption capacities of the adsorbent materials were computed. The MB was centered at a wavelength of 663 nm [26], and the MO was centered at a wavelength of 463 nm [27]. According to Formulas (1) and (2), the adsorption capacity q_e (mg·g^{-1}) and removal rate can be calculated, respectively. The calculation formula of the Langmuir model and Freundlich model was shown as Formulas (3) and (4). Additionally, Formulas (5) and (6) provided the expressions for the pseudo-first-order model and pseudo-second-order model. In order to ensure the uniformity of measurement and accuracy of dates, all experiments were repeated three times.

$$q_e = \frac{(C_0 - C_e) \times V}{W} \quad (1)$$

$$Removal\ rate = \frac{C_0 - C_e}{C_0} \times 100\% \quad (2)$$

where C_0 (mg·L^{-1}) was the initial concentration of dye, C_e (mg·L^{-1}) was the concentration of dye solutions after adsorption, V (L) was the volume, and W (g) was the quantity of the adsorbent.

$$\text{Langmuir model}: q_e = \frac{q_m \cdot K_L \cdot C_e}{1 + K_L \cdot C_e} \quad (3)$$

$$\text{Freundlich model}: q_e = K_L \cdot C_e^{\frac{1}{nF}} \quad (4)$$

where C_e (mg·L^{-1}) was the action's equilibrium content, q_e (mg·g^{-1}) was the target pollutant's absorption capacity, K_L (L·mg^{-1}) was the constant in the Langmuir model, K_F (mg^{1-n} L^n·g^{-1}) was the constant in the Freundlich model, and q_m (mg·g^{-1}) was the highest adsorbent capacity.

$$\text{Pseudo–first–order model}: q_t = q_e\left(1 - e^{-k_1 t}\right) \quad (5)$$

$$\text{Pseudo–sec ond–order model}: q_t = \frac{q_e^2 k_2 t}{1 + q_e k_{2t}} \quad (6)$$

The quantities of MB and MO that were adsorbed onto the CSHC600 at equilibrium and time t, respectively, are denoted by the variables q_e and q_t in these formulations, whereas k_1 (min^{-1}) and k_2 ($g \cdot (mg\ min)^{-1}$) were the constants for the pseudo-first-order and pseudo-second-order models, respectively. The fitting results are shown in Table 1. The final fitting results showed that during the adsorption of MB, the R^2 of the pseudo-second-order model was superior to that of the pseudo-first-order model, demonstrating that chemical adsorption dominated. Moreover, k_2 on MB (2.58×10^{-2}) was larger than k_2 on MO (1.87×10^{-4}), indicating that the equilibrium time of adsorption of MB by CSHC600 was shorter than that of MO.

Table 1. Pore structure analysis of four activated carbons.

Sample	Specific Area ($m^2 \cdot g^{-1}$)	Pore Volume ($m^3 \cdot g^{-1}$)	Pore Size (Å)
CSHC800	201.33	0.16	27.15
CSHC600	186.40	0.15	28.37
CSHC400	159.71	0.13	25.32
HC600	122.33	0.12	20.35

3. Findings and Discussion

3.1. Adsorbent Material Characterization

SEM directly displayed the morphologic characteristics of HC composites. The morphologies of unmodified HC600 and CSHC600 are shown in Figure 1. It was obvious to observe the different morphologies of HC600 and CSHC600. Compared with CSHC600, HC600 had a more smooth surface and dense structure with the unobvious distribution of micropores [28]. The edges of HC600 were mainly broken sheets, and a few fibrous structures could be detected, which was likely due to the pyrolysis of structural fibers in the SD [29]. There were metallic luster and white particles on the surface of CSHC600, which was most likely ZVI. The existence of ZVI could be attributed to the magnetic nanoparticles (Fe_3O_4) [30,31] and demonstrated that zero-valent iron effectively bonded to the surface of hydrochar [32]. In addition, some pores with different sizes were formed on the surface of CSHC600, which showed random and uneven characteristics in disorganized directions, a large number of irregular agglomeration spherical particles were attached, and a small number of micro and macro cracks appeared around the agglomeration particles, which may be the cross-linking process of multi-pore development [26]. It helped the dispersion and adsorption of dye. Compared with HC600, it was also noted that the pore diameters of CSHC600 were significantly reduced (Figure 1b). This phenomenon was due to the introduction of magnetite modifications, which blocked the pore.

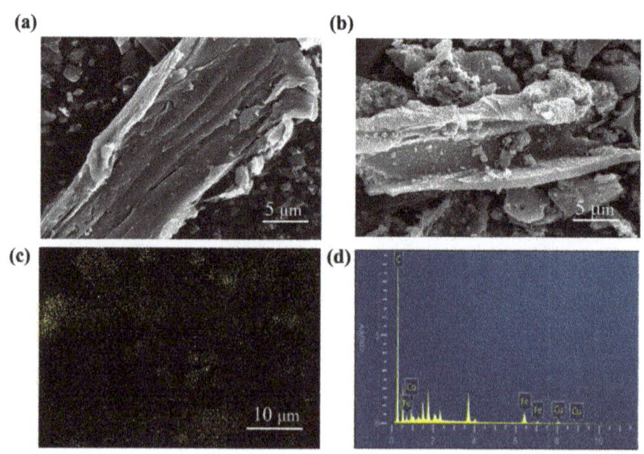

Figure 1. SEM images of HC600 (**a**) and SEM−EDS images of CSHC600 (**b**–**d**).

According to Figure 2, based on the IUPAC classification, CSHC is a typical Type I adsorption isotherm. When $P/P^0 \leq 0.48$, adsorption isotherms coincide with desorption isotherms, indicating that activated carbon has a microporous structure. With the increase of pressure, capillary condensation occurs in the pore structure, and the desorption rate of N_2 is higher than the adsorption rate, and H4-type hysteresis rings are produced, indicating the existence of a mesoporous structure. Such microporous mesopores not only provide more adsorption sites but also improve the adsorption rate. The formation of pores is caused by the dehydration of hydrochar. Surface materials begin to aggregate and form granular microcarbon balls. There are pores of different sizes inside and between the particles. Table 1 shows that the specific surface area of CSHC600 and HC600 are 186.40 $m^2 \cdot g^{-1}$ and 122.33 $m^2 \cdot g^{-1}$, respectively, and the total pore volume is 0.15 $m^3 \cdot g^{-1}$ and 0.12 $m^3 \cdot g^{-1}$, indicating that the pore structure and S_{BET} are improved by about 1.52 times after CS modification. V_{tot} improved by about 1.25 times. This shows that CS-modified hydrochar can be a very good preparation of high specific surface area adsorbent.

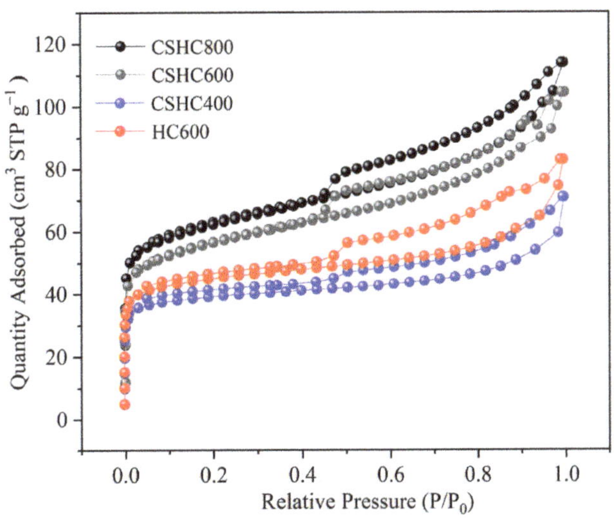

Figure 2. N_2 adsorption/desorption isotherms of four activated carbons.

The XRD diffractograms of HC600 and CS600 are shown in Figure 3. There was an intense peak corresponding to SiO_2 at 2θ of 26.66°, and there were diffraction peaks of $CaCO_3$ at 2θ of 20.95° and 29.56°. In the XRD diffractograms of CSHC, there were diffraction peaks of Fe^0 and Fe_3O_4 at 2θ of 44.57°, 35.47°, and 57.54°, respectively. The diffraction peak at 2θ = 46.81° represented Cu_5FeS_4, and the appearance of Cu_5FeS_4 may be the result of a high-temperature vulcanization reaction between copper, iron, and residual sulfur in CS during the pyrolysis process [33]. Compared HC600 to CSHC600, Fe^0 and Fe_3O_4 were successfully attached to the surface of the modified hydrochar [34]. In addition, when the temperature rose, the peak intensity of Fe^0 rose. This was because more reducing gases were created at high temperatures, which reduced Fe^{2+} and Fe^{3+} into Fe^0.

In the FTIR spectra (Figure 2b), the wide peaks at 3435 cm^{-1} and 877 cm^{-1} [35] represented the O–H and C–H, and the oscillation peak at 1647 cm^{-1} and 1008 cm^{-1} represented C=O [36,37], and the peak at 1065 cm^{-1} represented Si–O–Si [38]. The FTIR spectra showed the presence of functional groups such as –OH (3429 cm^{-1}), C=O (1008 cm^{-1}), C–H (1647 cm^{-1}), C=C (1426 cm^{-1}), and so on in these adsorbents.

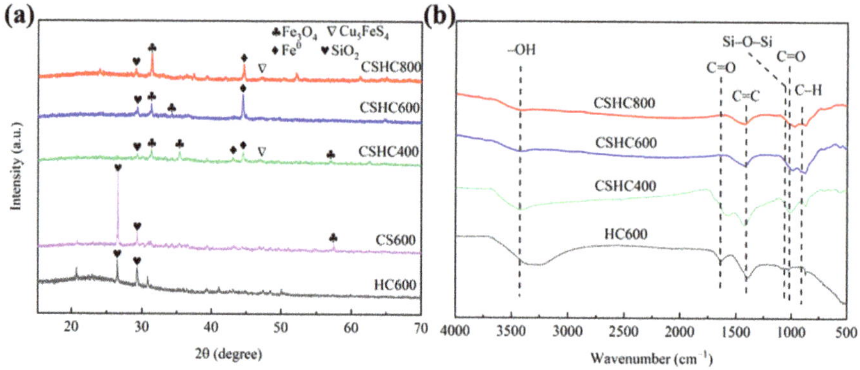

Figure 3. (**a**,**b**) XRD diffractograms and FTIR spectra of CSHC, CS600, and HC600.

In addition, it was observed that the oxygen-rich functional groups in the modified hydrochar prepared by pyrolysis at 600 °C and 800 °C were more than adsorbent prepared by pyrolysis at 400 °C. This indicated that pyrolysis at 600 °C stimulated and retained oxygen-rich functional groups and aromatic substances better [38], and thus CSHC600 had a better effect on the adsorption of MB and MO.

The XPS analysis is shown in Figure 4. The binding energy at 288.1 eV and 529.1 eV represented C 1s and O 1s, respectively, and the binding energy at 348.3 eV represented Ca 2p. The peaks at 710.13 eV and 725.14 eV represented Fe^{2+}, moreover, the peaks at 711.26 eV and 727.24 eV represented Fe^{3+}. The peaks at 720.10 eV represented Fe^0, indicating that Fe^0 was successfully loaded onto HC [39,40].

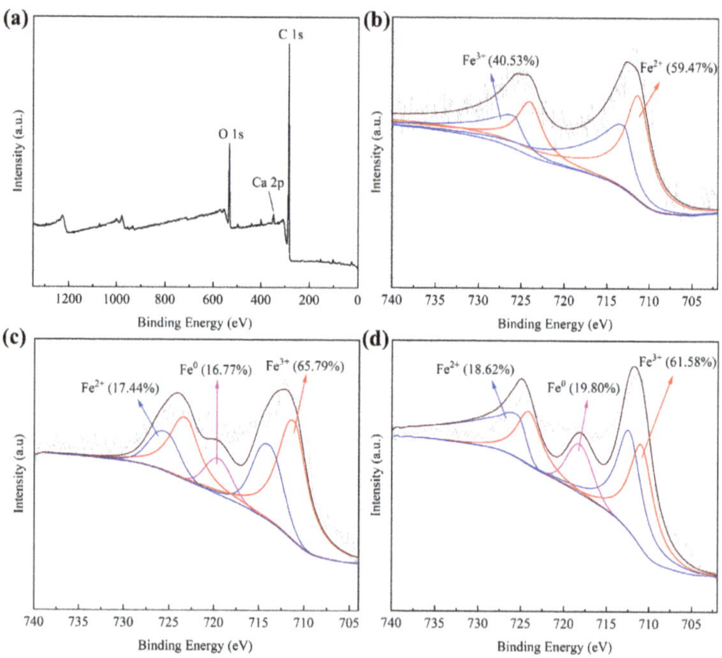

Figure 4. The full spectrum of CSHC600 (**a**) and XPS images of CSHC400 (**b**), CSHC600 (**c**), and CSHC800 (**d**).

The magnetic properties of CSHC400, CSHC600, and CSHC800 were determined by a vibrating sample magnetometer with a pole diameter of 5 cm under the condition of 300 K. The magnetic effect of zero-valent iron modified hydrochar was prepared from copper residue. The saturation magnetic strength of CSHC600 was 79.83 emu·g^{-1}, while that of CSHC400 and CSHC800 were 62.60 emu·g^{-1} and 60.92 emu·g^{-1}, respectively (Figure 5). Thus, the high-temperature pyrolysis magnetized the hydrochar [41], but the magnetization did not change with temperature. The intensity of magnetization was related to the generation of Fe_3O_4 during pyrolysis, and Fe_3O_4 had a high conductivity [42]. The adsorption properties were usually due to the electron transfer of ferromagnetism (ferrite magnets) between Fe^{2+} and Fe^{3+}. A secondary phase transition occurred above the Curie temperature to a paramagnetic substance. Due to the Curie temperature of Fe_3O_4 being 585 °C, the pyrolytic temperature of 600 °C can oxidize the iron part of CS to Fe_3O_4 to the greatest extent. The stronger the magnetic properties of the adsorbent, the more favorable the magnetization separation between the adsorbent and the dyeing wastewater [43]. Under the attraction of a permanent magnet, the adsorbent can be quickly separated from the adsorption liquid. Considering the adsorption capacity and magnetic properties, CSHC600 was considered the optimal one for further analysis.

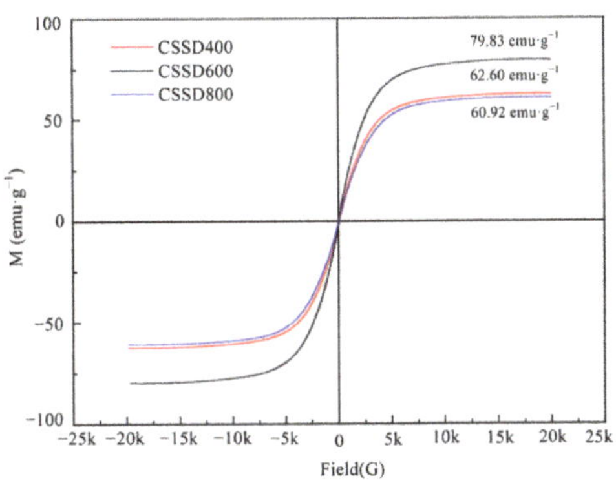

Figure 5. VSM spectra of CS600, CSHC400, CSHC600, and CSHC800.

3.2. Analysis of Batch Experiments

3.2.1. Effect of pH

Figure 6 shows the effect of pH value on the adsorption capacity of CSHC600. The initial concentration of MB was 50 mg·L^{-1}, the initial concentration of MO was 500 mg·L^{-1}, and the pH of the solution was 3–12. When pH > 3, the adsorption capacity of CSHS600 for MB was high, but the overall trend was stable. When pH = 10, the adsorption capacity increased slightly. The results showed that CSHS600 can remove MB over a wide pH range. For MO, the maximum adsorption capacity appealed when pH = 3, but the adsorption capacity evidently decreased with the increase of pH value. This was probably due to pHpzc 6.42. The positive charge of protonation of functional groups on the surface of CSHC600 led to positive zeta potential, and the anionic dye MO can be adsorbed by electrostatic adsorption. As can be seen from Figure 6, the adsorbent had a strong adsorption capacity for MB under an alkaline environment and MO under an acidic environment [44]. According to previous studies, the adsorption effect of CS-derived zero-valent iron-modified hydrochars on MB and MO solutions varies significantly at different pH levels [45]. In addition, the pH of the adsorbed solution was measured, and it was found that for MB, the pH level varied from 2.8 to 11.2. Similarly, the pH of MO varies from 3.5 to 11.8. The reason for this result may be that electrostatic attraction neutralizes part of the charge in the adsorption process, making the solution nearly neutral. However, the adsorbent material itself shows alkalinity, so most of the solution after adsorption is weak alkalinity, while the individual shows strong alkalinity.

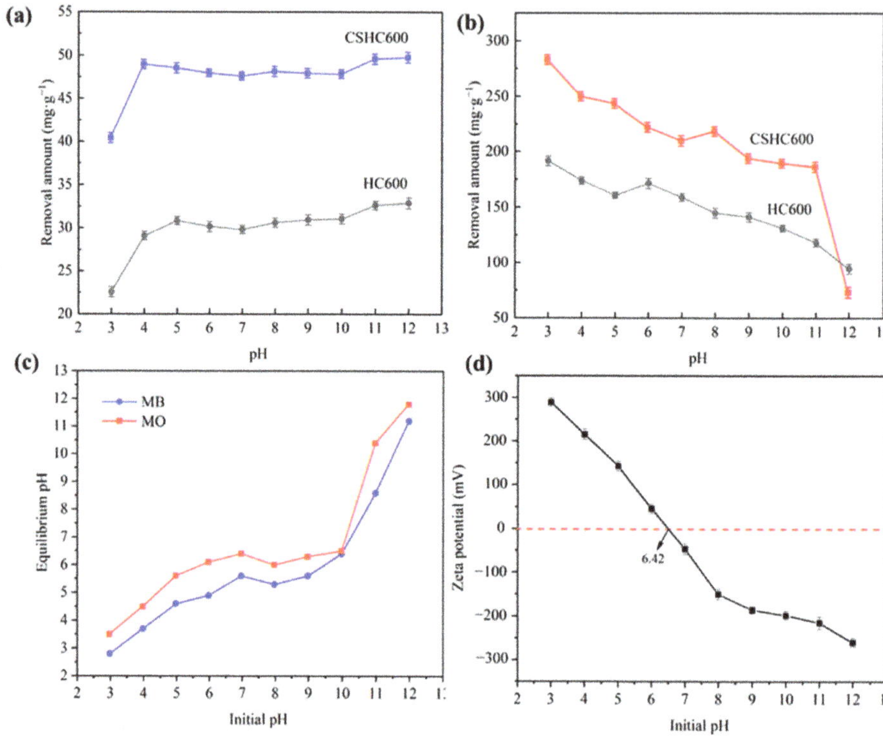

Figure 6. MB (**a**) and MO (**b**) were adsorbed onto CSHC600 and HC600 at various pH levels; Solution pH after adsorption of MB and MO (**c**); Zeta potential determination (**d**). Experiment conditions: [dosage = 1 g·L^{-1}], [temperature = 318 K], [MB = 50 mg·L^{-1}], [MO = 500 mg·L^{-1}], [reaction time = 3 h].

3.2.2. Analysis of Isothermal Adsorption

The adsorption efficiency of CSHC600 on MB and MO at various initial concentrations are shown in Figure 7c,d, and the similarities between MB and MO were obvious. The adsorption capacity gradually rose as the pollutant concentration rose. The adsorption capacity of CSHC600 tended to be steady when the concentration rose to a certain point. The experimental dates were fitted by the Langmuir and Freundlich models [46], and the results are shown in Table 2. For the adsorption MB by CSHC600, the R^2 of the Langmuir model was considerably higher (0.99) than that of the Freundlich model (0.94). For the adsorption MO by CSHC600, R^2 values were 0.97 and 0.88 for the Langmuir and Freundlich models, respectively. Thus, the MB and MO absorption process was better fitted by the Langmuir model [47]. The statement of Langmuir has four basic assumptions. There is no interaction between substrates, there is no adsorption site with the same energy, all adsorption sites are equivalent, and the substrate is uniformly distributed on the surface of the adsorbent monolayer film [48,49]. In other words, the adsorption preponderance of MB and MO by CSHC600 was mediated by chemisorption on a single molecular layer [50,51]. CSHC600 was an effective MB and MO adsorbent, which showed the best effect [52], as its greatest adsorption capacity respectively reached 278.21 mg·g^{-1} and 357.02 mg·g^{-1} for MB and MO (Table 2), which compared with other research data (Table 3). These results show that CSHC600 was an effective substance for removing these two kinds of organic dyes.

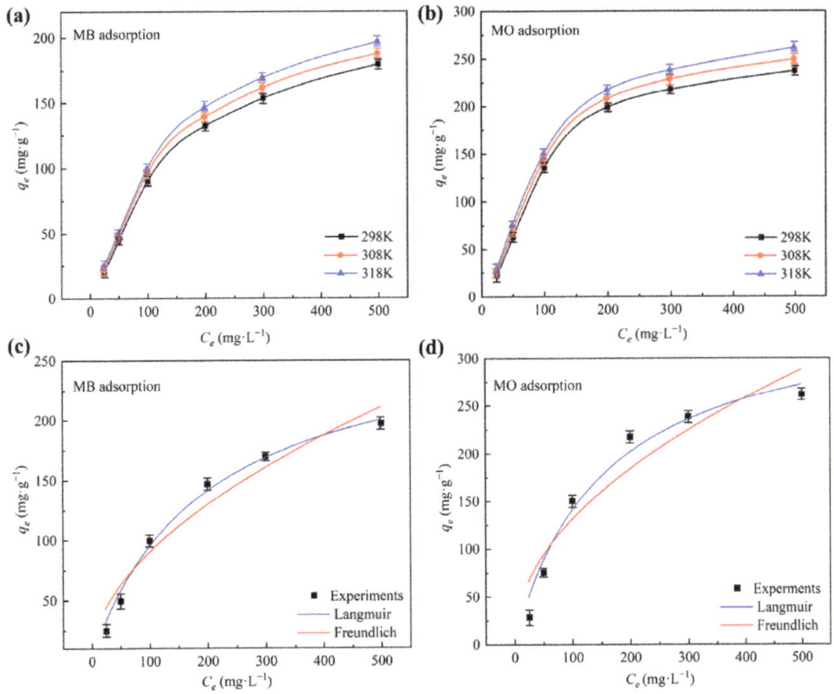

Figure 7. The MB (**a**) and MO (**b**) removal capacities of the isotherms, thermodynamics (**c**); Fitted curves of MB (**c**) and MO (**d**) for the Langmuir and Freundlich isotherms. Experiment conditions: [dosage = 1 g·L^{-1}], [temperature = 318 K], [MB = 50 mg·g^{-1}], [MO = 500 mg·g^{-1}], [reaction time = 3 h].

Table 2. Adsorption isotherms, kinetics, and related factors summarized.

Isotherms	**Target**	**Experimental Data**	**Parameters**		**R^2**
Langmuir			q_m (mg·g^{-1})	K_L (L·mg^{-1})	
	MB	196.89	278.21	0.005	0.99
	MO	261.78	357.02	0.006	0.97
Freundlich			K_F (mg^{1-n}L^{n}·g^{-1})	$1/n$	
	MB	196.89	7.80	0.53	0.94
	MO	261.78	13.68	0.49	0.88
Kinetics	Target	Experimental data	Parameters		
Pseudo-first-order			q_e (mg·g^{-1})	k_1 (min^{-1})	
	MB	49.95	48.56	0.43	0.99
	MO	254.38	250.65	0.04	0.99
Pseudo-second-order			q_e (mg·g^{-1})	k_2 (g·(mg·min)$^{-1}$)	
	MB	49.95	49.52	2.58×10^{-2}	0.99
	MO	254.38	287.14	1.87×10^{-4}	0.99

Table 2. *Cont.*

Isotherms	Target	Experimental Data	Parameters		R^2
Thermodynamics	T (K)	ΔG^0 (KJ·mol^{-1})	ΔS^0 (KJ·(mol·K)$^{-1}$)	ΔH^0 (KJ·(mol·K)$^{-1}$)	
MB	298	−2.74			
	308	−3.68			
	318	−4.10	1.96×10^{-2}	2.13	
MO	298	−4.85			
	308	−5.78			
	318	−6.02	3.11×10^{-2}	3.87	

Table 3. Comparison of different magnetic hydrochar or biomass—based adsorbents' abilities to bind MB and MO.

Sample	Dosage (g·L^{-1})	pH	Initial Concentration (mg·L^{-1})	Capacity (mg·g^{-1}) MB	MO	
Fe_3O_4-PAMH	0.8	11 for MB 5 for MO	100	148.84	202.02	[53]
BM	0.5	7	50	2	41.49	[54]
NFGPBC	0.7	7	50	67.54	-	[55]
Asph-Al	10	3	20	-	41.25	[56]
FE-LB	2	4	200	180.00	-	[57]
CSHC600	1	12 for MB 3 for MO	50 for MB 500 for MO	49.58	261.78	This study

3.2.3. Adsorption Kinetics

Response time was also a key parameter in adsorption, which reflected the adsorption speed. Response time contributed to the efficient removal of pollutants and provided more valuable information for the adsorption process. According to the existing experimental data, CSHC600 had the best adsorption capacities for MB and MO when pH = 12 and pH = 3, respectively. The results of the fitting are displayed in Figure 8a,b. The adsorption equilibrium for MO was 130 min, and the adsorption equilibrium for MB was 30 min. The adsorption capacity of MO and MB increased rapidly during the early stages of adsorption, increased slowly, and ultimately tended to a stable state. MB and MO rapidly diffused from the solution to the surface of the adsorbent material, after that, the adsorption tended to be slow. There may have been a lot of adsorption active sites on the surface of CSHC600, but when the reaction was going on, these active sites gradually filled with small molecules such as MO and MB, which weakened the adsorption.

3.2.4. Thermodynamic analysis

Thermodynamic analysis was carried out by evaluating the entropy and energy changes in the adsorption process. The formula for figuring out the Gibbs free energy and other factors during the adsorption process is shown in Formulas (7) and (8).

$$\ln\left(\frac{q_m}{C_e}\right) = \frac{\Delta S^0}{R} - \frac{\Delta H^0}{RT} \quad (7)$$

$$\Delta G^0 = \Delta H^0 - T\Delta S^0 \quad (8)$$

where ΔG^0 represented the change in free energy, ΔS^0 represented the change in entropy, and ΔH^0 represented the change in enthalpy. The isothermal adsorption data were used to calculate $\ln\left(\frac{q_m}{C_e}\right)$, which had been given a value of 8.314 J·(mol·K)$^{-1}$ [58], where T was the temperature and R was the universal gas constant.

Table 2 showed that ΔG^0 adsorbing MB by CSHC600 at 318 K was −4.10 KJ·mol^{-1}, while that of MO was −6.02 KJ·mol^{-1}. Both ΔG^0 values were negative when adsorbing MB and MO, indicating that the adsorption reactions may be totally irreversible [59]. The ΔH^0 was positive, indicating the reactions were endothermic.

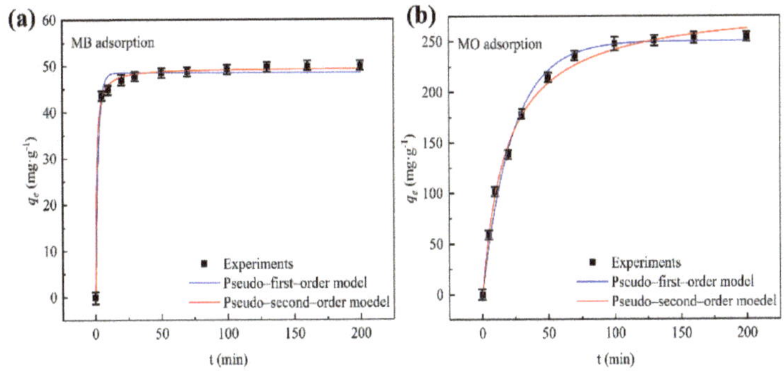

Figure 8. The adsorption kinetics of MB (**a**) and MO (**b**). Fitting of adsorption kinetics by linear relations of pseudo-first-order and pseudo-second-order models.

3.3. Regeneration Experiments

CSHC600 was a reusable adsorbent (Figure 9). Solvent desorption was used to evaluate the reproducibility of CSHC. The main method was achieved through five continuous adsorption-desorption processes [47]. To achieve pH neutrality, the attachments of the adsorbent were eluted, and the residue was repeatedly rinsed with deionized water. The recovered adsorbent was dried in a vacuum drying oven at 60 °C. To explore regeneration capabilities, the adsorption was placed into the same adsorption circumstances as previously, and the adsorption-desorption cycle was carried out five times. By using CSHC600, MB (50 mg·L^{-1}) and MO (500 mg·L^{-1}) can be reused with ease. As shown in Figure 10, MB could almost be desorbed completely. The removal rate of CSHC600 on MB was still achieved at 97% after five cycles. At pH = 3, the removal rate of MO by CSHC600 was 51% in the first cycle and more than 41% in the fifth cycle. The removal rates of CSHC600 for MB and MO were great even after five cycles. The decrease in adsorption efficiency may be related to the fact that some original adsorption sites of CSHC600 were occupied and did not regenerate after the preliminary test. The removal rates of CSHC600 for MB and MO were great even after five cycles. The outcomes demonstrate that CSHC600 was simple to recycle and has strong adsorption capability.

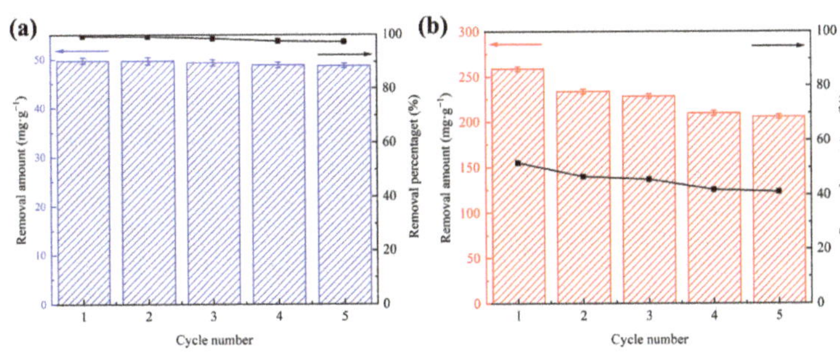

Figure 9. Adsorption of MB (**a**) and MO (**b**) using CSHC600 as an adsorbent that is reusable. Experiment conditions: [dosage = 1 g·L^{-1}], [temperature = 318 K], [MB = 50 mg·g^{-1}], [MO = 500 mg·g^{-1}], [reaction time = 3 h].

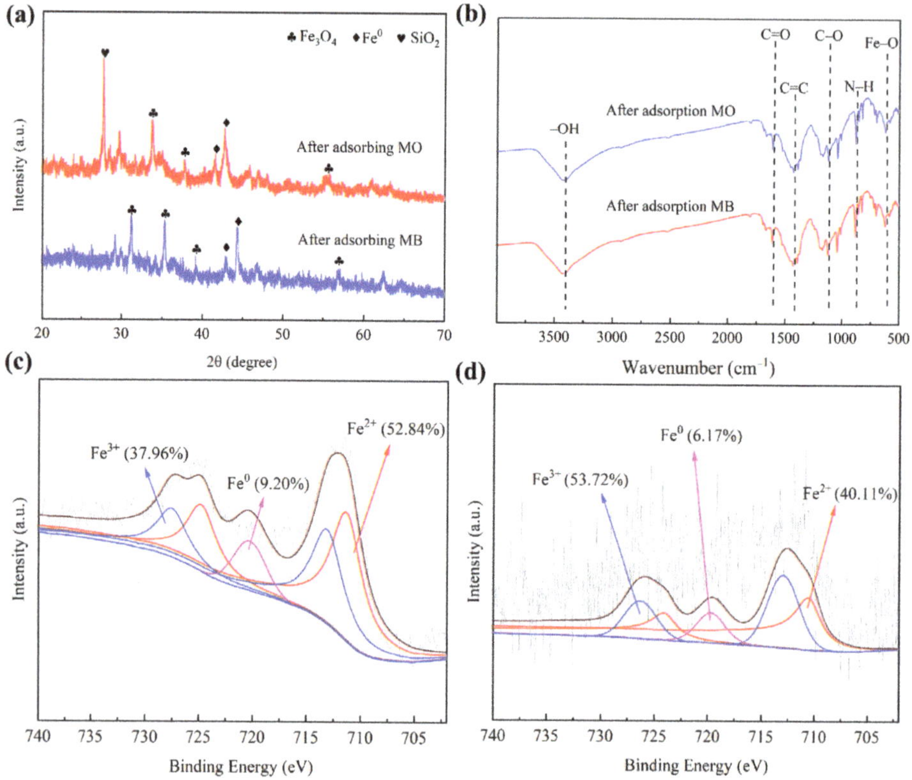

Figure 10. The images of XRD, FTIR, and XPS of CSHC600 after adsorption. (**a**) XRD diffractograms, (**b**) FTIR spectra, (**c**) XPS of CSHC600 adsorption of MB, (**d**) XPS of CSHC600 adsorption of MO.

3.4. Mechanisms

The removal mechanisms of MB and MO were studied. The SEM images showed that CSHC600 had a smooth surface and uniform pore distribution. In addition, the crystal structure of the adsorbent was discovered. The FTIR spectra after adsorption (Figure 10) showed that the interaction of these functional groups on the surface of CSHC600 with MB and MO, for example, –OH and –COOH functional groups shifted, the strength of C–C functional groups decreased, and additionally C–O and –COOH participated in the reaction. This indicated that hydrogen bonding existed in the adsorption process of CSHC600 [60]. By comparing the XRD diffractograms of CSHC before and after adsorption, it can be seen that Fe^0 decreased obviously after adsorption. It was also clear that the peak of Fe^0 significantly weakened after adsorption, which may be because ZVI loaded on CSHC600 reduced mineralized MB and MO. Eventually, the dye was degraded to small molecules such as CO_2 and H_2O, while ZVI was oxidized to Fe^{2+} and Fe^{3+}. As MO was a typical anionic compound, the positive and negative charges between Fe^{2+}, Fe^{3+}, and MO attracted each other, which attached MO to the surface of the adsorbent. According to the XPS of CSHC before and after adsorption, the peak of Fe^{3+} increased significantly after adsorption. A large amount of Fe^{3+} flocculated with the free –OH in solution to form $Fe(OH)_3$. Precipitation was the main mechanism of dye adsorption by hydrochar. In the same way, HC was usually alkaline and carried a negative charge, which can be well combined with cationic dye (MB) via complexation (Figure 11).

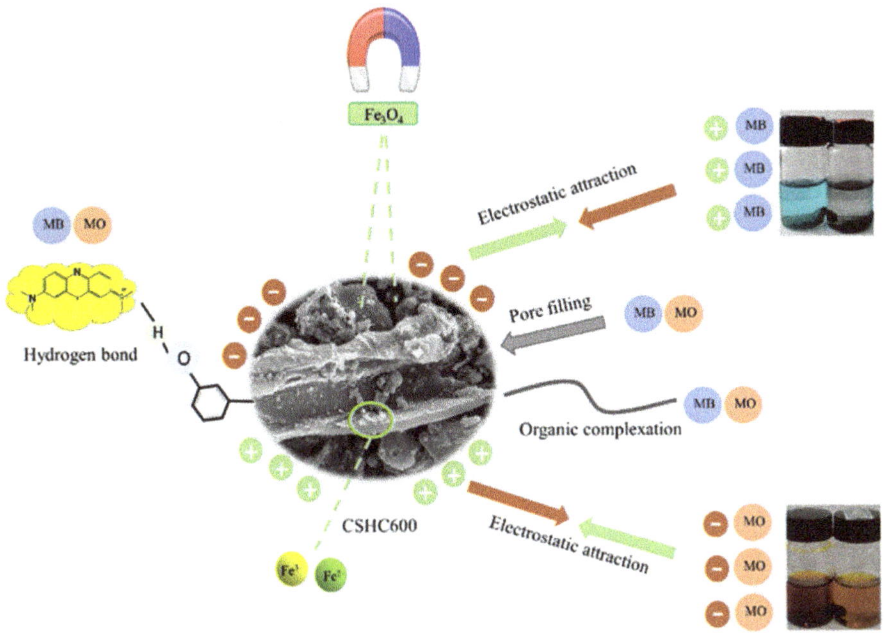

Figure 11. Adsorption mechanism of MO and MB by CSHC.

4. Conclusions

In this work, the green construction of CSHC was realized by the hydrothermal method with pine sawdust as a carrier and copper slag as an iron source. A variety of characterization methods have demonstrated the successful loading of ZVI. The adsorption process of CSHC600 on MB and MO was fitted by the pseudo-second-order kinetic equation and Langmuir model, respectively. The adsorption process is mainly controlled by chemisorption, and the maximum adsorption capacity is 278.21 mg·g^{-1} and 357.02 mg·g^{-1}, respectively. CSHC600 has good recovery performance. The experimental results prove the feasibility of copper slag and pine chip adsorbent for purifying industrial wastewater containing cationic dyes and anionic dyes.

Author Contributions: H.W.: Conceptualization; Y.W. (Yi Wu): Formal Analysis, Writing—Original Draft; Y.W. (Yi Wen): Data Curation, Software, Methodology; D.C.: Data Curation, Investigation, Methodology; J.P.: Supervision; Y.D.: Supervision; S.K.: Conceptualization; S.W.: Writing—Review & Editing; R.X.: Funding Acquisition, Resources, Supervision, Supervision, Writing—Review & Editing. All authors have read and agreed to the published version of the manuscript.

Funding: This research was supported by the Applied Basic Research Foundation of Yunnan Province (202201AS070020, 202201AU070061, 202101AU070008), Yunnan Province Education Department Project (2022J0136), Yunnan Academy of Experts Workstation (Xurui Shiping Workstation).

Institutional Review Board Statement: Not applicable.

Data Availability Statement: Not applicable.

Conflicts of Interest: This paper is an original work of the authors who have read and approved this version, and due care has been taken to ensure the integrity of the work. No part of this paper has been published or submitted elsewhere. No conflict of interest exists in the submission of this manuscript. A statement explaining why the manuscript is novel and significant is stated below.

References

1. Shao, Q.; Li, Y.; Wang, Q.; Niu, T.; Li, S.; Shen, W. Preparation of copper doped walnut shell-based biochar for efficiently removal of organic dyes from aqueous solutions. *J. Mol. Liq.* **2021**, *336*, 116314. [CrossRef]
2. Shaikh, W.A.; Chakraborty, S.; Islam, R.U.; Ghfar, A.A.; Naushad, M.; Bundschuh, J.; Maity, J.P.; Mondal, N.K. Fabrication of biochar-based hybrid Ag nanocomposite from algal biomass waste for toxic dye-laden wastewater treatment. *Chemosphere* **2022**, *289*, 133243. [CrossRef] [PubMed]
3. Sutar, S.; Patil, P.; Jadhav, J. Recent advances in biochar technology for textile dyes wastewater remediation: A review. *Environ. Res.* **2022**, *209*, 112841. [CrossRef] [PubMed]
4. Gao, Y.; Zhang, J.; Chen, C.; Du, Y.; Teng, G.; Wu, Z. Functional biochar fabricated from waste red mud and corn straw in China for acidic dye wastewater treatment. *J. Clean. Prod.* **2021**, *320*, 128887. [CrossRef]
5. Sun, Y.; Wang, T.; Han, C.; Bai, L.; Sun, X. One-step preparation of lignin-based magnetic biochar as bifunctional material for the efficient removal of Cr(VI) and Congo red: Performance and practical application. *Bioresour. Technol.* **2023**, *369*, 128373. [CrossRef]
6. Gautam, R.K.; Mandavi, G.M.; Mishra, R.K.; Chaturvedi, P.; Awashthi, M.K.; Singh, R.K.; Giri, B.S.; Pandey, P. Biochar for remediation of agrochemicals and synthetic organic dyes from environmental samples: A review. *Chemosphere* **2021**, *272*, 129917. [CrossRef]
7. Cheng, H.; Liu, Y.; Li, X. Adsorption performance and mechanism of iron-loaded biochar to methyl orange in the presence of Cr^{6+} from dye wastewater. *J. Hazard. Mater. J. Hazard. Mater.* **2021**, *415*, 125749. [CrossRef]
8. Cheng, C.; Liu, B.; Liu, C.; Shen, J.; Nurlan, J.; Khan, M.F.S.; Huang, Z.; Qian, Y.; Shen, F.; Wu, J. Tracking variation of fluorescent dissolved organic matter during full-scale printing and dyeing wastewater treatment. *Chemosphere* **2020**, *252*, 126559. [CrossRef]
9. Chen, Y.; Ma, R.; Pu, X.; Fu, X.; Ju, X.; Arif, M.; Yan, X.; Qian, J.; Liu, Y. The characterization of a novel magnetic biochar derived from sulfate-reducing sludge and its application for aqueous Cr(VI) removal through synergistic effects of adsorption and chemical reduction. *Chemosphere* **2022**, *308*, 136258. [CrossRef]
10. Li, X.; Xu, J.; Luo, X.; Shi, J. Efficient adsorption of dyes from aqueous solution using a novel functionalized magnetic biochar: Synthesis, kinetics, isotherms, adsorption mechanism, and reusability. *Bioresour. Technol.* **2022**, *360*, 127526. [CrossRef]
11. Madduri, S.; Elsayed, I.; Hassan, E.B. Novel oxone treated hydrochar for the removal of Pb(II) and methylene blue(MB) dye from aqueous solutions. *Chemosphere* **2020**, *260*, 127683. [CrossRef]
12. Tan, X.; Zhang, C.; Wei, H.; Shi, P.; Chang, H.; Ho, S. Versatile strategy of sulfanilamide antibiotics removal via microalgal biochar: Role of oxygen-enriched functional groups. *Chemosphere* **2022**, *304*, 135244. [CrossRef]
13. Duan, L.; Wang, Q.; Li, J.; Wang, F.; Yang, H.; Guo, B.; Hashimoto, Y. Zero valent iron or Fe_3O_4-loaded biochar for remediation of Pb contaminated sandy soil: Sequential extraction, magnetic separation, XAFS and ryegrass growth. *Environ. Pollut.* **2022**, *308*, 119702. [CrossRef]
14. Hou, Y.; Liang, Y.; Hu, H.; Tao, Y.; Zhou, J.; Cai, J. Facile preparation of multi-porous biochar from lotus biomass for methyl orange removal: Kinetics, isotherms, and regeneration studies. *Bioresour. Technol.* **2021**, *329*, 124877. [CrossRef]
15. Zhang, X.; Tran, H.N.; Liu, Y.; Yang, C.; Zhang, T.; Guo, J.; Zhu, W.; Ahmad, M.; Xiao, H.; Song, J. Nitrogen-doped magnetic biochar made with $K_3[Fe(C_2O_4)_3]$ to adsorb dyes: Experimental approach and density functional theory modeling. *J. Clean. Prod.* **2023**, *383*, 135527. [CrossRef]
16. Ren, Z.; Wang, Z.; Lv, L.; Ma, P.; Zhang, G.; Li, Y.; Qin, Y.; Wang, P.; Liu, X.; Gao, W. Fe-N complex biochar as a superior partner of sodium sulfide for methyl orange decolorization by combination of adsorption and reduction. *J. Environ. Manag.* **2022**, *316*, 115213. [CrossRef]
17. Chu, J.-H.; Kang, J.-K.; Park, S.-J.; Lee, C.G. Application of magnetic biochar derived from food waste in heterogeneous sono-Fenton-like process for removal of organic dyes from aqueous solution. *J. Water Process Eng.* **2020**, *37*, 101455. [CrossRef]
18. Zhang, H.; Xue, G.; Chen, H.; Li, X. Magnetic biochar catalyst derived from biological sludge and ferric sludge using hydrothermal carbon ization: Preparation, characterization and its circulation in Fenton process for dyeing wastewater treatment. *Chemosphere* **2018**, *191*, 64–71. [CrossRef]
19. Chanda, P.T.; Pritam, S.; Nikoloski, N.A. The potential for copper slag waste as a resource for a circular economy: A review—Part II. *Miner. Eng.* **2021**, *172*, 107150.
20. Shi, G.; Liao, Y.; Su, B.; Zhang, Y.; Wang, W.; Xi, J. Kinetics of copper extraction from copper smelting slag by pressure oxidative leaching with sulfuric acid. *Sep. Purif. Technol.* **2020**, *241*, 116699. [CrossRef]
21. Zhou, W.; Liu, X.; Lyu, X.; Gao, W.; Su, H.; Li, C. Extraction and separation of copper and iron from copper smelting slag: A review. *J. Clean. Prod.* **2022**, *368*, 133095. [CrossRef]
22. Chanda, P.T.; Pritam, S.; Nikoloski, N.A. The potential for copper slag waste as a resource for a circular economy: A review—Part I. *Miner. Eng.* **2022**, *180*, 107474.
23. Yan, J.; Zuo, X.; Yang, S.; Chen, R.; Cai, T.; Ding, D. Evaluation of potassium ferrate activated biochar for the simultaneous adsorption of copper and sulfadiazine: Competitive versus synergistic. *J. Hazard. Mater.* **2022**, *424*, 127435. [CrossRef] [PubMed]
24. Du, J.; Zhang, F.; Hu, J.; Yang, S.; Liu, H.; Wang, H. Direct reduction of copper slag using rubber seed oil as a reductant: Iron recycling and thermokinetics. *J. Clean. Prod.* **2022**, *363*, 132546. [CrossRef]
25. Wang, H.; Cai, J.; Liao, Z.; Jawad, A.; Ifthikar, J.; Chen, Z.; Chen, Z. Black liquor as biomass feedstock to prepare zero-valent iron embedded biochar with red mud for Cr(VI) removal: Mechanisms insights and engineering practicality. *Bioresour. Technol.* **2020**, *311*, 123553. [CrossRef]

26. Bashir, A.; Pandith, A.H.; Qureashi, A.; Malik, L.A.; Gani, M.; Perez, J. Catalytic propensity of biochar decorated with core-shell nZVI@Fe_3O_4: A sustainable photo-Fenton catalysis of methylene blue dye and reduction of 4-nitrophenol. *J. Environ. Chem. Eng.* **2022**, *10*, 107401. [CrossRef]
27. Ma, C.; Zhang, Y.; Yin, B.; Chen, J.; Guo, M.; Gao, X. Wood powder biochar in CdS-WPB-g-C_3N_4 heterojunction as an electron transfer medium for enhancing photocatalytic performance toward degradation methyl orange. *J. Environ. Chem. Eng.* **2023**, *11*, 109135. [CrossRef]
28. Zhu, Y.; Ji, S.; Liang, W.; Li, C.; Nie, Y.; Dong, J.; Shi, W.; Ai, S. A low-cost and eco-friendly powder catalyst: Iron and copper nanoparticles supported on biochar/geopolymer for activating potassium peroxymonosulfate to degrade naphthalene in water and soil. *Chemosphere* **2022**, *303 Pt 2*, 135185. [CrossRef]
29. Alshahrani, H.; Arun Prakash, V.R. Mechanical, fatigue and DMA behaviour of high content cellulosic corn husk fibre and orange peel biochar epoxy biocomposite: A greener material for cleaner production. *J. Clean. Prod.* **2022**, *374*, 133931. [CrossRef]
30. Li, Y.; Zimmerman, A.R.; He, F.; Chen, J.; Han, L.; Chen, H.; Hu, X.; Gao, B. Solvent-free synthesis of magnetic biochar and activated carbon through ball-mill extrusion with Fe_3O_4 nanoparticles for enhancing adsorption of methylene blue. *Sci. Total Environ.* **2020**, *722*, 137972. [CrossRef]
31. Wang, H.; Duan, R.; Zhou, X.; Wang, J.; Liu, Y.; Xu, R.; Liao, Z. Efficient removal of mercury and chromium from wastewater via biochar fabricated with steel slag: Performance and mechanisms. *Front. Bioeng. Biotechnol.* **2022**, *10*, 961907. [CrossRef]
32. Wei, A.; Ma, J.; Chen, J.; Zhang, Y.; Song, J.; Yu, X. Enhanced nitrate removal and high selectivity towards dinitrogen for groundwater remediation using biochar-supported nano zero-valent iron. *Chem. Eng. J.* **2018**, *353*, 595–605. [CrossRef]
33. Luo, Q.; Chen, D.; Cui, T.; Duan, R.; Wen, Y.; Deng, F.; Li, L.; Wang, H.; Zhang, Y.; Xu, R. Selenite elimination via zero-valent iron modified biochar synthesized from tobacco straw and copper slag: Mechanisms and agro-industrial practicality. *Front. Bioeng. Biotechnol.* **2022**, *10*, 1054801. [CrossRef]
34. Pozo, C.; Rego, F.; Yang, Y.; Puy, N.; Bartrolí, J.; Fàbregas, E.; Bridgwater, A.V. Converting coffee silverskin to value-added products by a slow pyrolysis-based biorefinery process. *Fuel Process Technol.* **2021**, *214*, 106708. [CrossRef]
35. Mahmoud, M.E.; Mohamed, A.K.; Salam, M.A. Self-decoration of N-doped graphene oxide 3-D hydrogel onto magnetic shrimp shell biochar for enhanced removal of hexavalent chromium. *J. Hazard. Mater.* **2021**, *408*, 124951. [CrossRef]
36. Xia, J.; Shen, Y.; Zhang, H.; Hu, X.; Mian, M.M.; Zhang, W.H. Synthesis of magnetic nZVI@biochar catalyst from acid precipitated black liquor and Fenton sludge and its application for Fenton-like removal of rhodamine B dye. *Ind. Crops Prod.* **2022**, *187*, 115449. [CrossRef]
37. Xu, Z.; Xu, X.; Zhang, Y.; Yu, Y.; Cao, X. Pyrolysis-temperature depended electron donating and mediating mechanisms of biochar for Cr(VI) reduction. *J. Hazard. Mater.* **2020**, *388*, 121794. [CrossRef]
38. Kumaraswamy, R.V.; Saharan, V.; Kumari, S.; Choudhary, R.C.; Pal, A.; Sharma, S.S.; Rakshit, S.; Raliya, R.; Biswas, P. Chitosan-silicon nanofertilizer to enhance plant growth and yield in maize (*Zea mays* L.). *Plant Physiol. Biochem* **2021**, *159*, 53–66. [CrossRef]
39. Wang, H.; Wen, Y.; Ding, Y.; Yue, Z.; Xu, D.; Liu, Y.; Zhang, Y.; Xu, R.; Zeng, W. Rapid and Effective Lead Elimination Using Cow Manure Derived Biochar: Balance between Inherent Phosphorus Release and Pollutants Immobilization. *Toxics* **2022**, *11*, 1.
40. Yang, X.; Zhu, W.; Chen, F.; Song, Y.; Yu, Y.; Zhuang, H. Modified biochar prepared from Retinervus luffae fructus for dyes adsorption and aerobic sludge granulation. *Chemosphere* **2023**, *322*, 138088. [CrossRef]
41. Wang, H.; Liu, Y.; Ifthikar, J.; Shi, L.; Khan, L.; Chen, Z.; Chen, Z. Towards a better understanding on mercury adsorption by magnetic bio-adsorbents with gamma-Fe_2O_3 from pinewood sawdust derived hydrochar: Influence of atmosphere in heat treatment. *Bioresour. Technol.* **2018**, *256*, 269–276. [CrossRef] [PubMed]
42. Truong, Q.-M.; Ho, P.-N.; Nguyen, T.-B.; Chen, W.-H.; Bui, X.-T.; Patel, A.K.; Singhania, R.R.; Chen, C.-W.; Dong, C.-D. Magnetic biochar derived from macroalgal Sargassum hemiphyllum for highly efficient adsorption of Cu(II): Influencing factors and reusability. *Bioresour. Technol.* **2022**, *361*, 127732. [CrossRef] [PubMed]
43. Yan, N.; Hu, B.; Zheng, Z.; Lu, H.; Chen, J.; Zhang, X.; Jiang, X.; Wu, Y.; Dolfing, J.; Xu, L. Twice-milled magnetic biochar: A recyclable material for efficient removal of methylene blue from wastewater. *Bioresour. Technol.* **2023**, *372*, 128663. [CrossRef] [PubMed]
44. Tian, X.; Yang, R.; Chen, T.; Cao, Y.; Deng, H.; Zhang, M.; Jiang, X. Removal of both anionic and cationic dyes from wastewater using pH-responsive adsorbents of L-lysine molecular-grafted cellulose porous foams. *J. Hazard. Mater.* **2022**, *426*, 128121. [CrossRef]
45. Yu, K.L.; Lee, X.J.; Ong, H.C.; Chen, W.H.; Chang, J.S.; Lin, C.S.; Show, P.L.; Ling, T.C. Adsorptive removal of cationic methylene blue and anionic Congo red dyes using wet-torrefied microalgal biochar: Equilibrium, kinetic and mechanism modeling. *Environ. Pollut.* **2021**, *272*, 115986. [CrossRef]
46. Tran, H.N.; Lima, E.C.; Juang, R.S.; Bollinger, J.C.; Chao, H.P. Thermodynamic parameters of liquid–phase adsorption process calculated from different equilibrium constants related to adsorption isotherms: A comparison study. *J. Environ. Eng.* **2021**, *9*, 106674. [CrossRef]
47. Mechnou, I.; Meskini, S.; El Ayar, D.; Lebrun, L.; Hlaibi, M. Olive mill wastewater from a liquid biological waste to a carbon/oxocalcium composite for selective and efficient removal of methylene blue and paracetamol from aqueous solution. *Bioresour. Technol.* **2022**, *365*, 128162. [CrossRef]

48. Raji, Y.; Nadi, A.; Mechnou, I.; Saadouni, M.; Cherkaoui, O.; Zyade, S. High adsorption capacities of crystal violet dye by low-cost activated carbon prepared from Moroccan Moringa oleifera wastes: Characterization, adsorption and mechanism study. *Diam. Relat. Mater.* **2023**, *135*, 109834. [CrossRef]
49. Mechnou, I.; Meskini, S.; Mourtah, I.; Lebrun, L.; Hlaibi, M. Use of phosphorus-doped microporous carbon from olive mill wastewater for effective removal of Crystal violet and Methylene blue. *J. Clean. Prod.* **2023**, *393*, 136333. [CrossRef]
50. Zhang, Z.; Li, Y.; Zong, Y.; Yu, J.; Ding, H.; Kong, Y.; Ma, J.; Ding, L. Efficient remov al of cadmium by salts modified-biochar: Performance assessment, theoretical calculation, and quantitative mechanism analysis. *Bioresour. Technol.* **2022**, *361*, 127717. [CrossRef]
51. Li, H.; Kong, J.; Zhang, H.; Gao, J.; Fang, Y.; Shi, J.; Ge, T.; Fang, T.; Shi, Y.; Zhang, R.; et al. Mechanisms and adsorption capacities of ball milled biomass fly ash/biochar composites for the adsorption of methylene blue dye from aqueous solution. *J. Water Process Eng.* **2023**, *53*, 103713. [CrossRef]
52. Lu, L.; Shan, R.; Shi, Y.; Wang, S.; Yuan, H. A novel TiO_2/biochar composite catalysts for photocatalytic degradation of methyl orange. *Chemosphere* **2019**, *222*, 391–398. [CrossRef]
53. Liu, J.L.; Qian, W.C.; Guo, J.Z.; Shen, Y.; Li, B. Selective removal of anionic and cationic dyes by magnetic Fe_3O_4-loaded amine-modified hydrochar. *Bioresour. Technol.* **2021**, *320 Pt A*, 124374. [CrossRef]
54. Nguyen, X.C.; Nguyen, T.T.H.; Nguyen, T.H.C.; Le, Q.V.; Vo, Y.B.; Tran, T.C.P.; La, D.D.; Kumar, G.; Nguyen, V.K.; Chang, S.; et al. Sustainable carbonaceous biochar adsorbents derived from agro-wastes and invasive plants for cation dye adsorption from water. *Chemosphere* **2021**, *282*, 131009. [CrossRef]
55. Rubangakene, N.O.; Elkady, M.; Elwardany, A.; Fujii, M.; Sekiguchi, H.; Shokry, H. Effective decontamination of methylene blue from aqueous solutions using novel nano-magnetic biochar from green pea peels. *Environ. Res.* **2023**, *220*, 115272. [CrossRef]
56. Eshraghian, A.; Yu, L.; Achari, G.; Sundararaj, U. Development of an effective asphaltene-derived adsorbent for wastewater treatment: Characterization and methyl orange removal study. *J. Environ. Chem. Eng.* **2023**, *11*, 109221. [CrossRef]
57. Sun, Y.; Wang, T.; Han, C.; Lv, X.; Bai, L.; Sun, X.; Zhang, P. Facile synthesis of Fe-modified lignin-based biochar for ultra-fast adsorption of methylene blue: Selective adsorption and mechanism studies. *Bioresour. Technol.* **2022**, *344 Pt A*, 126186. [CrossRef]
58. Zhang, P.; O'connor, D.; Wang, Y.; Jiang, L.; Xia, T.; Wang, L.; Tsang, D.C.; Ok, Y.S.; Hou, D. A green biochar/iron oxide composite for methylene blue removal. *J. Hazard. Mater.* **2020**, *384*, 121286. [CrossRef]
59. Mu, Y.; Du, H.; He, W.; Ma, H. Functionalized mesoporous magnetic biochar for methylene blue removal: Performance assessment and mechanism exploration. *Diam. Relat. Meterials* **2022**, *121*, 108795. [CrossRef]
60. Yuan, Y.; Zhang, C.; Zhao, C.; Wang, B.; Wang, X.; Gao, B.; Wang, B.; Rinklebe, J. One-step preparation of a novel graphitic biochar/Cu^0/Fe_3O_4 composite using CO_2-ambiance pyrolysis to activate peroxydisulfate for dye degradation. *J. Environ. Sci.* **2023**, *125*, 26–36. [CrossRef]

toxics

MDPI

Article

Mercury Biogeochemical Cycle in Yanwuping Hg Mine and Source Apportionment by Hg Isotopes

Xingang Jin [1], Junyao Yan [2], Muhammad Ubaid Ali [2], Qiuhua Li [1,*] and Ping Li [2,*]

[1] Key Laboratory for Information System of Mountainous Area and Protection of Ecological Environment of Guizhou Province, Guizhou Normal University, Guiyang 550001, China; jinxingang0725@foxmail.com

[2] State Key Laboratory of Environmental Geochemistry, Institute of Geochemistry, Chinese Academy of Sciences, Guiyang 550081, China; yanjunyao@mail.gyig.ac.cn (J.Y.); ubaid@mail.gyig.ac.cn (M.U.A.)

* Correspondence: qiuhua2002@126.com (Q.L.); liping@mail.gyig.ac.cn (P.L.)

Abstract: Although mercury (Hg) mining activities in the Wanshan area have ceased, mine wastes remain the primary source of Hg pollution in the local environment. To prevent and control Hg pollution, it is crucial to estimate the contribution of Hg contamination from mine wastes. This study aimed to investigate Hg pollution in the mine wastes, river water, air, and paddy fields around the Yanwuping Mine and to quantify the pollution sources using the Hg isotopes approach. The Hg contamination at the study site was still severe, and the total Hg concentrations in the mine wastes ranged from 1.60 to 358 mg/kg. The binary mixing model showed that, concerning the relative contributions of the mine wastes to the river water, dissolved Hg and particulate Hg were 48.6% and 90.5%, respectively. The mine wastes directly contributed 89.3% to the river water Hg contamination, which was the main Hg pollution source in the surface water. The ternary mixing model showed that the contribution was highest from the river water to paddy soil and that the mean contribution was 46.3%. In addition to mine wastes, paddy soil is also impacted by domestic sources, with a boundary of 5.5 km to the river source. This study demonstrated that Hg isotopes can be used as an effective tool for tracing environmental Hg contamination in typical Hg-polluted areas.

Keywords: mercury; mine wastes; surface water; paddy soil; Hg isotopes; source apportionment

Citation: Jin, X.; Yan, J.; Ali, M.U.; Li, Q.; Li, P. Mercury Biogeochemical Cycle in Yanwuping Hg Mine and Source Apportionment by Hg Isotopes. *Toxics* **2023**, *11*, 456. https://doi.org/10.3390/toxics11050456

Academic Editors: Junhao Qin, Peidong Su, Feng Zhu and Lin Ding

Received: 10 April 2023
Revised: 4 May 2023
Accepted: 11 May 2023
Published: 14 May 2023

1. Introduction

Mercury (Hg) is a highly toxic heavy metal that can travel a long distance in the atmosphere and is therefore considered a global pollutant [1]. The toxicity of Hg depends on its chemical form. The elevated levels of Hg in the air are mostly attributed to industrial emissions, such as coal burning, Hg mining, gold mining, wastes incinerators, and cement production [2]. Methylmercury (MeHg) is neurotoxic, and it has the ability to bioaccumulate and become ultimately biomagnified in the food web. Humans are exposed to MeHg mainly through the consumption of food [3–5]. The Minamata Convention went into effect in August 2017 to reduce the effects of Hg exposure on human health [6,7].

The Wanshan Hg Mine is considered the "capital of Hg" in China. Since 2002, mining activities have been banned at the site due to the depletion of Hg resources and the environmental implications [8,9]. However, long-term Hg mining activities have produced a large amount of mine wastes, which are an important source of Hg pollution in the surrounding atmosphere and surface water. Most Hg calcine piles are distributed at the source of the river. Under external forces, such as rainwater leaching, surface runoff, and wind erosion, the Hg from the mine wastes is released and enters the downstream water system [10–12]. Therefore, evaluating the ecological risks caused by Hg mines is crucial for local ecological restoration.

The mine wastes from Hg mines can diffuse into the surrounding environment through water and atmospheric transportation. The paddy soils are more heavily contaminated by Hg in Hg mining areas compared with other areas [13]. Among the crops grown in the

Wanshan Hg mining area, rice has been identified as significantly capable of bioaccumulating MeHg in its grain, and rice ingestion could be the main route of MeHg exposure for local residents [14–17], which can pose serious health risks [18]. To avoid persistent paddy soil Hg pollution and subsequent human MeHg exposure, the sources of the paddy soil Hg need to be identified, and the Hg emissions from these sources can then be controlled by optimizing the major emission processes.

The Hg stable isotopes are an effective tool to track pollution sources and environmental processes [19–22]. There are seven natural stable isotopes of Hg: ^{196}Hg, ^{198}Hg, ^{199}Hg, ^{200}Hg, ^{201}Hg, ^{202}Hg, and ^{204}Hg. Mercury isotopes not only have mass-dependent fractionation (MDF) (reported as $\delta^{202}Hg$), but they also have mass-independent fractionation (MIF, mainly reported as $\Delta^{199}Hg$ or $\Delta^{201}Hg$). Mass-dependent fractionation can occur in physical, chemical, and biological processes, while MIF only occurs in a few specific processes, such as the photochemical reduction of Hg^{2+} and the photodegradation of MeHg [19,23–26]. Mixing models based on Hg isotopic MDF and MIF values has been used to quantify the contribution of the primary Hg sources in sediments [27–29]. Song et al. [6] used binary and ternary mixed models to calculate soil Hg pollution sources and their contribution ratios at different polluted sites. Yan et al. [30] used ternary mixed models to analyze the contribution ratios of the main Hg pollution sources in river water in Hg mining areas, and Fu et al. [31] used them for the quantitative source apportionment of Hg in the atmosphere. These studies indicate that the binary and ternary mixed models have been helpful in tracing the sources and biogeochemical processes of Hg in the environment. However, the contribution ratio of Hg mine wastes to the paddy soil in Hg mining areas remains unclear.

This study had the following aims: (1) to study the impact of the Hg mine wastes on the surrounding soil, water, and atmosphere; (2) to use Hg isotopic mixed models for the source apportionment of the Hg pollution in the downstream river water and paddy soil; and (3) to provide a theoretical basis for the source control of soil Hg pollution in paddy fields.

2. Materials and Methods

2.1. Study Area

The Wanshan Hg Mine is located in Guizhou Province, southwest China (Figure 1a). Mineralization at the Wanshan Hg Mine is primarily associated with thin-layered, laminated, fine-grained, dolomite or limestone beds of the mid-Cambrian age. The wall rocks are intensively altered by silicification, dolomitization, calcification, subordinate bituminization, and pyritization [32]. The primary ore mineral in the Hg deposits is cinnabar, with less metacinnabar [33].The Yanwuping Hg Mine (YMM) is one of the largest Hg mines in the Wanshan area. The Yanwuping Hg Mine is hilly and karstic, and it is located at an altitude of 340–1010 m. The climate is subtropical humid, with an annual rainfall of 1200–1400 mm and an annual temperature of 15 °C [34].

The Yanwuping Hg Mine's historic Hg extraction facility and about 3.1×10^5 m^3 of mine wastes are located at the upper Wengman River [35]. In 2011, the government renovated the YMM and tailing dams, but 1.3×10^4 m^2 of the calcine deposits remained. The Wengman River (Figure 1b) originates in the YMM zone and belongs to the Yangtze River basin, which has an average summer depth of 1 m and is directly affected by upstream mine wastes [36].

Mine wastes, surface-layer soils, and deep-layer soils were collected at Yanwuping Mercury Mine. W1–16 and S1–6 were the water and paddy soils sampling sites, respectively.

The total gaseous Hg sampling sites were the same as those for mine wastes, soils, and paddy soils.

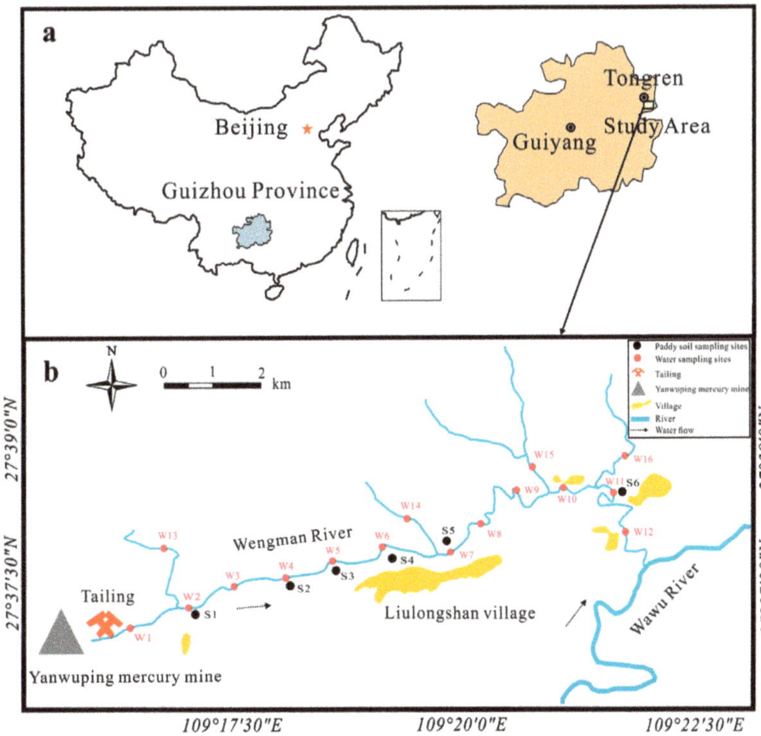

Figure 1. Location of study area (**a**) and distribution of sample sites (**b**).

2.2. Sample Collection

Water and atmospheric samples were collected and monitored twice, in December 2021 and August 2022, due to the high seasonal variability of the various indicators in the river water and atmosphere. The interannual variability in the soil Hg is not significant. Thus, soil and mine wastes samples were collected only once, in December 2021. There are two main types of mine wastes: calcines, the residues of Hg ore after high-temperature calcination, and waste rock, which is lower-grade surrounding rock [37]. Because most of the site has been restored, a total of 75 samples were collected from the surface layer and below 30 cm, and the difference between the restored area and bare area was compared and evaluated. Among them, 42 samples comprised surface soil, calcines, and waste rock, and 33 samples comprised deep soil, calcines, and waste rock. During the same period, the soil samples from the paddy fields downstream of the YMM were collected. For each site, a final sample composed of 3–5 subsamples was collected using the diagonal sampling method (15 paddy soil samples; Figure 1b). The collected soil, calcine, and waste rock samples were kept in clean polyethylene bags, air-dried, ground, and passed through a 200-mesh sieve, followed by total Hg (THg) and THg isotopic analysis.

The YMM downstream rainwater and surface water of the Wengman River were sampled for unfiltered THg, filtered dissolved Hg (DHg), DHg isotopes, particulate Hg (PHg) isotopes, anions, and cations. The THg and DHg isotopes samples were acidified with ultrapure hydrochloric acid, the cation samples with distilled nitric acid, and the anion samples without acid. The water samples were sealed in double-layer polyethylene bags, sent back to the laboratory, protected from light, and stored in a refrigerator at 4 °C. The analytical tests were completed within 28 days.

The total gaseous Hg (TGM) concentrations at the YMM and downstream paddy field sites were monitored 48 times using a portable RA-915+ Zeeman Hg Analyzer (Lumex,

Saint Petersburg, Russia). The Lumex instrument's detection limit was 0.5 ng/m^3. The instrument instantaneously displays the TGM concentrations per second, and each sampling point dataset represents an average monitoring time of at least 5 min in the field [37].

2.3. Analytical Methods

Approximately 0.1 g of the mine wastes and soil samples (dry weight) were digested with a mixture of HNO_3 and HCl (v:v = 1:3) for 2 h in a water bath at 95 °C. BrCl was added to the samples, and they were stored for 24 h for the conversion of all forms of Hg to Hg^{2+}, followed by the addition of acidic $SnCl_2$ to the solution to reduce the Hg ions to Hg^0. They were analyzed using cold-vapor atomic absorption spectrometry (CVAAS, F732-S, Shanghai Huaguang Instrument Factory, Shanghai, China). The detection limit of this method was 0.1 μg/L.

To determine the concentrations of THg and DHg in a water sample, BrCl was added to the sample and allowed to oxidize for 24 h. The Hg ions in the solution were then reduced to Hg^0 using acidic $SnCl_2$. The samples were preconcentrated into gold tubes and were later tested using a cold-vapor atomic fluorescence spectrophotometer (CVAFS, Tekran 2500, Tekran, Toronto, Ontario, Canada). The detection limit of this method was 0.1 μg/L. The THg in the water passing through a 0.45 μm filter is defined as DHg; subtracting the DHg from the THg yields the concentration of PHg in the water [34]. The anions and cations were analyzed by automated Dionex ICS-90 ion chromatography (Dionex, Sunnyvale, CA, USA) and an inductively coupled plasma optical emission spectrometer (ICP-OES, Varian, Palo Alto, CA, USA), respectively [38].

The Hg isotopic composition was analyzed using Neptune Plus MC-ICP-MS (Thermo Fisher Scientific, Waltham, MA, USA)at the State Key Laboratory of Environmental Geochemistry, the Institute of Geochemistry, the Chinese Academy of Sciences, following the method described by Yin et al. [39]. The total soluble Hg (TSHg) of the Hg mine wastes was extracted using a leaching experiment, as the Hg isotopes were to be tested along with the digested soil sample [37]. To ensure the minimum Hg concentration required for the DHg isotopes analysis of aqueous samples, each filtered water sample was pre-enriched into 5 mL of 40.0% aqua regia absorbent solution (v:v, HNO_3: HCl = 2:1), as shown in the method established by Li et al. [40]. For the Hg isotopes of PHg in the water samples, 2–5 L of water was filtered through a high-temperature purified Teflon membrane and freeze-dried. The Hg in the membrane was extracted into 5 mL of 40.0% anti-aqua regia absorbent solution using a tubular muffle furnace [41].

2.4. Hg Isotopes Analysis

The Hg isotopic composition was calculated using the formula presented by Blum and Bergquist (2007). Mass-dependent fractionation is expressed as delta (δ), and the results were calculated as follows:

$$\delta^{xxx}Hg_{sample}\ (‰) = [(^{xxx/198}Hg_{sample}/^{xxx/198}Hg_{NIST3133} - 1)] \times 1000 \quad (1)$$

where xxx is 199, 200, 201, 202, or 204. Mass-independent fractionation is expressed as "Δ", and it was calculated using the following equations:

$$\Delta^{199}Hg = \delta^{199}Hg - \delta^{202}Hg \times 0.252 \quad (2)$$

$$\Delta^{200}Hg = \delta^{200}Hg - \delta^{202}Hg \times 0.502 \quad (3)$$

$$\Delta^{201}Hg = \delta^{201}Hg - \delta^{202}Hg \times 0.752 \quad (4)$$

In this study, the binary mixed model was used to calculate the two sources of DHg of River Water No. 1. The calculations were performed using Equations (5) and (6) [1,42,43]:

$$\delta^{202}Hg_3 = \delta^{202}Hg_1 \times F_1 + \delta^{202}Hg_2 \times F_2 \quad (5)$$

$$1 = F_1 + F_2 \quad (6)$$

where F represents the percentage of the pollution source, subscript 1 represents the TSHg, subscript 2 represents the mountain spring water DHg, and subscript 3 represents the River Water No. 1 DHg. When calculating the two sources of River Water No. 1 PHg using a binary mixing model, subscript 1 represents the Hg mine wastes, subscript 2 represents the mountain spring water PHg, and subscript 3 represents the River Water No. 1 PHg.

The fractions of Hg in the paddy soil were derived from rainwater sources, river water sources, and geological background sources, and they were calculated using a triple-member mixing model as follows:

$$\delta^{202}Hg_{soil} = \delta^{202}Hg_{rain} \times F_{rain} + \delta^{202}Hg_{river} \times F_{river} + \delta^{202}Hg_{nat} \times F_{nat} \quad (7)$$

$$\Delta^{199}Hg_{soil} = \Delta^{199}Hg_{rain} \times F_{rain} + \Delta^{199}Hg_{river} \times F_{river} + \Delta^{199}Hg_{nat} \times F_{nat} \quad (8)$$

$$1 = F_{rain} + F_{river} + F_{nat} \quad (9)$$

where the subscripts rain, river, and nat represent rainwater sources, river water sources, and geological background sources, respectively, and F_{rain}, F_{river}, and F_{nat} represent the percentages of rainwater sources, river water sources, and geological background sources, respectively.

2.5. Quality Control

The quality control included blanks, duplicate samples, and certified reference materials (CRMs). The mean THg concentration in the method blanks was 0.026 ng/mL. Duplicate samples were measured after every 10 samples, and the mean relative standard deviations of the THg in the duplicate samples were 3.00% (n = 16). The low, medium, and high concentrations of the soil and mine wastes samples were controlled using GSS-5 (soil, 0.290 mg/kg), CRM021 (soil, 5.00 mg/kg), and CC580 (sediment, 132 mg/kg), with mean recovery ratios of 99.3% ± 3.25%, 91.7% ± 1.06%, and 99.3% ± 5.22%, respectively. The mean recovery ratio for the enrichment experiment of the water sample was 101% ± 4.31%, and the mean recovery ratio of the CRM in the PHg concentration experiment of the water sample was 99.0% ± 8.40%. The results of the UM-Almadén standard solution (δ^{202}Hg: −0.52‰ ± 0.05‰; Δ^{199}Hg: 0.00‰ ± 0.04‰; Δ^{201}Hg: −0.02‰ ± 0.04‰, n = 7) and the CC580 for sediment (δ^{202}Hg: −0.47‰ ± 0.04‰; Δ^{199}Hg: −0.06‰ ± 0.02‰; Δ^{201}Hg: −0.04‰ ± 0.02‰, n = 3) were consistent with previous studies (CC580, δ^{202}Hg: −0.51‰ ± 0.04‰; Δ^{199}Hg: 0.00‰ ± 0.03‰; Δ^{201}Hg: −0.02‰ ± 0.05‰, n = 2 [43–46]).

2.6. Data Analysis

Statistical analysis of the data, including means, standard deviations, and t-tests, was performed using IBM SPSS Statistics 26.0 (IBM, Armonk, NY, USA) and Microsoft Excel 2019 software (Microsoft, Redmond, WA, USA) (statistical significance = $p < 0.05$). Origin 2021 (OriginLab, Northampton, MA, USA) was used for the graphical demonstration of the data, and Arcmap 10.7 (ESRI, RedLands, CA, USA) was used to plot the spatial distributions by inverse distance weighting.

3. Results and Discussion

3.1. Hg Pollution in Mine Wastes

Considerable variation was observed in the THg concentrations of the YMM mine wastes. Except for the highest Hg concentration of 1.98×10^4 mg/kg found in surface mine wastes, the THg concentrations in the remaining samples showed a geometric mean of 38.4 mg/kg, with a range of 1.60–358 mg/kg. For the deep mine wastes, the THg concentrations showed a geometric mean of 46.8 mg/kg, with a range of 14.5–1.07×10^3 mg/kg. The considerable variations in the calcine THg concentrations may be attributed to the different retort furnaces used at the YMM. As the early smelting methods were not so advanced, inadequate ore burning resulted in low Hg recovery and a high concentration of calcines. The advancement in smelting technology led to adequate ore roasting, increased Hg recovery (≥95.0%), and a lower Hg concentration in the calcines [47]. The THg concentrations in the deep mine wastes analyzed in this study were higher than that of the local bedrock (0.35 mg/kg) [32], which was similar to the previous study also conducted at the Wanshan Hg Mine (geometric mean of THg concentrations: 49.0 mg/kg; THg concentration range: 4.15–825 mg/kg) [33]. The THg concentrations in 38.5% (15/39) of the YMM mine wastes samples exceeded the second-type soil pollution risk screening value (38.0 mg/kg) [48], and for 17.9% (7/39) of the samples, the THg concentrations exceeded the soil pollution risk control value of the second category of construction land (82.0 mg/kg) [48]. This demonstrates that a significant proportion of Hg persists even after the high-temperature melting of Hg ore [29].

The THg concentrations in the soils covered by restoration showed a geometric mean of 7.70 mg/kg, with a range of 1.68–139 mg/kg. The THg concentrations in the surface samples and deep samples correlated significantly ($p < 0.05$), indicating that the surface soil has been polluted by calcines in the lower layer. The soil THg concentrations were much higher than the agricultural land soil pollution risk control value (4.00 mg/kg, $6.5 < pH \leq 7.5$) [49], which is 70 times higher than the Guizhou Province soil background value of 0.110 mg/kg [50]. A comparison of the calcine area before and after the YMM restoration is shown in Figure 2a,b. The distribution of the Hg pollution in the surface and deep layers reveals that the most serious Hg pollution occurs in the exposed calcines areas. The mine wastes in the YMM are still the primary source of Hg pollution in the surrounding ecosystem. The exposed calcines seriously impact the local ecological environment by continually releasing Hg into the atmosphere, entering surface water bodies, and leaching into downstream farming soil [8,33].

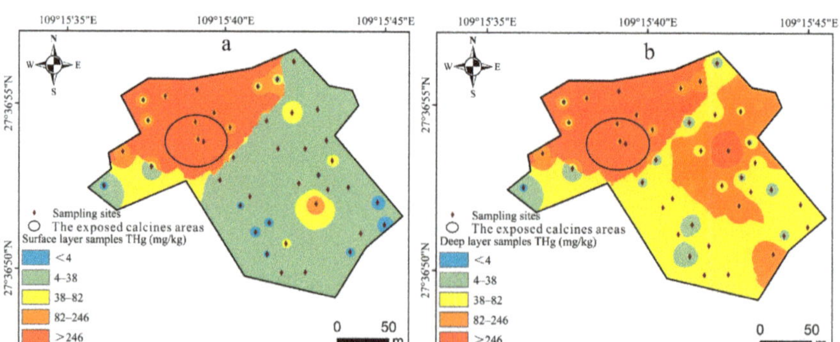

Figure 2. Spatial distribution of Hg pollution in mine wastes and soils at YMM: (**a**) surface layer; (**b**) deep layer.

3.2. Atmospheric Hg

The spatial distribution of the TGM at the YMM showed significant variations (Figure 3a,b). In wintertime, the TGM concentrations averaged 24.1 ± 6.90 ng/m^3, with a range of 10.1–45.0 ng/m^3. In summertime, the TGM concentrations averaged 153 ± 129 ng/m^3, with a range of 43.3–700 ng/m^3. The TGM concentrations at the exposed calcines areas were found to be the highest both in winter and summer, and the TGM concentrations in summer were much higher than those in winter (winter: 45.0 ng/m^3; summer: 700 ng/m^3), while the TGM concentrations at the restored sampling point were much lower (winter: 23.6 ng/m^3; summer: 153 ng/m^3). The results show that mine wastes are still an important emission source of atmospheric Hg pollution, and remediated measures could effectively reduce the Hg emission flux at the interface between the mine wastes and air [51]. Compared with other Hg mining areas in China, the TGM concentrations in the YMM (43.3–700 ng/m^3) were much higher than those in the Xunyang (7.40–410 ng/m^3) and Wanshan (13.5–309 ng/m^3) Hg mining areas [33,52]. However, the concentrations were much lower than those in the Xiushan Hg mining area (29.0–4.21 × 10^4 ng/m^3) [53]. The mean concentration of TGM was three times higher when compared with the air quality reference standard of 50.0 ng/m^3 set by the Ministry of Environmental Protection of China, and it may pose a potential risk to local residents [54]. Therefore, the Hg emissions from mine wastes should be strictly controlled to reduce the environmental risks.

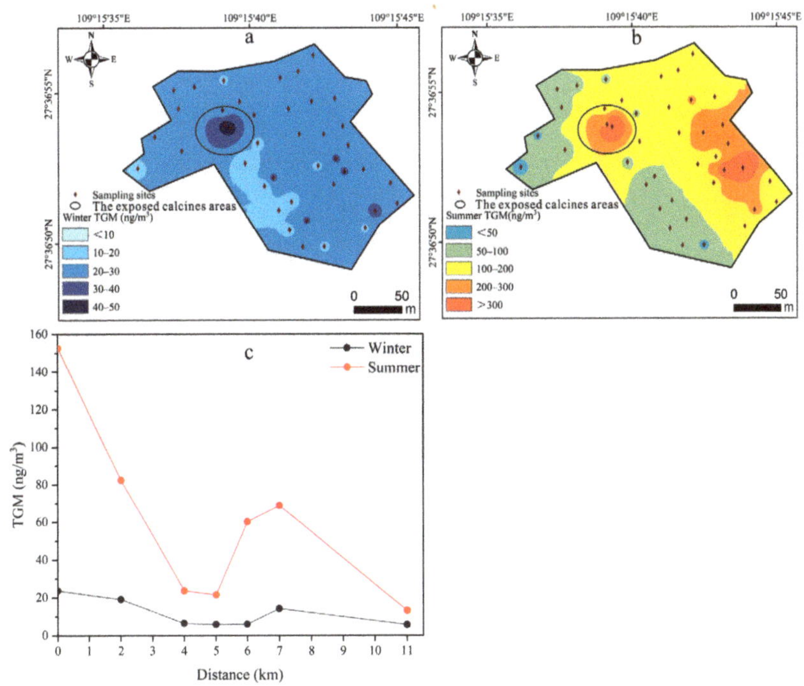

Figure 3. Spatial distribution of TGM at YMM and downstream paddy field: (**a**) spatial distribution of TGM at YMM in winter; (**b**) spatial distribution of TGM at YMM in summer; (**c**) variation in TGM with distance of paddy field downstream of YMM.

Previous studies have reported that the TGM released from pollution sources can settle into paddy soil after migration [55,56]. After the TGM monitoring in the downstream paddy fields, the following trends were noticed (Figure 3c). The results indicate that the TGM concentrations gradually decreased within 5.5 km of the YMM both in winter and summer (winter: 5.79 ng/m^3; summer: 21.4 ng/m^3). However, the TGM concentrations were still higher than the global background (1.50–1.60 ng/m^3) [33,57]. The TGM concentrations considerably increased at around 5.5 km, with the highest value of 68.7 ng/m^3 in summer, and then gradually decreased with the increasing distance. The increase in the TGM in the ambient air far from the YMM indicates the influence of nearby domestic Hg emission sources, such as Hg emissions from coal burning, waste incineration, traffic pollution, etc. [35,58,59].

3.3. Surface Water Hg and Source Apportionment

3.3.1. Surface Water Hg Pollution

The Hg concentrations in the surface water of the Wengman River flowing through the YMM varied greatly. The THg, PHg, and DHg concentrations in the winter water samples averaged 101 ng/L (5.34–1.25 $\times$ 10^3 ng/L), 86.9 ng/L (2.22–1.22 $\times$ 10^3 ng/L), and 14.9 ng/L (3.12–80.9 ng/L), respectively. In the summer water samples, the averages were 34.4 ng/L (2.14–241 ng/L) for THg, 22.9 ng/L (0.08–198 ng/L) for PHg, and 11.5 ng/L (1.75–42.7 ng/L) for DHg. The Hg mine wastes upstream are considered to be the main source of the surface water Hg pollution [37,60]. The highest Hg concentration was found at the W1 site near the YMM (Figure 4a,b). The THg concentration in the winter surface water at this site exceeded the standard limit of 1000 ng/L stipulated by China's Class V surface water environmental quality standard [61], indicating the direct influence of the Hg mine wastes. In order to reduce the impact of the upstream calcine leachate on the downstream water system, Xu et al. [36] designed and built a weir 1.5 km away from the YMM, which can intercept 40.4% of the THg per year and significantly reduce the THg concentrations in the river. The upstream calcine leachate is mainly composed of PHg. In this study, the proportions of the water PHg to THg in the winter and summer flowing through the weir decreased by 76.5% and 52.9%, respectively, indicating that the removal ratios were higher than those presented by Xu et al. [36]. The difference between the two studies might be attributed to the flow of the river water, as previous studies have reported that water flow is the main factor that influences the transport and migration of Hg [53,62]. In this study, the average THg concentrations in the summer samples (34.4 ng/L) were lower than those of the winter samples (101 ng/L). This might be because the summer samples were collected after a heavy rain event and the erosive, and leaching effect of the rainwater was not significant. The river flows in summer were higher than those in winter, which mainly showed that the dilution effect resulted in lower Hg concentrations in the summer samples. However, the highest concentration still exceeded the threshold limit of 100 ng/L set by China's Class III surface water environmental quality standard [61].

The above studies have proven that the weir can indeed cause the particulate matter to settle, which is because the water flow slows down and the suspension time increases, thereby reducing the Hg pollution downstream. However, Xu et al. [36] only monitored before and after the weir, and they did not set sampling points downstream of the Wengman River. The results of this study show that at 4–5 km downstream of the Wengman River, the THg concentrations decreased to 10.3–11.3 ng/L, as a large amount of PHg had settled. The proportion of DHg to THg increased to 55.4%–98.9%, indicating the PHg sedimentation effect. The Hg concentrations at 4–5 km were close to the mean concentration of 7.09 ng/L in the tributaries, which indicated the baseline concentration of the surface water in this area. Except for W14, the water Hg concentrations in the tributaries in this study were similar to those of Qiu et al. [35] (tributary: 3.00–17.0 ng/L). After 5.5 km, the Hg concentration in the river water gradually increased, but it did not exceed the limit of 100 ng/L stipulated by China's Class III surface water environmental quality standard [61]. The Hg concentration at 6.5 km of the tributary (W14 sample) in wintertime reached 85.0 ng/L, exceeding the 50.0 ng/L limited stipulated by China's Class II surface water environmental quality standard [61]. This indicated that the surface water after 5.5 km may be impacted by other external sources of Hg.

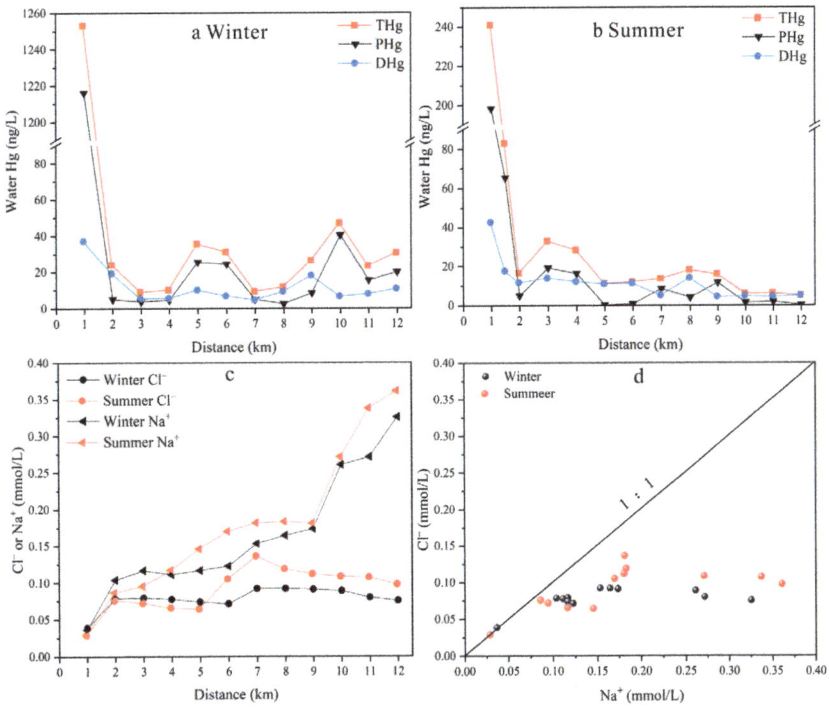

Figure 4. Water Hg and ion concentrations at Wengman River: (**a**) variation in different forms of Hg concentration with distance from YMM in winter; (**b**) variation in different forms of Hg concentration with distance from YMM in summer; (**c**) variation in Cl^-, Na^+ concentrations with distance from YMM; (**d**) evolution of Cl^- / Na^+ ratios.

This study analyzed the anions and cations of the Wengman River, and it found that the Na^+ and Cl^- concentrations increased significantly after 5.5 km in both winter and summer. The increase in the Na^+ concentration was much higher than that of Cl^- (Figure 4c). Figure 4d shows that the ratio of Cl^-:Na^+ in the sample at 1 km was 1:1, indicating that it was mainly derived from the dissolution of evaporite rock. However, the ratio of the water Cl^-:Na^+ downstream gradually fell below the 1:1 ratio line. The ratio of Cl^- and Na^+ ranged from 0.232 to 0.876, which indicates the contribution of sources other than the dissolution of evaporite rocks [63,64]. Because Liulongshan Township is 5.5 km away, with a dense population, domestic activities have significant impacts on the water chemistry [65]. Cl^- is not affected by physical, chemical, or biological processes, and it is a good indicator of anthropogenic activities, such as the use of agricultural fertilizers, animal manure, and domestic sewage [66]. The Wanshan area belongs to the karst landform, and the main rock type is carbonate rock. The Na^+ concentration in the surface water is relatively low. The increase in Na^+ relative to Cl^- may be due to silicate weathering (e.g., plagioclase) and the effect of the input from domestic pollution sources. In order to further elucidate the contribution of both to the excess Na^+, this study used the molar ratio bivariate plots of the Na^+-normalized Ca^{2+} and Mg^{2+} and Na^+-normalized Ca^{2+} and HCO_3^- distributions (Figure S1) to identify the contribution of rock weathering to the ion source in the river [67,68]. As shown in Figure S1, the ionic composition of the river water in the study area was mainly located near the weathered end element of the carbonate rock, with a trend towards the weathered end element of the silicate rock. This indicates that the contribution of silicate rocks to river water ions is small. Therefore, this study concluded that the increase in Na^+ concentration relative to the Cl^- concentration after 5.5 km from the

Wengman River may be caused by domestic pollution [69,70]. This domestic pollution may come from domestic sewage, domestic waste (such as batteries, thermometers, pigment and paint residues, fluorescent lamps, and so on), etc. [59,71].

3.3.2. Source Apportionment by Hg Isotopes

The river water THg increased significantly when it flowed through the Hg mining area (Figure 4a,b), which is consistent with previous literature [9,72]. In winter, the δ^{202}Hg and Δ^{199}Hg values in the river water samples downstream averaged $-0.29‰ \pm 0.30‰$ (-0.71~$0.11‰$, $n = 7$) and $-0.02‰ \pm 0.07‰$ (-0.12~$0.06‰$, $n = 7$), respectively (Table S1). Mercury isotopes were used to trace the source of Hg in the upstream water of the Wengman River. The results showed that the main contributing sources included Hg mine wastes and mountain spring water. This study assumes that the Hg from mine wastes and mountain spring water enters the River Water No. 1 sampling site in a rapid mixing process and that no significant MDF and MIF will occur during this process. The pollution sources of the DHg in the River Water No. 1 sample mainly include the TSHg from the Hg mine wastes and the DHg from the mountain spring water. For the TSHg of the Hg mine wastes, the DHg of the mountain spring water, and the DHg of River Water No. 1, the δ^{202}Hg values were $-0.90‰$, $-1.57‰$, and $-1.25‰$, respectively, and the Δ^{199}Hg values were all close to zero. Many previous studies have demonstrated the usefulness of end-member mixing models for Hg-source tracking in water environments [73,74]. In this study, the relative contributions of the different sources for the River Water No. 1 DHg were calculated using a binary mixing model. The relative contribution ratio of the mine waste TSHg to DHg was 48.6%, and that of the mountain spring water was 51.4%.

The pollution sources of the PHg in River Water No. 1 mainly included the Hg mine wastes and mountain spring water. The δ^{202}Hg values in the Hg mine wastes, the PHg of mountain spring water, and the River Water No. 1 PHg were $-0.35‰$, $-1.78‰$, and $-0.48‰$, respectively, and the Δ^{199}Hg values were all close to zero. The observed Δ^{199}Hg values in Hg mine wastes are consistent with previous studies [29,75,76]. In this study, the relative contributions of the two pollution sources to the River Water No. 1 PHg were calculated using a binary mixing model. The relative contribution ratio of the Hg mine wastes was 90.5%, and that of the mountain spring water was 9.50%.

The THg in the River Water No. 1 was 1.25×10^3 ng/L, while DHg accounted for 3.00%, and PHg for 97.0%. It was calculated that the Hg mine wastes directly contributed 89.3% to the river Hg pollution at the No. 1 site, indicating that the erosion of Hg mine wastes by runoff is the main process of the Hg pollution in the rivers near the Hg mine. This shows that the upstream water of the Wengman River is seriously polluted with Hg. The government needs to remediate the mine wastes left at the site and reinforce the tailings dam, which could reduce the Hg pollution in the downstream river and the health risk to local residents.

3.4. Paddy Soil Hg and Source Apportionment

3.4.1. Paddy Soil Hg Pollution

The THg concentrations in the paddy soil downstream of the YMM averaged 3.58 ± 1.82 mg/kg with a range of 1.49–8.51 mg/kg. Compared with other Hg mining areas, the paddy soil THg concentrations at the YMM were lower than those in China's Xunyang (1.30–750 mg/kg), Xiushan (0.45–68.0 mg/kg), and Wanshan (0.50–188 mg/kg) Hg mining areas [52,53,77]. However, they were still much higher than the agricultural soil pollution risk screening value (0.60 mg/kg, $6.5 < pH \leq 7.5$) and Guizhou soil Hg background value (0.110 mg/kg) [49,50]. The results showed that the downstream paddy soils are still seriously polluted by THg and indicated serious ecological risks. It is critical to identify the sources and contributions of Hg pollution in paddy soil. Therefore, preventive measures should be taken to control the Hg release. The obtained results provide a reliable theoretical and scientific basis for the treatment and safe utilization of Hg-contaminated soil.

The trends of the soil THg in the downstream paddy fields in this study were not consistent with those presented by Xu et al. [53], who reported that, with the increase in the distance from the Hg mining area, the soil THg concentrations tended to decrease. However, in this study, a different trend was noticed. A decreasing trend was observed prior to the 5.5 km distance; however, after 5.5 km, the THg concentrations at sites S4 and S5 significantly increased. The variation in the THg concentration in the downstream paddy soil is consistent with that of the TGM (Figure S2). Apart from this, a significant correlation was found between the TGM and paddy soil THg ($p < 0.05$). This indicates that atmospheric dry and wet depositions play a vital role in the Hg pollution of paddy soil [35]. The paddy fields are located along the banks of the Wengman River, and the local people have been using the Hg-contaminated river water for irrigation for a long time, which could also be a key source of the Hg pollution in the paddy soil [18,78].

3.4.2. Source Apportionment by Hg Isotopes

Previous studies have shown that soils can preserve the isotopic fingerprints of Hg pollution sources [42,79,80]. The paddy field downstream of the YMM are located on both sides of the Wengman River. There was a significant correlation between the TGM and THg in the paddy soil ($p < 0.05$), indicating that dry and wet atmospheric depositions are an important source of Hg pollution in the paddy soil. The contribution of wet deposition is of key importance as shown in Figure 5a,b. Pribil et al. [47] stated that the Hg in the soils to the north and east of the mining area may be the result of atmospheric deposition, geological background influence, and gaseous Hg emissions from calcines piles during Hg processing. The irrigation of paddy fields with Hg-contaminated river water is also one of the key sources [18].

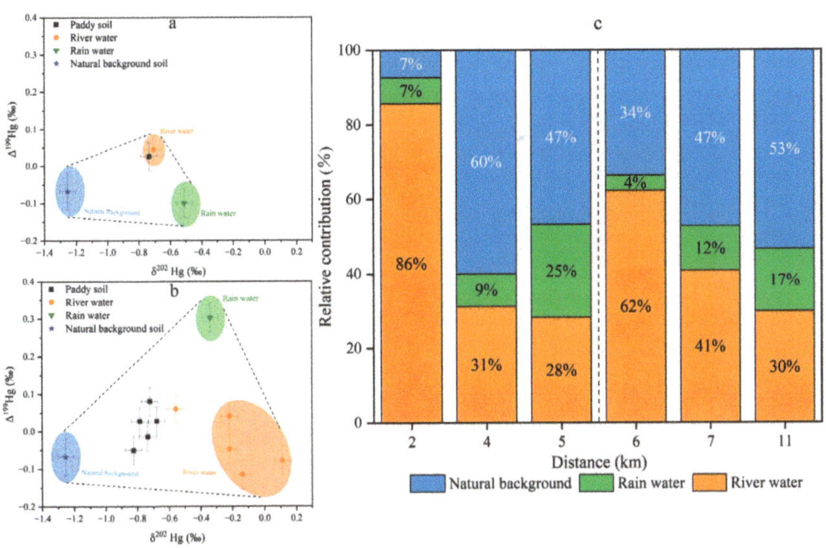

Figure 5. δ^{202}Hg and Δ^{199}Hg values from different sources and relative contributions to paddy soil: (**a**) soil Hg pollution source analysis before 2 km; (**b**) soil Hg pollution source analysis after 2 km; (**c**) three pollution sources of Hg in paddy soil. Natural background soil Hg isotopes values were adopted from Song et al. [6].

The Rain Water No. 1 sample was collected 2 km away from the mining area. Due to its proximity to the Hg mining area, the Hg mine wastes have a greater impact. As shown in Figure 5a, the Hg isotopes compositions were δ^{202}Hg = −0.51‰ ± 0.05‰ and Δ^{199}Hg = −0.10‰ ± 0.04‰. The Rain Water No. 2 sample was collected at a distance of 6 km from the mining area, and we noticed that the Hg mine wastes had less of an impact. The Δ^{199}Hg in the rainwater was positive, which was similar to the Hg isotopic value of the rainwater in Guiyang (δ^{202}Hg: −0.44~−4.27‰, Δ^{199}Hg: 0.19~1.16‰) [81]. The δ^{202}Hg (δ^{202}Hg = −0.34‰ ± 0.05‰) and Δ^{199}Hg (Δ^{199}Hg = 0.30‰ ± 0.04‰) values in the rainwaters are presented in Figure 5b. The mean values of the δ^{202}Hg and Δ^{199}Hg in the paddy soil were −0.73‰ ± 0.11‰ (−0.91~−0.56‰, n = 12) and 0.03‰ ± 0.05‰ (−0.05~0.10‰, n = 12), respectively. According to Song et al. [6], the mean values of the δ^{202}Hg and Δ^{199}Hg in the paddy soils in the Wanshan area were −1.26‰ ± 0.06‰ (−1.30~−1.21‰, n = 2) and −0.07‰ ± 0.10‰ (−0.14~0.00‰, n = 2), which were used as the background values in this study (Table S2).

As shown in Figure 5a,b, the combined characteristics of the Δ^{199}Hg and δ^{202}Hg indicate that the paddy soil is a ternary mixture of different sources, such as rainwater, river water, and geological background sources. Therefore, the δ^{202}Hg and Δ^{199}Hg of the corresponding point samples were used to trace the source of the Hg pollution in the paddy soil. This study calculated the relative contributions of the three sources using a ternary mixed model, as shown in Figure 5c. The results showed that the exogenous input of the Hg pollution in the paddy soil can be divided into two parts. The first part is the area contaminated by Hg mining activities. Within 5.5 km of the YMM, the river water is mainly polluted by Hg mine wastes, while the paddy soil pollution at 2 km is mostly attributed to river water, reaching 86.0%. The contribution range of river water was 28.0~86.0%, while in the case of rainwater, the range was 7.00~25.0%. The contribution range of the geological background was 7.00~60.0%. With the increase in distance, the Hg contribution of the river water gradually decreased, while the contribution ratio of the rainwater and geological background sources gradually increased. The paddy fields in this range were mainly contaminated by the Hg mine. The second part is the domestic Hg-polluted area. After 5.5 km from the YMM, the contribution of the river water to the paddy soil increased to 62.0% at 6 km, which is consistent with the increase in the Hg concentrations in the river water, indicating the influence of domestic pollution sources. The contribution ranges were 30.0~62.0% by river water, 4.00~17.0% by rainwater, and 34.0~53.0% by geological background. With the increase in distance, the contribution ratio of the river water gradually decreased, and the contribution ratio of the rainwater and geological background sources increased again, indicating that the paddy soil pollution in this range was mainly attributed to domestic pollution sources.

4. Conclusions

The YMM is still the primary source of Hg pollution in the surrounding ecosystem, and especially of the significant atmospheric Hg emissions in the exposed calcines area. Peak concentrations of THg were observed at the river upstream, and the source apportionment for the DHg and PHg in the river water using a binary mixing model demonstrated that mine wastes were the main source of Hg in the surface water. The TGM concentrations downstream showed specific spatial distributional characteristics, indicating a large amount of Hg emissions from Hg mine wastes and unidentified domestic pollution sources. This study calculated the contributions of the river water, atmospheric wet deposition, and geological background to the paddy soil Hg pollution by Hg isotopes, which were also verified by the spatial distributions of the river water Hg, river water anions and cations, TGM, and paddy soil Hg. The study shows that the paddy soil upstream at 5.5 km is mainly polluted by the Hg mine, while domestic sources are the main contributors after a distance of 5.5 km. This study provides an important scientific basis for the source control of Hg in the surface water and paddy fields in Hg mining areas. This is needed to control the Hg emissions from mine wastes to the river water and atmosphere, which would finally reduce the Hg bioaccumulation in agricultural crops and the associated human health risks in Hg-polluted areas.

Supplementary Materials: The following supporting information can be downloaded at: https://www.mdpi.com/article/10.3390/toxics11050456/s1, Figure S1: Molar ratio bivariate plots of (a) Na+-normalized Ca2+ and Mg2+ and (b) Na+-normalized Ca2+ and HCO3–; Figure S2: Variation of THg with distance in paddy field downstream of YMM; Table S1: Mercury isotopes composition of water samples; Table S2: Mercury isotopic composition in soil samples.

Author Contributions: Conceptualization, P.L.; Data curation, X.J. and M.U.A.; Funding acquisition, P.L.; Investigation, X.J. and J.Y.; Methodology, P.L.; Supervision, Q.L. and P.L.; Writing—original draft, X.J., J.Y., M.U.A. and Q.L.; Writing—review and editing, X.J., J.Y., M.U.A., Q.L. and P.L. All authors have read and agreed to the published version of the manuscript.

Funding: This research was funded by the National Key R&D Program of China (2020YFC1807600), Guizhou Provincial Science and Technology Subsidies, the CAS Interdisciplinary Innovation Team (JCTD-2020-20), and Youth Innovation Promotion Association of Chinese Academy of Sciences (Y2021106).

Institutional Review Board Statement: Not applicable.

Informed Consent Statement: Not applicable.

Data Availability Statement: Not applicable.

Conflicts of Interest: The authors declare no conflict of interest.

References

1. Feng, X.; Foucher, D.; Hintelmann, H.; Yan, H.; He, T.; Qiu, G. Tracing mercury contamination sources in sediments using mercury isotope compositions. *Environ. Sci. Technol.* **2010**, *44*, 3363–3368. [CrossRef] [PubMed]
2. Li, P.; Du, B.; Maurice, L.; Laffont, L.; Lagane, C.; Point, D.; Sonke, J.E.; Yin, R.; Lin, C.-J.; Feng, X. Mercury isotope signatures of methylmercury in rice samples from the Wanshan mercury mining area, China: Environmental implications. *Environ. Sci. Technol.* **2017**, *51*, 12321–12328. [CrossRef] [PubMed]
3. Wright, L.P.; Zhang, L.; Cheng, I.; Aherne, J.; Wentworth, G.R. Impacts and effects indicators of atmospheric deposition of major pollutants to various ecosystems—A review. *Aerosol Air Qual. Res.* **2018**, *18*, 1953–1992. [CrossRef]
4. Donovan, P.M.; Blum, J.D.; Singer, M.B.; Marvin-DiPasquale, M.; Tsui, M.T.K. Isotopic Composition of Inorganic Mercury and Methylmercury Downstream of a Historical Gold Mining Region. *Environ. Sci. Technol.* **2016**, *50*, 1691–1702. [CrossRef] [PubMed]
5. Gerson, J.R.; Topp, S.N.; Vega, C.M.; Gardner, J.R.; Yang, X.; Fernandez, L.E.; Bernhardt, E.S.; Pavelsky, T.M. Artificial lake expansion amplifies mercury pollution from gold mining. *Sci. Adv.* **2020**, *6*, 4953. [CrossRef]
6. Song, Z.; Wang, C.; Ding, L.; Chen, M.; Hu, Y.; Li, P.; Zhang, L.; Feng, X. Soil mercury pollution caused by typical anthropogenic sources in China: Evidence from stable mercury isotope measurement and receptor model analysis. *J. Clean. Prod.* **2021**, *288*, 125687. [CrossRef]
7. Brocza, F.M.; Biester, H.; Richard, J.-H.; Kraemer, S.M.; Wiederhold, J.G. Mercury Isotope Fractionation in the Subsurface of a Hg(II) Chloride-Contaminated Industrial Legacy Site. *Environ. Sci. Technol.* **2019**, *53*, 7296–7305. [CrossRef] [PubMed]
8. Li, P.; Feng, X.; Shang, L.; Qiu, G.; Meng, B.; Liang, P.; Zhang, H. Mercury pollution from artisanal mercury mining in Tongren, Guizhou, China. *Appl. Geochem.* **2008**, *23*, 2055–2064.
9. Qiu, G.L.; Feng, X.B.; Wang, S.F.; Fu, X.W.; Shang, L.H. Mercury distribution and speciation in water and fish from abandoned Hg mines in Wanshan, Guizhou province, China. *Sci. Total Environ.* **2009**, *407*, 5162–5168. [CrossRef]
10. Kim, K.-H.; Kabir, E.; Jahan, S.A. A review on the distribution of Hg in the environment and its human health impacts. *J. Hazard. Mater.* **2016**, *306*, 376–385. [CrossRef]
11. Li, P.; Feng, X.; Qiu, G.; Shang, L.; Li, G. Human hair mercury levels in the Wanshan mercury mining area, Guizhou Province, China. *Environ. Geochem. Health* **2009**, *31*, 683–691. [CrossRef] [PubMed]
12. Zhang, C.; Qiu, G.L.; Anderson, C.W.N.; Zhang, H.; Meng, B.; Liang, L.; Feng, X.B. Effect of Atmospheric Mercury Deposition on Selenium Accumulation in Rice (*Oryza sativa* L.) at a Mercury Mining Region in Southwestern China. *Environ. Sci. Technol.* **2015**, *49*, 3540–3547. [CrossRef]
13. Xia, J.C.; Wang, J.X.; Zhang, L.M.; Anderson, C.W.N.; Wang, X.; Zhang, H.; Dai, Z.H.; Feng, X.B. Screening of native low mercury accumulation crops in a mercury—Polluted mining region: Agricultural planning to manage mercury risk in farming communities. *J. Clean. Prod.* **2020**, *262*, 121324. [CrossRef]
14. Feng, X.; Li, P.; Qiu, G.; Wang, S.; Li, G.; Shang, L.; Meng, B.; Jiang, H.; Bai, W.; Li, Z. Human exposure to methylmercury through rice intake in mercury mining areas, Guizhou Province, China. *Environ. Sci. Technol.* **2008**, *42*, 326–332. [CrossRef] [PubMed]
15. Dai, Z.; Feng, X.; Sommar, J.; Li, P.; Fu, X. Spatial distribution of mercury deposition fluxes in Wanshan Hg mining area, Guizhou province, China. *Atmos. Chem. Phys.* **2012**, *12*, 6207–6218. [CrossRef]
16. Du, B.; Feng, X.; Li, P.; Yin, R.; Yu, B.; Sonke, J.E.; Guinot, B.; Anderson, C.W.; Maurice, L. Use of mercury isotopes to quantify mercury exposure sources in inland populations, China. *Environ. Sci. Technol.* **2018**, *52*, 5407–5416. [CrossRef] [PubMed]

17. Xing, Y.; Wang, J.; Xia, J.; Liu, Z.; Zhang, Y.; Du, Y.; Wei, W. A pilot study on using biochars as sustainable amendments to inhibit rice uptake of Hg from a historically polluted soil in a Karst region of China. *Ecotoxicol. Environ. Saf.* **2019**, *170*, 18–24. [CrossRef]
18. Zhang, Y.; Zhou, X.; Ma, W.; Yin, D.; Wang, Y.; Zhang, C.; Wang, D. Distribution of Mercury and Methylmercury in Farmland Soils Affected by Manganese Mining and Smelting Activities. *Int. J. Environ. Res. Public Health* **2022**, *19*, 10288. [CrossRef]
19. Kwon, S.Y.; Blum, J.D.; Yin, R.; Tsui, M.T.-K.; Yang, Y.H.; Choi, J.W. Mercury stable isotopes for monitoring the effectiveness of the Minamata Convention on Mercury. *Earth Sci. Rev.* **2020**, *203*, 103111. [CrossRef]
20. Jimenez-Moreno, M.; Barre, J.P.G.; Perrot, V.; Berail, S.; Rodriguez Martin-Doimeadios, R.C.; Amouroux, D. Sources and fate of mercury pollution in Almaden mining district (Spain): Evidences from mercury isotopic compositions in sediments and lichens. *Chemosphere* **2016**, *147*, 430–438. [CrossRef]
21. Wiederhold, J.G.; Smith, R.S.; Siebner, H.; Jew, A.D.; Brown, G.E., Jr.; Bourdon, B.; Kretzschmar, R. Mercury Isotope Signatures as Tracers for Hg Cycling at the New Idria Hg Mine. *Environ. Sci. Technol.* **2013**, *47*, 6137–6145. [CrossRef] [PubMed]
22. Stetson, S.J.; Gray, J.E.; Wanty, R.B.; Macalady, D.L. Isotopic Variability of Mercury in Ore, Mine-Waste Calcine, and Leachates of Mine-Waste Calcine from Areas Mined for Mercury. *Environ. Sci. Technol.* **2009**, *43*, 7331–7336. [CrossRef] [PubMed]
23. Bergquist, B.A.; Blum, J.D. Mass-dependent and-independent fractionation of Hg isotopes by photoreduction in aquatic systems. *Science* **2007**, *318*, 417–420. [CrossRef]
24. Blum, J.D.; Sherman, L.S.; Johnson, M.W. Mercury isotopes in earth and environmental sciences. *Annu. Rev. Earth Planet. Sci.* **2014**, *42*, 249–269. [CrossRef]
25. Zhang, H.; Feng, X.; Larssen, T.; Qiu, G.; Vogt, R.D. In inland China, rice, rather than fish, is the major pathway for methylmercury exposure. *Environ. Health Perspect.* **2010**, *118*, 1183–1188. [CrossRef]
26. Smith, R.S.; Wiederhold, J.G.; Jew, A.D.; Brown, G.E., Jr.; Bourdon, B.; Kretzschmar, R. Stable Hg Isotope Signatures in Creek Sediments Impacted by a Former Hg Mine. *Environ. Sci. Technol.* **2015**, *49*, 767–776. [CrossRef]
27. Liu, J.; Feng, X.; Yin, R.; Zhu, W.; Li, Z. Mercury distributions and mercury isotope signatures in sediments of Dongjiang, the Pearl River Delta, China. *Chem. Geol.* **2011**, *287*, 81–89. [CrossRef]
28. Yin, R.; Feng, X.; Chen, B.; Zhang, J.; Wang, W.; Li, X. Identifying the Sources and Processes of Mercury in Subtropical Estuarine and Ocean Sediments Using Hg Isotopic Composition. *Environ. Sci. Technol.* **2015**, *49*, 1347–1355. [CrossRef]
29. Yin, R.; Feng, X.; Wang, J.; Li, P.; Liu, J.; Zhang, Y.; Chen, J.; Zheng, L.; Hu, T. Mercury speciation and mercury isotope fractionation during ore roasting process and their implication to source identification of downstream sediment in the Wanshan mercury mining area, SW China. *Chem. Geol.* **2013**, *336*, 72–79. [CrossRef]
30. Yan, J.; Li, R.; Ali, M.U.; Wang, C.; Wang, B.; Jin, X.; Shao, M.; Li, P.; Zhang, L.; Feng, X. Mercury migration to surface water from remediated mine waste and impacts of rainfall in a karst area—Evidence from Hg isotopes. *Water Res.* **2023**, *230*, 119592. [CrossRef]
31. Fu, X.; Marusczak, N.; Wang, X.; Gheusi, F.; Sonke, J.E. Isotopic Composition of Gaseous Elemental Mercury in the Free Troposphere of the Pic du Midi Observatory, France. *Environ. Sci. Technol.* **2016**, *50*, 5641–5650. [CrossRef] [PubMed]
32. Hua, Y.; Cui, M. *Wanshan Mercury Deposit in Guizhou Province*; China Geological Press: Beijing, China, 1994. (In Chinese)
33. Yan, J.; Wang, C.; Wang, Z.; Yang, S.; Li, P. Mercury concentration and speciation in mine wastes in Tongren mercury mining area, southwest China and environmental effects. *Appl. Geochem.* **2019**, *106*, 112–119. [CrossRef]
34. Zhang, H.; Feng, X.; Zhu, J.; Sapkota, A.; Meng, B.; Yao, H.; Qin, H.; Larssen, T. Selenium in soil inhibits mercury uptake and translocation in rice (*Oryza sativa* L.). *Environ. Sci. Technol.* **2012**, *46*, 10040–10046. [CrossRef] [PubMed]
35. Qiu, G.; Feng, X.; Meng, B.; Zhang, C.; Gu, C.; Du, B.; Lin, Y. Environmental geochemistry of an abandoned mercury mine in Yanwuping, Guizhou Province, China. *Environ. Res.* **2013**, *125*, 124–130. [CrossRef]
36. Xu, X.; Gu, C.; Feng, X.; Qiu, G.; Shang, L.; Xu, Z.; Lu, Q.; Xiao, D.; Wang, H.; Lin, Y.; et al. Weir building: A potential cost-effective method for reducing mercury leaching from abandoned mining tailings. *Sci. Total Environ.* **2019**, *651*, 171–178. [CrossRef]
37. Li, P.; Feng, X.; Qiu, G.; Zhang, J.; Meng, B.; Wang, J. Mercury speciation and mobility in mine wastes from mercury mines in China. *Environ. Sci. Pollut. Res.* **2013**, *20*, 8374–8381. [CrossRef]
38. Shao, M.; Liu, Z.; Sun, H.; Lai, C.; Ma, Z.; He, X.; Fang, Y.; Chai, Q. C-N-P driven changes to phytoplankton community structure and gross primary productivity in river-fed reservoir ecosystems on the Chinese Loess Plateau. *J. Hydrol.* **2023**, *616*, 128781. [CrossRef]
39. Yin, R.; Krabbenhoft, D.P.; Bergquist, B.A.; Zheng, W.; Lepak, R.F.; Hurley, J.P. Effects of mercury and thallium concentrations on high precision determination of mercury isotopic composition by Neptune Plus multiple collector inductively coupled plasma mass spectrometry. *J. Anal. At. Spectrom.* **2016**, *31*, 2060–2068. [CrossRef]
40. Li, K.; Lin, C.-J.; Yuan, W.; Sun, G.; Fu, X.; Feng, X. An improved method for recovering and preconcentrating mercury in natural water samples for stable isotope analysis. *J. Anal. At. Spectrom.* **2019**, *34*, 2303–2313. [CrossRef]
41. Fu, X.; Feng, X.; Yin, R.; Zhang, H. Diurnal variations of total mercury, reactive mercury, and dissolved gaseous mercury concentrations and water/air mercury flux in warm and cold seasons from freshwaters of southwestern China. *Environ. Toxicol. Chem.* **2013**, *32*, 2256–2265. [CrossRef]
42. Estrade, N.; Carignan, J.; Donard, O.F. Tracing and quantifying anthropogenic mercury sources in soils of northern France using isotopic signatures. *Environ. Sci. Technol.* **2011**, *45*, 1235–1242. [CrossRef] [PubMed]
43. Qin, C.; Du, B.; Yin, R.; Meng, B.; Fu, X.; Li, P.; Zhang, L.; Feng, X. Isotopic Fractionation and Source Appointment of Methylmercury and Inorganic Mercury in a Paddy Ecosystem. *Environ. Sci. Technol.* **2020**, *54*, 14334–14342. [CrossRef] [PubMed]

44. Blum, J.D.; Bergquist, B.A. Reporting of variations in the natural isotopic composition of mercury. *Anal. Bioanal. Chem.* **2007**, *388*, 353–359. [CrossRef] [PubMed]
45. Estrade, N.; Carignan, J.; Sonke, J.E.; Donard, O.F. Measuring Hg isotopes in bio-geo-environmental reference materials. *Geostand. Geoanal. Res.* **2010**, *34*, 79–93. [CrossRef]
46. Janssen, S.E.; Johnson, M.W.; Blum, J.D.; Barkay, T.; Reinfelder, J.R. Separation of monomethylmercury from estuarine sediments for mercury isotope analysis. *Chem. Geol.* **2015**, *411*, 19–25. [CrossRef]
47. Pribil, M.J.; Rimondi, V.; Costagliola, P.; Lattanzi, P.; Rutherford, D.L. Assessing mercury distribution using isotopic fractionation of mercury processes and sources adjacent and downstream of a legacy mine district in Tuscany, Italy. *Appl. Geochem.* **2020**, *117*, 104600. [CrossRef]
48. Ministry of Ecology and Environment of China (MEE). *Soil Environmental Qualityrisk Control Standard for Soil Contamination of Development Land (GB3600–2018)*; Environmental Science Press of China: Beijing, China, 2018. (In Chinese)
49. Ministry of Ecology and Environment of China (MEE). *Soil Environmental Quality Risk Control Standard for Soil Contamination of Agricultural Land (GB15618–2018)*; Environmental Science Press of China: Beijing, China, 2018. (In Chinese)
50. China National Environmental Monitoring Center (CNEMC). *The Soil Environmental Background Value in the People's Republic of China*; China Environmental Press: Beijing, China, 1990. (In Chinese)
51. Dai, Z.; Feng, X.; Zhang, C.; Wang, J.; Jiang, T.; Xiao, H.; Li, Y.; Wang, X.; Qiu, G. Assessing anthropogenic sources of mercury in soil in Wanshan Hg mining area, Guizhou, China. *Environ. Sci. Pollut. Res.* **2013**, *20*, 7560–7569. [CrossRef]
52. Qiu, G.; Feng, X.; Meng, B.; Sommar, J.; Gu, C. Environmental geochemistry of an active Hg mine in Xunyang, Shaanxi Province, China. *Appl. Geochem.* **2012**, *27*, 2280–2288. [CrossRef]
53. Xu, X.; Lin, Y.; Meng, B.; Feng, X.; Xu, Z.; Jiang, Y.; Zhong, W.; Hu, Y.; Qiu, G. The impact of an abandoned mercury mine on the environment in the Xiushan region, Chongqing, southwestern China. *Appl. Geochem.* **2018**, *88*, 267–275. [CrossRef]
54. Ministry of Environment Protection of China (MEP). *Ambient Air Quality Standards (GB3095–2012)*; Environmental Science Press of China: Beijing, China, 2012. (In Chinese)
55. Xu, X.; Liu, N.; Landis, M.S.; Feng, X.; Qiu, G. Characteristics and distributions of atmospheric mercury emitted from anthropogenic sources in Guiyang, southwestern China. *Acta Geochim.* **2016**, *35*, 240–250. [CrossRef]
56. Zhao, L.; Qiu, G.; Anderson, C.W.; Meng, B.; Wang, D.; Shang, L.; Yan, H.; Feng, X. Mercury methylation in rice paddies and its possible controlling factors in the Hg mining area, Guizhou province, Southwest China. *Environ. Pollut.* **2016**, *215*, 1–9. [CrossRef] [PubMed]
57. Fu, X.; Zhang, H.; Yu, B.; Wang, X.; Lin, C.-J.; Feng, X. Observations of atmospheric mercury in China: A critical review. *Atmos. Chem. Phys.* **2015**, *15*, 9455–9476. [CrossRef]
58. Li, Z.G.; Feng, X.; Li, P.; Liang, L.; Tang, S.L.; Wang, S.F.; Fu, X.W.; Qiu, G.L.; Shang, L.H. Emissions of air-borne mercury from five municipal solid waste landfills in Guiyang and Wuhan, China. *Atmos. Chem. Phys.* **2010**, *10*, 3353–3364. [CrossRef]
59. Das, R.; Wang, X.; Khezri, B.; Webster, R.D.; Sikdar, P.K.; Datta, S. Mercury isotopes of atmospheric particle bound mercury for source apportionment study in urban Kolkata, India. *Elem. Sci. Anthr.* **2016**, *4*, 1–12. [CrossRef]
60. Zhang, G.; Liu, C.-Q.; Wu, P.; Yang, Y. The geochemical characteristics of mine-waste calcines and runoff from the Wanshan mercury mine, Guizhou, China. *Appl. Geochem.* **2004**, *19*, 1735–1744. [CrossRef]
61. Environmental Protection Administration of China (EPA). *Chinese Standards for Surface Water Quality (GB3838–2002)*; Environmental Protection Administration of China: Beijing, China, 2002. (In Chinese)
62. Qiu, G.; Feng, X.; Wang, S.; Xiao, T. Mercury contaminations from historic mining to water, soil and vegetation in Lanmuchang, Guizhou, southwestern China. *Sci. Total Environ.* **2006**, *368*, 56–68. [CrossRef]
63. Yokoo, Y.; Nakano, T.; Nishikawa, M.; Quan, H. Mineralogical variation of Sr–Nd isotopic and elemental compositions in loess and desert sand from the central Loess Plateau in China as a provenance tracer of wet and dry deposition in the northwestern Pacific. *Chem. Geol.* **2004**, *204*, 45–62. [CrossRef]
64. Zhang, F.; Jin, Z.; Yu, J.; Zhou, Y.; Zhou, L. Hydrogeochemical processes between surface and groundwaters on the northeastern Chinese Loess Plateau: Implications for water chemistry and environmental evolutions in semi-arid regions. *J. Geochem. Explor.* **2015**, *159*, 115–128. [CrossRef]
65. Xu, Z.; Liu, C.-Q. Water geochemistry of the Xijiang basin rivers, South China: Chemical weathering and CO_2 consumption. *Appl. Geochem.* **2010**, *25*, 1603–1614. [CrossRef]
66. Yin, C.; Yang, H.; Wang, J.; Guo, J.; Tang, X.; Chen, J. Combined use of stable nitrogen and oxygen isotopes to constrain the nitrate sources in a karst lake. *Agric. Ecosyst. Environ.* **2020**, *303*, 107089. [CrossRef]
67. Gaillardet, J.; Dupre, B.; Allegre, C.J.; Negrel, P. Chemical and physical denudation in the Amazon River basin. *Chem. Geol.* **1997**, *142*, 141–173. [CrossRef]
68. Gaillardet, J.; Dupre, B.; Louvat, P.; Allegre, C.J. Global silicate weathering and CO_2 consumption rates deduced from the chemistry of large rivers. *Chem. Geol.* **1999**, *159*, 3–30. [CrossRef]
69. Li, X.; Liu, C.; Harue, M.; Li, S.; Liu, X. The use of environmental isotopic (C, Sr, S) and hydrochemical tracers to characterize anthropogenic effects on karst groundwater quality: A case study of the Shuicheng Basin, SW China. *Appl. Geochem.* **2010**, *25*, 1924–1936. [CrossRef]
70. Liu, C.; Li, S.; Lang, Y.; Xiao, H. Using δ15N-and δ18O-values to identify nitrate sources in karst ground water, Guiyang, Southwest China. *Environ. Sci. Technol.* **2006**, *40*, 6928–6933. [CrossRef]

71. Mao, Y.; Cheng, L.; Ma, B.; Cai, Y. The fate of mercury in municipal wastewater treatment plants in China: Significance and implications for environmental cycling. *J. Hazard. Mater.* **2016**, *306*, 1–7. [CrossRef]
72. Zhang, H.; Feng, X.B.; Larssen, T.; Shang, L.H.; Vogt, R.D.; Lin, Y.; Li, P.; Zhang, H.I. Fractionation, distribution and transport of mercury in rivers and tributaries around Wanshan Hg mining district, Guizhou Province, Southwestern China: Part 2-Methylmercury. *Appl. Geochem.* **2010**, *25*, 642–649. [CrossRef]
73. Washburn, S.J.; Blum, J.D.; Kurz, A.Y.; Pizzuto, J.E. Spatial and temporal variation in the isotopic composition of mercury in the South River, VA. *Chem. Geol.* **2018**, *494*, 96–108. [CrossRef]
74. Wiederhold, J.G.; Skyllberg, U.; Drott, A.; Jiskra, M.; Jonsson, S.; Bjorn, E.; Bourdon, B.; Kretzschmar, R. Mercury isotope signatures in contaminated sediments as a tracer for local industrial pollution sources. *Environ. Sci. Technol.* **2015**, *49*, 177–185. [CrossRef]
75. Gehrke, G.E.; Blum, J.D.; Marvin-DiPasquale, M. Sources of mercury to San Francisco Bay surface sediment as revealed by mercury stable isotopes. *Geochim. Cosmochim. Acta* **2011**, *75*, 691–705. [CrossRef]
76. Sonke, J.E.; Schäfer, J.; Chmeleff, J.; Audry, S.; Blanc, G.; Dupré, B. Sedimentary mercury stable isotope records of atmospheric and riverine pollution from two major European heavy metal refineries. *Chem. Geol.* **2010**, *279*, 90–100. [CrossRef]
77. Yin, D.; He, T.; Yin, R.; Zeng, L. Effects of soil properties on production and bioaccumulation of methylmercury in rice paddies at a mercury mining area, China. *J. Environ. Sci.* **2018**, *68*, 194–205. [CrossRef] [PubMed]
78. Du, J.; Liu, F.; Zhao, L.; Liu, C.; Fu, Z.; Teng, Y. Mercury horizontal spatial distribution in paddy field and accumulation of mercury in rice as well as their influencing factors in a typical mining area of Tongren City, Guizhou, China. *J. Environ. Health Sci. Eng.* **2021**, *19*, 1555–1567. [CrossRef]
79. Feng, X.; Yin, R.; Yu, B.; Du, B. Mercury isotope variations in surface soils in different contaminated areas in Guizhou Province, China. *Chin. Sci. Bull.* **2013**, *58*, 249–255. [CrossRef]
80. Zhang, H.; Yin, R.-S.; Feng, X.-B.; Sommar, J.; Anderson, C.W.; Sapkota, A.; Fu, X.-W.; Larssen, T. Atmospheric mercury inputs in montane soils increase with elevation: Evidence from mercury isotope signatures. *Sci. Rep.* **2013**, *3*, 3322. [CrossRef] [PubMed]
81. Wang, Z.H.; Chen, J.B.; Feng, X.B.; Hintelmann, H.; Yuan, S.L.; Cai, H.M.; Huang, Q.; Wang, S.X.; Wang, F.Y. Mass-dependent and mass-independent fractionation of mercury isotopes in precipitation from Guiyang, SW China. *Comptes Rendus Geosci.* **2015**, *347*, 358–367. [CrossRef]

toxics MDPI

Article

Efficient Calibration of Groundwater Contaminant Transport Models Using Bayesian Optimization

Hao Deng [1,2,3], Shengfang Zhou [1], Yong He [1,2,3,*], Zeduo Lan [1], Yanhong Zou [1,2,3] and Xiancheng Mao [1,2,3]

1 School of Geosciences and Info-Physics, Central South University, South Lushan Road, Changsha 410083, China
2 Key Laboratory of Metallogenic Prediction of Nonferrous Metals and Geological Environment Monitoring, Central South University, Ministry of Education, Changsha 410083, China
3 Key Laboratory of Non-Ferrous Resources and Geological Hazard Detection, Changsha 410083, China
* Correspondence: heyong18@csu.edu.cn

Abstract: Numerical modeling is a significant tool to understand the dynamic characteristics of contaminants transport in groundwater. The automatic calibration of highly parametrized and computationally intensive numerical models for the simulation of contaminant transport in the groundwater flow system is a challenging task. While existing methods use general optimization techniques to achieve automatic calibration, the large numbers of numerical model evaluations required in the calibration process lead to high computing overhead and limit the efficiency of model calibration. This paper presents a Bayesian optimization (BO) method for efficient calibration of numerical models of groundwater contaminant transport. A Bayes model is built to fully represent calibration criteria and derive the objective function for model calibration. The efficiency of model calibration is made possible by the probabilistic surrogate model and the expected improvement acquisition function in BO. The probabilistic surrogate model approximates the computationally expensive objective function with a closed-form expression that can be computed efficiently, while the expected improvement acquisition function proposes the most promising model parameters to improve the fitness to the calibration criteria and reduce the uncertainty of the surrogate model. These schemes allow us to find the optimized model parameters effectively by using a small number of numerical model evaluations. Two case studies for the calibration of the Cr(VI) transport model demonstrate that the BO method is effective and efficient in the inversion of hypothetical model parameters, the minimization of the objective function, and the adaptation of different model calibration criteria. Specifically, this promising performance is achieved within 200 numerical model evaluations, which substantially reduces the computing budget for model calibration.

Keywords: model calibration; groundwater contaminant transport; Bayesian optimization

Citation: Deng, H.; Zhou, S.; He, Y.; Lan, Z.; Zou, Y.; Mao, X. Efficient Calibration of Groundwater Contaminant Transport Models Using Bayesian Optimization. *Toxics* **2023**, *11*, 438. https://doi.org/10.3390/toxics11050438

Academic Editor: Roberto Rosal

Received: 22 March 2023
Revised: 21 April 2023
Accepted: 4 May 2023
Published: 6 May 2023

1. Introduction

The contaminant transport in groundwater flow systems is a concealed, long-term, invertible process that has become a serious environmental problem and poses threats to sustainable development [1–4]. Understanding the dynamic characteristics of contaminant migration and retention is crucial for pollution assessment and contaminant remediation in soils, sites, and groundwater.

The numerical modeling techniques offer effective and robust tools to quantify, simulate, and predict the process of contaminants transport in groundwater flow systems. Over the past few decades, an enormous number of groundwater numerical modeling studies have been devoted to simulating the transportation and fate of contaminants [5–13]. The numerical modeling of contaminant transport is dependent on the dynamic governing equations, which include a series of model parameters that affect the model solutions, i.e., the simulation results. Thus, regardless of the dynamic governing equations and solution methods, numerical modeling is confronted with the non-uniqueness of simulation results, which limits the accuracy of simulation results. Aiming to generate accurate and meaningful simulation results, extensive efforts have been made in model calibration [2,14–24].

Initially, the model calibration is conducted manually. However, calibration of the model with only a few parameters would lead to a tedious and time-consuming "trial-and-error" process. Automatic model calibration, which is generally cast as an optimization problem, greatly alleviates the human workload to tune the model parameters and improves the efficiency of parameter calibration. To solve the highly nonlinear optimization problem derived from automatic model calibration, common optimization techniques such as gradient-based methods and global optimization methods are generally used. The gradient-based optimization is adopted in most model calibration approaches, such as those in the literature [25–27] and well-known software such as PEST [14], UCODE [28], and Dakota [29]. To find the minimum of objective functions, this category of methods involves multiple iterations to update the solution in terms of gradients. To estimate the gradient with respect to N model parameters, $N+1$(forward difference) or $2N+1$ (center difference) times of numerical modeling are required. Moreover, gradient-based methods can lead to local minimum solutions, hindering the accuracy of simulation results. The other line of research uses global optimization methods. The exemplary methods include genetic algorithms [20,30], simulated annealing [31], adaptive cluster covering [16], shuffled complex evolution [32], differential evolution [33], particle swarm optimization [34], and ant colony optimization [35]. While these methods are shown to be effective in finding the global minimum of the objective function and improving the model accuracy, the significantly increased numbers of numerical model evaluations compared with the gradient-based method are the major concern. For a high-dimensional and long-term simulation, only a single evaluation of numerical modeling would lead to a high computational workload. However, common optimization techniques are not tailored to fit this characteristic and include multiple objective function evaluations, which lead to extremely expensive and even prohibitive computation for the model calibration scenario.

Over the past few decades, the surrogate model has been introduced to hydrological modeling. The surrogate model has been widely used in the probabilistic approach for groundwater model calibrations. The probabilistic approaches [36–41], e.g., the well-known MCMC-based DRAME algorithms [42,43], generally entail hundreds or thousands of numerical model evaluations and would lead to prohibitive computational overhead. The surrogate model is an approximation to the computationally intensive numerical model but can be evaluated in a much more efficient fashion. A series of data-driven surrogate models have been proposed, such as Bagging Multivariate Adaptive Regression Splines [44], neural networks [45], Gaussian processes [46], kriging [47], radial basis functions [48], support vector machines [49], and genetic programming [50]. While these surrogate models solidly reduce the computational cost, the generation of a surrogate model that well fits the numerical model still entails a large number of numerical model evaluations, and little effort has been paid to the tailoring of an efficient optimization algorithm for surrogate models.

This paper proposes a novel model calibration method based on the Bayesian optimization (BO) method. The BO method is a global optimization method to optimize objective functions that are time-consuming to estimate. The BO method has shown its success in parameter tuning in machine learning [51,52], materials design [53], and recommender systems [54]. Compared with common optimization techniques that find the minimum by extensively exploring the objective function, the BO method uses a surrogate model to approximate the objective function and ensures computational efficiency. In terms of the surrogate model, specifically, an acquisition function is devised for the minimization of the objective function and the avoidance of unnecessary objective function evaluations at certain model parameter spaces. This allows us to find the optimal objective function using a very small number of numerical modeling evaluations. The proposed method is applied to model the transport and retention of chromium in groundwater flow systems. The results demonstrate that the BO method is effective and efficient in calibrating numerical models of groundwater flow and contaminant transport.

2. Materials and Methods

Figure 1 illustrates the BO framework for calibration of the numeric model of groundwater contaminant transport. Given the numeric models of groundwater flow and contaminant transport, we first choose the target parameters to be calibrated. Then, an objective function is formulated to quantify calibration criteria. We use the Bayes' rule to convert the calibration criteria into the objective function, in which we measure the fitness of simulated values to the observations and prior knowledge. To find the calibrated model parameters, the objective function is minimized by the BO method, in which the minimum of the objective function (i.e., optimized model parameters) is progressively searched in an iterated process. To reduce the expensive objective evaluations, the probabilistic surrogate model replaces the objective function in the BO method. In terms of the surrogate model, the acquisition function is defined to propose new model parameters for minimization of the objective function. Through specified numbers of BO iterations, the calibrated model parameters that well fit the calibration criteria can be obtained.

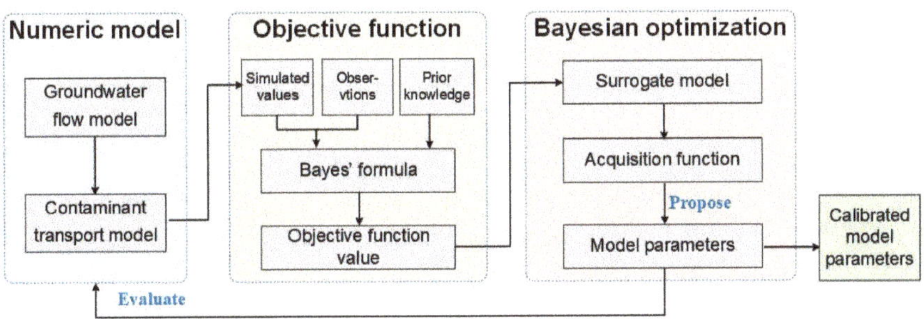

Figure 1. BO framework for calibration of the numeric model of groundwater contaminant transport.

In the following, we will first derive the object function in Section 2.1 and then elaborate on the BO in Section 2.2, while the numeric models are introduced in Section 2.3.

2.1. Objective Function

Formally, we can represent the contaminant transport in the groundwater flow system as follows:

$$\widetilde{C} = M(X, \Theta) + \epsilon, \tag{1}$$

where $\widetilde{C}$ is the observation of contaminant concentration, $M(X, \Theta)$ is a numerical model that simulates the contaminant transport with input variables X and model parameters $\Theta = \{\Theta\}_i^n$, and ϵ is the residual error. Here, the input X involves spatial-temporal locations of interest and boundary conditions. Given the set of observations $\widetilde{C} = \left\{\widetilde{c}_i\right\}_{i=1}^N$, the posterior probability of model parameters Θ that guides model calibration can be formulated in terms of Bayes' theorem:

$$p\left(\Theta \middle| \widetilde{C}\right) = \frac{p\left(\widetilde{C} \middle| \Theta\right) p(\Theta)}{p\left(\widetilde{C}\right)} \propto p\left(\widetilde{C} \middle| \Theta\right) p(\Theta), \tag{2}$$

where $p\left(\widetilde{C} \middle| \Theta\right)$ is the likelihood function that quantifies the fitness between simulation results and observations $\widetilde{C}$, $p(\Theta)$ is the prior distribution $p(\Theta)$ that represents prior knowledge on model parameters $p(\Theta)$, and $p\left(\tilde{C}\right) = \int p\left(\tilde{C}\middle|\Theta\right)p(\Theta)d\Theta$ is the evidence and normalizes $p\left(\Theta\middle|\tilde{C}\right)$.

To calibrate the numerical model, we solve the model parameter Θ in a maximum a posteriori fashion, i.e., given the observations $\tilde{C}$, Θ is optimized such that its posterior probability $p\left(\Theta\middle|\tilde{C}\right)$ is maximized:

$$\Theta = \underset{\Theta}{\mathrm{argmax}}\, p\left(\Theta\middle|\tilde{C}\right) = \underset{\Theta}{\mathrm{argmax}}\, p\left(\tilde{C}\middle|\Theta\right)p(\Theta). \tag{3}$$

Here, by assuming that the observation error ϵ is normally distributed with zero means, the likelihood function can be defined in the form of Gaussian probability density:

$$P\left(\tilde{C}\middle|\Theta\right) = \prod_{i=1}^{N} \frac{1}{\sqrt{2\pi\sigma^2}} \exp\left(-\frac{\left(M_i(X,\Theta) - \tilde{c}_i\right)^2}{2\sigma_i^2}\right), \tag{4}$$

where $M_i(X,\Theta)$, $\tilde{c}_i$, and σ_i are the numerical modeling result, the observation, and the standard deviation, respectively, at the i-th spatial-temporal location. Here, a standard deviation with a value of σ_i indicates that it is roughly a 95% confidence that the simulation error $M_i(X,\Theta) - \tilde{c}_i$ falls into the interval of $(-2\sigma_i, 2\sigma_i)$ according to the 68–95–99.7 rule [55]. Thus, σ_i is set according to a confidence level of the simulation error at i-th location and controls the fitness between $M_i(X,\Theta)$ and $\tilde{c}_i$.

On the other hand, we assume each parameter Θ_i in the model parameters Θ is independent of the other parameters in Θ. Then, the joint prior probability $p(\Theta)$ of model parameters Θ is factorized as:

$$p(\Theta) = \prod_{i=1}^{n} p(\Theta_i). \tag{5}$$

For the prior probability $p(\Theta_i)$, we can define it by using the kernel density estimation [56] if we have measurements on Θ_i. Otherwise, we define it as the uniform distribution over the interval $\left[\Theta_i^{inf}, \Theta_i^{sup}\right]$: $\Theta_i \sim U\left(\Theta_i^{inf}, \Theta_i^{sup}\right)$, where Θ_i^{inf} and Θ_i^{sup} define the valid range of Θ_i.

Combining Equations (2)–(5) to ensure numeric stability and following the convention of optimization, we use the negative log-posterior as the objective function of model calibration:

$$\min_{\Theta} f(\Theta) = \min_{\Theta}\left\{-\log p\left(\Theta\middle|\tilde{C}\right)\right\} \tag{6}$$

2.2. Bayesian Optimization

The optimization problem in Equation (6) involves a "black-box" and computationally expensive objective function. In terms of such characteristics, the BO method is a promising approach to solving the optimization problem in Equation (6). The BO method uses a probabilistic surrogate model to represent the "black-box" objective function and sequentially refines (updates) the model with new sampling to the objective function. In contrast to gradient-based optimization, the BO method does not require knowledge about the gradients of the objective function and allows finding a global minimum for the non-convex objective function.

The key components of the BO method are the surrogate model and the acquisition model. The surrogate model is the probabilistic model of the objective function that represents our assumptions and knowledge of the "black-box" and the expected observations of the objective functions. The acquisition function evaluates a sampling for the objective function concerning the surrogate model, and the sequence of the samples for the objective function evaluation is generated by maximizing the acquisition function. Ideally, the minimum objective function can be found by optimizing a sequence of samples. For our scenario, each sample of the objective function leads to a simulation with the numerical model. The sampling updates the posterior probability of the surrogate model of the objective function. Algorithm 1 summarizes the workflow of BO for model calibration. Interested readers are referred to the literature [53,57–59] for more details of the BO method.

Algorithm 1. Bayesian optimization for model calibration.

```
Initialize sample set Q_1
for t = 1, ..., T − 1
    Update probabilistic surrogate model p(f|Q_t, Θ) with Q_t
    Update acquisition function α(Θ; Q_t) with p(f|Q_t)
    Select the new model parameters Θ_t by maximizing the α(Θ; Q_t):
                    Θ_t ← argmax_Θ α(Θ; Q_t)
    Compute simulation results M(X, Θ_t)
    Compute the objective function f(Θ_t) with M(X, Θ_t)
    Q_{t+1} = Q_t ∪ {Θ_t, f(Θ_t)}
end for
Output model parameters Θ* with the smallest f(Θ)
```

Surrogate model. Given the sampling results $\mathcal{Q}$ to the objective function, Gaussian process (GP) regression is generally used as the probabilistic surrogate model $p(f|\mathcal{Q},\Theta)$ for the computationally intensive objective function. The GP regression is flexible and scalable to model an arbitrary nonlinear function and requires a relatively small number of data examples. This is well-suited to model to objective function calibration. The GP extends the multivariate Gaussian distribution over vectors (finite-dimension space) to a distribution over functions (infinite-dimension space), for which any finite collection of dimensions follows a multivariate Gaussian distribution. The distribution of the Gaussian process is the joint distribution of infinite random variables. To surrogate the objective function $f(\Theta)$, the distribution of the Gaussian process for f is defined by a mean function $m(\Theta)$ and a covariance function $k(\Theta,\Theta\prime)$:

$$f(\Theta) \sim GP(m(\Theta), k(\Theta,\Theta\prime)). \tag{7}$$

Following conventions [60], the prior mean function $m(\Theta)$ is assumed to be a mean function, while the covariance function is set to the squared exponential function. Then, given the known samples $\mathcal{Q}$ including t evaluations of objective function values: $\mathbf{f} = (f(\Theta_1), \cdots, f(\Theta_t))^T$, according to the Sherman–Morrison–Woodbury formula, a sample of objective function $f(\Theta)$ at Θ is derived to be Gaussian distributed [60]:

$$P(f|\mathcal{Q},\Theta) = \mathcal{N}\left(\mu(\Theta), \sigma^2(\Theta)\right), \tag{8}$$

where

$$\begin{aligned} \mu(\Theta) &= \mathbf{k}^T\mathbf{K}^{-1}\mathbf{f} \\ \sigma^2(\Theta) &= k(\Theta,\Theta) - \mathbf{k}^T\mathbf{K}^{-1}\mathbf{k}, \end{aligned} \tag{9}$$

$\mathbf{K} \in \mathbb{R}^{t\times t}$ is the covariance matrix of $\mathbf{f}$, with the entry at the i-th row, the j-th column being $k(\Theta_i,\Theta_j)$, and $\mathbf{k} \in \mathbb{R}^t$ is a vector with the i-th entry being $k(\Theta,\Theta_i)$. In realistic cases, an evaluation of the objective function can include the residential error. Assuming the residential error is independent and identically Gaussian distributed with the zero mean and the variance σ_n^2, the covariance matrix of the noisy version of $\mathbf{f}$ becomes $\mathbf{K} + \sigma_n^2\mathbf{I}$, which can also avoid overfitting to the noise-corrupted objective functions [60]. Thus, the predictive surrogate model $p(f_*|\mathcal{Q},\Theta)$ is:

$$P(f_*|\mathcal{Q},\Theta) = \mathcal{N}\left(\mu_*(\Theta),\sigma_*^2(\Theta)\right), \tag{10}$$

where

$$\begin{aligned} \mu_*(\Theta) &= \mathbf{k}^{\mathrm{T}}\left(\mathbf{K}+\sigma_n^2\mathbf{I}\right)^{-1}\mathbf{f} \\ \sigma_*^2(\Theta) &= k(\Theta,\Theta) - \mathbf{k}^{\mathrm{T}}\left(\mathbf{K}+\sigma_n^2\mathbf{I}\right)^{-1}\mathbf{k}. \end{aligned} \tag{11}$$

Note that in Equation (10), the surrogate model with GP returns the posterior distribution of $f(\Theta)$ rather than its scalar value.

Acquisition function. Given the probabilistic surrogate model, as shown in Algorithm 1, the acquisition function is updated for optimization of model parameters and for determining the model parameters for the next sampling. The acquisition function considers the tradeoff between exploitation (i.e., sampling regions with small values of the objective function) and exploration (i.e., sampling regions with high uncertainty). The expected improvement (EI) criterion [61] is used as the acquisition function, which is indicated to be effective theoretically and empirically [62,63]. For our model calibration scenario, given known sampling results $\mathcal{Q}$, the EI function $\alpha(\Theta)$ is defined as the expected amount of improvement the new model parameters can yield:

$$\alpha(\Theta) = \mathbb{E}\left(\max\{0, f(\Theta^+) - f(\Theta)\}\right) = \int_{-\infty}^{+\infty} \max\{0, f(\Theta^+) - f(\Theta)\} p(f|\mathcal{Q},\Theta) df, \tag{12}$$

where $f(\Theta^+)$ denotes the smallest value of objective function given current sampling results $\mathcal{Q}$. When $p(f|\mathcal{Q},\Theta)$ is defined as GP, the EI function has an analytic expression:

$$\alpha(\Theta) = \left(\mu(\Theta) - f(\Theta^+)\right)\Phi(Z) + \sigma_i(\Theta)\phi(Z), \tag{13}$$

where $\Phi(\cdot)$ and $\phi(\cdot)$ are the cumulated distribution function and the distribution function of standard normal distribution, and

$$Z = \frac{f(\Theta^+) - f(\Theta)}{\sigma_i(\Theta)}. \tag{14}$$

Unlike the objective function evaluation that includes numerical modeling, $\alpha(\Theta)$ only includes simple algebra calculations, such as $\mu(\Theta)$, $\Phi(Z)$, and $\phi(Z)$, and thus is inexpensive to estimate. Therefore, the gradient-based optimization can be used to maximize $\alpha(\Theta)$ in Algorithm 1, which can derive model parameters for the next sample efficiently.

2.3. Numerical Model

Contaminant transport model. The fate and transport of a contaminant (hexavalent chromium) in 3D transient groundwater flow systems are modeled by the Fickian advection-dispersion transport equation with retardation [64]:

$$\vartheta R\frac{\partial c}{\partial t} = \nabla\cdot(\vartheta D\nabla c) - \nabla\cdot(qc) + q_s c_s, \tag{15}$$

where c is the contaminant concentration (ML^{-3}), t is the time (T^{-1}), R is the retardation factor, ϑ is the porosity, D is the dispersion coefficient tensor (L^2T^{-1}), q is the Darcy velocity (LT^{-1}), q_s is the volume flow rate per unit volume of fluid sink/source (T^{-1}), c_s is the contaminant concentration of fluid sink/source (T^{-1}).

The transport of contaminant through advection in porous media, represented by Darcy velocity q in Equation (15), is related to the groundwater flow equation by Darcy's Law:

$$q = -K\nabla h, \tag{16}$$

where K is hydraulic conductivity tensor (LT^{-1}) and h is the hydraulic head (L).

Groundwater flow model. The hydraulic head in the 3D groundwater flow system is modeled by the groundwater flow equation:

$$S_s \frac{\partial h}{\partial t} = \nabla \cdot (K \nabla h) + q_s, \tag{17}$$

where S_s is the storage of aquifer (T^{-1}).

The finite difference method is adopted and discretizes and solves the above dynamic equation for groundwater flow and contaminant transport. In the finite difference method, the central difference scheme is used for spatial discretization, whereas the Crank-Nicolson scheme difference is used for time discretization. We developed a C++ program to implement the finite difference method, analogous to the implementations of MODFLOW [65] and MT3DMS [8].

3. Study Area

We applied the proposed model calibration method to simulate the transport of Cr(VI) in the groundwater flow system. Among the heavy metals, chromium, especially in its VI oxidation state, is extremely toxic and carcinogenic and has been added to the priority pollutant list by the United States Environmental Protection Agency (US EPA). The development of an accurate numerical model can provide in-depth insight into the migration and fate of Cr(VI) and guide post-remediation in the study area.

The study area is a Cr(VI)-contaminated site located in a city in Hunan province, south China (Figure 2). The region has a humid subtropical climate with an annual average temperature of 17.5 °C and an average annual precipitation of nearly 1400 mm. The site, with an area of 0.59 km^2 was originally an abandoned ferroalloy plant that produced 2000 tons of chromium metals annually. The production and processing of chromium lasted more than half a century at the site. It is estimated that overall 200,000 tons of chromium-bearing slags and tailings were produced during this period of history. Large amounts of chromium-bearing slags and tailings were directly dumped on the ground around the plant, leading to the leakage of chromium into the soil and groundwater.

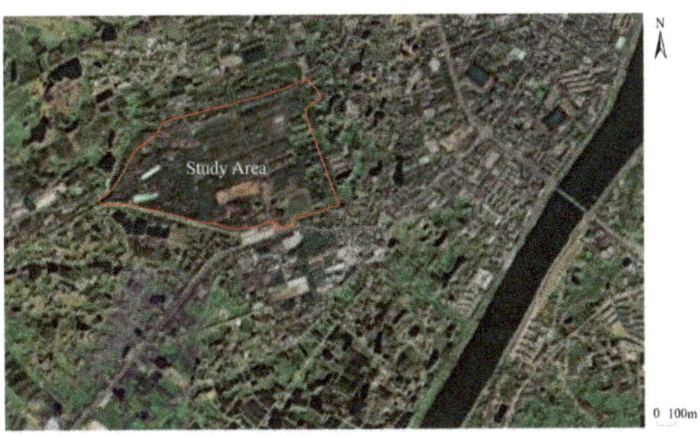

Figure 2. Arial image of the study area.

The Cr(VI)-contaminated site is located on an alluvial stratum with fairly flat terrain. According to the engineering geology survey of the site, the alluvial stratum comprises four layers, including mixed fills, silty clays, sandy gravels, and weathered argillaceous siltstones from surface to depth. Table 1 summarizes the settings for the four layers. The layers of sandy gravels, which contain the confined pore water, are treated as a confined aquifer. The aquifer is characterized by weak permeability at its top and bottom boundaries, which can be regarded as an aquiclude due to the low vertical permeability. The major aquifer recharge comes mainly from the lateral flow of groundwater and rainfall. The recharge is considered to be uniform over the boundary, and lateral flow is roughly from northwest to southeast across the site.

Table 1. Characteristics of soil layers in the study area.

Layer	Thickness	Water Permeability
Mixed fill	3~4 m	Strong permeability
Silty clay	2~4 m	Weak permeability
Sandy gravel	3~4 m	Medium permeability
Weathered argillaceous siltstone	9~10 m	Weak permeability

In terms of the environmental survey, Cr(VI), Zn(II), Mn(II), Sn (III), Pb(II), Cd(II), and Na(I) have been detected in the groundwater. The concentration of Cr(VI) (maximum mg L^{-1}) significantly exceeds the referenced environmental limit (1.5 mg L^{-1}). Figure 3 shows the overserved Cr(VI) plume in 3D, which is generated by interpolation of the concertation of Cr(VI). The high concentration of Cr(VI) chromium (0.0005 mg L^{-1}~5.006 mg L^{-1}) is located in the central north of the site and extends from 5 m to 15 m beneath the surface.

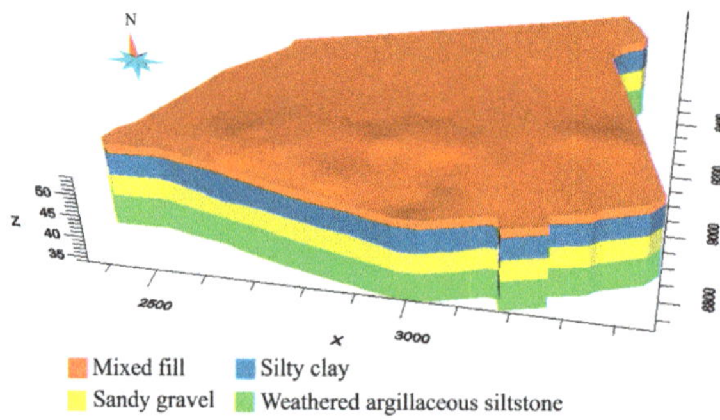

Figure 3. A 3D model of the study area.

4. Results

4.1. Setup

In our experiments, two cases are set up to investigate the effectiveness of the proposed method. The first one is a hypothetical case for testing the accuracy of model calibration and, especially, the identifiability of the hypothetical "true" values of model parameters. The second one is a real case for testing the fitness between simulated values and observations acquired in real fields.

Before carrying out model calibration in the hypothetical and real cases, we reconstruct 3D models of the site (Figure 4) using drill-hole data at the site. The Cr(VI)-contaminated aquifer is discretized and modeled by a high-resolution finite-difference grid comprising $120 \times 150 \times 60 = 1,080,000$ grids x, y, and z dimensions (Figure 4). Every grid is 10 m × 10 m × 0.5 m in size.

To simulate the groundwater flow, the boundary conditions for the flow model are shown in Figure 5. The northwest boundary condition specifies flux along the northwest boundaries, which represents the recharge of the lateral flow from the northwest of the site. The southeast boundary condition is a specified-head boundary that is set to a uniform constant. According to the hydrological survey for the site, a uniform recharge rate (3.0×10^{-7} ms^{-1}) is set for the northwest boundary, and 48 m is used as the water head for the outflow boundary (southeast boundary) according to the river stage in the vicinity of the site. The top boundary condition is also a flux-specific boundary condition to represent the recharge of rainfall at rates of 140 cm yr^{-1} according to the annual precipitation of the region. The bottom boundary condition is a no-flow boundary since the argillaceous siltstones at the bottom of the aquifer can be considered impermeable. To simulate the transport of Cr (VI), the boundary conditions for the transport model are set to zero mass flux over all boundaries except the outflow boundary (i.e., the southeast boundary), where the mass flux is determined by the flow rate of the outflow through the boundary. The Cr(VI) concentration illustrated in Figure 6 is used as the initial condition for transport simulation.

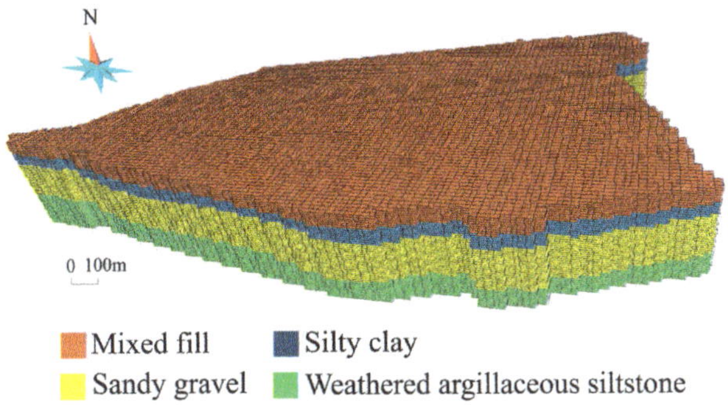

Figure 4. The 3D finite differential grids of the study area.

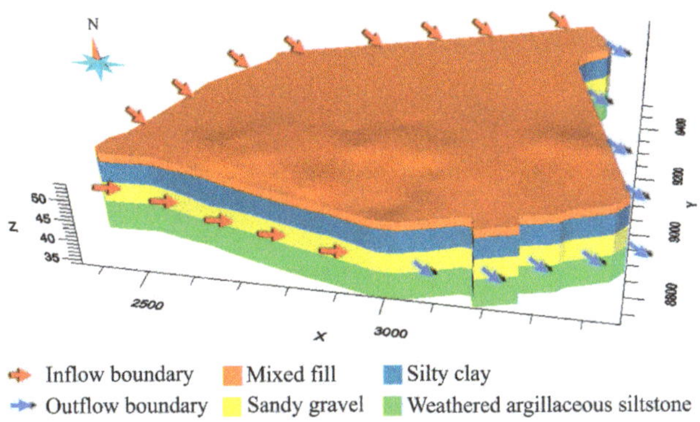

Figure 5. Boundary conditions for the groundwater flow model for the study area.

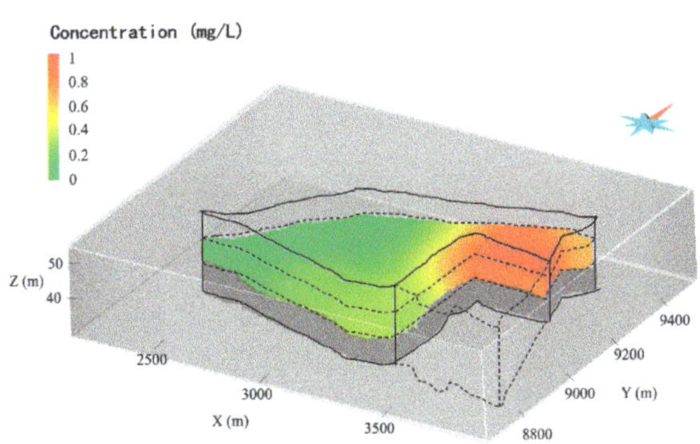

Figure 6. Initial concentration in the aquifer.

We used the BO method to calibrate the parameters of groundwater flow and Cr(VI) transport models for the study area. Overall, six model parameters involved in Equations (15) and (17) were taken for model calibration, and their feasible ranges are listed in Table 2. We implemented the calibration method with Python, in which numerical simulations are called with a function implemented in C++ for the evaluation of objective functions. The BO method was run on an ordinary PC equipped with an Intel 3.6 GHz CPU, and only one computing thread was used.

Table 2. Model parameters.

Model Parameter	Value
Releasing water (m^{-1})	0.00×10^{-3}~1.00×10^{-3}
Effective porosity	0.10~0.60
Dispersity ($m^2\ s^{-1}$)	0.00×10^{-3}~1.00×10^{-3}
Retardation factor	1.00~5.00
Horizontal permeability ($m\ s^{-1}$)	6.90×10^{-5}~8.30×10^{-5}
Vertical permeability ($m\ s^{-1}$)	6.90×10^{-7}~8.30×10^{-7}

In the following, we present the results of model calibration in the hypothetical case and the real case.

4.2. *Hypothetical Case*

Since real model parameters for the entire site are impossible to acquire, we set up a hypothetical case to validate the performance of the proposed method. Here, the "true" values of model parameters were specified beforehand, which are listed in Table 3. The "observations" were synthesized by the numerical model with the predefined model parameters (Table 3). To mimic the system stochastics and the measurement errors, Gaussian noises were added to the synthesized observations. The Gaussian noises are Gaussian distributed with a zero mean and the specified standard deviation (0.1 mg L^{-1} was used if not stated otherwise). The synthesized data consists of observations at 21 monitoring wells (dots in Figure 7) in 32 weeks.

Table 3. Target parameters.

Model Parameter	Value
Releasing water (m^{-1})	$1.0\,0 \times 10^{-4}$
Effective porosity	0.30
Dispersity ($m^2\ s^{-1}$)	1.00×10^{-4}
Retardation factor	2.00
Horizontal permeability ($m\ s^{-1}$)	6.97×10^{-10}
Vertical permeability ($m\ s^{-1}$)	4.37×10^{-12}

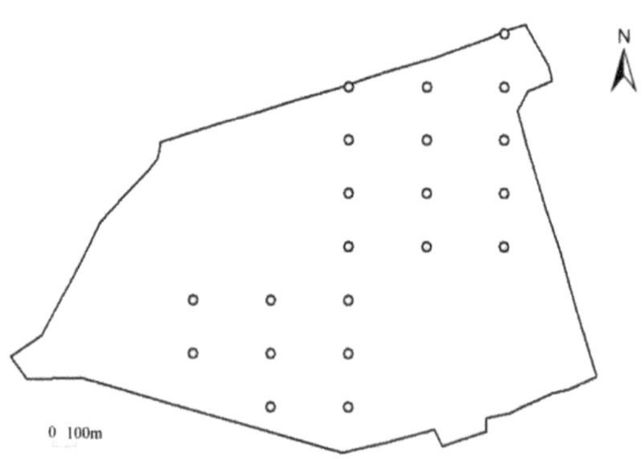

Figure 7. Spatial distribution of hypothetical monitoring wells.

Given the hypothetical observations, the BO method was used to optimize the objective function in Equation (3). Here, we set the maximum number of iterations to 200, which includes 200 times of numerical modeling. The overall model calibration process took 146 min. Figure 8 shows the variation of objective function values during the BO process. It is shown that the values of the objective function fluctuate significantly during iterations, reflecting that the BO process uses the acquisition function to probe the parameter space where the values of the objective function are relatively uncertain while exploiting the accumulated information to search for the minimum of the objective function. With the progression of BO iterations, the known minima of objective functions decrease substantially and gradually converge to zero (Table 4). The objective function considerably decreases after only 8 iterations (from an initial value >2000 to a value of 191.93). The BO process further searches for the minimum afterwards and finds the minimum (value = 1.03) at the 150th iteration.

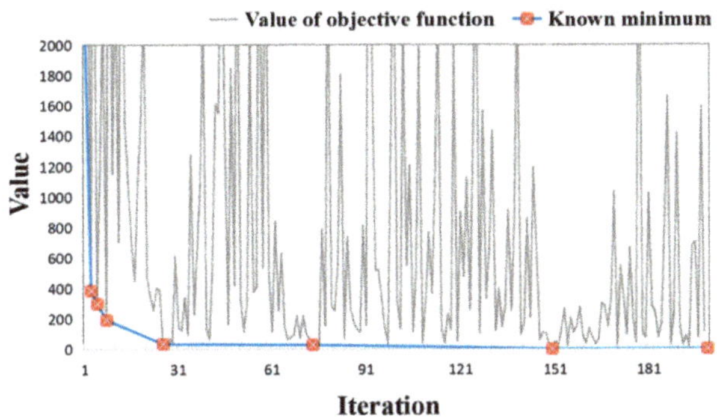

Figure 8. Variation of objective function values and the associated known minima during the BO process in hypothetical case.

Table 4. Statistics of iterations and minima when a better solution of the objective function minimum is found.

Iteration	**1**	**3**	**5**	**8**	**26**	**74**	**150**
Minimum	2284.01	383.84	297.55	191.93	34.18	28.09	1.31

To further investigate the model calibration accuracy, we compare the difference between simulated and observed concentrations. Especially, scatters of simulated concentrations versus observed concentrations, along with the 1:1 line and the associated 95% confidence interval lines, are visualized. To quantify the model calibration accuracy, the mean squared error (MAE) and the coefficient of determination (R^2) are used as the metrics, which are defined according to the notations in Equation (4):

$$MAE = \frac{1}{N}\sum_{i=1}^{N}\left|M_i(X,\Theta) - \tilde{c}_i\right|, \tag{18}$$

$$R^2 = \frac{\left(\sum_{i=1}^{N}\left(M_i(X,\Theta) - \bar{M}\right)\left(\tilde{c}_i - \bar{c}\right)\right)^2}{\sum_{i=1}^{N}\left(M_i(X,\Theta) - \bar{M}\right)^2 \sum_{i=1}^{N}\left(\tilde{c}_i - \bar{c}\right)^2}, \tag{19}$$

where $\bar{M}$ and $\bar{c}$ are the mean of simulated concentrations and observed concentrations, respectively.

Figure 8 illustrates the evolution of model performance during the BO iterations, in which the rows correspond to the iterations (except the first one) in Table 4 that find better solutions than the known minima. Analogous to the values of the objective function in

Figure 9, the simulation errors are remarkably decreased, which demonstrates the ability of the BO to eliminate the simulation errors for the contaminant transport model.

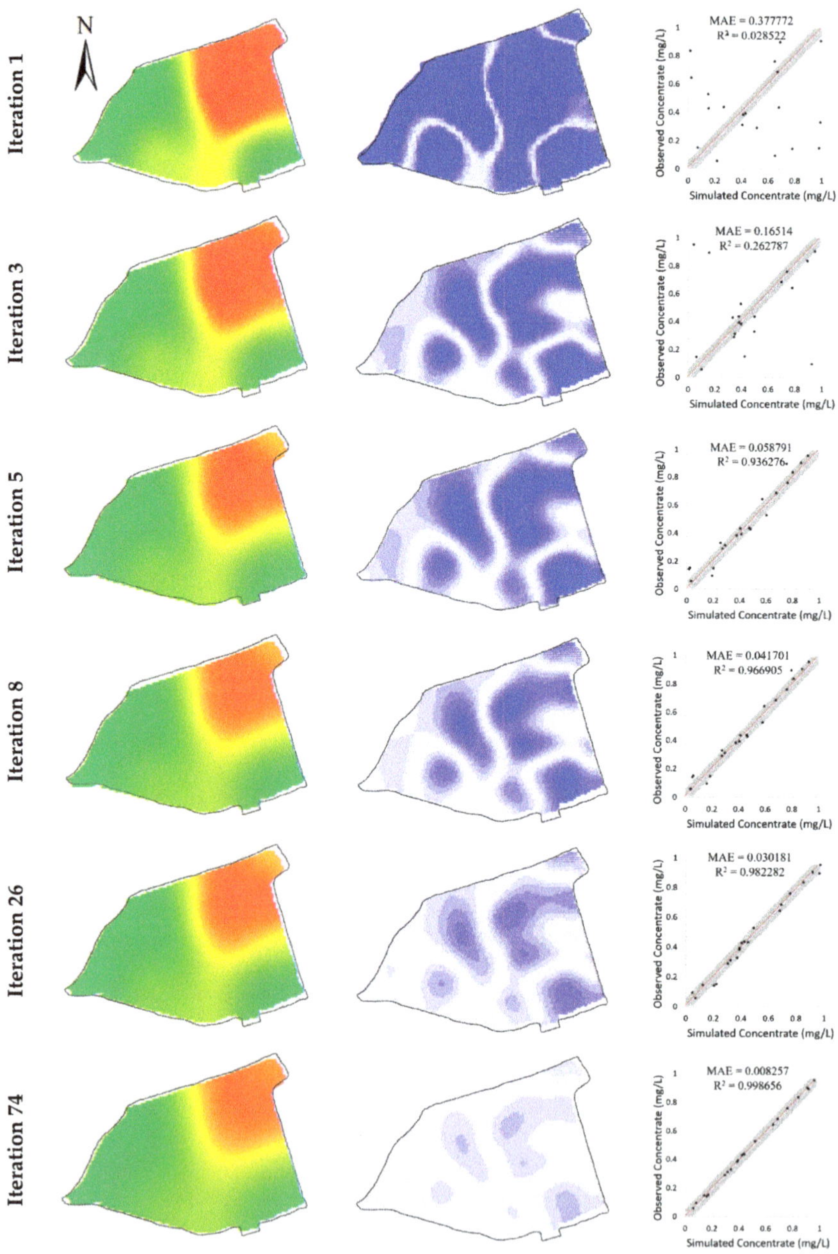

Figure 9. *Cont.*

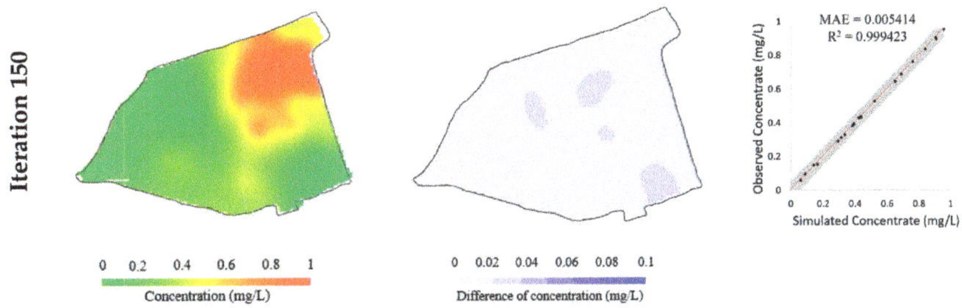

Figure 9. Evolution of simulated concentrations (**left column**), their differences from observed concentrations (**middle column**), and scatters of simulated versus observed concentrations (**right column**) during the BO process in hypothetical case.

Figure 10 illustrates the variations of model parameters during the BO process, which further shows, by probing the objective function, how the model parameter space is explored. As observed in Figure 10, with the accumulated evaluations of objective functions, the values of parameters gradually approach their "true" values. This demonstrates that, given a limited number of objective function evaluations, the BO method can reasonably identify the model parameters for a relatively ideal contaminant transport system.

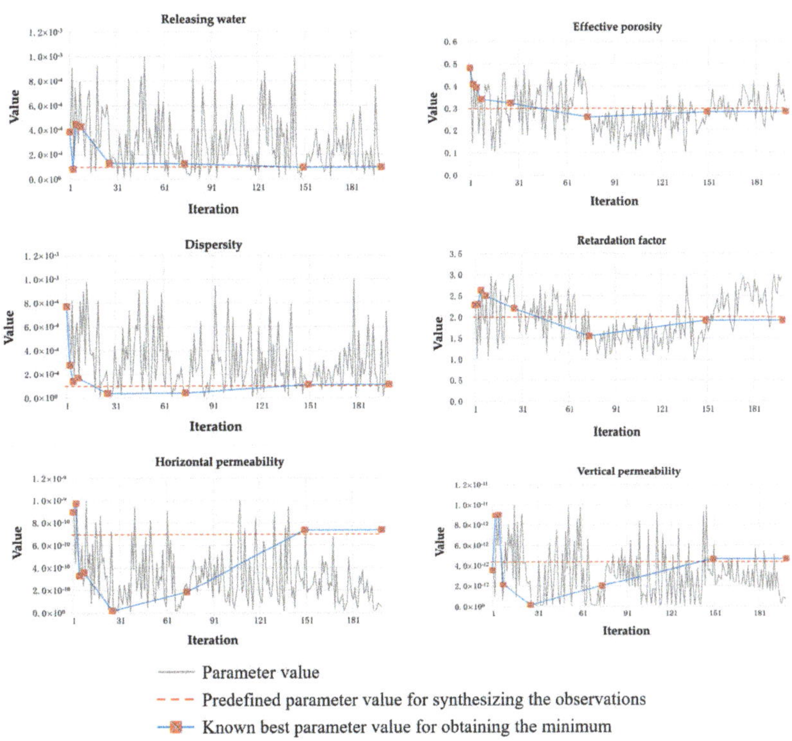

Figure 10. Variations model parameters during the BO process.

To validate the performance of the BO method in response to the observations, we designed an ablation study and a robustness study. In the ablation study, we removed the observations at the 13 monitoring wells in the north of the site and tested the impact on the model calibration. Figure 11 and Table 5 show that the simulation errors evidently increase for the removed 13 monitoring wells, whereas there is much less increase for the remaining 8 monitoring wells as calibration targets. This reflects that the BO method can be adaptive to the calibration targets. Table 5 summarizes the calibrated model parameters after removing the observations at the 13 monitoring wells. It is shown that the accuracy of the identified model parameters is slightly impacted, reflecting that the BO method is flexible to meet the set of observations. Note that the accuracy of horizontal and vertical permeability is evidently different. This may be attributed to the fact that the major transport mechanism of Cr(VI) in this study is advection, which is dominated by permeability. Given that the given concentrations at 8 monitoring wells in the south are much lower than the concentrations at 13 monitoring wells in the north, it is reasonable that the calibrated permeability is much deviated from its "true" values after removing the observations in the south. Thus, the calibrated model parameters suggest the ability of BO to explore the parameter space and fit the calibration targets.

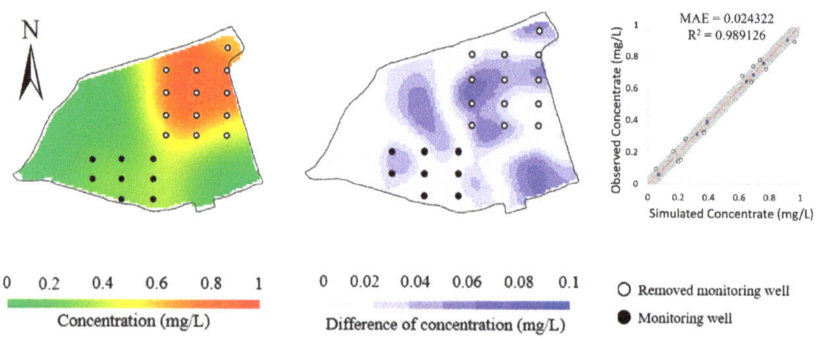

Figure 11. Simulated concentrations (**left column**), their differences from observed concentrations (**middle column**), and scatters of simulated versus observed concentrations (**right column**) obtained by removing the 13 monitoring wells in the north of the study area.

Table 5. Comparison of the model parameters identified with partial data and the full data shown in Figure 7. The partial data, compared with the full data, only include observations at eight monitoring wells in the south.

Model Parameter	Partial Data	Full Data	Target Parameters
Releasing water (m^{-1})	9.00×10^{-5}	1.00×10^{-4}	1.00×10^{-4}
Effective porosity	0.31	0.28	0.30
Dispersity ($m^2\ s^{-1}$)	1.10×10^{-4}	1.10×10^{-4}	1.00×10^{-4}
Retardation factor	1.90	1.92	2.00
Horizontal permeability ($m\ s^{-1}$)	3.58×10^{-10}	7.35×10^{-10}	6.97×10^{-10}
Vertical permeability ($m\ s^{-1}$)	1.82×10^{-12}	6.07×10^{-12}	4.37×10^{-12}
Objective function value	9.87	1.31	-

The robustness study is designed to validate the robustness of the BO method to measurement error and other unknown system interference. Here, we enhanced the Gaussian noises added to the hypothetical observations. By using different standard deviations of Gaussian noise, four datasets of hypothetical observations were generated. Figure 12 illustrates the simulation errors of the four calibrated models. It is observed that, with the increase in standard deviation, the simulation errors increase slightly, demonstrating that the BO method is able to obtain an acceptable calibrated model even given the calibration targets generated from an unideal contaminant transport system. This is further verified by the inversed model parameters listed in Table 6, which show the inversed model parameters are kept close to the "true" model parameters.

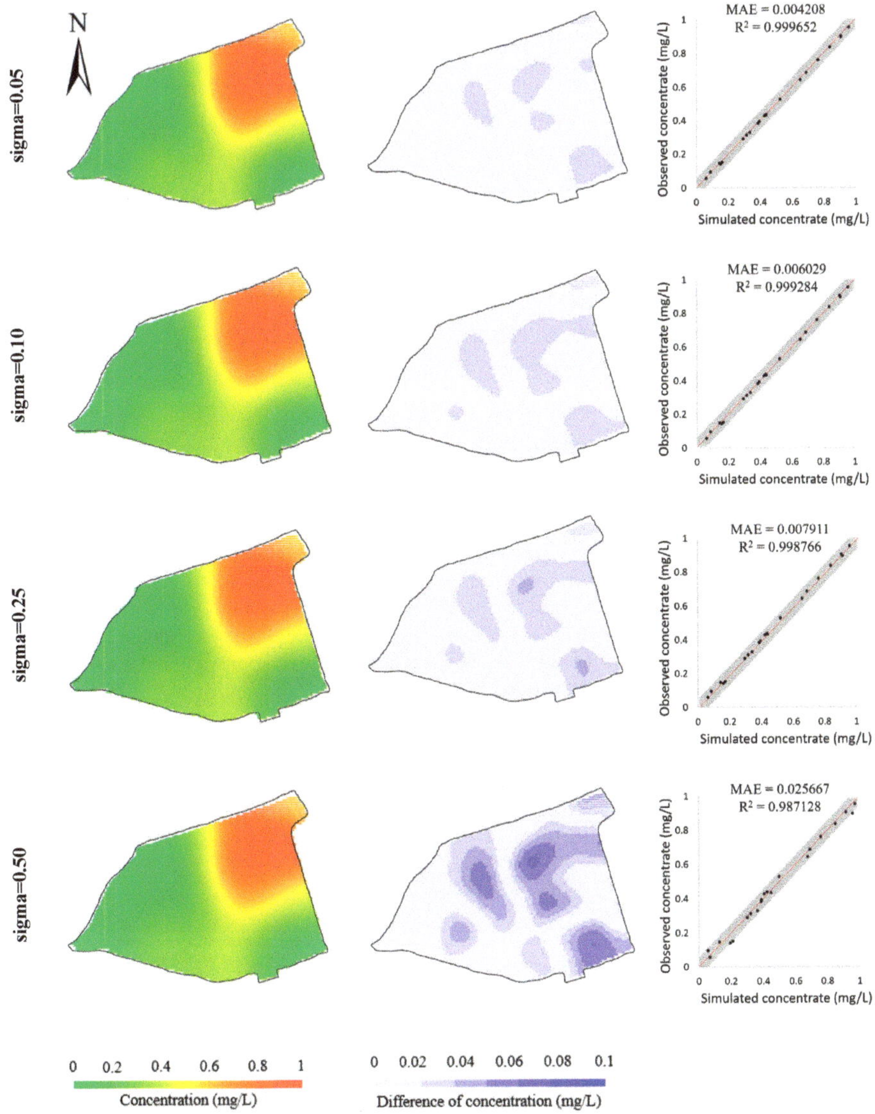

Figure 12. Comparisons of model calibration resulting from the calibration targets with different levels of Gaussian noise (Sigma/mgL^{-1}: standard deviation of the Gaussian noise added to calibration targets). Simulated concentrations (**left column**); differences from observed concentrations (**middle column**); scatters of simulated concentrations versus observed concentrations (**right column**) in hypothetical case.

Table 6. Calibrated model parameters resulting from calibration targets with different levels of noise (sigma/mgL^{-1} : standard deviation of the Gaussian noise added to calibration targets).

Model Parameter	Sigma = 0.05	Sigma = 0.10	Sigma = 0.25	Sigma = 0.50	Target Parameters
Releasing water (m^{-1})	1.10×10^{-4}	1.00×10^{-4}	1.10×10^{-4}	1.10×10^{-4}	1.00×10^{-4}
Effective porosity	0.30	0.31	0.30	0.31	0.30
Dispersity (m^2 s^{-1})	1.00×10^{-4}	1.20×10^{-4}	1.00×10^{-4}	1.00×10^{-4}	1.00×10^{-4}
Retardation factor	1.96	1.98	1.95	2.08	2.00
Horizontal permeability (m s^{-1})	5.71×10^{-10}	5.88×10^{-10}	7.84×10^{-10}	6.02×10^{-10}	6.97×10^{-10}
Vertical permeability (m s^{-1})	4.80×10^{-12}	6.06×10^{-12}	3.98×10^{-12}	4.87×10^{-12}	4.37×10^{-12}
Objective function value	1.79	2.64	1.63	1.92	-

To validate the efficiency of the BO method in model calibration, we compared it with Particle Swarm Optimization (PSO) [34], a well-known and effective model calibration method in the hydrological field due to its flexibility, ease of implementation, and efficiency [66,67]. The PSO method is a stochastic, population-based, global optimization technique inspired by swarm behavior observed in nature, such as fish and bird schooling. The number of particles is a significant hyperparameter in PSO. Following the setting in [34], an excessive number of particles (=40) was used for the efficiency of highly-parameterized model calibration. Here, an update of a particle's position would lead to a numerical modeling process. To compare the efficiency between the PSO and BO methods, we stop the PSO iteration when the minimum of the objective function is close to or smaller than the minimum found by the BO method. Figure 13 illustrates the variations in objective function values during the PSO iterations and the BO iterations. It is noted that the PSO method took overall 1520 objective function evaluations (1109 min versus 146 min resulting from the BO method) to gain an objective function value (1.70) close to the minimum (1.31) of objective function found by the BO method, which is much more than 200 evaluations in the BO method. Figure 14 illustrates the model calibration performance of the PSO method after 200 objective function evaluations, which is the same number of objective function evaluations as that of the BO method. Compared with the results of the BO method (bottom row of Figure 9), the simulation error of the PSO method is significantly higher after the same number of objective function evaluations, which demonstrates the efficiency of the BO method in the calibration of Cr(VI) transport models.

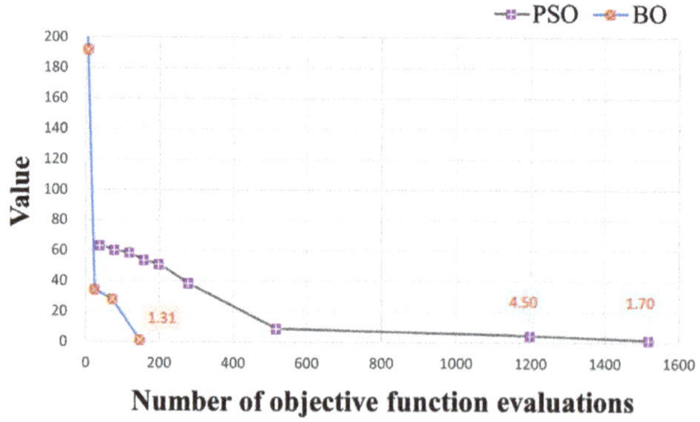

Figure 13. Comparison of objective function values and the associated known minima during the BO and the PSO iteration processes in hypothetical case.

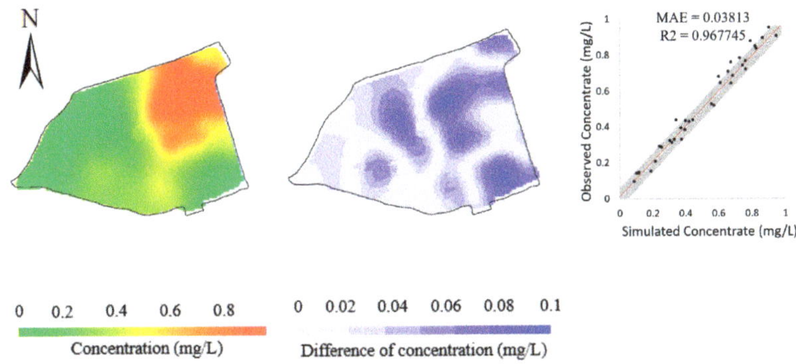

Figure 14. Simulated concentrations (**left column**), the differences from observed concentrations (**middle column**), and scatters of simulated versus observed concentrations (**right column**) resulting from the PSO method after 200 objective function evaluations.

4.3. Real Case

While the BO method attains promising performance in the hypothetical cases, we further verify its effectiveness and reliability in a real case. These real data come from chromium slag sites in Hunan Province, China. Here, the Cr(VI) concentrations measured at 53 monitoring wells were set as the calibration targets. Since time series observations at these 53 bores are withheld by the stakeholder at the site, only the observations obtained seven months after the time the simulation starts are used for model calibration. The model calibration took 148 min. Figure 15 shows the variations of objective functions during the model calibration process, whereas Table 7 lists the values of objective functions at the iterations when a better solution is found. Analogous to the results in Figure 8, the values of the optimization function are minimized dramatically within two decades of iterations, which took only 15 min in the experiment. The minimum is found in the 153rd iteration, taking a computational time of 111 min. Figure 16 illustrates the evolution of simulated concentrations during the BO iterations, which shows that the simulation errors are considerably minimized during the BO process. Thus, the numerical model calibrated by the BO method is promising to fit the observations in the real-field scenario.

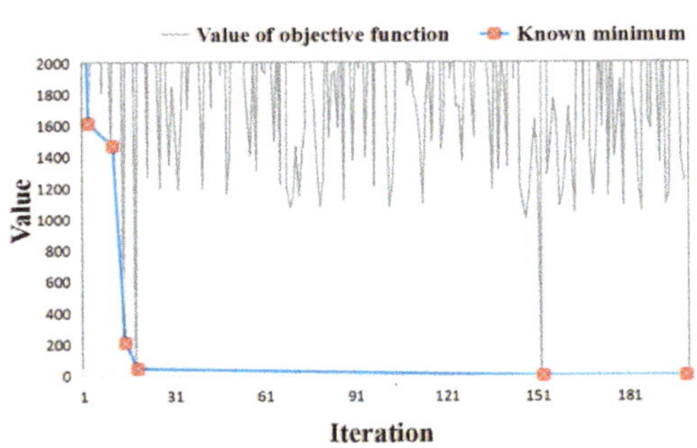

Figure 15. Objective function variation curve of the BO method process.

Table 7. Descending process of the optimal objective function.

Iteration	1	3	11	15	19	153
Known minimum	2975.10	1614.30	1470.65	210.01	46.03	1.58

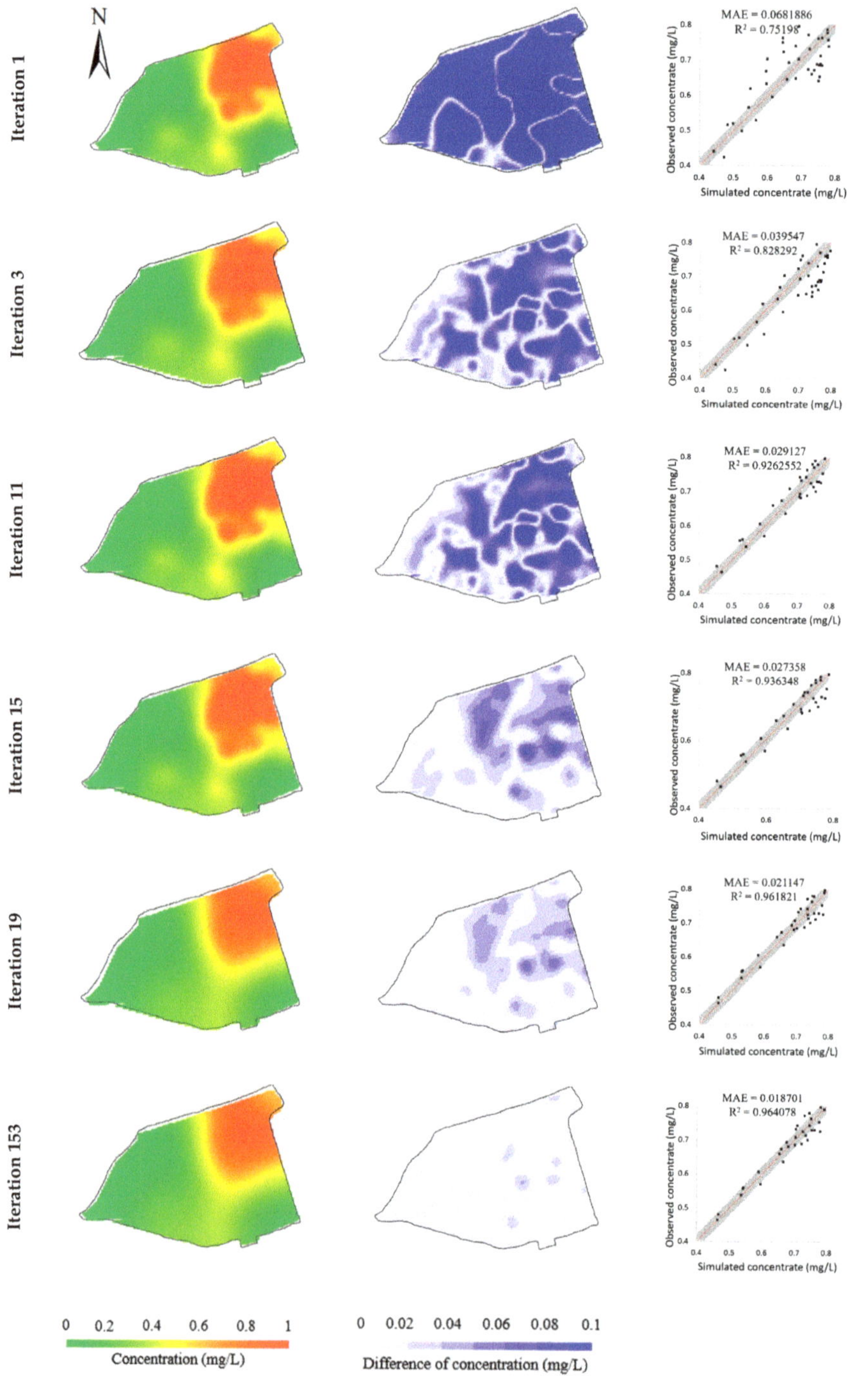

Figure 16. Evolution of simulated concentrations (**left column**), their differences from observed concentrations (**middle column**), and scatters of simulated versus observed concentrations (**right column**) during the BO process in real case.

Figure 16 illustrates the evolution of simulated concentrations during the BO iterations, which shows that the simulation errors are gradually minimized during the optimization process. When the optimization reaches the minimum, the simulated concentration is within the 95% confidence interval, which indicates the calibrated model is promising to fit the observations.

We further conducted the ablation study and the robustness study to verify the effectiveness and adaptivity of the BO. In the ablation study, we removed the observations at 23 monitoring wells in the north and only left the observations at 20 monitoring wells in the south as calibration targets. Figure 17 shows the model calibration performance. The simulation errors at the remaining 30 monitoring wells involved in model calibration are clearly smaller than those at the 23 removed monitoring wells. Combining the simulation error in Figure 16, it is indicated that the BO method is faithful to the input calibration targets and is capable of fitting the observations.

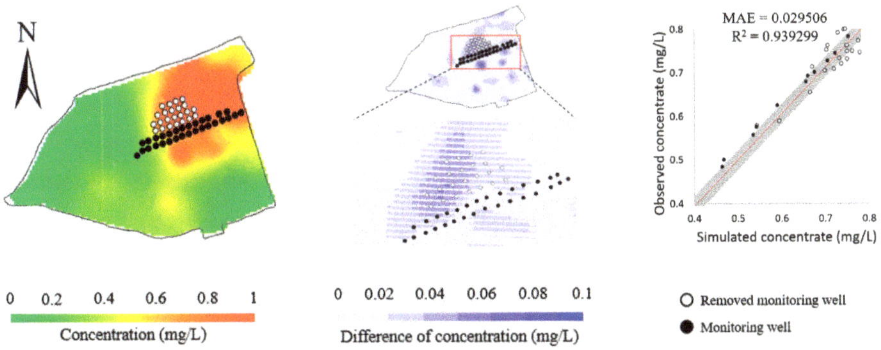

Figure 17. Simulated concentrations (**left column**), their differences from observed concentrations (**middle column**), and scatters of simulated versus observed concentrations (**right column**) obtained by removing the 23 monitoring wells in the north of the study area.

In the robustness study, similarly to the hypothetical case, we added different levels of Gaussian noise to the observations. Overall, four standard deviations were used to generate the Gaussian noise. As shown in Figure 18, while the simulation errors are elevated due to the increase in Gaussian noises, it can be seen that the BO method is less sensitive to noises added to observations. Even though the noise added to the observations increases to 0.5, the MAE of model calibration remains approximately 0.03. This reflects that the BO method is robust enough to handle observation errors and possibly other zero-mean system errors in real-field applications. The observations errors can essentially affect the values of the objective function. In the BO method, the objective function is approximated by a surrogate model that explicitly considers noise-corrupted objective functions. Thus, it is reasonable that the model calibration results are robust to the noise of the objective function in practice.

The BO method is further compared with the PSO method in the real case. Here, the setting of the PSO method is the same as that used in the hypothetical case. The PSO took 1720 object function evaluations (1255.6 min) to converge to an objective function value of 1.21 (Figure 19), the same order of magnitude as the objective function value (1.58) obtained by the BO method. When the number of numerical models is 200, the BO method evidently outperforms the PSO method (Figure 20). Combining the comparison results in both the hypothetic case and the real case, it is demonstrated that the BO method is evidently efficient in model calibration. In contrast to the stochastic PSO method, which is unaware of the structure of the objective function [68], the BO method uses the surrogate model to explicitly approximate the objective function and adopts an acquisition function to systematically search for the minimum objective function. Therefore, the BO method can efficiently search for the minimum of the objective function with fewer objective function evaluations.

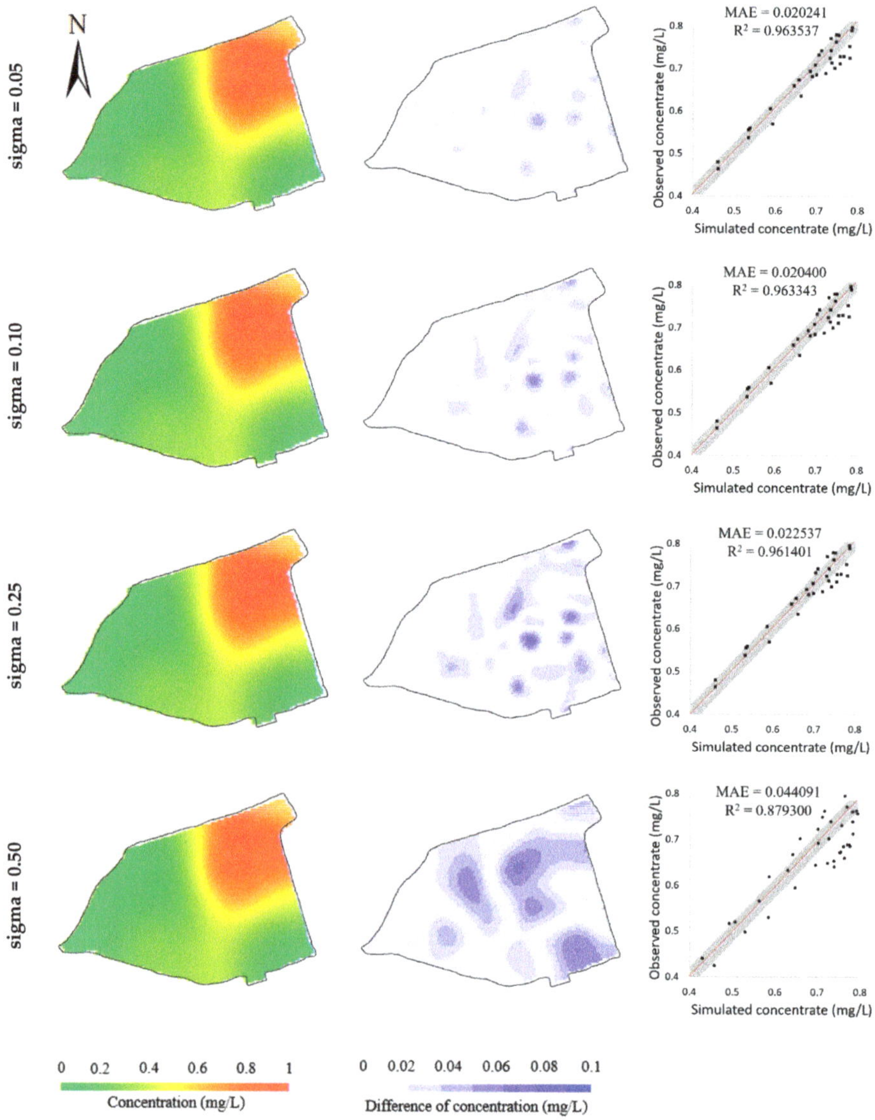

Figure 18. Comparisons of model calibration resulting from the calibration targets with different levels of Gaussian noise (Sigma/mgL^{-1}: standard deviation of the Gaussian noise added to calibration targets). Simulated concentrations (**left column**); differences from observed concentrations (**middle column**); scatters of simulated concentrations versus observed concentrations (**right column**) in real case.

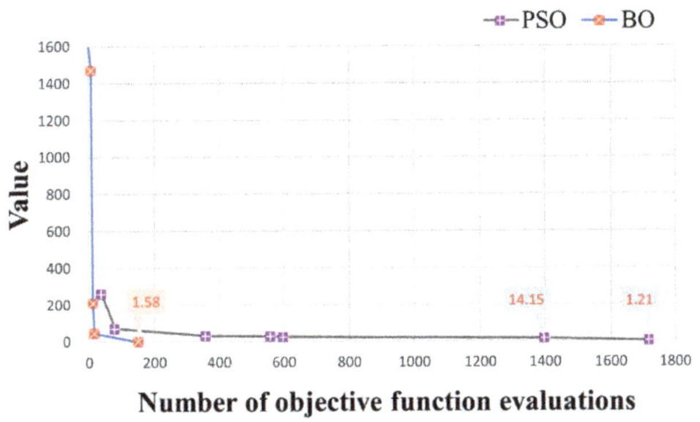

Figure 19. Comparison of objective function values and the associated known minima during the BO and the PSO iteration processes in real case.

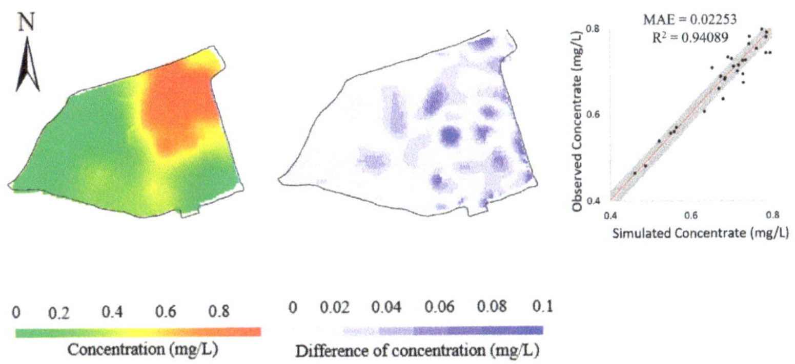

Figure 20. Simulated concentrations (**left column**), the differences from observed concentrations (**middle column**), and scatters of simulated versus observed concentrations (**right column**) resulting from PSO after 200 object function evaluations.

5. Conclusions

In this paper, we propose a BO method for the calibration of 3D numerical models of groundwater flow and contaminant transport. A Bayes model is built to fully represent calibration criteria and derive the objective function for model calibration. In contrast to the use of common optimization approaches for model calibration, the computationally expensive objective function is approximated by a computationally efficient probabilistic surrogate model in BO. An effective acquisition function is adopted to search for the most promising model parameters that facilitate minimizing the objective function while exploring the highly uncertain region of the objective function. This enables finding the optimized model parameters effectively and efficiently by minimizing the objective function with a small number of objective function evaluations.

Two case studies of model calibration for Cr(VI) transport simulation, which involve the *hypothetical case* and the *real case*, demonstrate that, as the BO process proceeds, the BO method is effective in identifying hypothetical model parameters, minimizing the objective function, and adapting different calibration criteria. The promising performance of the BO method is achieved within 200 objective function evaluations, which took 111 min as reported on an ordinary PC. Compared with the PSO model calibration method, the BO method shows superior efficiency in both the *hypothetical* and *real case*.

In the future, we will further validate the performance of the BO method in more challenging scenarios and expand its applicability and flexibility. The performance of the

BO method motivates the use of high-resolution numerical models for long-term and high-fidelity simulation. Due to the evidently lower number of objective function evaluations in BO, the high-resolution numerical simulations that were originally computationally prohibitive in the model calibration can now be allowed in applications. Another avenue is to scale the BO method to numerical models with hundreds of model parameters. This would allow us to comprehensively consider the spatial heterogeneity of model parameters in the numerical models. These potential improvements can further enforce the fitness of observations and result in more convincing numerical models for the simulation of contaminant transport in the groundwater flow system.

Author Contributions: Conceptualization, H.D. and Y.H.; methodology, H.D. and S.Z.; software, S.Z. and Z.L.; validation, H.D., S.Z. and Z.L.; formal analysis, H.D. and S.Z.; investigation, Y.H.; resources, Y.H. and Y.Z.; data curation, Y.H. and Y.Z.; writing—original draft preparation, S.Z.; writing—review and editing, H.D. and S.Z.; visualization, S.Z.; supervision, X.M.; project administration, Y.Z.; funding acquisition, Y.Z. All authors have read and agreed to the published version of the manuscript.

Funding: This research was funded by the National Key R&D Program of China (grant number 2019YFC1805905) and the National Natural Science Foundation of China (grant numbers 41972309, 42030809, 42272344 and 42072318).

Institutional Review Board Statement: Not applicable.

Informed Consent Statement: Not applicable.

Data Availability Statement: Not applicable.

Acknowledgments: We thank Wei Lou and Xing Li in Hunan HIKEE Environmental Technology CO., Ltd., Changsha for their assistance during the field investigation.

Conflicts of Interest: The authors declare no conflict of interest.

References

1. Mackay, D.M.; Roberts, P.V.; Cherry, J.A. Transport of organic contaminants in groundwater. *Environ. Sci. Technol.* **1985**, *19*, 384–392. [CrossRef] [PubMed]
2. He, Q.; He, Y.; Hu, H.-P.; Lou, W.; Zhang, Z.; Zhang, K.-N.; Chen, Y.-G.; Ye, W.-M.; Sun, J. Laboratory investigation on the retention performance of a soil–bentonite mixture used as an engineered barrier: Insight into the effects of ionic strength and associated heavy metal ions. *Environ. Sci. Pollut. Res.* **2023**, *30*, 50162–50173. [CrossRef] [PubMed]
3. United States Environmental Protection Agency; Office of Solid Waste, Emergency Response, & Environmental Management Support (Firm). *Cleaning Up the Nation's Waste Sites: Markets and Technology Trend*; US Environmental Protection Agency, Office of Solid Waste and Emergency Response: Washington, DC, USA, 1997.
4. He, Y.; Hu, G.; Wu, D.-Y.; Zhu, K.-F.; Zhang, K.-N. Contaminant migration and the retention behavior of a laterite–bentonite mixture engineered barrier in a landfill. *J. Environ. Manag.* **2022**, *304*, 114–338. [CrossRef] [PubMed]
5. Gorelick, S.M.; Voss, C.I.; Gill, P.E.; Murray, W.; Saunders, M.A.; Wright, M.H. Aquifer Reclamation Design: The Use of Contaminant Transport Simulation Combined with Nonlinear Programing. *Water Resour. Res.* **1984**, *20*, 415–427. [CrossRef]
6. Burnett, R.D.; Frind, E.O. Simulation of contaminant transport in three dimensions: 2. Dimensionality effects. *Water Resour. Res.* **1987**, *23*, 695–705. [CrossRef]
7. Konikow, L.F.; Goode, D.J.; Hornberger, G.Z. *A Three-Dimensional Method-of-Characteristics Solute-Transport Model (MOC3D)*; US Department of the Interior: Washington, DC, USA; US Geological Survey: Reston, VA, USA, 1996; Volume 96.
8. Zheng, C.; Wang, P.P. *MT3DMS: A Modular Three-Dimensional Multispecies Transport Model for Simulation of Advection, Dispersion, and Chemical Reactions of Contaminants in Groundwater Systems Documentation and User's Guide*; US Army Corps of Engineer: Washington, DC, USA, 1999.
9. Greskowiak, J.; Hay, M.B.; Prommer, H.; Liu, C.; Post, V.E.A.; Ma, R.; Davis, J.A.; Zheng, C.; Zachara, J.M. Simulating adsorption of U(VI) under transient groundwater flow and hydrochemistry: Physical versus chemical nonequilibrium model. *Water Resour. Res.* **2011**, *47*, 118–132. [CrossRef]
10. Steefel, C.I.; MacQuarrie, K.T. Approaches to modeling of reactive transport in porous media. *React. Transp. Porous Media* **2018**, *34*, 83–130.
11. Seyedpour, S.; Kirmizakis, P.; Brennan, P.; Doherty, R.; Ricken, T. Optimal remediation design and simulation of groundwater flow coupled to contaminant transport using genetic algorithm and radial point collocation method (RPCM). *Sci. Total Environ.* **2019**, *669*, 389–399. [CrossRef]

12. Locatelli, L.; Binning, P.J.; Sanchez-Vila, X.; Søndergaard, G.L.; Rosenberg, L.; Bjerg, P.L. A simple contaminant fate and transport modelling tool for management and risk assessment of groundwater pollution from contaminated sites. *J. Contam. Hydrol.* **2019**, *221*, 35–49. [CrossRef]
13. Masum, S.A.; Zhang, Z.; Tian, G.; Sultana, M. Three-dimensional fully coupled hydro-mechanical-chemical model for solute transport under mechanical and osmotic loading conditions. *Environ. Sci. Pollut. Res.* **2023**, *30*, 5983–6000. [CrossRef]
14. Doherty, J. PEST: A Unique Computer Program for Model-independent Parameter Optimisation. In *Water Down Under 94: Groundwater/Surface Hydrology Common Interest Papers*; Preprints of Papers: Groundwater/Surface Hydrology Common Interest Papers; Preprints of Papers; Institution of Engineers, Australia: Barton, Australia, 1994.
15. Doherty, J. Ground Water Model Calibration Using Pilot Points and Regularization. *Ground Water* **2003**, *41*, 170–177. [CrossRef] [PubMed]
16. Solomatine, D.P.; Dibike, Y.B.; Kukuric, N. Automatic calibration of groundwater models using global optimization techniques. *Hydrol. Sci. J.* **1999**, *44*, 879–894. [CrossRef]
17. Mugunthan, P.; Shoemaker, C.A.; Regis, R.G. Comparison of function approximation, heuristic, and derivative-based methods for automatic calibration of computationally expensive groundwater bioremediation models. *Water Resour. Res.* **2005**, *41*, 134–152. [CrossRef]
18. Hill, M.C.; Tiedeman, C.R. *Effective Groundwater Model Calibration*; John Wiley & Sons: Hoboken, NJ, USA, 2007.
19. Shoemaker, C.A.; Regis, R.G.; Fleming, R.C. Watershed calibration using multistart local optimization and evolutionary optimization with radial basis function approximation. *Hydrol. Sci. J.* **2007**, *52*, 450–465. [CrossRef]
20. Shafii, M.; De Smedt, F. Multi-objective calibration of a distributed hydrological model (WetSpa) using a genetic algorithm. *Hydrol. Earth Syst. Sci.* **2009**, *13*, 2137–2149. [CrossRef]
21. Tang, G.; D'azevedo, E.F.; Zhang, F.; Parker, J.C.; Watson, D.B.; Jardine, P.M. Application of a hybrid MPI/OpenMP approach for parallel groundwater model calibration using multi-core computers. *Comput. Geosci.* **2010**, *36*, 1451–1460. [CrossRef]
22. Gaganis, P.; Smith, L. A Bayesian approach to the quantification of the effect of model error on the predictions of groundwater models. *Water Resour. Res.* **2001**, *37*, 2309–2322. [CrossRef]
23. Xu, T.; Valocchi, A.J.; Ye, M.; Liang, F. Quantifying model structural error: Efficient Bayesian calibration of a regional groundwater flow model using surrogates and a data-driven error model. *Water Resour. Res.* **2017**, *53*, 4084–4105. [CrossRef]
24. Pang, M.; Shoemaker, C.A.; Bindel, D. Early termination strategies with asynchronous parallel optimization in application to automatic calibration of groundwater PDE models. *Environ. Model. Softw.* **2022**, *147*, 105–237. [CrossRef]
25. Dai, Z.; Samper, J. Inverse problem of multicomponent reactive chemical transport in porous media: Formulation and applications. *Water Resour. Res.* **2004**, *40*, 407–425. [CrossRef]
26. EL Harrouni, K.; Ouazar, D.; Wrobel, L.; Cheng, A.-D. Groundwater parameter estimation by optimization and DRBEM. *Eng. Anal. Bound. Elements* **1997**, *19*, 97–103. [CrossRef]
27. Lin, A.C.; Yeh, W.W.G. Identification of parameters in an inhomogenous aquifer by the use of the maximum priciple of optimal control and quasi-linearization. *Water Resour. Res.* **1974**, *10*, 829–838. [CrossRef]
28. Poeter, E.P.; Hill, M.C.; Lu, D.; Tiedeman, C.R.; Mehl, S. UCODE_2014, with new capabilities to define parameters unique to predictions, calculate weights using simulated values, estimate parameters with SVD, evaluate uncertainty with MCMC, and more. *Integr. Groundw. Model. Cent. Rep. Number GWMI* **2022**, *2*, 2014.
29. Adams, B.M.; Bohnhoff, W.J.; Dalbey, K.R.; Ebeida, M.S.; Eddy, J.P.; Eldred, M.S.; Hooper, R.W.; Hough, P.D.; Hu, K.T.; Jakeman, J.D.; et al. *Dakota, A Multilevel Parallel Object-Oriented Framework for Design Optimization, Parameter Estimation 2020, Uncertainty Quantification, and Sensitivity Analysis: Version 6.13 User's Manual (No. SAND2020-12495)*; Sandia National Lab. (SNL-NM): Albuquerque, NM, USA, 2020.
30. Giacobbo, F.; Marseguerra, M.; Zio, E. Solving the inverse problem of parameter estimation by genetic algorithms: The case of a groundwater contaminant transport model. *Ann. Nucl. Energy* **2002**, *29*, 967–981. [CrossRef]
31. Zheng, C.; Wang, P. Parameter structure identification using tabu search and simulated annealing. *Adv. Water Resour.* **1996**, *19*, 215–224. [CrossRef]
32. Duan, Q.Y.; Gupta, V.K.; Sorooshian, S. Shuffled complex evolution approach for effective and efficient global minimization. *J. Optim. Theory Appl.* **1993**, *76*, 501–521. [CrossRef]
33. Braak, C.J.T. A Markov Chain Monte Carlo version of the genetic algorithm Differential Evolution: Easy Bayesian computing for real parameter spaces. *Stat. Comput.* **2006**, *16*, 239–249. [CrossRef]
34. Zambrano-Bigiarini, M.; Rojas, R. A model-independent Particle Swarm Optimisation software for model calibration. *Environ. Model. Softw.* **2013**, *43*, 5–25. [CrossRef]
35. Wang, Z.; Lu, W.; Chang, Z.; Wang, H. Simultaneous identification of groundwater contaminant source and simulation model parameters based on an ensemble Kalman filter—Adaptive step length ant colony optimization algorithm. *J. Hydrol.* **2022**, *605*, 127–352. [CrossRef]
36. Laloy, E.; Rogiers, B.; Vrugt, J.A.; Mallants, D.; Jacques, D. Efficient posterior exploration of a high-dimensional groundwater model from two-stage Markov chain Monte Carlo simulation and polynomial chaos expansion. *Water Resour. Res.* **2013**, *49*, 2664–2682. [CrossRef]
37. Xu, T.; Valocchi, A.J. A Bayesian approach to improved calibration and prediction of groundwater models with structural error. *Water Resour. Res.* **2015**, *51*, 9290–9311. [CrossRef]

38. Xu, T.; Valocchi, A.J.; Ye, M.; Liang, F.; Lin, Y.F. Bayesian calibration of groundwater models with input data uncertainty. *Water Resour. Res.* **2017**, *53*, 3224–3245. [CrossRef]
39. Zhang, J.; Man, J.; Lin, G.; Wu, L.; Zeng, L. Inverse modeling of hydrologic systems with adaptive multifidelity Markov chain Monte Carlo simulations. *Water Resour. Res.* **2018**, *54*, 4867–4886. [CrossRef]
40. Sun, X.; Zeng, X.; Wu, J.; Wang, D. A Two-Stage Bayesian Data-Driven Method to Improve Model Prediction. *Water Resour. Res.* **2021**, *57*, e2021WR030436. [CrossRef]
41. Yang, H.Q.; Zhang, L.; Xue, J.; Zhang, J.; Li, X. Unsaturated soil slope characterization with Karhunen–Loève and polynomial chaos via Bayesian approach. *Eng. Comput.* **2019**, *35*, 337–350. [CrossRef]
42. Vrugt, J.A.; Ter Braak, C.J.; Clark, M.P.; Hyman, J.M.; Robinson, B.A. Treatment of input uncertainty in hydrologic modeling: Doing hydrology backward with Markov chain Monte Carlo simulation. *Water Resour. Res.* **2008**, *44*, 720–735. [CrossRef]
43. Vrugt, J.A.; Ter Braak, C.J.F. DREAM (D): An adaptive Markov Chain Monte Carlo simulation algorithm to solve discrete, noncontinuous, and combinatorial posterior parameter estimation problems. *Hydrol. Earth Syst. Sci.* **2011**, *15*, 3701–3713. [CrossRef]
44. Chen, M.; Izady, A.; Abdalla, O.A. An efficient surrogate-based simulation-optimization method for calibrating a regional MODFLOW model. *J. Hydrol.* **2017**, *544*, 591–603. [CrossRef]
45. Kourakos, G.; Mantoglou, A. Pumping optimization of coastal aquifers based on evolutionary algorithms and surrogate modular neural network models. *Adv. Water Resour.* **2009**, *32*, 507–521. [CrossRef]
46. Stone, N. Gaussian Process Emulators for Uncertainty Analysis in Groundwater Flow. Ph.D. Thesis, University of Nottingham, Nottingham, UK, 2011.
47. Garcet, J.D.P.; Ordonez, A.; Roosen, J.; Vanclooster, M. Metamodelling: Theory, concepts and application to nitrate leaching modelling. *Ecol. Model.* **2006**, *193*, 629–644. [CrossRef]
48. Regis, R.G.; Shoemaker, C.A. Constrained global optimization of expensive black box functions using radial basis functions. *J. Glob. Optim.* **2005**, *31*, 153–171. [CrossRef]
49. Yoon, H.; Jun, S.C.; Hyun, Y.; Bae, G.O.; Lee, K.K. A comparative study of artificial neural networks and support vector machines for predicting groundwater levels in a coastal aquifer. *J. Hydrol.* **2011**, *396*, 128–138. [CrossRef]
50. Fallah-Mehdipour, E.; Haddad, O.B.; Mariño, M.A. Prediction and simulation of monthly groundwater levels by genetic programming. *J. Hydro Environ. Res.* **2013**, *7*, 253–260. [CrossRef]
51. Hoffman, M.W.; Shahriari, B.; Freitas, N.D. On correlation and budget constraints in model-based bandit optimization with application to automatic machine learning. In Proceedings of the Seventeenth International Conference on Artificial Intelligence and Statistics, Reykjavik, Iceland, 22–25 April 2014; pp. 365–374.
52. Zhang, Y.; Sohn, K.; Villegas, R.; Pan, G.; Lee, H. Improving object detection with deep convolutional networks via Bayesian optimization and structured prediction. In Proceedings of the IEEE Conference on Computer Vision and Pattern Recognition (CVPR), Boston, MA, USA, 7–12 June 2015; pp. 249–258.
53. Frazier, P.I.; Wang, J. Bayesian optimization for materials design. In *Information Science for Materials Discovery and Design*; Springer International Publishing: Cham, Switzerland, 2015.
54. Vanchinathan, H.P.; Nikolic, I.; Bona, F.D.; Krause, A. Explore-Exploit in top-n recommender systems via Gaussian processes. In Proceedings of the 8th ACM Conference on Recommender systems, Foster City, CA, USA, 6–10 October 2014; Volume 31.
55. Nadarajah, S. A generalized normal distribution. *J. Appl. Stat.* **2005**, *32*, 685–694. [CrossRef]
56. Chen, Y.-C. A tutorial on kernel density estimation and recent advances. *Biostat. Epidemiology* **2017**, *1*, 161–187. [CrossRef]
57. Frazier, P.I. A tutorial on Bayesian optimization. *arXiv* **2018**, arXiv:1807.02811.
58. Pelikan, M.; Goldberg, D.E.; Cantú-Paz, E. BOA: The Bayesian optimization algorithm. In Proceedings of the genetic and evolutionary computation conference GECCO-99, Orlando, FL, USA, 13–17 July 1999; Volume 1, pp. 525–532.
59. Snoek, J.; Larochelle, H.; Adams, R.P. Practical bayesian optimization of machine learning algorithms. *Adv. Neural Inf. Process. Syst.* **2012**, *25*, 1–9.
60. Williams, C.K.; Rasmussen, C.E. *Gaussian Processes for Machine Learning*; MIT Press: Cambridge MA, USA, 2006.
61. Mockus, J. The application of Bayesian methods for seeking the extremum. *Towards Glob. Optim.* **1998**, *2*, 117.
62. Bull, A.D. Convergence rates of efficient global optimization algorithms. *J. Mach. Learn. Res.* **2011**, *12*, 2879–2904.
63. Snoek, J.R. Bayesian Optimization and Semiparametric Models with Applications to Assistive Technology. Ph.D. Thesis, University of Toronto, Toronto, Canada, 2013.
64. Zheng, C.; Bennett, G.D. *Applied Contaminant Transport Modeling*; Wiley-Interscience: New York, NY, USA, 2002.
65. Harbaugh, A.W.; Banta, E.R.; Hill, M.C.; McDonald, M.G. *Modflow-2000, the U. S. Geological Survey Modular Ground-Water Model-User Guide to Modularization Concepts and the Ground-Water Flow Process*; USGS: Reston, VA, USA, 2000.
66. Kennedy, J.; Eberhart, R. Particle swarm optimization. In Proceedings of the ICNN'95—International Conference on Neural Networks, Perth, WA, Australia, 27 November–1 December 1995; Volume 4, pp. 1942–1948.

67. Poli, R. Analysis of the Publications on the Applications of Particle Swarm Optimisation. *J. Artif. Evol. Appl.* **2008**, *2008*, 685175. [CrossRef]
68. Asher, M.J.; Croke, B.F.W.; Jakeman, A.J.; Peeters, L.J. A review of surrogate models and their application to groundwater modeling. *Water Resour. Res.* **2015**, *51*, 5957–5973. [CrossRef]

toxics

MDPI

Article

Water Management Impacts on Chromium Behavior and Uptake by Rice in Paddy Soil with High Geological Background Values

Zeting Guan [1], Ran Wei [1], Ting Liu [1,*], Jingjing Li [1], Ming Ao [2], Shengsheng Sun [2], Tenghaobo Deng [3], Shizhong Wang [2], Yetao Tang [2], Qingqi Lin [1], Zhuobiao Ni [1] and Rongliang Qiu [1,2]

[1] Guangdong Laboratory for Lingnan Modern Agriculture, Guangdong Provincial Key Laboratory of Agricultural & Rural Pollution Abatement and Environmental Safety, College of Natural Resources and Environment, South China Agricultural University, Guangzhou 510642, China

[2] Guangdong Provincial Key Laboratory of Environmental Pollution Control and Remediation, School of Environmental Science and Engineering, Sun Yat-sen University, Guangzhou 510006, China

[3] Institute of Quality Standard and Monitoring Technology for Agro-Products of Guangdong Academy of Agricultural Sciences, Guangzhou 510640, China

* Correspondence: tingliu@scau.edu.cn

Abstract: Chromium (Cr) is an expression toxic metal and is seriously released into the soil environment due to its extensive use and mining. Basalt is an important Cr reservoir in the terrestrial environment. Cr in paddy soil can be enriched by chemical weathering. Therefore, basalt-derived paddy soils contain extremely high concentrations of Cr and can enter the human body through the food chain. However, the water management conditions' effect on the transformation of Cr in basalt-derived paddy soil with high geological background values was less recognized. In this study, a pot experiment was conducted to investigate the effects of different water management treatments on the migration and transformation of Cr in a soil–rice system at different rice growth stages. Two water management treatments of continuous flooding (CF) and alternative wet and dry (AWD) and four different rice growth stages were set up. The results showed that AWD treatment significantly reduced the biomass of rice and promoted the absorption of Cr in rice plants. During the four growth periods, the root, stem and leaf of rice increased from 11.24–16.11 mg kg^{-1}, 0.66–1.56 mg kg^{-1} and 0.48–2.29 mg kg^{-1} to 12.43–22.60 mg kg^{-1}, 0.98–3.31 mg kg^{-1} and 0.58–2.86 mg kg^{-1}, respectively. The Cr concentration in roots, stems and leaves of AWD treatment was 40%, 89% and 25% higher than CF treatment in the filling stage, respectively. The AWD treatment also facilitated the potential bioactive fractions conversion to the bioavailable fraction, compared with the CF treatment. In addition, the enrichment of iron-reducing bacteria and sulfate-reducing bacteria with AWD treatment also provided electron iron for the mobilization of Cr, thus affecting the migration and transformation of Cr in the soil. We speculated that the reason for this phenomenon may be the bioavailability of Cr was affected by the biogeochemical cycle of iron under the influence of alternating redox. This indicates that AWD treatment may bring certain environmental risks in contaminated paddy soil with high geological background, and it is necessary to be aware of this risk when using water-saving irrigation to plant rice.

Keywords: chromium; paddy soil; rice; water management; high geological background

Citation: Guan, Z.; Wei, R.; Liu, T.; Li, J.; Ao, M.; Sun, S.; Deng, T.; Wang, S.; Tang, Y.; Lin, Q.; et al. Water Management Impacts on Chromium Behavior and Uptake by Rice in Paddy Soil with High Geological Background Values. *Toxics* **2023**, *11*, 433. https://doi.org/10.3390/toxics11050433

Academic Editors: Junhao Qin, Peidong Su, Feng Zhu and Lin Ding

Received: 2 April 2023
Revised: 22 April 2023
Accepted: 28 April 2023
Published: 5 May 2023

1. Introduction

Heavy metals-contaminated agricultural soil has become a global and widespread issue [1,2], which poses a serious threat to food security. Chromium (Cr) is one of the most phytotoxic metals even at trace levels, which is released into agricultural soils through anthropogenic and geological backgrounds [3]. Cr can easily enter the food chain, and its excessive accumulation in soil–plant systems poses a risk to human health [3]. In many countries, the main natural source of Cr in agricultural soils is weathering from Cr-rich ultramafic salts (e.g., serpentine), such as France, the United States, Brazil and New

Caledonia [4–7]. In China, the average content of Cr in arable land soil is 78.94 mg kg^{-1}, significantly higher than the background content 57.30 mg kg^{-1}. About 1.39% of arable land grain production has a high risk of Cr pollution [8]. According to the screening and control values of soil pollution risk on agricultural land, as specified in the "Soil Environmental Quality Soil Pollution Risk Control Standard (Trial)" (GB15618-2018) [9], the standard Cr concentration in paddy soil limit is 250 mg kg^{-1} ($pH \leq 5.5$, $5 < pH \leq 6.5$), 300 mg kg^{-1} ($6.5 < pH \leq 7.5$) and 350 mg kg^{-1} ($pH > 7.5$), depending on the soil pH, respectively. In basalt watersheds (such as the Yunnan and Leiqiong areas in China), the background value of Cr exceeds the national standard of China. Rice (*Oryza sativa* L.) is a staple food crop for half of the world's population and 60% of China's residents [10,11]. Therefore, it is necessary to focus on the Cr behavior in the paddy-rice system.

Rice is mostly grown under continuous flooding (CF) conditions, but due to climate change and limited development, the agronomic measure of alternative wet and dry (AWD) has been derived for the lack of water in the dry season [12]. AWD is the aerobic and anaerobic environment that manipulates rice by alternating unflooded and flooded conditions during the growing season and is an alternative to traditional continuous flooding management [13,14], which is considered to be an effective rice cultivation method. Water management in rice planting has greatly affected the redox environment of paddy soil, which can affect the bioavailability of arsenic in paddy soil and its accumulation in rice [15,16]. Similarly, as a variable valence element, Cr may lead to dynamic redox state changes under different water management conditions, thus affecting the absorption of Cr by plants [17]. It has been reported that Cr(VI) is reduced by 69–71.8% under continuous flooding conditions(CF), while only 33.3–38.6% under intermittent flooding conditions(AWD) [17]. Different chemical forms of Cr in soil can produce different biogeochemical behaviors, which is crucial for assessing the hazards of Cr pollution in soil [18,19]. Cr in soil contains exchangeable, carbonate binding, oxidizable, reducible and residual forms of Cr [20]. Oxidizable Cr and reducible Cr can also be subdivided into Fe/Mn oxides-bound fraction, organic matter-bound fraction, amorphous iron oxides-bound fraction and crystalline iron oxides bound fraction [21]. Among them, the exchangeable fraction and carbonate binding fraction of Cr are the bioavailable mobile components, and the oxidizable fraction and the reducible fraction are the potential mobile components. Although Cr in residual fraction cannot be directly absorbed by plants, it can be activated and released through hydrolysis, oxidation and reduction [22,23]. The availability and morphological changes of Cr in soil systems are also related to soil properties such as iron oxides [20,24]. Iron has a variety of oxidation and reducing states, so it can be an electron donor or acceptor in soil, and it mainly plays a role in reducing and adsorbing Cr in the environment [25]. It has been reported that Cr may be precipitated in the form of $Cr_xFe_{1-x}(OH)_3$ in soil [26]. Thus, the reductive dissolution of Fe(III) oxides promotes the release of structural binding Cr [27].

Microorganisms play an important role in the immobilization of heavy metals because they can promote the redox process of iron and affect the migration of metals [28]. For example, iron-reducing bacteria (FeRB) promote iron reduction to release related metals [29], so FeRB activity may affect the mobility and absorption of Cr [30]. Microorganisms also play a crucial role in the transformation and migration of Cr. Iron and sulfur-reducing bacteria are dominant in paddy soil [31]. *Bacillus* and *Anaeromyxobacter* have shown to be associated with Fe-Cr reaction [32,33]. Some scholars have pointed out that the reduction of Cr was essentially related to the microbial-mediated Fe(II)/Fe(II) cycle [34]. Understanding the role of rhizosphere microorganisms in Cr transformation and bioavailability is necessary to alleviate Cr pollution in rice. Therefore, it is necessary to determine the effect of water management on the composition and activity of rhizosphere soil microbial structure community.

At present, most of the research was mainly focused on the artificially added soil with high mobility and high toxicity of Cr(VI), but there are few studies on the effect of Cr availability on natural soil with high Cr geological background. In this study, we investigated the effects of different water management conditions on (1) the effect of Cr uptake by rice; (2) the distribution of Cr and Fe fractions in rhizosphere soil at different rice growth stages; and (3) the diversity and activity of microbial structure community in the rhizosphere soil. The results provide a scientific basis for the agronomic management practices of Cr-contaminated paddy fields in high geological background areas.

2. Materials and Methods

2.1. Pot Experiment

The experimental soil was collected from a paddy field in Maichen Town, Xuwen County, Zhanjiang City, Guangdong Province. The soil texture was loamy clay (40% sand, 28% silt, 32% clay), and the total Cr content was 228 mg kg^{-1}. Other soil physical and chemical properties are shown in Table 1.

Table 1. The basic physical and chemical properties of soil.

Properties	Soil
pH	5.74
Total Cr (mg kg^{-1})	228
Total Fe (g kg^{-1})	76.84
Organic matter (g kg^{-1})	30.21
Cation exchange capacity (cmol kg^{-1})	25.16
Total N (g kg^{-1})	2.121
Soil texture	40% sand, 28% silt, and 32% clay

Pot experiments were carried out in the net room of the South China Agricultural University in Guangdong Province. We had chosen the local variety seeds of hybrid rice named Guang 8 You 305 for cultivation. Rice seeds were sterilized with 10% H_2O_2 for 15 min and then soaked in warm water at 30 °C for 24 h. Then, the soaked seeds were evenly spread on a large petri dish with wet filter paper and germinated in the dark. About 2 days later, the germinated seeds were transplanted into quartz sand matrix for culture. The Hoagland nutrient solution was used for culture, and the nutrient solution was replaced every 3 days. When the seedlings grew to 3 leaves, the rice was transplanted into the pure nutrient solution system for a week and finally transplanted into the soil pot.

The pot experiment included four growth stages (tillering stage, jointing stage, booting stage and filling stage) and two water management systems (continuous flooding (CF)) and alternating wet and dry (AWD)). Each treatment set up three repeats, a total of 24 pots. The water management methods are as follows: CF is to keep flooding to the liquid level of 5 cm during the whole growth period of rice; AWD means that the soil surface is flooded to 3 cm, and then the water is naturally dried, and the soil surface is dried for one day before flooding, so repeatedly. Each pot (diameter 25 cm, height 30 cm, bottom diameter 16.5 cm) had 5 kg soil added to it; 5 g compound fertilizer (C/N/P = 15/15/15) was added as base fertilizer; and 2 L water was added to submerge. Then, rice was transplanted on the 35th day of germination; 3 rice seedlings were transplanted in each pot.

2.2. Plants Analyses

Rice was collected at the tillering stage, jointing stage, booting stage and filling stage on the 27th, 63rd, 80th and 90th day after transplanting, respectively. The roots, stems and leaves of rice were separated, rinsed with water and then frozen in a refrigerator, followed by freeze-drying, crushing and digestion (HNO_3:$HClO_4$ = 4:1, v/v). The concentrations of total Cr and total Fe in plant samples were determined by flame atomic absorption spectrophotometer (FAAS, Z-2300, HITACHI, Tokyo, Japan).

2.3. Soil Analyses

Rhizosphere soil samples were collected with plants at different periods. One part of the collected soil was used for the determination of physical and chemical properties such as

Cr content and morphological characteristics, and the other part was stored at −80 °C in a refrigerator waiting for the determination of microbial community structure. The first part of the soil samples was dried, crushed and screened with 10 mesh and 200 mesh sieves. Cr fractions were extracted according to the sequentially extraction method proposed by Rinklebe [21]. In brief, weigh accurately 2 g soil and add the corresponding reagent in turn according to the order of extraction, then shake at room temperature (120rpm) and centrifuge (3000 rpm, centrifuge for 10 min). Finally, the filtrate was filtered with 0.45 μm membrane and stored in a 4 °C refrigerator for test. After that, the extracted soil was washed with 10 mL ultrapure water and then shaken and centrifuged at room temperature for 15 min. After the washing solution was poured out, the next extraction was performed. The fractions and extraction of Cr are as follows: exchangeable fraction (F1) was extracted with 0.2 mol L^{-1} $CaCl_2$ for 24 h; carbonate-bound fraction (F2) was extracted with 1.0 mol L^{-1} NH_4OAc (pH 6.0) for 24 h; Fe/Mn oxide-bound fraction (F3) was extracted with 0.1 mol L^{-1} NH_2OH-HCl + 1.0 mol L^{-1} NH_4OAc (pH 6.0) for 0.5 h; organic matter-bound fraction (F4) was extracted with 0.025 mol L^{-1} NH_4EDTA (pH 4.6) for 1.5 h; amorphous iron oxide-bound fraction (F5) was extracted with 0.2 mol L^{-1} ammonium oxalate (pH 3.25) for 4 h; crystalline iron oxide-bound fraction (F6) was extracted with 0.1 mol L^{-1} ascorbic acid +0.2 mol L^{-1} NH_4-oxalate (pH 3.25) and shaken in a 95 °C water bath for 0.5 h; and the residual fraction (F7) was digested with HNO_3-$HClO_4$-HF [35,36]. The Cr concentration of the above digestion solution was determined by inductively coupled plasma optical emission spectrometer (ICP-OES, 710-ES, VARIAN, Palo Alto, CA, USA).

In addition, the extraction of different fractions of iron referred to the method proposed by Yu et al. [37]. The fraction includes dissolved-Fe, hydrochloric acid extractable iron (HCl-Fe), sulfite-citric acid-sodium bicarbonate extractable iron (DCB-Fe) and amorphous iron (Oxalate-Fe). The extraction method of each component iron is as follows: add 20 mL ultrapure water (pH = 7.0) to 1 g soil and shake for 16 h, then centrifuge at 4500 rpm for 30 min, filter, and collect filtrate to determine the concentration of dissolved-Fe; add 25 mL 0.5 M HCl to 0.5 g soil, shake for 4 h, filter and collect the filtrate to determine the concentration of HCl-Fe. The extraction method of disulfite-citric acid-sodium bicarbonate (DCB) extractable Fe is by adding 1.0 g soil to a 50 mL polyethylene centrifuge tube and adding 20 mL 0.3 M trisodium citrate and 0.1 M sodium bicarbonate. The mixed solution was shaken at 120 rpm for 10 min in a 75 °C water bath, and then 1.0 g of sodium disulfite was added. After shaking for 5 min, 1.0 g of sodium disulfite was added again and then shaken for 10 min. Finally, the mixture was centrifuged at 2500 rpm for 5 min, and then the supernatant was collected for determination. Amorphous iron (oxalate-Fe) was added to 0.5 g soil in a 50 mL polyethylene centrifuge tube; 25 mL oxalic acid-ammonium oxalate buffer solution (0.2 M ammonium oxalate pH 3.0) was added then shaken at 25 °C for 4 h and then filtered to determine Fe concentration. Finally, about 0.1 g soil samples were weighed and digested with HNO_3-$HClO_4$-HF to determine the total iron concentration. The iron content in the above extract were analyzed by atomic absorption spectrometer (FAAS, Z-2300, HITACHI, Tokyo, Japan).

2.4. Microbial Diversity Analyses

The rhizosphere soil samples stored at −80 °C were used for microbial community diversity analysis. The formal PCR test used transgenic AP22-02: Trans Start Fastpfu DNA Polymerase, 20 μL reaction system.

A set of primers (forward primer: 338F (ACTCCTACGGGAGGCAGCAG) and reverse primer 806R (GGACTACHVGGGTWTCTAAT)) were used for amplification. According to the preliminary quantitative results of electrophoresis, the PCR products were detected and quantified by QuantiFluor TM-ST blue fluorescence quantitative system (Promege, Paris, France), and then the Illumina library was constructed and sequenced. The final data were analyzed on the online platform of Majorbio cloud platform (www.majorbio.com, accessed on 5 October 2022).

2.5. Quality Control and Statistical Analyses

Quality control and assurance measurements for all analytes were performed using method blanks, triplicates and certified reference materials. GBW07405(GSS-5) (Beijing, China) and GBW07443(GSF-3) (Beijing, China), obtained from the National Standard Material Centre of China, were used as the standard materials for metals and fraction of Cr analyses. Recovery rates of reference materials ranged from 91% to 106% and from 95 to 109% for total Cr and fraction of Cr in soil samples, respectively. The sums of relative proportions of the 7 fractions were in the range of 89–105%, indicating that the Cr contents of sequential extraction were in an acceptable range. GBW10020(GSB-11) served as standard sample for plant samples. Recovery rates were ranged from 95% to 101%.

The data were analyzed with IBM SPSS Statistics 25.0 (IBM Corp., Armonk, NY, USA) software. The significant differences among the different treatment means were determined by one-way ANOVA analysis at $p < 0.05$. All data are presented as means ± standard error (SE) ($n = 3$). Redundancy analysis (RDA) between microbial community samples and soil environmental factors was performed on the microbial Majorbio cloud platform (software: R language vegan package), and spearman correlation coefficient was also used for correlation Heatmap analysis (software: R (version 3.3.1) pheatmap package).

3. Results and Discusses

3.1. Biomass of Rice Plant

Dry biomass reflects the resistance of plants to different soil conditions. The dry weight of each part of plant increased with the growth of rice (Table 2). At the same time, the dry weight of each part of the rice in CF was larger than AWD. The dry weight of rice roots, stems and leaves in CF treatment increased from 1.12 g, 1.72 g and 1.82 g to 4.6 g, 23.71 g and 13.2 g, respectively. The dry weight of roots, stems and leaves treated with AWD ranged from 1.08–2.27 g, 1.5–14.0 g and 1.49–9.38 g, respectively. This indicated that AWD treatment inhibited the growth of rice. This is consistent with the results of a previous study that the grain yield, plant height and root biomass of rice under AWD conditions are lower than CF conditions [38]. Studies have shown that AWD treatment at low Cr rates may promote rice growth by increasing the content of available nitrogen and organic matter in the soil during AWD, while inhibiting rice growth at high Cr rates [17]. Thus, the growth inhibition may be due to the increase in Cr toxicity [39]. In contrast, other literature reported that AWD irrigation increased the grain yield of rice. The differences between these studies may be due to differences in the timing of irrigation methods and basic physical and chemical properties of soils.

Table 2. Biomass of rice plants in different water management at four growth stages.

Dry Weight (g)		CF	AWD
root	Tillering	1.12 ± 0.05 a	1.08 ± 0.01 a
	Jointing	2.06 ± 0.35 a	1.78 ± 0.26 b
	Booting	3.07 ± 0.42 a	1.89 ± 0.18 b
	Filling	4.60 ± 0.57 a	2.27 ± 0.38 b
stem	Tillering	1.72 ± 0.43 a	1.500 ± 0.02 a
	Jointing	6.42 ± 0.45 a	5.66 ± 0.13 b
	Booting	3.07 ± 0.42 a	10.74 ± 0.77 b
	Filling	23.71 ± 1.44 a	14.00 ± 0.85 b
leaf	Tillering	1.82 ± 0.56 a	1.49 ± 0.05 a
	Jointing	5.04 ± 0.47 a	4.67 ± 0.43 b
	Booting	5.67 ± 1.34 a	4.68 ± 0.56 b
	Filling	13.20 ± 1.05 a	9.38 ± 0.46 b

The lowercase letters indicate the difference between different treatments in the same period, $p < 0.05$. CF refers to continuous flooding. AWD refers to alternate wet and dry irrigation.

3.2. The Uptake and Translocation of Cr by Rice Plants

As shown in Figure 1, the content of Cr in rice plants under CF treatment showed root (11.24–18.91 mg kg^{-1}) > stem (0.75–1.56 mg kg^{-1}) > leaf (0.48–2.29 mg kg^{-1}), and AWD treatment also showed root (12.43–22.60 mg kg^{-1}) > stem (0.81–3.31 mg kg^{-1}) > leaf (0.58–2.86 mg kg^{-1}), similarly. This is consistent with the distribution order of Cr in different parts of most plants: root > stem > leaf [40–42]. The Cr content of rice at the tillering stage and filling stage was higher than that at the jointing stage and booting stage, and there were significant differences among different growth stages. The high Cr content of rice at the tillering stage may be due to the rapid absorption of nutrients and relatively small biomass. With the growth of the plant, the biomass of rice is no longer growing rapidly, but the accumulation of Cr in rice continues. During the filling stage, there is a drainage period and an obvious oxidation process occurs. When the environmental conditions change from anaerobic to aerobic, Fe(II) oxidation can form to Fe(III)(hydride) oxides, which are the ubiquitous adsorbents of heavy metals (such as Cr). During oxidation, a large number of protons are released and the pH is significantly reduced. The complexation of Cr(III) and Fe(III/II) with NOM may result in the formation of Cr(III)-NOM-Fe colloids that enhance the fluidity of Cr(III). In addition, oxidation of S^{2-} to SO_4^{2-} may contribute to Cr release [21].

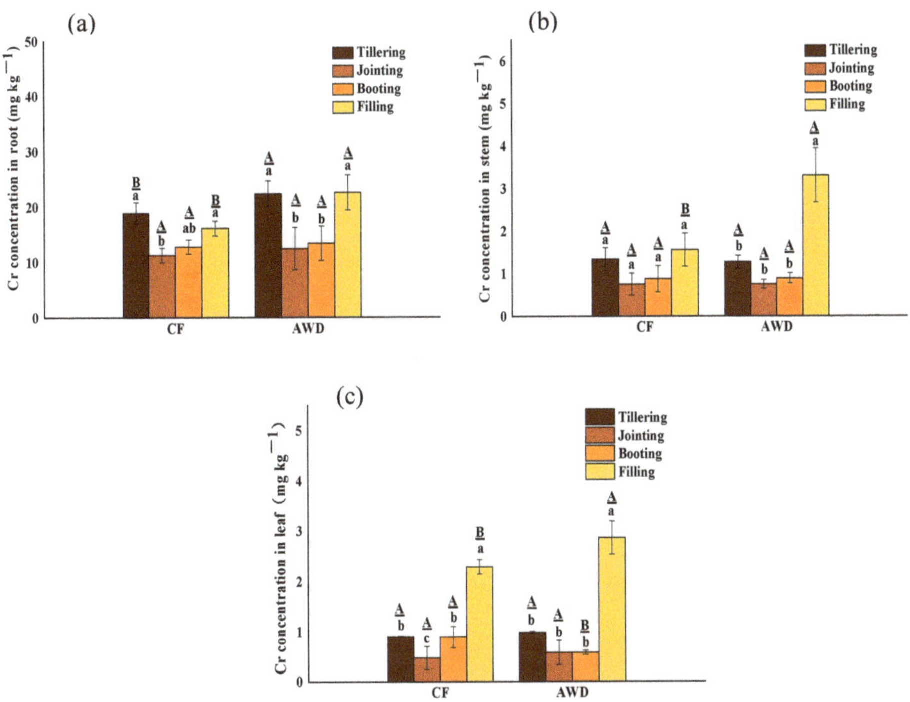

Figure 1. Effects of different water managements on Cr concentrations in roots (**a**), stems (**b**) and leaves (**c**) of rice at different growth stages. The lowercase letters indicate the differences among the same treatment in different periods; the uppercase letters indicate the differences between different treatments in the same period, $p < 0.05$.

There was a significant difference in rice tissues between different water managements. The Cr content in AWD treatment was higher than CF; the roots, stems and leaves were exceeded by 40%, 89% and 25% in the filling stage, respectively. This indicated that AWD treatment promoted the absorption and translocation of Cr by rice plants, which was consistent with the previous conclusion that AWD inhibited the biomass of rice plants. Studies had shown that irrigation has a significant effect on Cr content in rice tissues under

AWD treatment with high soil Cr concentration [17]. Compared with CF treatment, soil conditions changed alternately from reduction to oxidation during rice growth. The AWD treatment reduced the reduction rate of Cr (VI) and increased the content [43]. Therefore, despite the high Cr geological background area with low Cr(VI) content, AWD could also increase the absorption of Cr by rice. However, the situation is different between different elements. In contrast, arsenic (As) and Cr have the opposite results. For As, maintaining aerobic conditions for rice growth can significantly reduce As uptake by rice [44]. The water in the soil was usually dominated by As(V) rather than the highly soluble and toxic As(III) under oxidizing conditions [45]. The phenomenon may be affected by the transformation of heavy metal bioavailable fraction in the soil.

3.3. *Effects of Water Managements on Concentration and Fractions of Cr and Fe in Soil*

In order to better understand the mobility of Cr in paddy fields, the changes of Cr fractions in different periods and different water management were sequentially extracted (Figure 2). In general, the Cr fraction distribution is F7 (residual fraction) > F6 (crystalline iron oxidation fraction) > F5 (amorphous iron oxide fraction) > F4 (organic matter-bound fraction) > F3 (Fe/Mn oxide-bound fraction) > F2 (carbonate-bound fraction) > F1 (exchangeable fraction). Residual fraction Cr (F7) is the main component of Cr in soil, accounting for more than 90%, indicating that the mobility of Cr is low [46]. In the four growth stages, the concentrations of bioavailable fractions ∑(F1–F2) in CF and AWD were 0.19–0.20 mg kg^{-1} and 0.20–0.20 mg kg^{-1}, respectively, accounting for only 0.06–0.07% of total Cr. Interestingly, the CF and AWD treatments had significant differences between the bioavailable fractions ∑(F1-F2) and potential bioavailable fractions ∑(F3–F6), respectively. For CF treatment, the ∑(F1-F2) fraction and Fe/Mn oxide-bound fraction concentrations were decreased significantly with rice growth, and organic-bound Cr and residual Cr were increased significantly. This result indicated that the Cr fraction in soil was affected by Fe-Mn oxides in a strong redox state under flooding conditions. This result was in accordance with the latest research [47] that newly released Cr may be re-adsorbed on the surface of minerals and organic matter through electrostatic or physical complexes, resulting in a decrease in bioavailable Cr. However, it is different in AWD treatment; the ∑(F1–F2) fraction and Fe/Mn oxide-bound fraction showed a significant increase with rice growth. In contrast, the organic-bound and residual Cr fraction concentration were significantly reduced with rice growth. It is indicated that organic-bound Cr and residual Cr are significantly reduced and converted into Fe/Mn oxide-bound Cr and bioavailable Cr. The reason could be due to the oxidation of iron, because the oxidation of Fe(II) produces amorphous iron hydrates with poor crystallinity [48]. In summary, it can be seen that compared with CF treatment, AWD treatment increased the content of bioavailable and potential bioavailable component Cr in the soil and increased the absorption of Cr by rice.

CF treatment usually results in oxygen depletion, leading to anoxic fermentation of organic matter using various alternative electron acceptors (e.g., iron oxide, NO_3^- and SO_4^{2-}) [49]. At the same time, a large number of protons are consumed, and pH is significantly increased, which may promote the adsorption of Cr(III) on the soil surface. In addition, organic matter (OM) may significantly affect the morphologic transformation of Cr. OM can be adsorbed on iron oxides to form organic-mineral iron oxides [50]. During flooding, reductive dissolution of Fe(III) (hydr) oxides results in the release of dissolved organic matter (DOM) from Fe(III) (hydr) oxides [6,51]. DOM contains rich active functional groups, which can form complexes with heavy metals [52,53]. Thus, DOM, such as low molecular weight DOM, may increase the mobility of Cr in paddy soils, resulting in significant dissolution of Cr(III) from Cr(III)-loaded goethite [54]. In addition, under flooding conditions, SO_4^{2-} reduces to S^{2-}, which may contribute to the fixation of Cr, as metals combine with S^{2-} to form metal sulfide precipitates, resulting in reduced solubility [55–57].

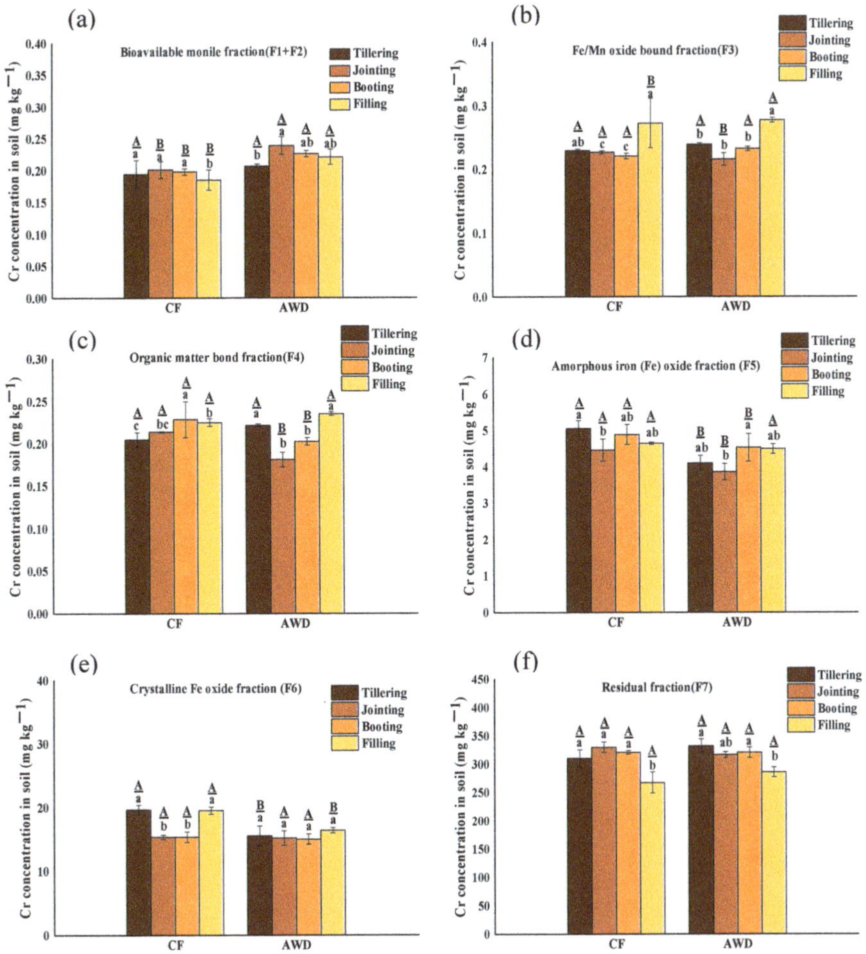

Figure 2. The concentration distribution of bioavailable fraction (F1+F2) (**a**), Fe/Mn oxide-bound fraction F3 (**b**), organic matter-bond fraction F4 (**c**), amorphous iron oxide fraction F5 (**d**), crystalline Fe oxide fraction F6 (**e**) and residual fraction F7 (**f**) of Cr in soil under different water management treatments at different growth stages. The lowercase letters indicate the differences among the same treatment in different periods; the uppercase letters indicate the differences between different treatments in the same period, $p < 0.05$.

The distribution of Fe fraction concentration in soil is shown in Figure 3. There was a significant difference in soil-dissolved Fe among different periods, and it decreased significantly with the growth period. The concentration of dissolved-Fe in AWD treatment was higher than that in CF treatment at the tillering stage and booting stage but lower in the other two stages. It may be the reason that dissolved-Fe was actively utilized by microorganisms under AWD treatment. There was no significant effect on the change of other Fe fraction concentration with CF treatment and AWD treatment. HCl-Fe is thought to be mainly derived from microbial reduction of poorly crystalline ferric hydroxide(III) or poorly ordered ferric hydroxide(III). Thus, the amount of extractable Fe(II) with hydrochloric acid can be used to assess the ability of microorganisms to reduce low crystalline Fe(III) hydroxide. The concentration of extractable Fe(II) in hydrochloric acid increased significantly in both treatments. For DCB-Fe, it is considered to be free iron oxides, both crystalline and low-crystalline. Similar to HCl-Fe, the concentration of DCB-Fe increased significantly after tillering. The reason may be that less O_2 was released from the roots after the tillering stage of rice compared to the previous stage, resulting in reduced oxidation of Fe(II), and strong reducing conditions led to reductive dissolution of Fe(III) (hydrodr) oxides [58].

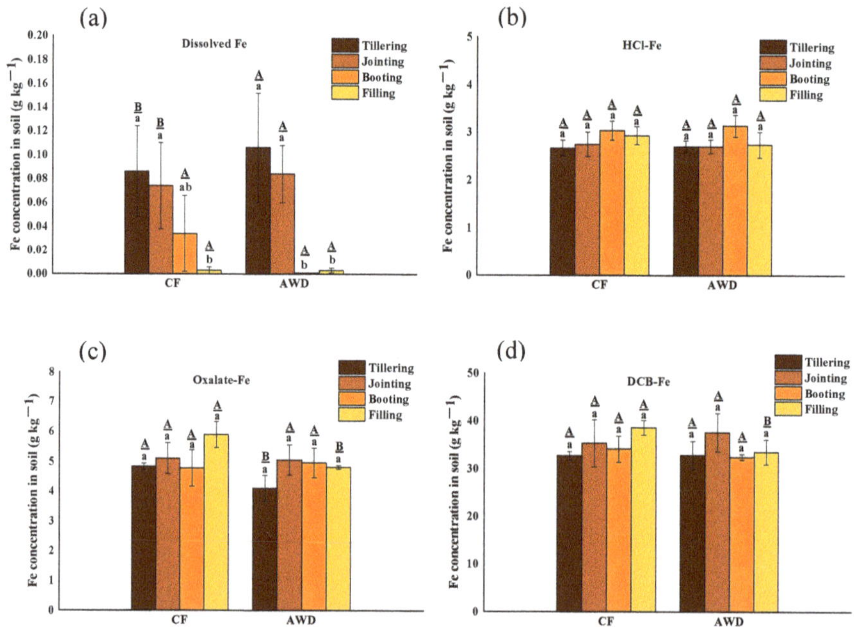

Figure 3. The concentration distribution of different fractions of Fe in soil under different water management treatments at different growth stages of rice. The lowercase letters indicate the differences among the same treatment in different periods; the uppercase letters indicate the differences between different treatments in the same period, $p < 0.05$.

During different redox conditions, iron biogeochemical cycles in paddy soils, including Fe(III) reduction and Fe(II) catalyzing recrystallization of Fe(III)(hydr)oxides, have a significant influence on the immobilization of heavy metals [58]. Strong reducing conditions lead to reductive dissolution of Fe(III) (hydr)oxides, which play a crucial role in controlling the mobility of metals and may have a large effect on the release of Cr. The presence of Fe(II) can promote the recrystallization of metastable Fe(III)(hydr) oxides, resulting in the incorporation of Cr. For example, ferrites can be converted into more stable goethite [59]. These transformations of Fe(III)(hydr) oxides may significantly affect the mobility and availability of Cr in paddy soils.

3.4. Microbial Community Analyses

The composition and abundance of microbial community are important factors affecting the migration and bioavailability of heavy metals in paddy soils. Rhizosphere soil microorganisms can maintain soil fertility and plant yield by assisting nutrient fixation, mineralization, decomposition and flow [60,61]. In this study, a total of 50 phyla, 163 classes, 373 orders, 566 families, 1052 genera and 2084 species were identified. Figure 4 describes the relative community abundance of microbial groups at the genus level in different water managements at different growth stages of rice. At genus level, the microbial community composition was similar, but the relative abundance was different. Among them, Bacillus (3.7–11.6%) accounted for the largest proportion, followed by Intrasporangium and Gaiella.

Compared with other growth stages, the soil at the filling stage significantly increased the abundance of Bacillus and decreased the abundance of Intrasporangium. Bacillus is known as Fe (III)-reducing genus, which is speculated to be related to Fe-Cr reaction in previous studies [62]. It indicates that the transformation of Cr in soil is closely related to microbial-mediated iron cycle [34]. Clostridium was found to exist in CF treatment (1.5–1.8%) and AWD treatment (1.2–1.6%) at the genus level, which is a typical sulfate-reducing bacteria related to sulfur metabolism [63,64]. It can reduce sulfate to sulfide and provide an electron donor for the conversion of Cr.

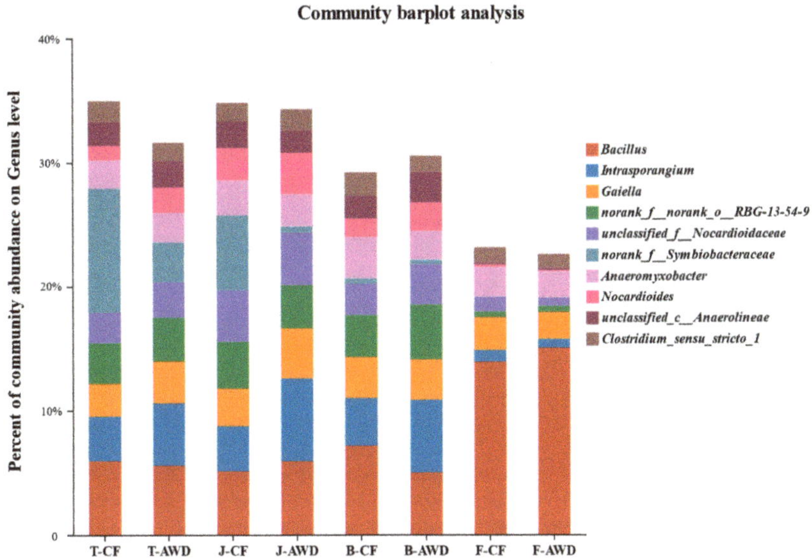

Figure 4. The relative community abundance of the top 10 soil bacteria on genus level in different treatments. T represents the Tillering stage; J represents the Jointing stage; B represents the Booting stage; F represents the Filling stage. CF refers to continuous flooding. AWD refers to alternate wet and dry irrigation.

In this paper, redundancy analysis (RDA) was used to determine the correlation between soil environmental variables and soil microbial community composition (Figure 5). RDA analysis showed that at the genus level, the relevant environmental variables explained 69.04% and 64.43% of the community composition of CF treatment (Figure 5a) and AWD treatment (Figure 5b), respectively. Compared with other environmental variables, dissolved-Fe (p = 0.048), Cr-F7 (p = 0.015) and Cr-∑(F3–F6) (p = 0.096) had significant effects on microbial community composition. Dissolved-Fe (RDA1 = −99.85%, r^2 = 0.50, p = 0.048) and Cr-F7 (RDA1 = −85.21%, r^2 = 0.64, p = 0.015) were significantly negative related with RDA1, while TFe (RDA1 = −99.07%, r^2 = 0.15, p = 0.484), TCr (RDA1 = −87.01%, r^2 = 0.14, p = 0.533) and DCB-Fe (RDA1 = −88.22%, r^2 = 0.01, p = 0.989) were critical for explaining the variations in community structure in CF treatment. The soil pH (RDA1 = 99.18%, r^2 = 0.17, p = 0.455), Cr-∑(F3–F6) (RDA1 = 99.48%, r^2 = 0.45, p = 0.096), TCr (RDA1 = −98.53%, r^2 = 0.12, p = 0.571) and Cr-F7 (RDA1 = −99.99%, r^2 = 0.35, p = 0.140) were critical factors explaining the variations communities in AWD treatment. Some scholars have shown that soil moisture has the greatest impact on the composition of microbial communities in the 0 cm surface soil, and pH may be the main reason for the stratification of the 20 cm surface soil. In addition, the composition of microbial communities below 40 cm is mainly significantly affected by total Cr [53]. In the RDA plot, the angle between the 2 factor vectors is less than 90°, indicating a positive correlation between them. It can be seen from Figure 5a that TCr is positively correlated with TFe and dissolved-Fe under CF condition, while TCr is positively correlated with TFe, Cr-F7 and dissolved-Fe under AWD condition. It shows that the Cr content in the soil is greatly affected by the effective state of iron. Similarly, the RDA results of Xiao et al. [65] also showed a close correlation between microbial community composition and soil pH and Fe(II) concentration.

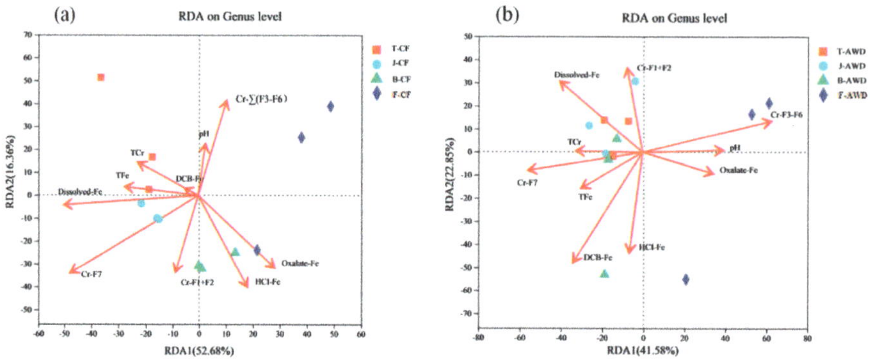

Figure 5. The redundancy analysis (RDA) between bacterial community and related environmental variables of continuous flooding treatment (CF) (**a**) and alternative wet and dry treatment (AWD) (**b**) at genus level. TCr and TFe represent the Total Cr and the Total Fe, respectively; Cr-F1+F2 represent the bioavailable fractions of Cr; Cr-∑(F3–F6) represent potentially bioavailable fractions of Cr; Cr-F7 represents the residual fractions of Cr. (T represents the Tillering stage; J represents the Jointing stage; B represents the Booting stage; F represents the Filling stage.).

Spearman correlation analysis was further used to study the relationship between dominant species and soil environmental factors (Figure 6). The results showed that Cr-F7 ($p < 0.001$), dissolved-Fe ($p < 0.01$) and Cr-∑(F3-F6) ($p < 0.001$) were significantly and specifically correlated with multiple soil microbial communities in CF treatment, followed by Cr-F1+F2, TCr, TFe, DCB-Fe and Oxalate-Fe ($p < 0.05$), and the dominant species had little correlation with pH. It shows that different forms of Cr and different forms of iron have obvious ecological effects on soil microbial communities. In AWD treatment, pH ($p < 0.01$), TCr ($p < 0.001$), Cr-F7 ($p < 0.01$) and dissolved-Fe ($p < 0.001$) were significantly correlated with multiple microbial communities, respectively. Except for Cr-F1+F2, there was significant correlation between microbial communities and other environmental factors ($p < 0.05$). Among these bacteria, *Anaeromyxobacter*, another functional Fe(III)-reducing bacterium, has also been speculated to be related to Fe-Cr reactions in previous studies. *Anaeromyxobacter* was positively correlated with Cr-∑(F3–F6) in CF treatment. In contrast, *Anaeromyxobacter* was significantly negatively correlated with HCl-Fe in AWD treatment. In addition to Fe-reducing bacteria, the relative abundance of sulfate-reducing bacteria was considerably higher in the AWD-treated soils than CF, with *Clostridium* being the most dominant. In CF treatment, *Clostridium* was significantly positively correlated with dissolved-Fe and Cr-F7. In contrast, *Clostridium* was significantly positively correlated with Cr-∑(F3–F6) and dissolved-Fe in AWD treatment. It is particularly noteworthy that there is a significant difference in the correlation between pH value and microbial community between AWD treatment and CF treatment, indicating that AWD treatment may affect the difference of microbial community by affecting soil pH, thus affecting the migration and transformation of Cr. Previous studies have also shown that soil pH can affect the composition and activity of soil microbial communities [66]. Changes in pH affect the physiological and biochemical processes of heavy metals and cell bodies and ultimately affect the survival and growth of specific bacteria, resulting in changes in microbial community structure [67]. In conclusion, it can be clear that the effect of AWD treatment on Cr in a rice–paddy soil system has microbial factors.

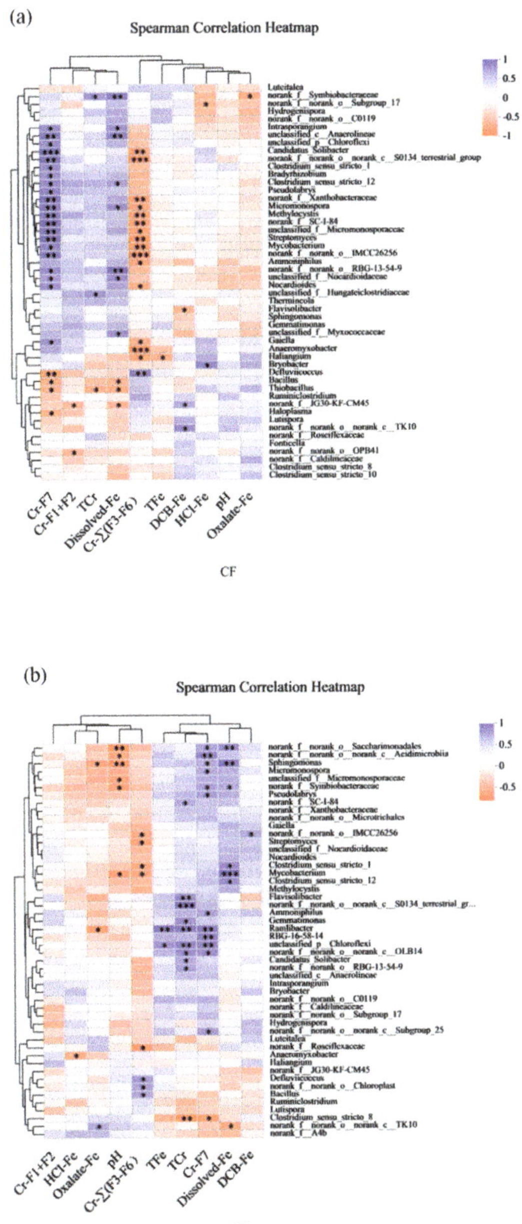

Figure 6. The spearman correlation on genus level between bacterial community and related environmental variables of continuous flooding treatment (CF) (**a**) and alternative wet and dry treatment (AWD) (**b**). TCr and TFe represent the Total Cr and the Total Fe, respectively; Cr-F1+F2 represent bioavailable fractions of Cr; Cr-∑(F3–F6) represent the potentially bioavailable fractions of Cr; Cr-F7 represents the residual fractions of Cr. (* means $0.01 \leq p \leq 0.05$, ** means $0.001 \leq p \leq 0.01$, and *** means $p \leq 0.001$).

4. Conclusions and Environmental Implications

Different water managements had significantly different effects on the migration and transformation of Cr in a rice–paddy soil system. Alternative wet and dry (AWD) treatment made Cr bioavailable in the soil and also affected the soil microbial community of iron-reducing microorganisms and sulfate-reducing bacteria, which provided electron donors for Cr transformation, thus affecting the migration and transformation of Cr in soil. AWD treatment promoted the absorption of Cr by rice and affected its biomass. The reason for this result may be that Cr is released through reduction of Fe(III) (hydr) oxides, and Cr is likely to enter the structure of Fe(III) (hydr) oxides through Fe(II)-catalyzed recrystallization. Basalt is an important Cr reservoir in the terrestrial environment. Cr in paddy soil can be enriched by chemical weathering. Basalt-derived paddy soils contain very high concentrations of Cr and can enter the human body through the food chain. Therefore, the fluidity and availability of Cr are critical to the risk assessment in basaltic paddy soils. This study revealed the variation of Cr availability and mobility in different water management processes of basalt-type paddy soils, which provided a basis for formulating effective water management strategies to control Cr conversion in basalt-type paddy soils.

Author Contributions: Conceptualization, Z.G.; Data curation, Q.L.; Formal analysis, Z.G., R.W., T.L., M.A. and S.S.; Investigation, Z.G., R.W., T.L., J.L., M.A., T.D. and Z.N.; Methodology, T.L., J.L. and S.S.; Software, Z.G., R.W. and S.S.; Supervision, S.W., Y.T. and R.Q.; Validation, Z.N.; Writing—original draft, Z.G.; Writing—review and editing, T.L., T.D., S.W., Y.T., Q.L., Z.N. and R.Q. All authors have read and agreed to the published version of the manuscript.

Funding: This research was funded by the Key Realm Research and Development Program of Guangdong Province, grant number 2020B0202080001, the National Natural Science Foundation of China, grant number 42230707, 42207263, the Guangdong Provincial Science and Technology Project, grant number 2021B1212040008, the Guangdong Laboratory for Lingnan Modern Agriculture Project, grant number NT2021010-2, and the Guangdong Special Support Program, grant number 2021TQ060163.

Institutional Review Board Statement: Not applicable.

Informed Consent Statement: Not applicable.

Data Availability Statement: Not applicable.

Acknowledgments: This work was co-supported by the Key Realm Research and Development Program of Guangdong Province (2020B0202080001), National Natural Science Foundation of China (42230707, 42207263), Guangdong Provincial Science and Technology Project (2021B1212040008), Guangdong Laboratory for Lingnan Modern Agriculture Project (NT2021010-2) and Guangdong Special Support Program (2021TQ060163).

Conflicts of Interest: The authors declare no conflict of interest.

References

1. He, Z.; Shen, J.; Ni, Z.; Tang, J.; Song, S.; Chen, J.; Zhao, L. Electrochemically Created Roughened Lead Plate for Electrochemical Reduction of Aqueous CO_2. *Catal. Commun.* **2015**, *72*, 38–42. [CrossRef]
2. Khalid, S.; Shahid, M.; Niazi, N.K.; Murtaza, B.; Bibi, I.; Dumat, C. A Comparison of Technologies for Remediation of Heavy Metal Contaminated Soils. *J. Geochem. Explor.* **2017**, *182*, 247–268. [CrossRef]
3. Coetzee, J.J.; Bansal, N.; Chirwa, E.M.N. Chromium in Environment, Its Toxic Effect from Chromite-Mining and Ferrochrome Industries, and Its Possible Bioremediation. *Expos. Health* **2020**, *12*, 51–62. [CrossRef]
4. Oze, C.; Fendorf, S.; Bird, D.K.; Coleman, R.G. Chromium Geochemistry of Serpentine Soils. *Int. Geol. Rev.* **2004**, *46*, 97–126. [CrossRef]
5. Becquer, T.; Quantin, C.; Rotte-Capet, S.; Ghanbaja, J.; Mustin, C.; Herbillon, A.J. Sources of Trace Metals in Ferralsols in New Caledonia. *Eur. J. Soil Sci.* **2006**, *57*, 200–213. [CrossRef]
6. Garnier, J.; Quantin, C.; Guimarães, E.; Garg, V.K.; Martins, E.S.; Becquer, T. Understanding the Genesis of Ultramafic Soils and Catena Dynamics in Niquelândia, Brazil. *Geoderma* **2009**, *151*, 204–214. [CrossRef]
7. Cheng, C.H.; Jien, S.H.; Iizuka, Y.; Tsai, H.; Chang, Y.H.; Hseu, Z.Y. Pedogenic Chromium and Nickel Partitioning in Serpentine Soils along a Toposequence. *Soil Sci. Soc. Am. J.* **2011**, *75*, 659–668. [CrossRef]
8. Zhang, X.; Zhong, T.; Liu, L.; Zhang, X.; Cheng, M.; Li, X.; Jin, J. Chromium Occurrences in Arable Soil and Its Influence on Food Production in China. *Environ. Earth Sci.* **2016**, *75*, 257. [CrossRef]
9. Wu, W.; Qu, S.; Nel, W.; Ji, J. The Influence of Natural Weathering on Tthe Behavior of Heavy Metals in Small Basaltic Watersheds: A Comparative Study From Different Regions In China. *Chemosphere* **2021**, *262*, 127897. [CrossRef]

10. Hu, Y.; Cheng, H.; Tao, S. The Challenges and Solutions for Cadmium-contaminated Rice in China: A Critical Review. *Environ. Int.* **2016**, *92–93*, 515–532. [CrossRef]
11. Chen, D.; Guo, H.; Li, R.; Li, L.; Pan, G.; Chang, A.; Joseph, S. Low Uptake Affinity Cultivars with Biochar to Tackle Cd-Tainted Rice—A Field Study over Four Rice Seasons in Hunan, China. *Sci. Total Environ.* **2016**, *541*, 1489–1498. [CrossRef]
12. Cabangon, R.J.; Tuong, T.P.; Castillo, E.G.; Bao, L.X.; Lu, G.; Wang, G.; Cui, Y.; Bouman, B.A.M.; Li, Y.; Chen, C.; et al. Effect of Irrigation Method and N-Fertilizer Management on Rice Yield, Water Productivity and Nutrient-Use Efficiencies in Typical Lowland Rice Conditions in China. *Paddy Water Environ.* **2004**, *2*, 195–206. [CrossRef]
13. Linquist, B.A.; Anders, M.M.; Adviento-Borbe, M.A.A.; Chaney, R.L.; Nalley, L.L.; Da Rosa, E.F.F.; Van Kessel, C. Reducing Greenhouse Gas Emissions, Water Use, and Grain Arsenic Levels in Rice Systems. *Glob. Chang. Biol.* **2015**, *21*, 407–417. [CrossRef]
14. Li, C.; Carrijo, D.R.; Nakayama, Y.; Linquist, B.A.; Green, P.G.; Parikh, S.J. Impact of Alternate Wetting and Drying Irrigation on Arsenic Uptake and Speciation in Flooded Rice Systems. *Agric. Ecosyst. Environ.* **2019**, *272*, 188–198. [CrossRef]
15. Talukder, A.S.M.H.M.; Meisner, C.A.; Sarkar, M.A.R.; Islam, M.S.; Sayre, K.D.; Duxbury, J.M.; Lauren, J.G. Effect of Water Management, Arsenic and Phosphorus Levels on Rice in a High-Arsenic Soil–Water System: Ii. Arsenic Uptake. *Ecotoxicol. Environ. Saf.* **2012**, *80*, 145–151. [CrossRef] [PubMed]
16. Zia, Z.; Bakhat, H.F.; Saqib, Z.A.; Shah, G.M.; Fahad, S.; Ashraf, M.R.; Hammad, H.M.; Naseem, W.; Shahid, M. Effect of Water Management and Silicon on Germination, Growth, Phosphorus and Arsenic Uptake in Rice. *Ecotoxicol. Environ. Saf.* **2017**, *144*, 11–18. [CrossRef]
17. Xiao, W.D.; Ye, X.Z.; Yang, X.; Li, T.Q.; Zhao, S.P.; Zhang, Q. Effects of Alternating Wetting and Drying Versus Continuous Flooding on Chromium Fate in Paddy Soils. *Ecotoxicol. Environ. Saf.* **2015**, *113*, 439–445. [CrossRef] [PubMed]
18. Austruy, A.; Shahid, M.; Xiong, T.; Castrec, M.; Payre, V.; Niazi, N.K.; Sabir, M.; Dumat, C. Mechanisms of Metal-Phosphates Formation in The Rhizosphere Soils of Pea and Tomato: Environmental and Sanitary Consequences. *J. Soil Sediment.* **2014**, *14*, 666–678. [CrossRef]
19. Rafiq, M.; Shahid, M.; Abbas, G.; Shamshad, S.; Khalid, S.; Niazi, N.K.; Dumat, C. Comparative Effect of Calcium and Edta on Arsenic Uptake and Physiological Attributes of *Pisum sativum*. *Int. J. Phytoremediat.* **2017**, *19*, 662–669. [CrossRef] [PubMed]
20. Demir, A. Speciation, Risk Assessment and Bioavailability of Metals in the Agricultural Soils of the Goksu Delta, Turkey. *Soil Sediment Contam.* **2021**, *30*, 292–313. [CrossRef]
21. Rinklebe, J.; Shaheen, S.M.; Schroeter, F.; Rennert, T. Exploiting Biogeochemical and Spectroscopic Techniques to Assess the Geochemical Distribution and Release Dynamics of Chromium and Lead in a Contaminated Floodplain Soil. *Chemosphere* **2016**, *150*, 390–397. [CrossRef]
22. Banks, M.K.; Schwab, A.P.; Henderson, C. Leaching and Reduction of Chromium in Soil As Affected by Soil Organic Content and Plants. *Chemosphere* **2006**, *62*, 255–264. [CrossRef]
23. Xiao, W.D.; Ye, X.Z.; Zhu, Z.Q.; Zhang, Q.; Zhao, S.P.; Chen, D.; Gao, N.; Hu, J. Continuous Flooding Stimulates Root Iron Plaque Formation and Reduces Chromium Accumulation in Rice (*Oryza sativa* L.). *Sci. Total Environ.* **2021**, *788*, 147786. [CrossRef]
24. Xu, Z.R.; Cai, M.L.; Chen, S.H.; Huang, X.Y.; Zhao, F.J.; Wang, P. High-Affinity Sulfate Transporter Sultr1;2 Is a Major Transporter for Cr(VI) Uptake in Plants. *Environ. Sci. Technol.* **2021**, *55*, 1576–1584. [CrossRef] [PubMed]
25. Suda, A.; Makino, T. Functional Effects of Manganese and Iron Oxides on the Dynamics of Trace Elements in Soils With a Special Focus on Arsenic and Cadmium: A review. *Geoderma* **2016**, *270*, 68–75. [CrossRef]
26. Li, X.M.; Liu, T.X.; Zhang, N.M.; Ren, G.; Li, F.B.; Li, Y.T. Effect of Cr(VI) on Fe(III) Reduction in Three Paddy Soils from the Hani Terrace Field at High Altitude. *Appl. Clay Sci.* **2012**, *64*, 53–60. [CrossRef]
27. Davranche, M.; Bollinger, J.C. Release of Metals from Iron Oxyhydroxides under Reductive Conditions: Effect of Metal/Solid Interactions. *J. Colloid Interf. Sci.* **2000**, *232*, 165–173. [CrossRef]
28. Somenahally, A.C.; Hollister, E.B.; Yan, W.; Gentry, T.J.; Loeppert, R.H. Water Management Impacts on Arsenic Speciation and Iron-Reducing Bacteria in Contrasting Rice-Rhizosphere Compartments. *Environ. Sci. Technol.* **2011**, *45*, 8328–8335. [CrossRef] [PubMed]
29. Wang, X.; Chen, X.; Yang, J.; Wang, Z.; Sun, G. Effect of Microbial Mediated Iron Plaque Reduction on Arsenic Mobility in Paddy Soil. *J. Environ. Sci.* **2009**, *21*, 1562–1568. [CrossRef]
30. Wang, X.; Tam, N.F.Y.; He, H.; Ye, Z. The Role of Root Anatomy, Organic Acids and Iron Plaque on Mercury Accumulation in Rice. *Plant Soil* **2015**, *394*, 301–313. [CrossRef]
31. Liesack, W.; Schnell, S.; Revsbech, N.P. Microbiology of Flooded Rice Paddies. *FEMS Microbiol. Rev.* **2000**, *24*, 625–645. [CrossRef]
32. Zhang, M.; Zhang, L.; Riaz, M.; Xia, H.; Jiang, C. Biochar Amendment Improved Fruit Quality and Soil Properties and Microbial Communities at Different Depths in Citrus Production. *J. Clean. Prod.* **2021**, *292*, 126062. [CrossRef]
33. Liu, K.; Li, F.; Tian, Q.; Nie, C.; Ma, Y.; Zhu, Z.; Fang, L.; Huang, Y.; Liu, S. A Highly Porous Animal Bone-Derived Char with a Superiority of Promoting nZVI for Cr(VI) Sequestration in Agricultural Soils. *J. Environ. Sci.* **2021**, *104*, 27–39. [CrossRef] [PubMed]
34. Zhang, K.; Li, N.; Liao, P.; Jin, Y.; Li, Q.; Gan, M.; Chen, Y.; He, P.; Chen, F.; Peng, M.; et al. Conductive Property of Secondary Minerals Triggered Cr(VI) Bioreduction by Dissimilatory Iron Reducing Bacteria. *Environ. Pollut.* **2021**, *286*, 117227. [CrossRef] [PubMed]

35. Sun, S.S.; Ao, M.; Geng, K.R.; Chen, J.Q.; Deng, T.H.B.; Li, J.J.; Guan, Z.T.; Mo, B.L.; Liu, T.; Yang, W.J.; et al. Enrichment and Speciation of Chromium During Basalt Weathering: Insights from Variably Weathered Profiles in the Leizhou Peninsula, South China. *Sci. Total Environ.* **2022**, *822*, 153304. [CrossRef] [PubMed]
36. Ao, M.; Sun, S.S.; Deng, T.H.B.; Zhang, F.; Liu, T.; Tang, Y.; Li, J.J.; Wang, S.Z.; Qiu, R.L. Natural Source of Cr(VI) in Soil: The Anoxic Oxidation of Cr(III) by Mn Oxides. *J. Hazard. Mater.* **2022**, *433*, 128805. [CrossRef]
37. Yu, H.Y.; Wang, X.; Li, F.; Li, B.; Liu, C.; Wang, Q.; Lei, J. Arsenic Mobility and Bioavailability in Paddy Soil under Iron Compound Amendments at Different Growth Stages of Rice. *Environ. Pollut.* **2017**, *224*, 136–147. [CrossRef]
38. Seyfferth, A.L.; Amaral, D.; Limmer, M.A.; Guilherme, L.R.G. Combined Impacts of Si-Rich Rice Residues and Flooding Extent on Grain As and Cd in Rice. *Environ. Int.* **2019**, *128*, 301–309. [CrossRef]
39. Athar, R.; Ahmad, M. Heavy Metal Toxicity: Effect on Plant Growth and Metal Uptake by Wheat, and on Free Living Azotobacter. *Water Air Soil Pollut.* **2002**, *138*, 165–180. [CrossRef]
40. Zhang, X.H.; Liu, J.; Huang, H.T.; Chen, J.; Zhu, Y.N.; Wang, D.Q. Chromium Accumulation by the Hyperaccumulator Plant Leersia Hexandra Swartz. *Chemosphere* **2007**, *67*, 1138–1143. [CrossRef]
41. Cipriani, H.N.; Ribeiro Bastos, A.R.; de Carvalho, J.G.; da Costa, A.L.; Oliveira, N.P. Chromium Toxicity in Hybrid Eucalyptus (*Eucalyptus Urophylla* S. T. Blake X *Grandis* W. Hill Ex. Maiden) Cuttings. *J. Plant Nutr.* **2012**, *35*, 1618–1638. [CrossRef]
42. Usman, K.; Al Jabri, H.; Abu-Dieyeh, M.H.; Alsafran, M.H.S.A. Comparative Assessment of Toxic Metals Bioaccumulation and the Mechanisms of Chromium (Cr) Tolerance and Uptake in *Calotropis procera*. *Front. Plant Sci.* **2020**, *11*, 883. [CrossRef] [PubMed]
43. Loyaux-Lawniczak, S.; Lecomte, P.; Ehrhardt, J.J. Behavior of Hexavalent Chromium in a Polluted Groundwater: Redox Processes and Immobilization in Soils. *Environ. Sci. Technol.* **2001**, *35*, 1350–1357. [CrossRef]
44. Li, R.Y.; Stroud, J.L.; Ma, J.F.; McGrath, S.P.; Zhao, F.J. Mitigation of Arsenic Accumulation in Rice with Water Management and Silicon Fertilization. *Environ. Sci. Technol.* **2009**, *43*, 3778–3783. [CrossRef] [PubMed]
45. Roberts, L.C.; Hug, S.J.; Voegelin, A.; Dittmar, J.; Kretzschmar, R.; Wehrli, B.; Saha, G.C.; Badruzzaman, A.B.M.; Ali, M.A. Arsenic Dynamics in Porewater of an Intermittently Irrigated Paddy Field in Bangladesh. Environ. *Sci. Technol.* **2011**, *45*, 971–976. [CrossRef]
46. Xu, M.; da Silva, E.B.; Gao, P.; Liao, R.; Wu, J.; Ma, J.; Yang, G.; Zhang, X.; Xiao, Y.; Long, L. Biochar Impact on Chromium Accumulation by Rice Through Fe Microbial-Induced Redox Transformation. *J. Hazard. Mater.* **2020**, *388*, 121807. [CrossRef]
47. Zhang, K.; Yang, Y.; Chi, W.; Chen, G.; Du, Y.; Hu, S.; Li, F.; Liu, T. Chromium Transformation Driven by Iron Redox Cycling in Basalt-Derived Paddy Soil With High Geological Background Values. *J. Environ. Sci.* **2023**, *125*, 470–479. [CrossRef] [PubMed]
48. Karimian, N.; Johnston, S.G.; Burton, E.D. Iron and sulfur cycling in acid sulfate soil wetlands under dynamic redox conditions: A review. *Chemosphere* **2018**, *197*, 803–816. [CrossRef] [PubMed]
49. Kögel-Knabner, I.; Amelung, W.; Cao, Z.; Fiedler, S.; Frenzel, P.; Jahn, R.; Kalbitz, K.; Kölbl, A.; Schloter, M. Biogeochemistry of Paddy Soils. *Geoderma* **2010**, *157*, 1–14. [CrossRef]
50. Bao, Y.; Bolan, N.S.; Lai, J.; Wang, Y.; Jin, X.; Kirkham, M.B.; Wu, X.; Fang, Z.; Zhang, Y.; Wang, H. Interactions between Organic Matter and Fe (Hydr)Oxides and Their Influences on Immobilization and Remobilization of Metal(Loid)s: A review. *Crit. Rev. Environ. Sci. Technol.* **2022**, *52*, 4016–4037. [CrossRef]
51. Song, X.X.; Wang, P.; Van, Z.L.; Bolan, N.; Wang, H.L.; Li, X.M.; Cheng, K.; Yang, Y.; Wang, M.L.; Liu, T.X.; et al. Towards a Better Understanding of the Role of Fe Cycling in Soil for Carbon Stabilization and Degradation. *Carbon Res.* **2022**, *1*, 5. [CrossRef]
52. Huang, M.; Zhou, M.; Li, Z.; Ding, X.; Wen, J.; Jin, C.; Wang, L.; Xiao, L.; Chen, J. How do Drying-Wetting Cycles Influence Availability of Heavy Metals in Sediment? A Perspective From Dom Molecular Composition. *Water Res.* **2022**, *220*, 118671. [CrossRef]
53. Liu, B.; Su, G.; Yang, Y.; Yao, Y.; Huang, Y.; Hu, L.; Zhong, H.; He, Z. Vertical Distribution of Microbial Communities in Chromium-Contaminated Soil and Isolation of Cr(VI)-Reducing Strains. *Ecotoxicol. Environ. Saf.* **2019**, *180*, 242–251. [CrossRef]
54. Stewart, A.G.; Hudson-Edwards, K.A.; Dubbin, W.E. Effect of Desferrioxamine B and Suwannee River Fulvic Acid on Fe(III) Release And Cr(III) Desorption from Goethite. *Geochim. Cosmochim. Acta* **2016**, *178*, 62–75. [CrossRef]
55. Armienta, M.A.; Rodríguez, R.; Ceniceros, N.; Juárez, F.; Cruz, O. Distribution, Origin and Fate of Chromium in Soils in Guanajuato, Mexico. *Environ. Pollut.* **1996**, *91*, 391–397. [CrossRef]
56. Furuya, M.; Hashimoto, Y.; Yamaguchi, N. Time-Course Changes in Speciation and Solubility of Cadmium in Reduced and Oxidized Paddy Soils. *Soil Sci. Soc. Am. J.* **2016**, *80*, 870–877. [CrossRef]
57. Zawadzka, A.M.; Crawford, R.L.; Paszczynski, A.J. Pyridine-2,6-bis(thiocarboxylic acid) Produced by *Pseudomonas stutzeri* KC Reduces Chromium(VI) and Precipitates Mercury, Cadmium, Lead and Arsenic. *BioMetals* **2007**, *20*, 145–158. [CrossRef] [PubMed]
58. Dong, H.; Huang, L.; Zhao, L.; Zeng, Q.; Liu, X.; Sheng, Y.; Shi, L.; Wu, G.; Jiang, H.; Li, F.; et al. A Critical Review Of Mineral-Microbe Interaction And Co-Evolution: Mechanisms And Applications. *Natl. Sci. Rev.* **2022**, *9*, nwac128. [CrossRef] [PubMed]
59. Huang, J.; Jones, A.; Waite, T.D.; Chen, Y.; Huang, X.; Rosso, K.M.; Kappler, A.; Mansor, M.; Tratnyek, P.G.; Zhang, H. Fe(II) Redox Chemistry in the Environment. *Chem. Rev.* **2021**, *121*, 8161–8233. [CrossRef]
60. Chai, L.; Huang, S.; Yang, Z.; Peng, B.; Huang, Y.; Chen, Y. Cr (VI) Remediation by Indigenous Bacteria in Soils Contaminated by Chromium-Containing Slag. *J. Hazard. Mater.* **2009**, *167*, 516–522. [CrossRef]

61. Desai, C.; Parikh, R.Y.; Vaishnav, T.; Shouche, Y.S.; Madamwar, D. Tracking the Influence Of Long-Term Chromium Pollution On Soil Bacterial Community Structures By Comparative Analyses of 16S rRNA Gene Phylotypes. *Res. Microbiol.* **2009**, *160*, 1–9. [CrossRef] [PubMed]
62. Dong, G.; Han, R.; Pan, Y.; Zhang, C.; Liu, Y.; Wang, H.; Ji, X.; Dahlgren, R.A.; Shang, X.; Chen, Z.; et al. Role of MnO_2 in Controlling Iron and Arsenic Mobilization from Illuminated Flooded Arsenic-Enriched Soils. *J. Hazard. Mater.* **2021**, *401*, 123362. [CrossRef]
63. Ravikumar, K.V.G.; Argulwar, S.; Sudakaran, S.V.; Pulimi, M.; Chandrasekaran, N.; Mukherjee, A. Nano-Bio Sequential Removal of Hexavalent Chromium Using Polymer-nZVI Composite Film and Sulfate Reducing Bacteria under Anaerobic Condition. *Environ. Technol. Innov.* **2018**, *9*, 122–133.
64. Kumar, S.S.; Kumar, V.; Gude, V.G.; Malyan, S.K.; Pugazhendhi, A. Alkalinity and Salinity Favor Bioelectricity Generation Potential of *Clostridium, Tetrathiobacter* and *Desulfovibrio consortium* in Microbial Fuel Cells (MFC) Treating Sulfate-laden Wastewater. *Bioresour. Technol.* **2020**, *306*, 123110. [CrossRef] [PubMed]
65. Xiao, W.D.; Ye, X.Z.; Ye, Z.Q.; Zhang, Q.; Zhao, S.P.; Chen, D.; Gao, N.; Huang, M.J. Responses of Microbial Community Composition and Function to Biochar and Irrigation Management and the Linkage to Cr Transformation in Paddy Soil. *Environ. Pollut.* **2022**, *304*, 119232. [CrossRef]
66. Montebello, A.M.; Bezerra, T.; Rovira, R.; Rago, L.; Lafuente, J.; Gamisans, X.; Campoy, S.; Baeza, M.; Gabriel, D. Operational Aspects, pH Transition and Microbial Shifts of a H2S Desulfurizing Biotrickling Filter with Random Packing Material. *Chemosphere* **2013**, *93*, 2675–2682. [CrossRef]
67. Zhang, H.; Wan, Z.; Ding, M.; Wang, P.; Xu, X.; Jiang, Y. Inherent Bacterial Community Response to Multiple Heavy Metals in Sediment from River-Lake Systems in the Poyang Lake, China. *Ecotoxicol. Environ. Saf.* **2018**, *165*, 314–324. [CrossRef]

toxics

Review

A Review of Antibiotics, Antibiotic Resistant Bacteria, and Resistance Genes in Aquaculture: Occurrence, Contamination, and Transmission

Xia Yuan [1,2], Ziqing Lv [1,2], Zeyu Zhang [1,2], Yu Han [1,2], Zhiquan Liu [1,2,3] and Hangjun Zhang [1,2,3,*]

1 School of Life and Environmental Sciences, Hangzhou Normal University, Hangzhou 311121, China
2 Zhejiang Provincial Key Laboratory of Urban Wetlands and Regional Change, Hangzhou 311121, China
3 School of Engineering, Hangzhou Normal University, Hangzhou 310018, China
* Correspondence: 13819172516@163.com

Abstract: Antibiotics are commonly used to prevent and control diseases in aquaculture. However, long-term/overuse of antibiotics not only leaves residues but results in the development of antibiotic resistant bacteria (ARB) and antibiotic resistance genes (ARGs). Antibiotics, ARB, and ARGs are widespread in aquaculture ecosystems. However, their impacts and interaction mechanisms in biotic and abiotic media remain to be clarified. In this paper, we summarized the detection methods, present status, and transfer mechanisms of antibiotics, ARB, and ARGs in water, sediment, and aquaculture organisms. Currently, the dominant methods of detecting antibiotics, ARB, and ARGs are UPLC−MS/MS, 16S rRNA sequencing, and metagenomics, respectively. Tetracyclines, macrolides, fluoroquinolones, and sulfonamides are most frequently detected in aquaculture. Generally, antibiotic concentrations and ARG abundance in sediment are much higher than those in water. Yet, no obvious patterns in the category of antibiotics or ARB are present in organisms or the environment. The key mechanisms of resistance to antibiotics in bacteria include reducing the cell membrane permeability, enhancing antibiotic efflux, and structural changes in antibiotic target proteins. Moreover, horizontal transfer is a major pathway for ARGs transfer, including conjugation, transformation, transduction, and vesiculation. Identifying, quantifying, and summarizing the interactions and transmission mechanisms of antibiotics, ARGs, and ARB would provide useful information for future disease diagnosis and scientific management in aquaculture.

Keywords: aquaculture; antibiotics; antibiotic resistant bacteria (ARB); antibiotic resistance genes (ARGs); contamination; transmission

check for updates

Citation: Yuan, X.; Lv, Z.; Zhang, Z.; Han, Y.; Liu, Z.; Zhang, H. A Review of Antibiotics, Antibiotic Resistant Bacteria, and Resistance Genes in Aquaculture: Occurrence, Contamination, and Transmission. *Toxics* **2023**, *11*, 420. https://doi.org/10.3390/toxics11050420

Academic Editors: Junhao Qin, Peidong Su, Feng Zhu and Lin Ding

Received: 31 March 2023
Revised: 21 April 2023
Accepted: 28 April 2023
Published: 30 April 2023

1. Introduction

Aquaculture is one of the fastest growing industries in the world, helping solve food shortages and promote economic development. The Food and Agriculture Organization of the United Nations (FAO) predicted that the fish production of aquaculture could reach 109 million tons in 2030, with Asia as the dominant sector, accounting for 89% of total production [1]. Water quality in aquaculture is the key factor determining fish health. However, in practice, high-density breeding, pollution, or eutrophication of water bodies often lead to the decline of water quality and outbreaks of fish diseases [2]. Therefore, antibiotics have been used globally as aquaculture drugs or feed mixes to control fish diseases. On the one hand, antibiotics protect fish from infectious diseases to some extent [3]; however, on the other hand, the wide application or improper use of antibiotics creates problems of bacterial resistance, promoting the occurrence of antibiotic resistant bacteria (ARB) and antibiotic resistance genes (ARGs) in various aquatic environments.

Antibiotic pollution and the resulting contamination by ARB and ARGs have aroused wide concern [4]. ARGs contamination and their distribution in microbial populations throughout the biosphere is the result of continuous selection by the heavy anthropogenic use of antibiotics over the past few decades [5,6]. The high market demand for aquaculture products means that antibiotics are used in farmed products more than in humans, leading to increasing antibiotic resistance in pathogens [7]. As a new class of environmental pollutants, the spread of ARB and ARGs in aquaculture environmental media might be more harmful than antibiotics themselves [8]. ARGs and ARB could pose a threat to human health when ingested by humans via cultured products [9]. Moreover, prolonged abuse of antibiotics, as well as their recalcitrance, inevitably lead to large amounts of antibiotic residues in aquaculture environments. Antibiotic residues remaining in the environment could be further redistributed in water and sediments, having adverse effects on the ecological environment. Therefore, although many countries (including China, Japan, America, and the European Union) have banned the use of antibiotics, antibiotic contamination in aquaculture environments and production still exists. In addition, there is a very low probability that resistance could be selected and transferred by bacteria when the antibiotic selection pressure is high [10]. Overall, it is necessary to further detect and analyze the virulence and transmission mechanisms of antibiotic residues and the subsequent or emerging pollutants in aquaculture. Micro-organisms have developed efficient mechanisms that enable them to tolerate antibiotics, of which ARGs are one. ARGs are initially present in the genome of environmental bacteria or are produced by bacterial mutations [11] and can be transmitted by vertical gene transfer (VGT) [12] and horizontal gene transfer (HGT) [13]. HGT is the main route of ARG transmission, including conjugation, transformation, and transduction [14]. VGT involves the transfer of genetic material from the parent to the offspring [15]. Many studies have focused on antibiotics, ARB, and ARGs; however, an overview and determination of the mechanisms of antibiotic, ARB, and ARG interactions to date are also necessary.

This review carried out an integrated assessment of the threats to aquaculture associated with antibiotics, ARB, and ARGs in different media, including water, sediment, and organisms. We mainly examined the presence of major antibiotics, resistant bacteria, and ARGs in aquaculture water, sediment, and aquaculture organisms, exploring the connection and interaction mechanism of antibiotics, resistant bacteria, and ARGs in these three media, and provided a scientific basis for further in-depth research into the control of ARGs in the aquaculture environment in the future.

2. Methodology for the Study of Antibiotics, ARB, and ARGs

Understanding the variety, content, and abundance of antibiotics in the environment can help researchers precisely control and remove antibiotics [16]. To investigate the contamination status of antibiotics, multiple methods were designed to detect residual antibiotics in water, sediment, and aquaculture organisms. The main processes for analyzing antibiotics include extraction and detection. Currently, the commonly used detection methods are physicochemical, including liquid chromatography–tandem mass spectrometry (LC-MS/MS), high-performance liquid chromatography–tandem mass spectrometry (HPLC-MS/MS), and ultra-performance liquid chromatography–electrospray tandem mass spectrometry (UPLC-MS/MS). For example, nine antibiotics were detected using LC-MS/MS after filtering water samples through filter paper and using solid-phase extraction for sediment [17]. A study used solid-phase extraction to extract antibiotics and high-performance liquid chromatography–electrospray tandem mass spectrometry (HPLC-ESI-MS/MS) to analyze the concentrations of 10 antibiotics [18]. The recoveries were 83–117% for antibiotics in water and 52–105% for antibiotics in sediment, with relative standard deviations less than 10%. High-performance liquid chromatography–electrospray tandem mass spectrometry was also used to identify antibiotics in fish tissues, and the recoveries of the four antibiotics were 92.3%, 91.6%, 94.1%, and 93.7%, respectively [19]. All three methods have proved to be highly accurate and sensitive. Moreover, several other methods, such as microbial assays or immunochromatographic methods, have also been used to detect antibiotics. The microbial assay can qualitatively determine the antibiotic residues in a sample and has the advantage of being easy to perform; however, the detection time is long, and the method is not quantitative [20]. Tetracycline drug residues in Qinghai yak meat were assessed using a fluorescence immunochromatographic method, and the recovery ranged from 80 to 120%, indicating the high accuracy and sensitivity of the method [21]. However, one method can only detect one or one class of antibiotics, and the range of detection is extremely limited. In summary, for a highly efficient and accurate detection of antibiotics, UPLC-MS/MS or similar methods could be given priority in practical work and in research.

The detection of ARB includes the screening and identification of resistant strains. The initial screening of ARB is generally performed using drug sensitivity tests and passage culture [22]. Subsequently, 16S rRNA sequencing is performed on the cultured colonies to explore the taxonomy of the isolated strains [23]. Meanwhile, PCR amplification is performed to detect the ARGs. The drug sensitivity assay for ARB is also a phenotypic assay for ARGs [24]. Drug sensitivity assays include the paper slice method, the agar/broth dilution method, and the concentration gradient method, which require a highly pure culture, and the process is time consuming and laborious [25]. A relatively quicker method is to perform PCR amplification of the extracted DNA and 16S rRNA sequencing after purification. The abundance of operational taxonomic units (OTUs) is obtained from the sequencing results, and the abundance of bacteria carrying ARGs is also called the ARB abundance, which is verified through gene annotation [26].

The detection of ARGs allows clarification of their distribution characteristics and propagation patterns in the environment for follow-up research. There are numerous methods to study ARGs, such as fluorescent quantitative PCR, high-throughput fluorescent quantitative PCR, and macro-genomics. Most studies have used traditional fluorescence quantitative PCR. However, this method can only quantify a few or dozens of resistance genes [27]. The high-throughput fluorescence quantitative PCR technique is capable of quantifying hundreds of genes or multiple samples of ARGs in water; using high-throughput fluorescence quantitative PCR can identify 211 ARGs [28]. In recent years, many studies began to apply the metagenomics approach to detect ARGs, which is capable of detecting all ARGs in the environment and can help uncover novel resistance genes [29]. For instance, metagenomics has been used for detecting the abundance of ARGs in activated sludge and predicting ARG hosts [30] and screening multiple antibiotic resistant clones from soil non-pathogenic bacteria [31]. The above three methods have their own advantages, whereby one comprehensive method could be chosen according to a study's requirements.

3. The Presence of Antibiotics, ARB, and ARGs in Aquaculture Environments

3.1. Type and Abundance of Antibiotics, ARB, and ARGs in Aquaculture Water

3.1.1. Antibiotics in Aquaculture Water

At present, the level of antibiotics detected in the water bodies of aquaculture areas in China is generally at the level of ng L^{-1} to μg L^{-1}. There are geographical differences in the level of antibiotics in farmed water because of the different farming habits and composition of farming species in different regions; however, the most commonly detected antibiotic types are basically the same. The commonly detected antibiotics in aquaculture water are tetracyclines, macrolides, fluoroquinolones, and sulfonamides (as shown in Table 1). The solubility, frequency, and usage of antibiotics are the main factors affecting antibiotic residues in water [32]. In general, antibiotic residues are also associated with the growth stages of cultured organisms, i.e., younger animals usually have less resistance and possibly need more antibiotics for disease control and growth promotion [33]. The concentration of antibiotics in water bodies is also influenced by environmental factors. For example, the average concentration of antibiotics in Zhuhai rivers and coastal waters was significantly higher in the dry season than in the rainy season [34]. This could be explained by the dilution effect of precipitation on antibiotics. Moreover, extreme temperatures could promote the biodegradation and photolysis of antibiotics [35], further reducing the level of antibiotic residues.

Table 1. The average amount of antibiotics detected in the aquaculture environment.

Antibiotics		Water	Sediment	Origin Location
Major Categories	Category	Concentration (ng·L^{-1}) (Average/Range)	Concentration (μg·kg^{-1}) (Average/Range)	
Tetracyclines	Oxytetracycline (OTC)	10.69	0.56	An aquaculture farm in Dongli District, Tianjin [36]
	Tetracycline (TC)	ND	ND~4.3	Trout farms along the Nera River [37]
	Chlortetracycline (CTC)	ND~50.32	-	Surface water of East Dongting Lake [38]
Macrolides	Roxithromycin (ROM)	1.22~110.66	-	Wusongkou, Yangtze River basin [38]
	Erythromycin (ERM)	ND	ND	Three forks of the Yangtze River basin [38]
Quinolones	Ciprofloxacin (CIP)	ND~2	ND~32.8	Trout farms along the Nera River [37]
	Norfloxacin (NOR)	ND~75.1	-	A trout aquaculture system located in the north of Portugal [39]
	Enrofloxacin (ENR)	0.5×10^3	45.4	Tha Chin River in Thailand [40]
Sulfonamides	Sulfadiazine (SDZ)	ND~571	3–553	Trout farms along the Nera River [37]
	Sulfadimethoxine (SDM)	0.14×10^3~ 0.88×10^3	7.7	Aquaculture systems of northwestern Germany [41]
	Sulfamethoxazole (SMX)	0.04×10^6~ 2.39×10^6	4.77×10^3~ 820.49×10^3	Tiger shrimp farm in mangrove area of Vietnam [42]
	Methicillin (TMP)	0.93	0.89	Beibu Gulf marine farm [43]

Note: - represents data not explicitly reported.

3.1.2. Occurrence of ARB and ARGs in Aquaculture Water

The excessive use of antibiotics in aquaculture has resulted in the emergence or development of bacterial resistance [44]. For example, in water samples from tilapia farms, more than nine ARB species were detected, which were *Cyanobacteria* (26.5%), *Proteobacteria* (26.4%), *Actinobacteria* (11.0%), *Bacteroidetes* (9.8%), *Planctomycetes* (8.3%), *Verrucomicrobia* (6.1%), *Chloroflexi* (2.1%), *Chlorobi* (1.7%), *Firmicutes* (1.0%), and Other (4%). The ARB species in water varied from one location to another. The common bacterial groups in the water of Danjiangkou Reservoir were *Actinobacteria* (40.2%), *Proteobacteria* (21.1%), *Bacteroidetes* (17.7%), and *Cyanobacteria* (13.7%) [45]. Among them, the main types of ARGs were bacitracin (0.047 copies cell^{-1}) and multidrug (0.016 copies cell^{-1}) resistance genes [26]. Aspergillus, Trichoderma, and Bacteroidetes were the main phyla detected in a shrimp farm in Thailand [46]. Notably, the antibiotics used differed from one aquaculture area to another, and the above results indicated that bacteria usually have the ability to resist multiple antibiotics.

Bacterial resistance is associated with carried ARGs, which are commonly detected in the environment as an emerging contaminant [47]. To date, the research on ARGs in aquatic environments has mainly focused on sulfa drugs (e.g., trimethoprim), tetracycline, quinolones, and macrolides. The most frequently detected ARGs were sulfonamide ARGs, including sul1, sul2, and sul3; tetracycline ARGs, including tetA, tetB, tetC, tetD, tetE, tetG, tetH, tetM, tetO, tetQ, tetS, tetW, tetX, and tetZ; quinolone ARGs, including qnrA, qnrD, and qnrS; and macrolide ARGs, including ermC, ermA, emf (C), and mph (G) (Table 2). Their relative abundance in aquaculture waters ranges between 8.57×10^{-7} and 3.45×10^{-2} copies/16S rRNA (Table 2). In most studies, sulfa resistance genes and tetracycline resistance genes usually have high relative abundance in cultured waters. For example, a study detected the highest levels of sul2 in mariculture farms, with relative expression ranging from 0.029 to 0.075 [48]. Researchers detected abundant tetracycline resistance genes in fish culture environments in South Jeolla province and Jeju Island, among which tetB and tetD accounted for 74.8–98.0% of the total ARGs [49]. TetA, sul1, and sul2 were also frequently detected in aquaculture effluents [50].

Table 2. ARGs abundance in aquaculture water.

Antibiotics	ARGs	Abundance	Origin Location
Tetracycline	tetB/tetD/tetE/tetH/tetX/tetZ/tetQ	4.24×10^{-3}~1.46×10^{-2} a	Fish culture environments in South Jeolla province and Jeju Island [49]
	tetA/tetC/tetG/tetM	-/-/-/1.16×10^{2} b	Hainan Dongzhai Port [51]
	tetA/tetB/tetC/tetH/tetM/tetO	5.46×10^{-4}~1.61×10^{-3} a/2.05×10^{-5}~6.35×10^{-5} a/4.71×10^{-4}~1.68×10^{-2} a/8.47×10^{-6}~2.37×10^{-5} a/8.84×10^{-5}~4.97×10^{-3} a/5.90×10^{-6}~6.99×10^{-5} a	West Coast of Pearl River Estuary [52]
	tetW/tetG/tetX	-	Jiangsu Province Baima Lake Aquaculture Farm [53]
	tetA/tetB/tetM/tetS/	-	Aquaculture farms in Sri Lanka [54]
Macrolides	ermC	-	Coastal farms in Jeollanam-do Province and Jeju Island, Korea [49]
	ermA	-	Pearl River Delta South China Zhongshan [55]
	emf(C)/mph(G)	-	A fish farm in Japan [56]
Quinolones	qnrA/qnrD/qnrS	-	Shuidongwan, Maoming City, Guangdong [57]
	qnrS	(100%); 9.97×10^{-3} a	Hangzhou Bay, Xiaoshan, Shaoxing, Cixi, Pinghu [58]
	qnrS	(100%); 8.57×10^{-7}~3.45×10^{-2} a; 8.68~1.37×10^{6} b	Hainan Dongzhai Port [51]
	qnrA/qnrD/qnrS	-	Guangzhou Pearl River Delta Estuarine Aquaculture Zone [59]
Sulfonamides	sul1	$3.29 \times 10^{2} \pm 4.81 \times 10^{2}$ b; $2.72 \times 10^{2} \pm 3.57 \times 10^{2}$ b; $4.08 \times 10^{2} \pm 2.06 \times 10^{2}$ b	Waste lagoon of JFRC of the Ministry of Agriculture in Saudi Arabia [60]
	sul1/sul2/sul3	(100%/100%/25%); 1.94×10^{-2} a/-/-	Hangzhou Bay, Xiaoshan, Shaoxing, Cixi, Pinghu [58]
	sull/sul2	1.21×10^{5} b/5.13×10^{5} b	Hainan Dongzhai Port [51]

Note: a is relative abundance (copies/16S rRNA); b is absolute abundance (copies mL^{-1}; copies g^{-1}); - is data not explicitly reported; the detection rate is shown in brackets.

3.2. *Type and Abundance of Antibiotics, ARB, and ARGs in Aquaculture Sediments*

3.2.1. Occurrence of Antibiotics in Sediments

Sediment is a reservoir of antibiotics in cultured water bodies. Antibiotics in the water column can be adsorbed on suspended particulate matter and settle into the sediment [61]. In addition, antibiotics carried by unused bait entering natural water bodies would eventually enter the sediment [62]. Thus, although antibiotics are susceptible to photolysis and hydrolysis in water, the decay process of antibiotics is slowed down when they enter the sediment, eventually leading to long-term accumulation of antibiotics [40].

Currently, the most commonly mentioned categories of antibiotics in sediment are tetracyclines, sulfonamides, fluoroquinolones, and macrolides [63–65]. For example, tetracyclines and macrolide antibiotics were the main antibiotics detected in the sediments of the Liaohe River basin, with concentrations ranging from not detected (ND) to 405 ng g^{-1} and ND to 375 ng g^{-1}, respectively. Fluoroquinolones and sulfonamides concentrations ranged from ND to 117 ng g^{-1} and ND to 2.63 ng g^{-1}, respectively [66]. Fluoroquinolones were the main antibiotics in Lake Baiyangdian, accounting for 80.9% of the total concentration [67]. Similarly, fluoroquinolones accounted for the largest proportion of antibiotics detected in North Bay sediments, followed by sulfonamides, with these two compounds accounting for 91.0–100% of the total [43]. These studies also showed that the concentrations of antibiotics in sediments are generally 3~4 orders of magnitude higher than that in water, suggesting that sediments are a more stable site for the presence of antibiotics.

3.2.2. Status of ARB and ARGs in Aquaculture Sediments

In aquaculture, antibiotic residues, feces, as well as feed with farmed organisms could gradually accumulate in the sediment, likely driving antibiotic-resistant pathogens to evolve [55,68]. Additionally, the dominant bacterial groups are commonly varied in different sedimentary environments. For example, sediments from shrimp culture ponds were studied for three culture years, and the five most abundant phyla were *Proteobacteria*, *Chlo-*

roflexi, *Firmicutes*, *Bacteroidetes*, and *Planctomycetes* [69]. In the sediment of carp farm ponds, *Proteobacteria*, *Bacteroidetes*, and *Actinobacteria* were the dominant bacterial groups [70]. Moreover, in the sediment of a cultured pond during the suspension of bait [71], *Proteobacteria* were the dominant phylum, followed by *Chloroflexi*, *Bacteroidetes*, and *Cyanobacteria*. Potentially, different bacterial community structures in sediments with antibiotic residues could further induce ARGs contamination.

The types of ARGs in sediment were similar to those in the water body: sulfonamide ARGs, mainly including sul1, sul2, and sul3; tetracycline ARGs, including tetA, tetB, tetC, tetG, tetH, tetM, tetO, tetT, tetW, and tetX; quinolone ARGs, such as qnrA, qnrB, qnrD, and qnrS; and macrolide ARGs ermB and ermA (Table 2). The species of ARGs are different in different culture areas. The abundance of sulfonamide *sul1* and *sul2* in aquaculture sediments of the Pearl River Estuary was relatively high, with *sul1* considered a potential indicator of ARG [59]. Multidrug resistance genes were the most common ARGs in sediments from the Shatian Lake aquaculture area, with a detection rate of 24.5%. Fluoroquinolone ARGs represented 8.8%, tetracyclines ARGs represented 6.8%, and the detection rate of macrolides ARGs was 2.9% [63]. The relative abundance of ARGs in aquaculture environmental sediments was 1.25×10^{-6}–1.08×10^{-1} copies/16S rRNA, which was higher than that in the water column. Yuan et al. [58] found that the spread of ARGs over time led to their accumulation in sediments, and sediments likely enhanced the spread of antibiotic resistance. The abundance of ARGs in sediments is also influenced by the culture pattern. It is shown that the abundance of ARGs in sediments from different modes of culture ponds (duck–fish mixed and separate culture), and the absolute abundance of ARGs, was significantly higher in integrated mode ponds than in non-integrated mode ponds [55]. Similarly, the abundance of ARGs in the sediment of an integrated duck and shrimp farm was more than one times higher than that of the shrimp monoculture ponds [72]. These findings implied that the abundance of ARGs was higher in the sediments of integrated mode ponds.

3.3. Type and Abundance of Antibiotics, ARB, and ARGs in Aquaculture Organisms

3.3.1. Retention of Antibiotics in Aquaculture Organisms

Antibiotics are widely used to treat fish diseases and promote the output of aquatic products. However, intensive research has shown that only a small fraction of the antibiotics in aquaculture organisms could be broken down, with most of them being retained in the body of the aquaculture organism or excreted into the external environment. For example, a study found that only 20–30% of the antibiotics added are absorbed by fish, while most are input into water [73]. Taking the Haihe River as an example, the highest level of sulfamethoxazole in aquatic organisms was 68 ng g^{-1}, and the average level of sulfamethoxazole in the water was 2.0×10^2 ng L^{-1} [74]. The level of antibiotic residues in aquatic organisms is influenced by several factors. It is clear that the amount of antibiotics input artificially during the farming process is an important influencing factor. Secondly, antibiotics in the pristine environment might also affect antibiotic levels in aquatic organisms. Antibiotic residues were detected in wild marine organisms that had not been exposed to artificially added antibiotics, which was most likely caused by the uptake and enrichment of antibiotics in the pristine environment by aquatic animals [75]. In addition, fishmeal is also a potential factor for the continuous accumulation of antibiotics in aquatic animal products [76]. Approximately 25% of the harvest from global commercial marine fisheries is used for the production of fishmeal and fish oil. Fishmeal is produced from fish tissue and is often used as feed for aquaculture organisms [77]. About 30–70% of fish tissues are further processed and made into bait in China's aquaculture industry [78], enriching the antibiotics in aquatic organisms from generation to generation, possibly resulting in greater levels of resistance [79]. Moreover, other pollutants (e.g., microplastics) could also aggravate the bioaccumulation of antibiotics in mollusks (such as *Mytilus* spp.).

The types of antibiotic residues in aquatic products are diverse. Intestinal antibiotics were detected in fish from five fish farms, revealing the presence of one sulfonamide, one quinolone, two tetracyclines, and three macrolides, among which the macrolide azithromycin was detected at a high rate in Tianjin, China [80]. Analysis of the intestines of South American white shrimp revealed the presence of sulfadiazine, ciprofloxacin, and norfloxacin in all samples, which belong to the sulfonamides and quinolones [81]. Moreover, antibiotics in the intestines of squid ginseng were detected, showing high levels of amikacin, gentamicin, ampicillin, and ribostamycin, which belong to the aminoglycoside and β-lactam antibiotic families [82]. Among them, the aminoglycoside amikacin was the most abundant antibiotic. Overall, in view of the abundance of diverse residual antibiotics in aquatic products, over 30 countries have banned the use of antibiotic growth promoters because of concerns over human health and environmental conservation.

3.3.2. Status of ARB and ARGs in Aquaculture Organisms

The presence of ARB commonly varies among aquaculture organisms. In the intestinal flora of spotted catfish, the dominant phyla were *Bacteroidetes, Fusobacteria, Firmicutes* [83]. *Aeromonas hydrophila* and *Pseudomonas aeruginosa* were the most abundant bacteria in Nile tilapia [84]. The representative resistant bacteria in the intestines of South American shrimp were *Aeromonas, Kinetobacter, Escherichia, Sphingomonas, Enterobacter, Lactobacillus, Pantotrichum, Bacillus, Leuconostoc*, and *Vibrio* [85]. Differently, *Acinetobacter, Bacillus*, and *Psychrobacter* species were most prevalent in snow crab [86]. Thus, the dominant ARB would vary in different host organisms.

Exogenous addition of antibiotics in aquaculture would induce one or a variety of resistance genes produced by bacteria in aquaculture organisms, leading to antibiotic resistance development. In aquaculture organisms, there are four major classes of ARGs, comprising those conferring resistance to quinolones, macrolides, tetracyclines, and sulfonamides (Table 2), among which sul1, sul2, qnrA, qnrS, macB, tetA, and tetG were more likely to be detected [82,87]. Currently, metagenomics is increasingly used to detect ARGs in aquaculture organisms or an environmental medium, in addition to PCR assays or gene chip technology. Moreover, studies on ARGs in aquaculture organisms have mainly focused on intestinal flora because it is the dominant reservoir of ARGs, although other studies have focused on liver or hepatopancreas micro-organisms. The type and abundance of resistance genes in aquaculture organisms are also related to the commonly used local antibiotics. In areas where the use of sulfonamides is more prevalent than tetracyclines, sulfonamide resistance genes were prevalent and were present in the highest concentrations [88]. Notably, antibiotic misuse promotes the prevalence of ARGs, contributes to the formation of multidrug resistant bacteria, and poses a significant threat to human health and the environment [87].

4. Association between Antibiotics, ARB, and ARGs in Aquaculture Ecosystems

4.1. Mechanism of Interaction between Antibiotics, ARB, and ARGs

Generally, antibiotics act as bactericides by blocking the synthesis of bacterial cell membranes or cell walls to kill or inhibit pathogenic bacteria [89]. Research has shown that quinolone antibiotics inhibit bacterial growth by inhibiting bacterial DNA topoisomerase [90]. Macrolides bind to large ribosome subunits near peptidyl transferase centers and cause cell growth to stall by inhibiting protein synthesis [91]. Tetracyclines inhibit protein synthesis by preventing the attachment of aminoacyl-tRNA to the ribosomal receptor (A) site, thus exerting an antibacterial effect [92]. Sulfonamides interact with dihydrofolate synthase and prevent the formation of tetrahydrofolate. Tetrahydrofolic acid is a precursor of folic acid. Bacteria cannot grow and reproduce properly without folic acid [93]. Some pathogens could be killed quickly under antibiotic stress; yet, some also survive via corresponding resistance mechanisms related to protein or gene changes (Figure 1). There are two main mechanisms of antibiotic resistance in bacteria. One mechanism involves physically preventing the antibiotic from binding to the target protein, and the other involves altering the structure of the target protein on which the antibiotic acts to resist the antibiotic. The first mechanism includes reducing the permeability of the cell membrane to antibiotics and export of the antibiotic via efflux pumps. The second mechanism causes mutation of the target structure or temporary modification of the target structure [94].

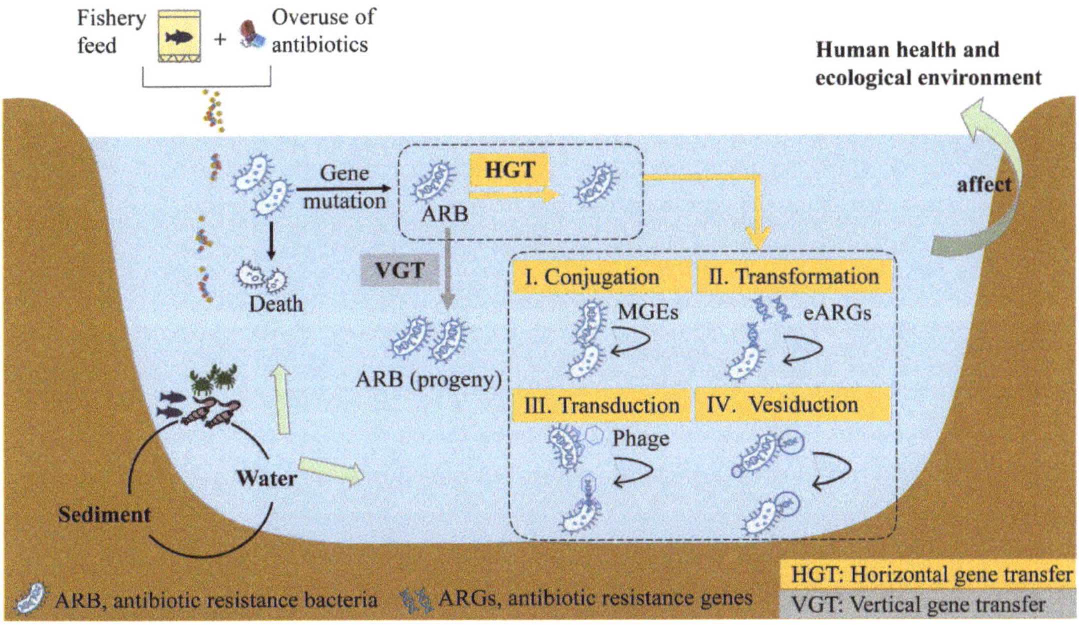

Figure 1. ARGs and ARB occurrence and transfer patterns in the aquaculture environment. MGE, mobile genetic element; eARGs, extracellular ARGs.

The acquisition or spreading of ARGs in resistant bacteria occurs mainly through gene mutation and horizontal and vertical transfer (Figure 1). Residues of antibiotics in an aquaculture environment can exert antibiotic selection pressure on bacterial communities. Under this pressure, bacteria could acquire antibiotic resistance (via ARGs) through random chromosomal mutations [53]. ARGs are transmitted by two main processes: VGT, attributed to bacterial host propagation, and HGT, attributed to the transfer of ARGs between different bacterial cells through mobile elements [95]. HGT is an important pathway for the spread of ARGs, where mobile genetic elements (MGEs) transfer ARGs to other host bacteria, whether they are in intercellular contact or not [96]. Currently, the horizontal transfer includes four pathways: classical conjugation, transformation, transduction, and vesiduction [97]. Conjugation is the process of transferring ARGs from donor to recipient cells through direct cellular contact of MGEs, which can carry multiple ARGs [98]. Transformation is the uptake of extracellular plasmid or chromosomal DNA by competent cells and its stable integration into the bacterial genome to functionally express antibiotic resistance [99]. In addition, adsorbed extracellular ARGs (eARGs) tend to transfer to competent cells more than free eARGs [100]. Transduction is achieved by phage infection of donor cells. The DNA containing ARGs in the donor cell is encapsulated in the capsid of the phage. When the donor cell is lysed by the phage, the ARGs are transferred to the recipient cell along with phage infection. Vesiduction is a recently discovered mode of horizontal transfer, which refers to the spread of ARGs through extracellular vesicles. The membrane vesicles produced by the cells are able to carry the DNA of the source bacterial strain through the environment and protect the DNA from degradation [101].

In most cases, gene mutation, horizontal transfer, and vertical transfer work together at the same time. Briefly, the resistance to an antibiotic obtained by mutation of an antibiotic target gene can be transferred by a plasmid or phage. MGEs carrying ARGs are able to insert into bacterial chromosomes, resulting in chromosomally encoded resistance, which in turn allows such bacteria to transmit ARGs by vertical transfer. Some MGEs provide integrase genes that enable ARGs to bind to the host chromosome, while some MGEs are able to utilize integrase genes located on the host bacterial chromosome that allow them to excise themselves from the chromosome after they have completed their role [102]. After binding to the host chromosome, ARGs are selectively retained only when they confer an advantage on the host strain [103].

4.2. Ecological Risks of Antibiotics Residues, ARB, and ARGs in Aquatic Ecosystems

Early studies reported the adverse environmental effects of antibiotic residues in aquaculture environments, and subsequently, many studies have focused on the risk assessment of antibiotics in aquaculture environments [104]. The ecological risks are assessed using Formula (1) [105].

$$\text{RQs} = \text{PEC or MEC/PNEC} \quad (1)$$

RQs, the risk quotients;
PEC, the predicted environmental concentration;
MEC, measured environmental concentration;
PNEC, predicted no-effect concentration.

The PNEC is computed using Formula (2).

$$\text{PNECs} = \text{LC}_{50} \text{ or } \text{EC}_{50}/\text{AF} \quad (2)$$

LC_{50}, half lethal concentration;
EC_{50}, half maximal effective concentration;
AF, assessment factor.

The classification method of the RQ proposed by Norton et al. [106] can evaluate whether drugs pose risks to the ecosystem. If RQ > 1, the regional pollution will be considered to be in the high-risk state [107]; if 0.1 < RQ < 1, the pollution in this region is in a state of medium risk [108]; and if RQ < 0.1, the region is at a low risk of contamination [109]. Ashbolt et al. [110] proposed a threshold assessment method for the occurrence of ARB based on the minimum selective concentration (MSC). The pathogenic risk of a pathogen was assessed based on problem formulation (hazard, risk setting, and pathway), hazard exposure assessment (ARB and ARGs), and the dose–response relationship. This microbiological risk assessment (MRA) is used to assess the level of pathogen exposure and its subsequent risk to human health qualitatively or quantitatively. However, in vitro databases and computational toxicology studies are key factors for minimizing the uncertainty of ARG risk assessment. Compared with that for antibiotics, the overall assessment system for emerging contaminants is not very robust and needs further study.

5. Conclusions and Prospects

The detection method, concurrence, and contamination status of antibiotics, ARB, and ARGs in aquaculture ecosystems were summarized. Currently, the major methods for detecting antibiotics, ARB, and ARGs are UPLC-MS/MS, 16S rRNA sequencing, and metagenomics, respectively. Antibiotic concentrations in sediment are generally 3 to 4 orders of magnitude higher than those in water. Similarly, the relative abundance of ARGs in sediment is also much higher than that in aquaculture water. However, in aquaculture organisms, no obvious patterns in changes in the category or concentration of antibiotics were detected. There were also no clear patterns in the categories of ARB in the aquaculture environment. ARB in aquaculture mainly belonged to the *Cyanobacteria*, *Proteobacteria*, *Actinobacteria*, *Chloroflexi*, *Bacteroidetes*, and *Planctomycetes*. The interactions between antibiotics, ARB, and ARGs were also clarified. The key mechanisms of resistance to antibiotics in bacteria include reducing the permeability of the cell membrane to antibiotics, enhancing antibiotic efflux, and changing the structure of antibiotic target proteins. Furthermore, genetic mutations, horizontal transfer, and vertical transfer are the common methods of inducing and spreading ARGs. Horizontal transfer is an important pathway for the spread of ARGs, which includes conjugation, transformation, transduction, and vesiduction.

Based on the problems in current research works on ARGs, the following research questions and future outlook should be taken into consideration in future research studies: (1) Our current understanding of multidrug resistant ARGs is still relatively shallow, and the relevant databases need to be enriched; (2) eARGs contribute to the spread of antibiotic resistance in the environment, and their persistence is detrimental to the environment; however, this persistence in the environment is more complicated, still requiring further in-depth study; (3) the antibiotic resistome is extremely important for the study of ARG pollution control. Moreover, it is necessary to monitor antibiotics regularly to determine whether silent genes are expressed, such that corresponding measures can be taken for timely control of pollution.

Author Contributions: Conceptualization, X.Y., Z.L. (Ziqing Lv) and H.Z.; Writing—Original draft, X.Y., Z.L. (Ziqing Lv) and H.Z.; Writing—Review and editing, X.Y., Z.L. (Ziqing Lv), Z.Z., Y.H., Z.L. (Zhiquan Liu) and H.Z. All authors have read and agreed to the published version of the manuscript.

Funding: This work was supported by Key Research & Development Project of the Science and Technology Department of Zhejiang Province (Grant No. 2021C02048) and Natural Science Foundation of Zhejiang Province, China (Grant No. LY22C030004).

Institutional Review Board Statement: Not applicable.

Informed Consent Statement: Not applicable.

Data Availability Statement: Not applicable.

Acknowledgments: The authors wish to express gratitude to Timothy R. Seastedt for improving the English quality. We are also grateful to anonymous reviewers and editors for their constructive comments, which greatly improved the manuscript.

Conflicts of Interest: The authors declare no conflict of interest.

References

1. Wenning, R. The state of world fisheries and aquaculture (sofia) 2020 report. *Integr. Environ. Assess. Manag.* **2020**, *16*, 800–801.
2. Yılmaz, S.; Ergün, S.; Yiğit, M.; Yılmaz, E. An extensive review on the use of feed additives against fish diseases and improvement of health status of fish in turkish aquaculture sector. *Aquac. Stud.* **2022**, *22*, AQUAST710. [CrossRef]
3. Assefa, A.; Abunna, F. Maintenance of fish health in aquaculture, review of epidemiological approaches for prevention and control of infectious disease of fish. *Vet. Med. Int.* **2018**, *2018*, 5432497. [CrossRef]
4. Dong, H.; Chen, Y.; Wang, J.; Zhang, Y.; Zhang, P.; Li, X.; Zou, J.X.; Zhou, A.J. Interactions of microplastics and antibiotic resistance genes and their effects on the aquaculture environments. *J. Hazard. Mater.* **2021**, *403*, 123961. [CrossRef]
5. Davies, J.; Davies, D. Origins and Evolution of Antibiotic Resistance. *Microbiol. Mol. Biol. Rev.* **2010**, *74*, 417–433. [CrossRef]
6. Knapp, C.W.; Dolfing, J.; Ehlert, P.A.I.; Graham, D.W. Evidence of Increasing Antibiotic Resistance Gene Abundances in Archived Soils since 1940. *Environ. Sci. Technol.* **2010**, *44*, 580–587. [CrossRef] [PubMed]
7. Tang, K.L.; Caffrey, N.P.; Nobrega, D.B.; Cork, S.C.; Ronksley, P.E.; Barkema, H.W.; Polachek, A.J.; Ganshorn, H.; Sharma, N.; Kellner, J.D.; et al. Restricting the use of antibiotics in food-producing animals and its associations with antibiotic resistance in food-producing animals and human beings, a systematic review and meta-analysis. *Lancet Planet. Health* **2017**, *1*, e316–e327. [CrossRef]
8. Yang, F.; Mao, D.; Luo, Y.; Wang, Q.; Mu, Q.H. Horizontal transfer of antibiotic resistance genes in the environment. *Chin. J. Appl. Ecol.* **2013**, *24*, 2993–3002.
9. Amarasiri, M.; Sano, D.; Suzuki, S. Understanding human health risks caused by antibiotic resistant bacteria (ARB) and antibiotic resistance genes (ARG) in water environments, Current knowledge and questions to be answered. *Crit. Rev. Environ. Sci. Technol.* **2020**, *50*, 2016–2059. [CrossRef]
10. Martinez, J.L. Antibiotics and antibiotic resistance genes in natural environments. *Science* **2008**, *321*, 365–367. [CrossRef]
11. Martinez, J.L.; Baquero, F.; Andersson, D.I. Predicting antibiotic resistance. *Nat. Rev. Microbiol.* **2007**, *5*, 958–965. [CrossRef] [PubMed]

12. Li, B.; Qiu, Y.; Song, Y.; Lin, H.; Yin, H. Dissecting horizontal and vertical gene transfer of antibiotic resistance plasmid in bacterial community using microfluidics. *Environ. Int.* **2019**, *131*, 105007. [CrossRef] [PubMed]
13. Su, H.C.; Pan, C.G.; Ying, G.G.; Zhao, J.L.; Zhou, L.J.; Liu, Y.S.; Tao, R.; Zhang, R.Q.; He, L.Y. Contamination profiles of antibiotic resistance genes in the sediments at a catchment scale. *Sci. Total Environ.* **2014**, *490*, 708–714. [CrossRef] [PubMed]
14. Huddleston, J.R. Horizontal gene transfer in the human gastrointestinal tract, potential spread of antibiotic resistance genes. *Infect. Drug Resist.* **2014**, *7*, 167–176. [CrossRef]
15. Li, S.; Zhang, C.; Li, F.; Hua, T.; Zhou, Q.; Ho, S.H. Technologies towards antibiotic resistance genes (ARGs) removal from aquatic environment, a critical review. *J. Hazard. Mater.* **2021**, *411*, 125148. [CrossRef]
16. Chen, J.; Huang, L.; Wang, Q.; Zeng, H.; Xu, J.; Chen, Z. Antibiotics in aquaculture ponds from Guilin, South of China, Occurrence, distribution, and health risk assessment. *Environ. Res.* **2022**, *204*, 112084. [CrossRef]
17. Xiong, W.; Sun, Y.; Zhang, T.; Ding, X.; Li, Y.; Wang, M.; Zeng, Z.L. Antibiotics, antibiotic resistance genes, and bacterial community composition in fresh water aquaculture environment in China. *Microb. Ecol.* **2015**, *70*, 425–432. [CrossRef]
18. Kim, S.-C.; Carlson, K. Occurrence of ionophore antibiotics in water and sediments of a mixed-landscape watershed. *Water Res.* **2006**, *40*, 2549–2560. [CrossRef]
19. Rezk, M.; Riad, S.; Khattab, F.; Marzouk, H. Multi-residues determination of antimicrobials in fish tissues by HPLC-ESI-MS/MS method. *J. Chromatogr. B* **2015**, *978*, 103–110. [CrossRef]
20. Park, Y.K.; Fox, L.K.; Hancock, D.D.; McMahan, W.; Park, Y.H. Prevalence and antibiotic resistance of mastitis pathogens isolated from dairy herds transitioning to organic management. *J. Vet. Sci.* **2012**, *13*, 103–105. [CrossRef]
21. Zhang, Z.; Cui, N.; Wang, Y.; Zhang, H.; Zhao, W.; Ma, L. Determination of tetracycline residues in Qinghai yak by time-resolved fluorescence immunochromatography. *Vet. Guide* **2021**, *2021*, 135–136.
22. Liu, Y.; Li, B.; Qiu, Y.; Ding, H.; Wang, Y.; Wang, S.; Li, J. Distribution and removal characteristics of antibiotic resistant bacteria in wastewater treatment plants at Wuxi. *Acta Sci. Circumstantiae* **2017**, *37*, 2114–2121.
23. Le, T.H.; Ng, C.; Chen, H.; Yi, X.Z.; Koh, T.H.; Barkham, T.M.S.; Zhou, Z.; Gin, K.Y.-H. Occurrences and characterization of antibiotic-resistant bacteria and genetic determinants of hospital wastewater in a tropical country. *Antimicrob. Agents Chemother.* **2016**, *60*, 7449–7456. [CrossRef] [PubMed]
24. Zhu, G.; Xue, C.; Fan, H.; Wu, N. Research progress of antibiotic resistance genes in environment. *Prev. Med.* **2020**, *32*, 1121–1125+1129.
25. Che, J.; Chen, X.; Li, J.; Lu, J. Application of different detection techniques for bacterial drug resistance test. *Dis. Surveill.* **2017**, *32*, 757–763.
26. Dang, C.; Xia, Y.; Zheng, M.; Liu, T.; Liu, W.; Chen, Q.; Ni, J.R. Metagenomic insights into the profile of antibiotic resistomes in a large drinking water reservoir. *Environ. Int.* **2020**, *136*, 105449. [CrossRef]
27. Su, J.; Huang, F.; Zhu, Y. Antibiotic resistance genes in the environment. *Biodivers. Sci.* **2013**, *21*, 481–487.
28. An, X.L.; Su, J.Q.; Li, B.; Ouyang, W.Y.; Zhao, Y.; Chen, Q.L.; Cui, L.; Chen, H.; Gillings, M.R.; Zhang, T. Tracking antibiotic resistome during wastewater treatment using high throughput quantitative PCR. *Environ. Int.* **2018**, *117*, 146–153. [CrossRef]
29. Hu, H.Y.; Liu, H.M.; Meng, L.; Dong, L.; Lan, T.; Zheng, N.; Chen, J.B.; Wang, J.Q. Application of metagenomics in the detection of microbial antibiotic resistance genes. *Microbiol. Chain.* **2019**, *46*, 3110–3123.
30. Sun, Y.; Clarke, B.; Clarke, J.; Li, X. Predicting antibiotic resistance gene abundance in activated sludge using shotgun metagenomics and machine learning. *Water Res.* **2021**, *202*, 117384. [CrossRef]
31. Forsberg, K.J.; Reyes, A.; Wang, B.; Selleck, E.M.; Sommer, M.; Dantas, G. The Shared Antibiotic Resistome of Soil Bacteria and Human Pathogens. *Science* **2012**, *337*, 1107–1111. [CrossRef]
32. Garcia-Galan, M.J.; Villagrasa, M.; Diaz-Cruz, M.S.; Barcelo, D. LC-QqLIT MS analysis of nine sulfonamides and one of their acetylated metabolites in the Llobregat River basin. Quantitative determination and qualitative evaluation by IDA experiments. *Anal. Bioanal. Chem.* **2010**, *397*, 1325–1334. [CrossRef]
33. Hedberg, N.; Stenson, I.; Pettersson, M.N.; Warshan, D.; Nguyen-Kim, H.; Tedengren, M.; Kautsky, N. Antibiotic use in Vietnamese fish and lobster sea cage farms; implications for coral reefs and human health. *Aquaculture* **2018**, *495*, 366–375. [CrossRef]
34. Li, S.; Shi, W.Z.; Li, H.M.; Xu, N.; Zhang, R.J.; Chen, X.J.; Sun, W.L.; Wen, D.H.; He, S.L.; Pan, J.G.; et al. Antibiotics in water and sediments of rivers and coastal area of Zhuhai City, Pearl River estuary, south China. *Sci. Total Environ.* **2018**, *636*, 1009–1019. [CrossRef]
35. Kim, S.-C.; Carlson, K. Temporal and spatial trends in the occurrence of human and veterinary antibiotics in aqueous and river sediment matrices. *Environ. Sci. Technol.* **2007**, *41*, 50–57. [CrossRef]
36. Ruan, Y.; Chen, J.; Guo, C.; Chen, S.; Wang, S.; Wang, Y. Distribution characteristics of typical antibiotics in surface water and sediments from freshwater aquaculture water in Tianjin suburban areas, China. *J. Agro-Environ. Sci.* **2011**, *30*, 2586–2593.
37. Sargenti, M.; Bartolacci, S.; Luciani, A.; Biagio, K.D.; Baldini, M.; Galarini, R.; Giusepponi, D.; Capuccella, M. Investigation of the correlation between the use of antibiotics in aquaculture systems and their detection in aquatic environments, a case study of the Nera River aquafarms in Italy. *Sustainability* **2020**, *12*, 5176. [CrossRef]
38. Song, R.; Guo, X.; Lu, S.; Liu, X.; Wang, X. Occurrence and source analysis of antibiotics and antibiotic resistance genes in surface water of East Dongting Lake basin. *Res. Environ. Sci.* **2021**, *34*, 2143–2153.
39. Pereira, A.M.; Silva, L.J.; Meisel, L.M.; Pena, A. Fluoroquinolones and tetracycline antibiotics in a Portuguese aquaculture system and aquatic surroundings, occurrence and environmental impact. *J. Toxicol. Environ. Health A* **2015**, *78*, 959–975. [CrossRef]

40. Rico, A.; Oliveira, R.; McDonough, S.; Matser, A.; Khatikarn, J.; Satapornvanit, K.; Nogueira, A.J.A.; Soares, A.M.V.M.; Domingues, I.; Van den Brink, P.J. Use, fate and ecological risks of antibiotics applied in tilapia cage farming in Thailand. *Environ. Pollut.* **2014**, *191*, 8–16. [CrossRef]
41. Hamscher, G.; Priess, B.; Nau, H. A survey of the occurrence of various sulfonamides and tetracyclines in water and sediment samples originating from aquaculture systems in Northern Germany in summer 2005. *Arch. Lebensm.* **2006**, *57*, 97–101.
42. Le, T.X.; Munekage, Y. Residues of selected antibiotics in water and mud from shrimp ponds in mangrove areas in Viet Nam. *Mar. Pollut. Bull.* **2004**, *49*, 922–929. [CrossRef]
43. Zhang, R.; Pei, J.; Zhang, R.; Wang, S.; Zeng, W.; Huang, D.; Wang, Y.; Zhang, Y.Y.; Wang, Y.H.; Yu, K.F. Occurrence and distribution of antibiotics in mariculture farms, estuaries and the coast of the Beibu Gulf, China, Bioconcentration and diet safety of seafood. *Ecotoxicol. Environ. Saf.* **2018**, *154*, 27–35. [CrossRef]
44. Patangia, D.V.; Ryan, C.A.; Dempsey, E.; Stanton, C.; Ross, R.P. Vertical transfer of antibiotics and antibiotic resistant strains across the mother/baby axis. *Trends Microbiol.* **2021**, *30*, 47–56. [CrossRef]
45. Zhou, A.; Xie, S.; Junaid, M.; Sun, D.; Tang, H.; Chuan, J.; Li, X.; Xu, G.H.; Zou, J.X. First insight into the environmental microbial communities associated with potentially pathogenic strains in pond cultured tilapia (*Oreochromis niloticus*) at various growth stages based on 16S, 18S, and ITS2 rRNA gene amplicons sequencing. *Aquaculture* **2021**, *532*, 736007. [CrossRef]
46. Suyamud, B.; Lohwacharin, J.; Yang, Y.; Sharma, V.K. Antibiotic resistant bacteria and genes in shrimp aquaculture water, Identification and removal by ferrate (VI). *J. Hazard. Mater.* **2021**, *420*, 126572. [CrossRef]
47. Pruden, A.; Pei, R.; Storteboom, H.; Carlson, K.H. Antibiotic resistance genes as emerging contaminants, Studies in northern Colorado. *Environ. Sci. Technol.* **2006**, *40*, 7445–7450. [CrossRef]
48. Chen, C.; Zheng, L.; Zhou, J.; Zhao, H. Persistence and risk of antibiotic residues and antibiotic resistance genes in major mariculture sites in Southeast China. *Sci. Total Environ.* **2017**, *580*, 1175–1184. [CrossRef]
49. Jang, H.; Kim, Y.; Choi, S.; Lee, Y.; Shin, S.; Unno, T.; Kim, Y.M. Prevalence of antibiotic resistance genes from effluent of coastal aquaculture, South Korea. *Environ. Pollut.* **2018**, *233*, 1049–1057. [CrossRef]
50. Makowska, N.; Koczura, R.; Mokracka, J. Class 1 integrase, sulfonamide and tetracycline resistance genes in wastewater treatment plant and surface water. *Chemosphere* **2016**, *144*, 1665–1673. [CrossRef]
51. Jiang, C.; Li, P.; Li, S.; Diiao, X.; Huang, W.; Wang, D.; Ye, C.X. Pollution characteristics of antibiotic resistance genes in seawater and sediment of Dongzhai Harbor, Hainan Province. *Ecol. Environ. Sci.* **2019**, *28*, 128–135.
52. Liang, X.; Nie, X.; Shi, Z. Preliminary studies on the occurrence of antibiotic resistance genes in typical aquaculture area of the Pearl River Estuary. *Environ. Sci.* **2013**, *34*, 4073–4080.
53. Shen, X.; Jin, G.; Zhao, Y.; Shao, X. Prevalence and distribution analysis of antibiotic resistance genes in a large-scale aquaculture environment. *Sci. Total Environ.* **2020**, *711*, 134626. [CrossRef] [PubMed]
54. Liyanage, G.; Manage, P.M. Occurrence and distribution of tetracycline resistance determinants and their pollution profile in selected aquaculture environments in Sri Lanka. *J. Natl. Sci. Found. Sri Lanka* **2019**, *47*, 455–465. [CrossRef]
55. Huang, L.; Xu, Y.; Xu, J.; Ling, J.; Chen, J.; Zhou, J.; Zheng, L.; Du, Q.P. Antibiotic resistance genes (ARGs) in duck and fish production ponds with integrated or non-integrated mode. *Chemosphere* **2017**, *168*, 1107–1114. [CrossRef]
56. Nonaka, L.; Maruyama, F.; Suzuki, S.; Masuda, M. Novel macrolide-resistance genes, *mef* (C) and *mph* (G), carried by plasmids from Vibrio and *Photobacterium* isolated from sediment and seawater of a coastal aquaculture site. *Lett. Appl. Microbiol.* **2015**, *61*, 1–6. [CrossRef]
57. Su, H.; Hu, X.; Wang, L.; Xu, W.; Xu, Y.; Wen, G.; Li, Z.J.; Cao, Y.C. Contamination of antibiotic resistance genes (ARGs) in a typical marine aquaculture farm, source tracking of ARGs in reared aquatic organisms. *J. Environ. Sci. Health B* **2020**, *55*, 220–229. [CrossRef]
58. Yuan, J.; Ni, M.; Liu, M.; Zheng, Y.; Gu, Z. Occurrence of antibiotics and antibiotic resistance genes in a typical estuary aquaculture region of Hangzhou Bay, China. *Mar. Pollut. Bull.* **2019**, *138*, 376–384. [CrossRef]
59. Su, H.C.; Liu, S.; Hu, X.J.; Xu, X.R.; Xu, W.J.; Xu, Y.; Li, Z.J.; Wen, G.L.; Liu, Y.S.; Cao, Y.C. Occurrence and temporal variation of antibiotic resistance genes (ARGs) in shrimp aquaculture, ARGs dissemination from farming source to reared organisms. *Sci. Total Environ.* **2017**, *607*, 357–366. [CrossRef]
60. Sanawar, H.; Xiong, Y.; Alam, A.; Croué, J.P.; Hong, P.Y. Chlorination or monochloramination, balancing the regulated trihalomethane formation and microbial inactivation in marine aquaculture waters. *Aquaculture* **2017**, *480*, 94–102. [CrossRef]
61. Hu, Y.; Wang, J.; Ma, D. Research progress on environmental effect of antibiotic agents in marine aquaculture. *Mar. Environ. Sci.* **2004**, *23*, 76–80.
62. Chen, H.; Liu, S.; Xu, X.; Liu, S.; Zhou, G.; Sun, K.; Zhao, J.L.; Ying, G.G. Antibiotics in typical marine aquaculture farms surrounding Hailing Island, South China, Occurrence, bioaccumulation and human dietary exposure. *Mar. Pollut. Bull.* **2015**, *90*, 181–187. [CrossRef] [PubMed]
63. Yang, H.; Xu, M.; Wang, L.; Wang, X.; Jeppesen, E.; Zhang, W. Metagenomic analysis to determine the characteristics of antibiotic resistance genes in typical antibiotic-contaminated sediments. *J. Environ. Sci.* **2023**, *128*, 12–25. [CrossRef] [PubMed]
64. Czekalski, N.; Gascon Diez, E.; Burgmann, H. Wastewater as a point source of antibiotic-resistance genes in the sediment of a freshwater lake. *ISME J.* **2014**, *8*, 1381–1390. [CrossRef]

65. Siedlewicz, G.; Borecka, M.; Białk-Bielińska, A.; Sikora, K.; Stepnowski, P.; Pazdro, K. Determination of antibiotic residues in southern Baltic Sea sediments using tandem solid-phase extraction and liquid chromatography coupled with tandem mass spectrometry. *Oceanologia* **2016**, *58*, 221–234. [CrossRef]
66. Bai, Y.; Meng, W.; Xu, J.; Zhang, Y.; Guo, C. Occurrence, distribution and bioaccumulation of antibiotics in the Liao River Basin in China. *Environ. Sci. Process Impacts* **2014**, *16*, 586–593. [CrossRef]
67. Li, W.; Shi, Y.; Gao, L.; Liu, J.; Cai, Y. Occurrence of antibiotics in water, sediments, aquatic plants, and animals from Baiyangdian Lake in North China. *Chemosphere* **2012**, *89*, 1307–1315. [CrossRef]
68. Chen, J.; Yang, Y.; Jiang, X.; Ke, Y.; He, T.; Xie, S. Metagenomic insights into the profile of antibiotic resistomes in sediments of aquaculture wastewater treatment system. *J. Environ. Sci.* **2022**, *113*, 345–355. [CrossRef]
69. Feng, Y.; Hu, J.; Chen, Y.; Xu, J.; Yang, B.; Jiang, J. Ecological effects of antibiotics on aquaculture ecosystems based on microbial community in sediments. *Ocean Coast. Manag.* **2022**, *224*, 106173. [CrossRef]
70. Cui, B.; Gao, T.; Chen, L. Typical pathogenic microorganisms detection and bacterial community analysis of gibel carp (Carassius auratus gibelio) breeding environment. *Microbiol. Chain.* **2019**, *46*, 3363–3377.
71. Li, X.; Li, B.; Dong, Y.; Zhu, J. Analysis of sediment microbial communities in *Megalobrama amblycephala* intensive rearing pond. *J. Fish. China* **2014**, *38*, 218–227.
72. Su, H.; Hu, X.; Xu, W.; Xu, Y.; Wen, G.; Cao, Y. Metagenomic analysis of the abundances, diversity, and distribution of antibiotic resistance genes and their potential bacterial hosts in two types of shrimp-rearing farms in South China. *Ecotoxicol. Environ. Saf.* **2022**, *241*, 113801. [CrossRef] [PubMed]
73. Samuelsen, O.B. Degradation of oxytetracycline in seawater at two different temperatures and light intensities, and the persistence of oxytetracycline in the sediment from a fish farm. *Aquaculture* **1989**, *83*, 7–16. [CrossRef]
74. Liu, X.; Lu, S.; Guo, W.; Xi, B.; Wang, W. Antibiotics in the aquatic environments, a review of lakes, China. *Sci. Total Environ.* **2018**, *627*, 1195–1208. [CrossRef] [PubMed]
75. Liu, S.; Bekele, T.; Zhao, H.; Cai, X.; Chen, J. Bioaccumulation and tissue distribution of antibiotics in wild marine fish from Laizhou Bay, North China. *Sci. Total Environ.* **2018**, *631*, 1398–1405. [CrossRef]
76. Han, Y.; Wang, J.; Zhao, Z.; Chen, J.; Lu, H.; Liu, G. Combined impact of fishmeal and tetracycline on resistomes in mariculture sediment. *Environ. Pollut.* **2018**, *242*, 1711–1719. [CrossRef] [PubMed]
77. Thiele, C.J.; Hudson, M.D.; Russell, A.E.; Saluveer, M.; Sidaoui-Haddad, G. Microplastics in fish and fishmeal, an emerging environmental challenge? *Sci. Rep.* **2021**, *11*, 2045. [CrossRef] [PubMed]
78. Cao, L.; Naylor, R.; Henriksson, P.; Leadbitter, D.; Metian, M.; Troell, M.; Zhang, W.B. China's aquaculture and the world's wild fisheries. *Science* **2015**, *347*, 133–135. [CrossRef]
79. Han, Y.; Wang, J.; Zhao, Z.; Chen, J.; Lu, H.; Liu, G. Fishmeal application induces antibiotic resistance gene propagation in mariculture sediment. *Environ. Sci. Technol.* **2017**, *51*, 10850–10860. [CrossRef]
80. Ding, Q.; Wang, J.; Feng, Y.; Wu, Z.; Ma, L.; Liu, J.; Wang, Y.; Jia, L.; Gao, L.J.; Shao, P.; et al. Determination of 32 kinds of antibiotic residues in fish intestinal content by high performance liquid chromatography-tandem mass spectrometry. *J. Food Saf. Qual.* **2022**, *13*, 1141–1149.
81. Li, W.; Li, Y.; Zheng, N.; Ge, C.; Yao, H. Occurrence and distribution of antibiotics and antibiotic resistance genes in the guts of shrimp from different coastal areas of China. *Sci. Total Environ.* **2022**, *815*, 152756. [CrossRef]
82. Siew, S.W.; Choo, M.Y.; Marshall, I.P.G.; Hamid, H.A.; Kamal, S.S.; Nielsen, D.S.; Ahmad, H.F. Gut microbiome and metabolome of sea cucumber (*Stichopus ocellatus*) as putative markers for monitoring the marine sediment pollution in Pahang, Malaysia. *Mar. Pollut. Bull.* **2022**, *182*, 114022.
83. Xiong, X.; Zhao, Y.; Wang, Z. Analysis of microbial community structure of channel catfish ictalurus punctatus intestine and culture water. *Fish. Sci.* **2022**, *41*, 589–596.
84. Sherif, A.H.; Gouda, M.; Darwish, S.; Abdelmohsin, A. Prevalence of antibiotic-resistant bacteria in freshwater fish farms. *Aquac. Res.* **2021**, *52*, 2036–2047. [CrossRef]
85. Liu, K.; Han, J.; Li, S.; Liu, L.; Lin, W.; Luo, J. Insight into the diversity of antibiotic resistance genes in the intestinal bacteria of shrimp *Penaeus vannamei* by culture-dependent and independent approaches. *Ecotoxicol. Environ. Saf.* **2019**, *172*, 451–459. [CrossRef] [PubMed]
86. Kim, M.; Kwon, T.H.; Jung, S.M.; Cho, S.H.; Jin, S.Y.; Park, N.H.; Kim, C.G.; Kim, J.S. Antibiotic Resistance of Bacteria Isolated from the Internal Organs of Edible Snow Crabs. *PLoS ONE.* **2013**, *8*, e70887. [CrossRef]
87. Zhao, Y.; Zhang, X.; Zhao, Z.; Duan, C.; Chen, H.; Wang, M.; Ren, H.Q.; Yin, Y.; Ye, L. Metagenomic analysis revealed the prevalence of antibiotic resistance genes in the gut and living environment of freshwater shrimp. *J. Hazard. Mater.* **2018**, *350*, 10–18. [CrossRef]
88. Gao, P.; Mao, D.; Luo, Y.; Wang, L.; Xu, B.; Xu, L. Occurrence of sulfonamide and tetracycline-resistant bacteria and resistance genes in aquaculture environment. *Water Res.* **2012**, *46*, 2355–2364. [CrossRef]
89. Li, J.; Yang, Y.; Li, T. Research Progress on Antibiotic Tolerance Mechanism and Treatment of Bacteria in VBNC State. *World Notes Antibiot.* **2022**, *43*, 1–9.
90. Bush, N.; Diez-Santos, I.; Abbott, L.R.; Maxwell, A. Quinolones, mechanism, lethality and their contributions to antibiotic resistance. *Molecules* **2020**, *25*, 5662. [CrossRef]

91. Weisblum, B. Erythromycin resistance by ribosome modification. *Antimicrob. Agents Chemother.* **1995**, *39*, 577–585. [CrossRef] [PubMed]
92. Chopra, I.; Roberts, M. Tetracycline antibiotics, mode of action, applications, molecular biology, and epidemiology of bacterial resistance. *Microbiol. Mol. Biol. Rev.* **2001**, *65*, 232–260. [CrossRef] [PubMed]
93. Christensen, S. Drugs that changed society: History and current status of the early antibiotics: Salvarsan, sulfonamides, and β-lactams. *Molecules* **2021**, *26*, 6057. [CrossRef] [PubMed]
94. Zhang, Z.; Song, X.; Huang, C.; Jin, A.; Tian, C.; Zhang, D.; Liu, M.C. Progress on antimicrobial resistance mediated by changes of target protein in bacteria. *Prog. Vet. Med.* **2017**, *38*, 74–77.
95. Tian, Z.; Zhang, Y.; Yu, B.; Yang, M. Changes of resistome, mobilome and potential hosts of antibiotic resistance genes during the transformation of anaerobic digestion from mesophilic to thermophilic. *Water Res.* **2016**, *98*, 261–269. [CrossRef]
96. Li, Y.; Gao, Q.; Zhang, C.; Shi, Z.; Peng, S.; Wang, J. Current status and prospect of antibiotic resistance genes (ARGs) pollution in the aquaculture. *Mar. Fish.* **2017**, *39*, 351–360.
97. Soler, N.; Forterre, P. Vesiduction, the fourth way of HGT. *Environ. Microbiol.* **2020**, *22*, 2457–2460. [CrossRef]
98. McInnes, R.S.; McCallum, G.E.; Lamberte, L.E.; Schaik, W. Horizontal transfer of antibiotic resistance genes in the human gut microbiome. *Curr. Opin. Microbiol.* **2020**, *53*, 35–43. [CrossRef]
99. Jiang, Q.; Feng, M.; Ye, C.; Yu, X. Effects and relevant mechanisms of non-antibiotic factors on the horizontal transfer of antibiotic resistance genes in water environments: A review. *Sci. Total Environ.* **2022**, *806*, 150568. [CrossRef]
100. Dong, P.; Wang, H.; Fang, T.; Wang, Y.; Ye, Q. Assessment of extracellular antibiotic resistance genes (eARGs) in typical environmental samples and the transforming ability of eARG. *Environ. Int.* **2019**, *125*, 90–96. [CrossRef]
101. Soler, N.; Forterre, P. Vesicles for DNA exchange, A new mechanism called vesiduction. *Med. Sci.* **2021**, *37*, 583–585.
102. Springael, D.; Top, E.M. Horizontal gene transfer and microbial adaptation to xenobiotics, new types of mobile genetic elements and lessons from ecological studies. *Trends Microbiol.* **2004**, *12*, 53–58. [CrossRef] [PubMed]
103. Lu, J.; Zhang, Y.; Wu, J.; Wang, J. Intervention of antimicrobial peptide usage on antimicrobial resistance in aquaculture. *J. Hazard. Mater.* **2022**, *427*, 128154. [CrossRef] [PubMed]
104. Sun, M.; Chang, Z.; Van den Brink, P.J.; Li, J.; Zhao, F.; Rico, A. Environmental and human health risks of antimicrobials used in Fenneropenaeus chinensis aquaculture production in China. *Environ. Sci. Pollut. Res.* **2016**, *23*, 15689–15702. [CrossRef] [PubMed]
105. Hossain, A.; Habibullah-Al-Mamun, M.; Nagano, I.; Masunaga, S.; Kitazawa, D.; Matsuda, H. Antibiotics, antibiotic-resistant bacteria, and resistance genes in aquaculture, risks, current concern and future thinking. *Environ. Sci. Pollut. Res. Int.* **2022**, *29*, 11054–11075. [CrossRef]
106. Norton, S.B.; Rodier, D.J.; van der Schalie, W.H.; Wood, W.P.; Slimak, M.W.; Gentile, J.H. A framework for ecological risk assessment at the EPA. *Environ. Toxicol. Chem.* **1992**, *11*, 1663–1672. [CrossRef]
107. Han, Q.; Zhao, S.; Zhang, X.; Wang, X.; Song, C.; Wang, S. Distribution, combined pollution and risk assessment of antibiotics in typical marine aquaculture farms surrounding the Yellow Sea, North China. *Environ. Int.* **2020**, *138*, 105551. [CrossRef]
108. Zhang, J.; Zhang, X.; Zhou, Y.; Han, Q.; Wang, X.; Song, C.; Wang, S.G.; Zhao, S. Occurrence, distribution and risk assessment of antibiotics at various aquaculture stages in typical aquaculture areas surrounding the Yellow Sea. *J. Environ. Sci.* **2023**, *126*, 621–632. [CrossRef]
109. Hossain, A.; Nakamichi, S.; Habibullah-Al-Mamun, M.; Tani, K.; Masunaga, S.; Matsuda, H. Occurrence, distribution, ecological and resistance risks of antibiotics in surface water of finfish and shellfish aquaculture in Bangladesh. *Chemosphere* **2017**, *188*, 329–336. [CrossRef]
110. Ashbolt, N.; Amezquita, A.; Backhaus, T.; Borriello, P.; Brandt, K.; Collignon, P.; Coors, A.; Finley, R.; Gaze, W.H.; Heberer, T.; et al. Human health risk assessment (hhra) for environmental development and transfer of antibiotic resistance. *Environ. Health Perspect.* **2013**, *121*, 993–1001. [CrossRef]

toxics

Article

The Adsorption Behaviors and Mechanisms of Humic Substances by Thermally Oxidized Graphitic Carbon Nitride

Hongxin Li [1,2], Jianlong Wang [1], Dongbei Yue [2,*], Jianchao Wang [3], Chu Tang [2] and Lingyue Zhang [4]

1 School of Environment and Energy Engineering, Beijing University of Civil Engineering and Architecture, Beijing 100044, China
2 School of Environment, Tsinghua University, Beijing 100084, China
3 School of Chemical and Environmental Engineering, China University of Mining and Technology (Beijing), Beijing 100083, China
4 School of Department of Civil Engineering, The University of Hong Kong, Pokfulam 999077, Hong Kong SAR, China
* Correspondence: yuedb@tsinghua.edu.cn; Tel./Fax: +86-10-62771931

Abstract: Thermal oxidation is efficient for enhancing the photocatalysis performance of graphitic carbon nitride (g-C_3N_4), while its effect on adsorption performance has not been fully studied, which is crucial to the application of g-C_3N_4 as adsorbents and photocatalysts. In this study, thermal oxidation was used to prepare sheet-like g-C_3N_4 (TCN), and its application for adsorption of humic acids (HA) and fulvic acids (FA) was evaluated. The results showed that thermal oxidation clearly affected the properties of TCN. After thermal oxidation, the adsorption performance of TCN was enhanced significantly, and the adsorption amount of HA increased from 63.23 (the bulk g-C_3N_4) to 145.35 mg/g [TCN prepared at 600 °C (TCN-600)]. Based on fitting results using the Sips model, the maximum adsorption amounts of TCN-600 for HA and FA were 327.88 and 213.58 mg/g, respectively. The adsorption for HA and FA was markedly affected by pH, alkaline, and alkaline earth metals due to electrostatic interactions. The major adsorption mechanisms included electrostatic interactions, π-π interactions, hydrogen bonding, along with a special pH-dependent conformation (for HA). These findings implied that TCN prepared from environmental-friendly thermal oxidation showed promising prospects for humic substances (HSs) adsorption in natural water and wastewater.

Keywords: humic substances; adsorption removal; π-π interactions; electrostatic interactions

Citation: Li, H.; Wang, J.; Yue, D.; Wang, J.; Tang, C.; Zhang, L. The Adsorption Behaviors and Mechanisms of Humic Substances by Thermally Oxidized Graphitic Carbon Nitride. *Toxics* **2023**, *11*, 369. https://doi.org/10.3390/toxics11040369

Academic Editors: Junhao Qin, Peidong Su, Feng Zhu and Lin Ding

Received: 17 March 2023
Revised: 6 April 2023
Accepted: 10 April 2023
Published: 12 April 2023

1. Introduction

Humic substances (HSs), including humic acid (HA) and fulvic acid (FA), are commonly found in various water bodies, such as surface water, underground water, wastewater, and landfill leachate [1]. In chemical compositions, HSs are mixtures of aliphatic and aromatic organic compounds and have abundant oxygen-containing groups, such as phenolic alcohol (-OH), methoxycarbonyl (-C-O), and carboxylic acid (-COOH) groups [2]. Naturally occurring HSs (5–20 mg/L) are not hazardous to human health, while due to abundant chemical functional groups, HSs have strong bonding affinity to heavy metals, pesticides, microplastics, and antibiotics, resulting in serious environmental pollution and health risks [3]. Particularly in chlorination disinfection, HSs can react with chlorine, generating highly carcinogenic disinfection by-products, such as trihalomethanes and haloacetic acids, which are highly toxic to humans and cause serious crisis for the sustainable supply of drinking water [4]. Recently, HSs have been regarded as ultraviolet quenching substances (UVQS), which can lower the UV transmittance and strongly decrease the efficacy of UV disinfection [5]; at the same time, as an emerging environmental problem, UVQS has attracted increasing attention from researchers and engineers. In addition, the accumulation of HSs in water bodies can cause serious organoleptic issues, such as unpleasant colors (yellowish to black) and undesirable odors. Consequently, the removal of HSs from water bodies is of important significance to control the environmental problems induced by HSs.

Several methods for HSs elimination from the aqueous solutions have been developed, such as electrocoagulation, advanced oxidation processes, membrane filtration, biological treatment, and adsorption [6]. Of these, adsorption is a widely used method due to its high removal efficiency, convenient operation, low investment, and reusability. In previous studies, several adsorbents, such as fly ash, clay minerals, and activated carbon, have been used for adsorption of HSs [7,8]. However, these adsorbents were not efficient for adsorption of HSs due to the inherent structural properties. For example, the micropore of activated carbon was not available for HSs adsorption due to the strong size exclusion effect. In addition, nanosized adsorbents, such as zero-valent iron [9] and layered double hydroxides-Fe_3O_4 nano-composites [10], were used for adsorption of HSs due to their large surface area. However, the nanosized adsorbents suffer from high cost and aggregation, limiting the practical application of these adsorbents [11]. Moreover, most of the nanosized adsorbents use toxic raw materials in the synthesis process, and thus the post-consumer counterparts have high eco-toxicity [8]. Therefore, the fabrication of cost-efficient and environmentally friendly adsorbents for adsorption of HSs has attracted more and more attention [12].

Recently, graphitic carbon nitride (g-C_3N_4), an emerging material, has been widely used for water splitting and photodegradation of organic pollutants due to its abundant precursors, simple synthesis, thermal/chemical stability, and environmental friendliness [13]. For example, the efficient mineralization of trichloroethylene using S-scheme heterojunctions of g-C_3N_4 was reported [14]. In the literature focusing on g-C_3N_4, most of the studies pay more attention to the photocatalysis performance, while there are few studies using g-C_3N_4 as adsorbents [15]. The current studies using g-C_3N_4-based adsorbents mainly focused on the adsorption of heavy metals, since pyridine-like nitrogen can efficiently capture cations via ligands due to the rich electron lone pairs. For example, Xiao et al. and Zhu et al. studied the adsorption capacity for lead (II) onto g-C_3N_4, which was synthesized from melamine via the pyrolysis method [16,17]. Tan et al. and Guo et al. reported the adsorption of cadmium (II) on virgin g-C_3N_4 and magnetic g-C_3N_4 [18,19]. However, there are few studies about g-C_3N_4 focusing on the adsorption of organic matters. Zhu et al. and Santoso et al. [20] reported the adsorption of methylene blue on g-C_3N_4. Yan et al. [21] reported that g-C_3N_4 showed great adsorption performance for perfluorooctane sulfonate. More recently, we revealed that g-C_3N_4 showed promise for adsorption of HSs, but the small specific surface area of g-C_3N_4 strongly restricts its application [22]. This is because the g-C_3N_4 materials are prepared via the pyrolysis method using precursors with earth-abundant carbon and nitrogen elements. Although the theoretical surface area for ideal monolayer g-C_3N_4 can be as high as 2500 m^2/g, the g-C_3N_4 materials prepared using pyrolysis exhibited a stacked structure, particularly for g-C_3N_4 derived from thiourea, melamine, and cyanamide [23]. As reported by Wang et al. [22], the specific surface area of bulk g-C_3N_4 derived from melamine, thiourea, and cyanamide was 10.35, 11.24, and 9.32 m^2/g, respectively, and the bulk g-C_3N_4 showed a very low performance for HSs adsorption. This makes the g-C_3N_4 materials with large surface area or with nanoscale structures highly desired for application as adsorbents.

Some methods, such as soft and hard templating methods, have been reported to prepare porous g-C_3N_4 with a high specific surface area. Yan [24] reported that using Pluronic P123 as a soft template, the worm-like porous g-C_3N_4 with a high surface area and great photocatalytic activity was prepared. Nevertheless, the soft template has undesirable residues of template polymers, which affects the performance of the materials, and the hard templating consumes hazardous hydrogen fluoride [14]. In this regard, a simple, efficient, and environmentally friendly thermal oxidation was proposed to prepare g-C_3N_4 with a high surface area [25]. This method remarkably affects the surface area and porosity of g-C_3N_4, leading to an improved photocatalysis. For instance, thermal oxidation was used to prepare porous g-C_3N_4 nanosheets, and the prepared g-C_3N_4 showed an enhanced visible light photocatalytic activity, and the radiative lifetime of charge carriers (τ_1 and τ_2) increased from 4.13 and 26.23 ns to 5.36 and 36.57 ns after thermal exfoliation, respectively [26]. The mechanisms are that thermal oxidation can destroy the layer structures of g-C_3N_4 bonded by hydrogen bonding and Van der Waals force. However, the effect of thermal oxidation on the adsorption performance of g-C_3N_4 has not been fully investigated, particularly for the effect of porosity and surface chemistry of g-C_3N_4 on its adsorption performance. Particularly, given that melamine was a most-used precursor of g-C_3N_4, while the effect of thermal oxidation on its adsorption performance for organic pollutants has not been studied. On the other hand, in heterogeneous photocatalysis using g-C_3N_4, the organics are adsorbed on the surface of g-C_3N_4 and then photodegraded, since free radicals cannot leave the surface of g-C_3N_4 due to a short half-life [27]. The adsorption of g-C_3N_4 for organics is a critical process, while this critical process is rarely elucidated, and only some reports studied the adsorption of methylene blue, a typical pollutant during photocatalysis [20]. Consequently, studies on the adsorption performance of g-C_3N_4 derived from melamine after thermal oxidation is important for the application of this emerging material as an adsorbent and photocatalyst.

Therefore, this study investigated the application of melamine-derived g-C_3N_4 after thermal oxidation for adsorption of HSs. First, the changes in the textural properties of g-C_3N_4 after thermal oxidation (TCN) were studied. Second, the adsorption kinetics and isotherms for HA and FA to TCN were investigated. The effect of contact time, initial pH, initial concentration, metal ions, and temperature on adsorption of HSs was further studied. Third, the adsorption mechanisms of HSs on TCN and the changes in the surface properties of TCN after adsorption were elucidated.

2. Materials and Methods

2.1. Preparation of Adsorbents and HS Materials

HSs materials were obtained from the Macklin Co., Ltd. (Shanghai, China). The precursor of g-C_3N_4, analytical grade melamine, was purchased from Sinopharm Chemical Reagent Co., Ltd., (Shanghai, China). The leachate nanofiltration concentrate was collected from the treatment plant located in Beijing, China, which was filtered through a 0.45 μm membrane and stored at 4 °C. The chemical properties of landfill leachate concentrate were: pH = 9.06, dissolved organic carbon (DOC) = 1122.92 mg C/L, conductivity = 26,487.83 μs/cm, and UV_{254} = 36.00 (/cm). The bulk g-C_3N_4 was synthesized via the pyrolysis of precursor at 500 °C for 2 h in air. After milling, the bulk g-C_3N_4 was then subjected to thermal oxidation. The thermal oxidation was conducted at 500, 550, and 600 °C for 3 h in air, and the resultant g-C_3N_4 samples were denoted as TCN-500, TCN-550, and TCN-600, respectively.

2.2. Characterization of Adsorbents

X-ray diffraction (XRD, Zürich, Switzerland) was used to characterize the crystalline structure of the adsorbents. The chemical composition of the adsorbents was analyzed using Fourier transforms infrared spectra spectroscopy (FT-IR, Bartlett, IL USA) and X-ray photoelectron spectroscopy (XPS, Hopkinton, MA, USA). The morphology of the adsorbents was recorded with field emission scanning electron microscopy (SEM, Hitachi, Tokyo, Japan). The Brunauer–Emmett–Teller (BET) surface area and pore-size distribution of the adsorbents were determined with Quanta Chrome ASAP 2460 analyzer (Micromeritics, Norcross, GA, USA). The zeta potential of the adsorbents was measured with an instrument at 25 °C with suspensions containing 1 g/L of solids (Delsa Nano C, Beckman Coulter, Brea, CA, USA). The pH of the solutions was adjusted using 0.1 M NaOH and HCl.

2.3. Adsorption Procedures

For each batch adsorption experiment, 0.04 g of g-C_3N_4 was added to 100 mL of HA or FA solution, and the adsorption was performed at 298.5 K in the dark using an instrument (Figure S1). The adsorption of HA or FA was systematically investigated according to the following parameters: initial concentration of HSs (25–200 mg/L), contact time (0–240 min), initial pH (2.0–10.0), ionic strength (I, NaCl, 0–0.07 M), temperature (298.15–318.15 K), alkali metals (K^+, 0–0.07 M), and alkali earth metals (Ca^{2+}, Mg^{2+}, 0–1.05 mM) (Table S1). HSs concentration was measured using a UV-visible spectrometer (UV-1800, Shimadzu, Kyoto, Japan) at a wavelength of 254 nm. The experiments were conducted in duplicate at least, and the average values were reported. At equilibrium, the removal rate and adsorption capacity were calculated as the following equations:

$$R = \frac{C_0 - C_e}{C_0} \times 100\% \quad (1)$$

$$q_e = \frac{(C_0 - C_e)V}{m} \quad (2)$$

where R (%) and q_e (mg/g) are the removal rate and adsorption capacity, respectively. m (g) is the mass of absorbents. C_0 is the initial concentration (mg/L). C_e is the equilibrium concentration (mg/L). V (L) is the volume of HSs solution.

The kinetics data were simulated with the pseudo-first-order (PFO) model, the pseudo-second-order (PSO) model, and the Elovich kinetic model according to the following equations, respectively:

$$q_t = q_e\left(1 - e^{-k_1 t}\right) \quad (3)$$

$$q_t = \frac{q_e^2 k_2 t}{(1 + q_e k_2 t)} \quad (4)$$

$$q_t = \frac{1}{\beta} \ln(1 + \alpha\beta t) \quad (5)$$

where q_e (mg/g) and q_t (mg/g) are the equilibrium and adsorption capacities at time t, respectively. k_1 (g/mg/min) and k_2 (g/mg/min) are the rate constants of the PFO and PSO models, respectively. α is the initial adsorption rate (mg/g/min). β is the desorption constant (g/mg) and t (min) is the contact time in the Elovich equation.

The intraparticle diffusion model was used to study the intraparticle diffusion process, as described by the following equation:

$$q_t = k_{id} t^{1/2} + C \quad (6)$$

where k_{id} [(mg/g/min$^{1/2}$)] and C (mg/g) are the rate constant and the boundary layer thickness constant, respectively.

The Freundlich, Langmuir, Sips, and Temkin models were used to describe the adsorption isotherms as follows:

$$q_e = k_f C_0^{\frac{1}{n}} \quad (7)$$

$$\frac{C_e}{q_e} = \frac{C_e}{q_m} + \frac{1}{k_l q_m} \quad (8)$$

$$q_e = \frac{q_m (k_s C_e)^{1/n}}{1 + (k_s C_e)^{1/n}} \quad (9)$$

$$q_e = \frac{RT}{bY} ln A_T C_e \quad (10)$$

where k_f [(mg/g) (L/mg)$^{1/n}$], k_l (L/mg), and k_s (L/mg) are the constants for the Langmuir, Freundlich, and Sips models, respectively. R is 8.314 J/mol·K. T (K) is temperature. A_T (L/g) and b_T are the constants of Temkin. C_e (mg/L) is the equilibrium concentration of HSs. q_m (mg/g) is the maximum adsorption capacity. $1/n$ is the heterogeneity factor.

The thermodynamic parameters, including the Gibbs free energy (ΔG^0, kJ/M), enthalpy (ΔH^0, kJ/M), and entropy (ΔS^0, J/M/K) were calculated as follows:

$$ln\frac{1}{C_e} = lnk_0 - \frac{\Delta H^0}{RT} \tag{11}$$

$$\Delta G^0 = -nRT \tag{12}$$

$$\Delta S^0 = \frac{\Delta H^0 - \Delta G^0}{T} \tag{13}$$

where T is temperature (K). K_0 is the thermodynamic constant. The values of ΔH^0 and ΔS^0 were calculated from the slope and intercept of the linear regression of lnK_0 versus T^{-1}.

2.4. Adsorption Mechanisms

The adsorption mechanisms were ascertained using dynamic light scattering (DLS), XPS, nitrogen adsorption–desorption isotherms, along with fluorescence excitation-emission matrix (EEM). DLS was utilized to evaluate the changes in the conformation of HSs with respect to pH using a Nanoparticle Size Analyzer (Delsa Nano C, Indianapolis, IN, USA) for a special adsorption mechanism. XPS was used to determine the differences in the binding energy before and after adsorption. The porosity of the adsorbents after HSs adsorption was investigated using nitrogen adsorption–desorption isotherms. EEM was recorded with a spectrofluorometer (F-7000, Hitachi) to investigate the changes in the fluorescence features of HSs before and after adsorption on TCN-600.

3. Results and Discussion

3.1. Characterization of Adsorbents

The crystal structure of the bulk g-C_3N_4 and TCN samples was characterized by XRD patterns. As shown in Figure 1a, there were two diffraction peaks at approximately 27.5° and 13.0°. The stronger peak at 27.5° was ascribed to the interlayer superposition of reflections in graphite, corresponding to the 002 crystal face of g-C_3N_4 [28]. The 002 peak was consistent with the known accumulation peak of aromatic systems. With the increasing of thermal oxidation temperature, the intensity of the 002 diffraction peak of TCN (27.55°) samples was higher than that of bulk g-C_3N_4 (27.35°). This indicated that the crystallinity of TCN samples increased. The reflection peak at 13.0° was attributed to the interlayer tri-s-triazine units stacking, which was indexed as (100) [28].

The functional groups of the bulk g-C_3N_4 and TCN samples were studied using FT-IR. As shown in Figure 1b, there were no differences in the FT-IR of the adsorbents. There were three distinct groups of bands, which were consistent with that of the typical g-C_3N_4 materials, in agreement with the previous reports [29]. The sharp peak at 800 cm^{-1} was related to the rings of triazine or heptazine, the absorption bands in the range of 1100–1600 cm^{-1} were the asymmetric stretching of C-N and C=N heterocycles, and the broad bands in the range of 3000–3500 cm^{-1} were the stretching vibrations of the -N-H or -N-H_2 groups derived from the uncondensed amino groups [30].

The surface compositions and chemical states of the adsorbents were investigated using XPS (Figure 1c,d, Tables S2 and S3). As shown in Figure 1c, there were four peaks at binding energies of 398.8, 399.6, 400.9, and 404.5 eV, respectively. The peaks of 398.8 and 399.6 eV were attributed to the sp^2-hybridized nitrogen bonded to carbon (C-N=C) and bridging N atoms (N-$(C)_3$), respectively. The peaks of 400.9 and 404.5 eV related to the primary or secondary amino groups (-NH_2 or =NH) and charging effect, respectively [31]. As shown in Figure 1d, there were three peaks at 284.8, 288.3, and 293.8 eV in the C 1s region; the first peak was attributed to adventitious carbon, the second was the sp^2-hybridized carbon, and the third peak was π-π^* excitation [32].

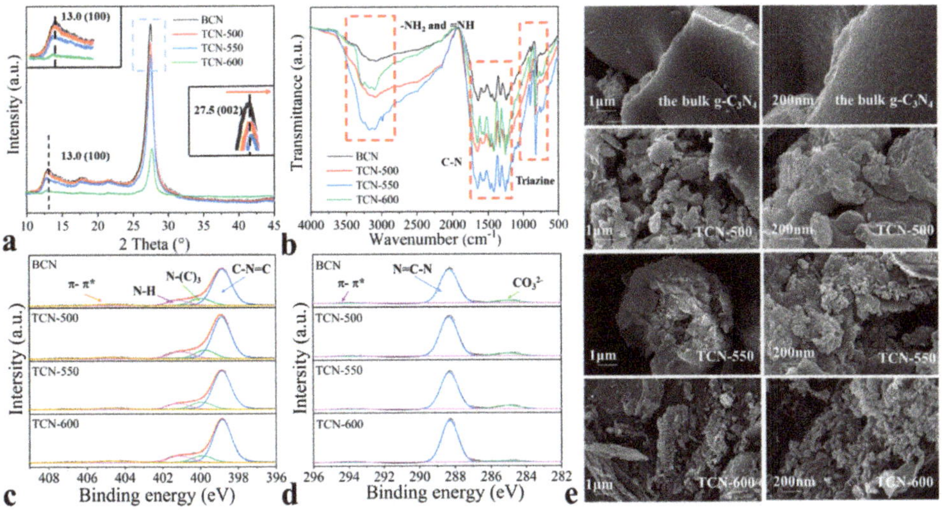

Figure 1. (**a**) The XRD patterns, (**b**) FT-IR spectra, (**c**) N 1s core region, (**d**) C 1s core region, and (**e**) SEM images of the bulk g-C_3N_4 and TCN samples.

As shown in Figure 1e, the morphological structure and interlayer stacking of the adsorbents were observed. The bulk g-C_3N_4 had a nonporous structure, whereas the TCN samples exhibited porous structure. The TCN-500 samples displayed a layered structure composed of large lumps with a few pores and exhibited a reduced thickness with more pores. When the thermal temperature was increased to 600 °C, the TCN-600 samples showed a thinner sheet-like structure with abundant pores. As the temperature of thermal oxidation increased, the thickness of TCN decreased and possessed more abundant pores, indicating that the thermal oxidation induced morphological changes, in agreement with the previous studies [25]. As previously reported [33], the energy of van der Waals forces and of hydrogen bonding in the interlayers of g-C_3N_4 are fragile and can be broken after thermal oxidation. The formation of pores was ascribed to the decomposition of g-C_3N_4 into gaseous products [25]. The yield of the adsorbents decreased with increasing thermal temperature, which was consistent with the above findings.

The nitrogen adsorption–desorption and pore-size distribution curves of the adsorbents are shown in Figure 2. The surface area of the TCN samples increased after thermal oxidation. Notably, the surface area of TCN-550 and TCN-600 was 68.57 and 60.39 m^2/g, respectively, which were much higher than those of the bulk g-C_3N_4. In addition, the pore volume of the TCN samples steadily increased with the increased thermal oxidation temperature, and that of TCN-600 reached 0.64 cm^3/g. All isotherms exhibited a H3-type hysteresis loop, suggesting the formation of slit-shaped pores [34]. In contrast to the bulk g-C_3N_4, the hysteresis loop of TCN samples shifted to lower pressure, and the area of the hysteresis loop grew with the increased thermal temperature, indicating the formation of a relatively large mesopore [35]. The above findings can be explained by the reduced layer thickness and size of the TCN samples, which were consistent with the results of morphological features. This was consistent with the surface area and pore-size distribution for the bulk g-C_3N_4 and TCN samples. These variations in the porosity of the TCN samples would affect their adsorption performance for HSs. The increased surface area and pore volume provided more active adsorption sites for HSs adsorption. For macromolecular HSs, however, the pore size also affects adsorption, besides the specific surface area.

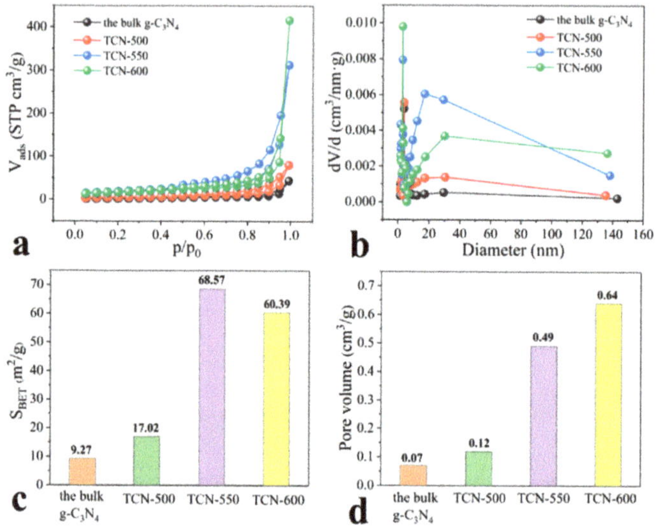

Figure 2. (**a**) The nitrogen adsorption–desorption isotherms, (**b**) pore-size distribution, (**c**) surface areas, and (**d**) pore volume of the bulk g-C_3N_4 and TCN.

3.2. Comparison of HA Adsorption Using Different Adsorbents

The adsorption capacities and removal rates of HA (as an example) by the bulk g-C_3N_4 and TCN are shown in Figure S2. It was found that TCN adsorbed greater amounts of HA with higher removal rates than the bulk g-C_3N_4. The amount of HA adsorbed by TCN-600 was 145.35 mg/g (more than twice that of the bulk g-C_3N_4). This was consistent with the changes in surface area and pore volume, suggesting that TCN-600 had greater potential for HSs removal because its large surface area and pore volume provide more active sites. Moreover, it was found that, although the surface area of TCN-550 was similar to that of TCN-600, the amount of HA adsorbed to TCN-600 was greater than that to TCN-550. This was due to the higher pore volume of TCN-600.

3.3. Adsorption Kinetics

The adsorption kinetics of HA and FA on the bulk g-C_3N_4 and TCN are shown in Figures 3 and S3. It was observed that the adsorption amounts of HA and FA increased until reaching equilibrium for both the bulk g-C_3N_4 and TCN. Within the initial 30 min, the adsorption amounts of HA and FA showed a significant increase. This phenomenon was attributed to abundant active sites within the initial stage. The active sites became more completely bound and eventually saturated with the contact time increased, resulting in the decreased adsorption rate. In terms of HA adsorption, TCN-600 showed the greatest adsorption amount, which was consistent with the results of adsorption capacities. In addition, TCN-600 exhibited greater adsorption capacity for HA than for FA, indicating a higher adsorption affinity towards HA.

For the adsorption of HA and FA on the bulk g-C_3N_4 and TCN, the adsorption kinetic data were fitted according to the PFO, PSO, and Elovich models (Figure 3 and Table S4). It was found that the correlation coefficient (R^2) of the PSO-fitted data was higher than that of the PFO-fitted data, suggesting that the adsorption of HA and FA is related to chemisorption [7]. Among these models, the Elovich model showed the best fitting, as indicated by the highest R^2 value. The Elovich model is often used to describe chemisorption on heterogeneous adsorbing surfaces [36]. Collectively, it was speculated that HSs adsorption is related to chemisorption.

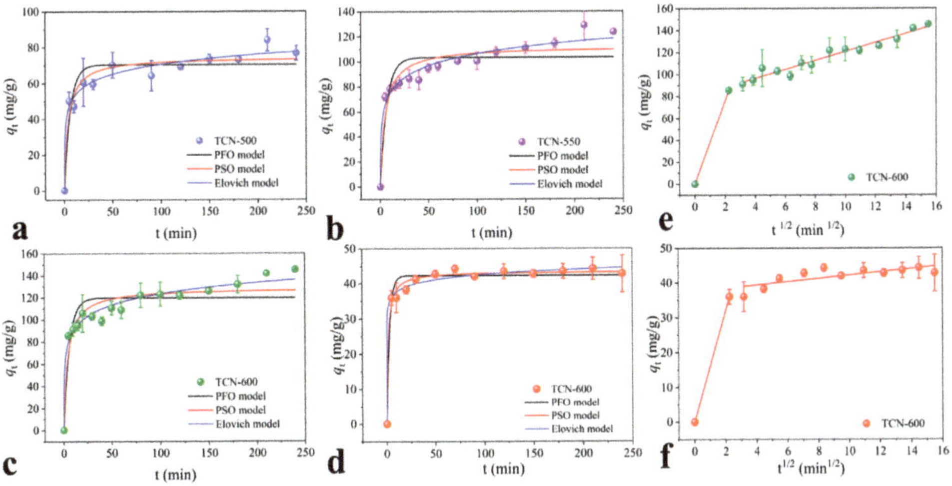

Figure 3. (**a–c**) The adsorption kinetics of HA on TCN (C_{HSs} = 100 mg/L, pH = 3.0, T = 298.5 K, C_{TCN} = 0.4 g/L, and I = 0.01 M); (**d**) The adsorption kinetics of FA on TCN-600 (C_{HSs} = 50 mg/L, pH = 3.0, T = 298.5 K, $C_{TCN\text{-}600}$ = 0.4 g/L, and I = 0.01 M); (**e**,**f**) The intraparticle diffusion of HA and FA on TCN-600 [C_{HSs} = 100 (HA) or 50 (FA) mg/L, pH = 3.0, T = 298.5 K, $C_{TCN\text{-}600}$ = 0.4 g/L, and I = 0.01 M].

As shown in Figure 3e,f, the diffusion process for adsorption of HSs to TCN-600 was ascertained based on the intraparticle diffusion model. It was found that the correlation curves of q_t and $t^{1/2}$ were multilinear, indicating that boundary layer diffusion and interior adsorption were involved in the adsorption of HSs to TCN-600. In the first stage, HA and FA molecules migrated from the solution to the external surface of TCN-600 with a high mass transfer rate. However, in the second stage, the steric effect slowed the mass transfer rate of macromolecular HA and FA, resulting in less intraparticle diffusion. A similar phenomenon was reported for adsorption of HSs on carbon nanotubes [37].

3.4. Effect of pH on Adsorption of HSs

The effect of pH on the adsorption of HA and FA on TCN-600 was investigated at pH 2.0–10.0. The adsorption amounts of HA and FA decreased rapidly as pH increased (Figure S4). This was consistent with that of a previous study about adsorption of HSs on activated carbon [38]. When pH was increased from 2.0 to 10.0, the adsorption amounts of HA and FA decreased from 113.55 to 9.08 mg/g and from 56.99 to 15.75 mg/g, respectively. Thus, the acidic condition apparently favored adsorption of HSs to TCN-600. This indicated that electrostatic interactions played an essential role in HSs adsorption. Moreover, the zeta potential of TCN-600 was measured as a function of pH (Figure S5). It was found that the surface of TCN-600 was positive when pH of the solution was below 3.0. The effect of pH on adsorption of HSs can be explained from two points. On the one hand, the surface of TCN contained abundant amino groups, which were protonated or deprotonated, resulting in negatively or positively charged surfaces. The carboxylic and phenolic groups of HSs could be protonated or deprotonated [39]. When pH was increased, the surface charge of TCN-600 became more negative, rendering TCN-600 more repulsive towards the negatively charged HSs. Accordingly, the decreased adsorption with pH was attributed to enhanced electrostatic repulsion. On the other hand, by contrast to FA adsorption, HA adsorption was more dependent on pH. This was possibly due to the pH-dependent conformation (a non-interaction adsorption mechanism), and HA could become more compact with a higher adsorption affinity at low pH values [40]. In order to verify this, DLS analysis of HA and FA with respect to pH was conducted (Figure S6), and the DLS results indicated that as pH increased, the size of HA increased due to the decoiling of macromolecular HA, in agreement with the studies reported by Wells and Stretz [41]. By contrast, the size of FA kept stable as pH increased. Accordingly, the adsorption of HA was affected by pH via electrostatic interactions and the special pH-dependent conformation.

3.5. Effect of the Common Ions on Adsorption of HSs

The alkali metals (Na^+ and K^+) and alkali earth metals (Ca^{2+} and Mg^{2+}) are ubiquitous in water. Thus, the effect of these ions on the adsorption of HA and FA to TCN-600 was investigated. When the concentration of the alkali and alkali earth metals was increased, the adsorption amounts of HA and FA increased significantly, as shown in Figure 4. This further elucidated electrostatic interactions as a primary mechanism for adsorption of HSs to TCN-600, consistent with the effect of pH. Thus, the increase in adsorption of HSs with the increase of alkali and alkali earth metals was due to the resulted compressed electrical double layer, reducing electrostatic repulsion between HSs and TCN [42]. As the concentration of ionic increased, the thickness of the electrical double layer was compressed, causing HSs to adhere to TCN-600 more closely. Moreover, the precipitation or aggregation of HSs at high ionic strengths could counteract electrostatic repulsion. At higher ionic strengths, HA formed a smaller configuration, thus occupying fewer adsorption sites and leading to a greater adsorption capacity to TCN-600 [43]. As shown in Figure 4b,d, the alkali earth metals (Ca^{2+} and Mg^{2+}) were favorable for the adsorption of HA and FA to TCN-600. This can be attributed to the role of Ca^{2+} and Mg^{2+} in neutralizing the negative charges, reducing electrostatic repulsions between HSs and TCN-600. Moreover, the effect of Ca^{2+} on the adsorption HA and FA to TCN-600 was greater than that of Mg^{2+} due to the differences in ionic size. This was consistent with the adsorption of HA by nano-amorphous calcium phosphate [44].

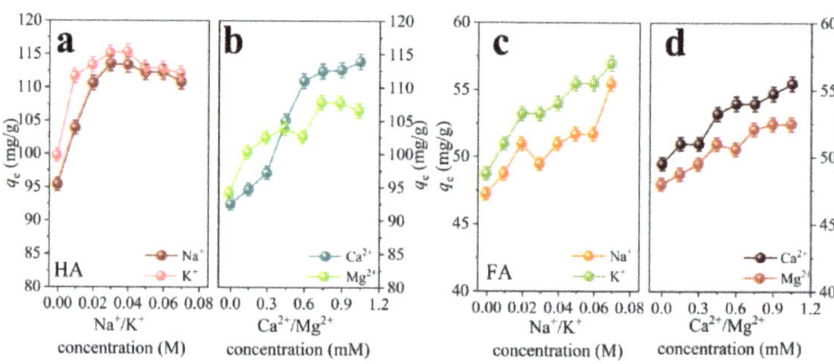

Figure 4. The effects of (**a**,**c**) alkali metals and (**b**,**d**) alkali earth metals [C_{HSs} = 50 mg/L, pH = 3.0, T = 298.5 K, $C_{TCN-600}$ = 0.4 g/L, I = 0.01 M (for K^+, Ca^{2+}, and Mg^{2+})] on the adsorption of HA and FA to TCN-600.

3.6. Adsorption Isotherms and Thermodynamic Analysis

Temperature was an important factor affecting adsorption of HSs, and thus the adsorption experiments were carried out at different temperatures. As shown in Figure 5, at the same temperature, the adsorption amounts of HA and FA increased with increasing initial concentration due to the concentration gradient. Apparently, the temperature had no noticeable effect on FA adsorption. By contrast, the adsorption amounts of HA increased with the increasing of temperature. The isotherm data were fitted using the two-parameter models (Freundlich, Langmuir, and Temkin) and a single three-parameter model (Sips). As shown in Table S5, all four models well described the adsorption isotherms of HSs to TCN, and the correlation coefficients R^2 were greater than 0.950. However, the Langmuir and Sips models fitted better compared to the other models. The Langmuir model is generally used to describe the uniform distribution of surface-active centers [45]. The Sips model combines the Langmuir and Freundlich isotherm models and is used to analyze the heterogeneous adsorption systems and monolayer adsorption characteristics at high concentration of adsorbate [46]. In our case, the k_s and $1/n$ values of the Sips model were close to zero and one, respectively, suggesting that the adsorption of HA and FA to TCN-600 was heterogeneous, and the surface of TCN-600 possessed heterogeneous adsorption sites for adsorption of HSs.

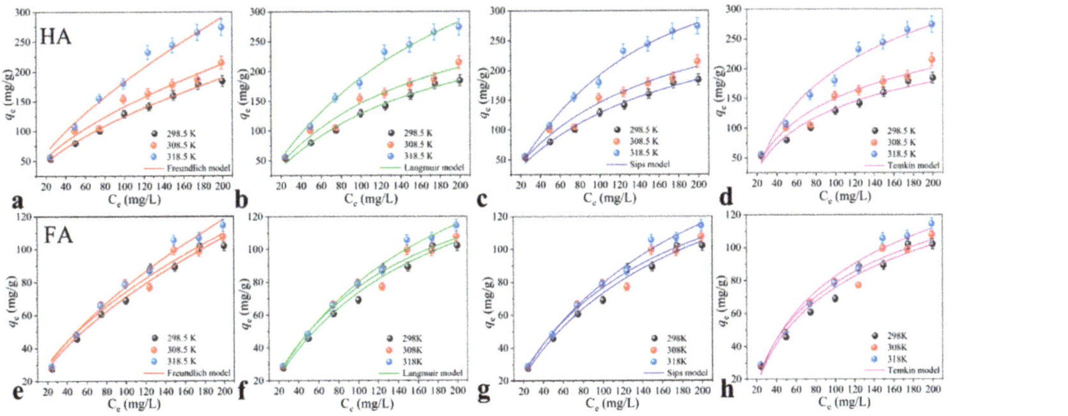

Figure 5. The adsorption isotherms of (**a**–**d**) HA and (**e**–**h**) FA to TCN-600 (C_{HSs} = 25–200 mg/L, pH = 3.0, $C_{TCN-600}$ = 0.4 g/L, and I = 0.01 M, T = 298.5, 308.5, and 318.5 K).

In addition, according to the Sips model, the maximum adsorption amounts of HA and FA at 298.5 K were 327.88 and 213.58 mg/g, respectively. The adsorption amount of HA was greater than that of FA, consistent with that of previous reports for adsorption of HSs to graphene oxides and carbon nanotubes [47]. This phenomenon was attributed to the abundance of aromatic rings and the polar nature of HA, which induced more robust π-π stacking interactions between HA and TCN-600 [34]. Compared to that of previously reported adsorbents (Table S6), the adsorbed amounts of both HA and FA by TCN-600 were greater. For example, palygorskite had a notable adsorption capacity for HA, with the maximum adsorption amount of 17 mg/g for HA at 298.5 K [48]. Thus, TCN-600 was highly effective for adsorbing HA. The thermodynamic parameters of HSs are summarized in Table S7. The very low ΔH^0 values for adsorption of HSs to TCN-600 indicated that temperature was not a very important factor, suggesting a physisorption process. The negative ΔG^0 values revealed that the adsorption progress was spontaneous.

In order to investigate the practical application of TCN, we studied the adsorption of HA and FA at low concentration (for simulating real surface waters) and adsorption of a landfill leachate concentrate on TCN-600 (Figures S7 and S8). It was found that the removal rates of HA and FA at a concentrate of 5~20 mg/L were above 90% and 60%, respectively, indicating that HSs can be removed effectively after adsorption using TCN. In addition, the removal rate measured as UV_{254} of landfill leachate concentrate reached 60.86%. EEM spectra showed that the fluorescent region of landfill leachate concentrate at Ex > 250 nm and Em > 380 nm was effectively eliminated, which was attributed to the humic-like components [49,50]. The above results indicated that HSs in surface waters and wastewaters can be effectively removed by adsorption using TCN.

3.7. Adsorption Mechanisms

The adsorption mechanisms for HSs on TCN were elucidated using nitrogen adsorption–desorption isotherms, XPS, and fluorescence EEM. As shown in Figure 6a, it was observed that after adsorption of HA and FA, and the hysteresis loops of isotherms shifted to higher relative pressures. This clearly indicated that HA and FA molecules diffused into the pore systems of TCN-600, which was consistent with the results of the intraparticle diffusion model. In addition, although the adsorption amounts of HA were greater than that of FA, the specific surface area and pore volume of TCN-600 both decreased more significantly after FA adsorption than HA (Figure 6b,c). This was because the average size of FA was less than that of HA, and as reported the average sizes of FA and HA molecules were 3.2 and 9.3 nm, respectively [51]. In this regard, more FA molecules could diffuse into the pore systems of TCN-600 than HA molecules. The similar diffusion mechanism was reported by Ateia et al. [52] for the adsorption of HA on carbon nanotubes.

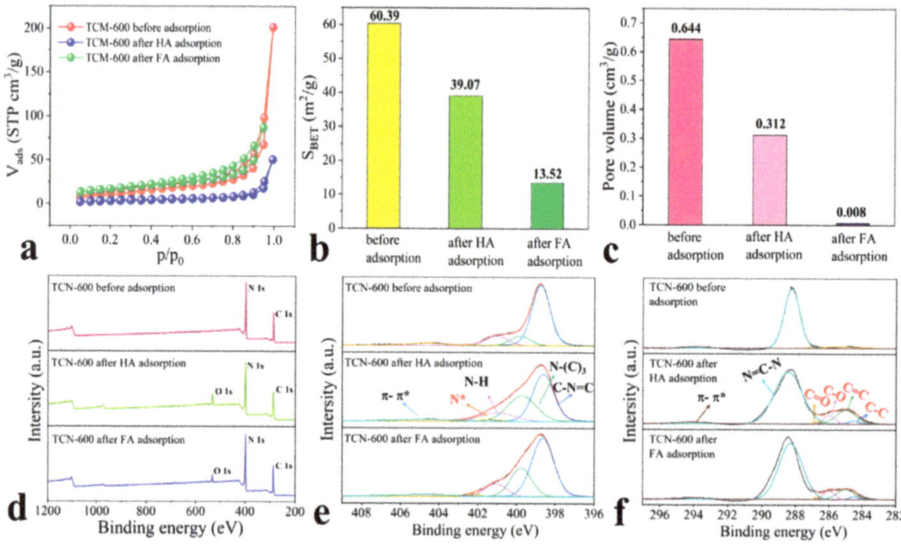

Figure 6. (**a**) The nitrogen adsorption–desorption isotherms, (**b**) specific surface area, (**c**) pore volume, (**d**) XPS spectra, (**e**) N 1s core region, and (**f**) C 1s core region before and after adsorption of HA and FA (C_{HSs} = 50 mg/L, pH = 3.0, $C_{TCN\text{-}600}$ = 0.4 g/L t = 120 min, and I = 0.01 M).

As shown in Figure 6d and Table S8, after HA and FA adsorption, a new O 1s peak appeared at approximately 532.0 eV in the XPS spectra of TCN-600, indicating that the molecules of HSs were adsorbed on the surface of TCN-600. The N 1s and C 1s regions of TCN-600 after HSs adsorption were also decomposed. As shown in Figure 6e, a new N 1s peak appeared at 402.2 eV, which may be attributed to protonated or polarized nitrogen (N*) in the -NH_2 and -NH- groups [22]. As reported in a previous study, amino groups have superior adhesion affinity towards HA [53]. As a result, the protonation and deprotonation of amine groups on the surface of TCN-600 played a critical role in HSs adsorption via electrostatic interactions (as discussed above). Collectively, we can conclude that electrostatic interactions were a major mechanism for HSs adsorption. As shown in Figure 6f, after HSs adsorption, four new peaks appeared in the C 1s region of TCN-600. The new peaks at 284.5 and 285.0 eV were attributed to carbonyl C=O and ether carbon C-O bonds, respectively, which may be the oxygen groups of HSs. The percentage of surface oxygen of TCN-600 increased from 1.41 to 7.90% and 4.66% after the adsorption of HA and FA, respectively (Table S2). The increases in the surface oxygen content after adsorption were consistent with the above discussions. The new peaks at 286.0 and 286.7 eV were attributed to aliphatic C-C and aromatic C=C bonds, respectively, which may be the carbon atom of HSs or to the combination with the carbon atom of TCN-600 [51]. In addition, HSs contain large quantities of carboxylic and phenolic moieties, which could act as hydrogen-bond donors [54]. The amino groups of TCN-600 could bind with the carboxylic and phenolic moieties of HSs via hydrogen bonding [55].

In addition, HSs contain large quantities of aromatic rings, which promote π-π interactions with adsorbents containing aromatic rings. As shown in Figure 7, EEM was used to characterize HSs before and after adsorption. As previously reported [56], the fluorescence of HSs included five regions. Among them, FA-like components at Ex/Em = 200–250/380–550 nm and HA-like substances at Ex/Em = 250–400/380–550 nm. Thus, it was clear that HA-like and FA-like components were preferentially adsorbed at higher Ex/Em wavelengths. In particular, the II and V regions in the HA EEM showed decreased fluorescence intensity after adsorption; however, decreases in the II regions in the FA EEM were not clear. In general, the components in these regions have great aromaticity [57]. Thus, such a phenomenon can be explained by π-π interactions, indicating the presence of π-π interactions in HSs adsorption on TCN-600. Liu et al. [58] reported a similar finding in previous study. To sum up, adsorption of HSs to TCN-600 was mainly mediated by electrostatic interactions, hydrogen bonding, and π-π interactions. Particularly, the adsorption of HA was also affected by its variable conformation.

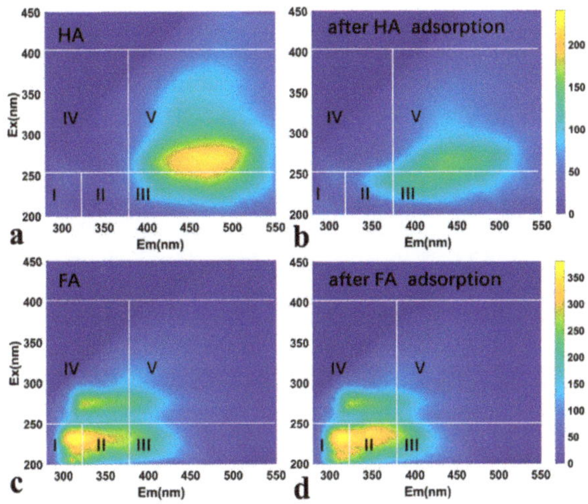

Figure 7. The EEM of HA and FA before (**a**,**c**) and after (**b**,**d**) adsorption by TCN-600 (C_{HSs} = 50 mg/L, pH = 3.0, t = 120 min, and I = 0.01 M).

4. Conclusions

In this study, the synthesis of TCN using thermal oxidation was investigated, and the adsorption of HA and FA on TCN was evaluated. After thermal oxidation, the textural properties of TCN were remarkably affected, particularly in specific surface areas and pore size. As thermal oxidation temperature increased, the specific surface area of TCN at 600 °C (TCN-600) was 60.39 m^2/g, much greater than that of bulk g-C_3N_4, enhancing the adsorption performance for HSs. The adsorption capacity of TCN-600 for HA was 145.35 mg/g, two times greater than that of bulk g-C_3N_4. The kinetics data were fitted well with the PSO and Elovich models, indicating HSs adsorption is related to chemisorption. Based on the Sips model, the maximum adsorption amounts of HA and FA at 298.5 K were quite substantial, calculated to be 327.88 and 213.58 mg/g, respectively. The adsorption capacity of TCN-600 was higher than that of most previously reported adsorbents. Moreover, TCN-600 could effectively remove HSs with low concentration. According to the intraparticle diffusion model, boundary layer diffusion and interior adsorption were involved in HSs adsorption. The adsorption of HSs was greatly affected by pH, alkali metals (Na^+ and K^+), and alkali earth metals (Ca^{2+} and Mg^{2+}), which was due to electrostatic interactions and hydrogen bonding. HA-like and FA-like components were preferentially adsorbed at higher Ex/Em wavelengths, indicating the presence of π-π interactions. In particular, the adsorption of HA was also affected by the pH-dependent conformation. The findings are valuable for designing of HSs-target adsorbents and elimination of HSs from aqueous solutions.

Supplementary Materials: The following supporting information can be downloaded at: https://www.mdpi.com/article/10.3390/toxics11040369/s1, Figure S1: The experimental instrument used in this study; Figure S2: The adsorption capacities and removal rate (%) of HA by the bulk g-C3N4 and TCNs; Figure S3: The adsorption kinetics of HA on the bulk g-C3N4; Figure S4: The effect of pH for adsorption of HSs on TCN-600; Figure S5: The zeta potential of TCN-600 as a function of pH; Figure S6: The particle size distributions of HA and FA as a function of pH; Figure S7: The removal rate of HA and FA at low initial concentration; Figure S8: The removal rates measured as UV254 of landfill leachate concentrate on the TCN-600; Table S1: The detailed experimental parameters adopted in this study; Table S2: The relative content of surface elements (%) of XPS spectra peaks of the bulk g-C3N4 and TCNs before and after adsorption of HSs; Table S3: The relative contents (%) of peaks in the N 1s and C 1s core region of the bulk g-C3N4 and TCNs; Table S4: Kinetics parameters for adsorption of HSs on the bulk g-C3N4 and TCNs; Table S5: The parameters of adsorption isotherms for adsorption of HSs on TCN-600; Table S6: A comparison for adsorption of HSs on various adsorbents; Table S7: Thermodynamic parameters for adsorption of HSs on TCN-600; Table S8: The relative contents (%) of peaks in the N 1s and C 1s core region of TCN-600 before and after adsorption of HSs. References [59–67] are cited in the Supplementary Materials.

Author Contributions: H.L.: Conceptualization, Experimental work, Methodology, Writing-original draft, Formal analysis; J.W. (Jianlong Wang): Supervision, Writing—Review and editing; D.Y.: Supervision, Resources, Funding acquisition, Writing—Review and editing. J.W. (Jianchao Wang), C.T., L.Z.: Methodology and Writing—Review and editing. All authors have read and agreed to the published version of the manuscript.

Funding: This research received no external funding or This research was funded by Dongbei Yue grant number 2018YFC1901405. And The APC was funded by Dongbei Yue. Information regarding the funder and the funding number should be provided.

Institutional Review Board Statement: Not applicable.

Informed Consent Statement: Not applicable.

Data Availability Statement: The data is confidential.

Acknowledgments: This study was funded by the National Key Research and Development Program of China (2018YFC1901405).

Conflicts of Interest: The authors declare that they have no known competing financial interest or personal relationship that could have appeared to influence the work reported in this paper.

References

1. Wu, H.; Ai, Z.H.; Zhang, L.Z. Anoxic and oxic removal of humic acids with Fe@Fe_2O_3 core-shell nanowires: A comparative study. *Water Res.* **2014**, *52*, 92–100. [CrossRef] [PubMed]
2. Mohanraj, J.; Durgalakshmi, D.; Saravanan, R. Water-soluble graphitic carbon nitride for clean environmental applications. *Environ. Pollut.* **2021**, *269*, 116172. [CrossRef] [PubMed]
3. Adusei-Gyamfi, J.; Ouddane, B.; Rietveld, L.; Cornard, J.-P.; Criquet, J. Natural organic matter-cations complexation and its impact on water treatment: A critical review. *Water Res.* **2019**, *160*, 130–147. [CrossRef] [PubMed]
4. Goslan, E.H.; Seigle, C.; Purcell, D.; Henderson, R.; Parsons, S.A.; Jefferson, B.; Judd, S.J. Carbonaceous and nitrogenous disinfection by-product formation from algal organic matter. *Chemosphere* **2017**, *170*, 1–9. [CrossRef]
5. Chen, W.M.; Li, Q.B. Elimination of uv-quenching substances from mbr- and saarb-treated mature landfill leachates in an ozonation process: A comparative study. *Chemosphere* **2020**, *242*, 125256. [CrossRef] [PubMed]
6. Dong, C.L.; Chen, W.; Liu, C. Preparation of novel magnetic chitosan nanoparticle and its application for removal of humic acid from aqueous solution. *Appl. Surf. Sci.* **2014**, *292*, 1067–1076. [CrossRef]
7. Liu, Z.J.; Zhou, S.Q. Removal of humic acid from aqueous solution using polyacrylamide/chitosan semi-ipn hydrogel. *Water Sci. Technol.* **2018**, *2017*, 16–26. [CrossRef]
8. Bhatnagar, A.; Sillanpää, M. Removal of natural organic matter (NOM) and its constituents from water by adsorption—A review. *Chemosphere* **2017**, *166*, 497–510. [CrossRef]

9. Giasuddin, A.B.M.; Kanel, S.R.; Choi, H. Adsorption of humic acid onto nanoscale zerovalent iron and its effect on arsenic removal. *Environ. Sci. Technol.* **2007**, *41*, 2022–2027. [CrossRef]
10. Wang, R.X.; Wen, T.; Wu, X.L.; Xu, A.W. Highly efficient removal of humic acid from aqueous solutions by mg/al layered double hydroxides-Fe_3O_4 nanocomposites. *RSC Adv.* **2014**, *4*, 21802–21809. [CrossRef]
11. Jiang, L.; Li, Y.; Shao, Y.; Zhang, Y.; Han, R.; Li, S.; Wei, W. Enhanced removal of humic acid from aqueous solution by novel stabilized nano-amorphous calcium phosphate: Behaviors and mechanisms. *Appl. Surf. Sci.* **2018**, *427*, 965–975. [CrossRef]
12. Fronczak, M. Adsorption performance of graphitic carbon nitride-based materials: Current state of the art. *J. Environ. Chem. Eng.* **2020**, *8*, 104411. [CrossRef]
13. Wang, S.; Li, D.; Sun, C.; Yang, S.; Guan, Y.; He, H. Synthesis and characterization of g-C_3N_4/Ag_3VO_4 composites with significantly enhanced visible-light photocatalytic activity for triphenylmethane dye degradation. *Appl. Catal. B: Environ.* **2014**, *144*, 885–892. [CrossRef]
14. Mkhalid, I.A.; Mohamed, R.M.; Alhaddad, M.; Basaleh, A.; Al-Hajji, L.A.; Ismail, A.A. S-scheme mesoporous Li_2MnO_3/g-C_3N_4 heterojunctions as efficient photocatalysts for the mineralization of trichloroethylene in aqueous media. *J. Colloid Interface Sci.* **2022**, *614*, 160–171. [CrossRef]
15. Li, X.; Xing, J.; Zhang, C.; Han, B.; Zhang, Y.; Wen, T.; Leng, R.; Jiang, Z.; Ai, Y.; Wang, X. Adsorption of lead on sulfur-doped graphitic carbon nitride nanosheets: Experimental and theoretical calculation study. *ACS Sustain. Chem. Eng.* **2018**, *6*, 10606–10615. [CrossRef]
16. Zhu, L.; You, L.J.; Wang, Y.; Shi, Z.X. The application of graphitic carbon nitride for the adsorption of Pb^{2+} ion from aqueous solution. *Mater. Res. Express* **2017**, *4*, 075606. [CrossRef]
17. Xiao, G.; Wang, Y.Q.; Xu, S.N.; Li, P.F.; Yang, C.; Jin, Y.; Sun, Q.F.; Su, H.J. Superior adsorption performance of graphitic carbon nitride nanosheets for both cationic and anionic heavy metals from wastewater. *Chin. J. Chem. Eng.* **2019**, *27*, 305–313. [CrossRef]
18. Tan, J.Z.Y.; Nursam, N.M.; Xia, F.; Sani, M.A.; Li, W.; Wang, X.D.; Caruso, R.A. High-performance coral reef-like carbon nitrides: Synthesis and application in photocatalysis and heavy metal ion adsorption. *ACS Appl. Mater. Interfaces* **2017**, *9*, 4540–4547. [CrossRef]
19. Guo, S.Z.; Wu, K.L.; Gao, Y.; Liu, L.H.; Zhu, X.X.; Li, X.L.; Zhang, F. Efficient removal of Zn (II), Pb (II), and Cd (II) in waste water based on magnetic graphitic carbon nitride materials with enhanced adsorption capacity. *J. Chem. Eng. Data* **2018**, *63*, 3902–3912. [CrossRef]
20. Zhu, B.; Xia, P.; Ho, W.; Yu, J. Isoelectric point and adsorption activity of porous g-C_3N_4. *Appl. Surf. Sci.* **2015**, *344*, 188–195. [CrossRef]
21. Yan, T.; Chen, H.; Wang, X.; Jiang, F. Adsorption of perfluorooctane sulfonate (PFOS) on mesoporous carbon nitride. *RSC Adv.* **2013**, *3*, 22480–22489. [CrossRef]
22. Wang, J.; Yue, D.; Cui, D.; Zhang, L.; Dong, X. Insights into adsorption of humic substances on graphitic carbon nitride. *Environ. Sci. Technol.* **2021**, *55*, 7910–7919. [CrossRef]
23. Cao, S.; Low, J.; Yu, J.; Jaroniec, M. Polymeric photocatalysts based on graphitic carbon nitride. *Adv. Mater.* **2015**, *27*, 2150–2176. [CrossRef]
24. Yan, H. Soft-templating synthesis of mesoporous graphitic carbon nitride with enhanced photocatalytic h2 evolution under visible light. *Chem. Commun.* **2012**, *48*, 3430–3432. [CrossRef]
25. Niu, P.; Zhang, L.L.; Liu, G.; Cheng, H.M. Graphene-like carbon nitride nanosheets for improved photocatalytic activities. *Adv. Funct. Mater.* **2012**, *22*, 4763–4770. [CrossRef]
26. Dong, F.; Li, Y.; Wang, Z.; Ho, W.-K. Enhanced visible light photocatalytic activity and oxidation ability of porous graphene-like g-C_3N_4 nanosheets via thermal exfoliation. *Appl. Surf. Sci.* **2015**, *358*, 393–403. [CrossRef]
27. Xu, B.; Ahmed, M.B.; Zhou, J.L.; Altaee, A.; Xu, G.; Wu, M. Graphitic carbon nitride based nanocomposites for the photocatalysis of organic contaminants under visible irradiation: Progress, limitations and future directions. *Sci. Total Environ.* **2018**, *633*, 546–559. [CrossRef]
28. Fina, F.; Callear, S.K.; Carins, G.M.; Irvine, J.T.S. Structural investigation of graphitic carbon nitride via xrd and neutron diffraction. *Chem. Mat.* **2015**, *27*, 2612–2618. [CrossRef]
29. Zhao, Z.W.; Sun, Y.J.; Dong, F. Graphitic carbon nitride based nanocomposites: A review. *Nanoscale* **2015**, *7*, 15–37. [CrossRef]
30. Lotsch, B.V.; Doblinger, M.; Sehnert, J.; Seyfarth, L.; Senker, J.; Oeckler, O.; Schnick, W. Unmasking melon by a complementary approach employing electron diffraction, solid-state nmr spectroscopy, and theoretical calculations-structural characterization of a carbon nitride polymer. *Chem. Eur. J.* **2007**, *13*, 4969–4980. [CrossRef]
31. Tong, J.H.; Li, Q.; Li, W.Y.; Wang, W.H.; Ma, W.M.; Su, B.T.; Bo, L.L. MoS_2 thin sheet growing on nitrogen self-doped mesoporous graphic carbon derived from ZIF-8 with highly electrocatalytic performance on hydrogen evolution reaction. *ACS Sustain. Chem. Eng.* **2017**, *5*, 10240–10247. [CrossRef]
32. Bhunia, K.; Chandra, M.; Khilari, S.; Pradhan, D. Bimetallic ptau alloy nanoparticles-integrated g-C_3N_4 hybrid as an efficient photocatalyst for water-to-hydrogen conversion. *ACS Appl. Mater. Interfaces* **2019**, *11*, 478–488. [CrossRef]
33. Dong, F.; Ou, M.Y.; Jiang, Y.K.; Guo, S.; Wu, Z.B. Efficient and durable visible light photocatalytic performance of porous carbon nitride nanosheets for air purification. *Ind. Eng. Chem. Res.* **2014**, *53*, 2318–2330. [CrossRef]
34. Wang, J.C.; Li, H.X.; Yue, D.B. Enhanced adsorption of humic/fulvic acids onto urea-derived graphitic carbon nitride. *J. Hazard. Mater.* **2022**, *424*, 127643. [CrossRef]

35. Li, W.F.; Wang, Y.; Yang, X.Y.; Liu, F.Q.; Li, W.H. Graphitic carbon nitride prepared by rapid recrystallization for photoelectrochemical anticorrosion. *ACS Appl. Nano Mater.* **2019**, *2*, 7559–7565. [CrossRef]
36. Wu, F.C.; Tseng, R.L.; Juang, R.S. Characteristics of elovich equation used for the analysis of adsorption kinetics in dye-chitosan systems. *Chem. Eng. J.* **2009**, *150*, 366–373. [CrossRef]
37. Wang, X.; Liang, S.; Wang, Y.; Xu, B.; Bai, Y.; Shu, T.; Xing, B. Sorption of peat humic acids to multi-walled carbon nanotubes. *Environ. Sci. Technol.* **2011**, *45*, 9276–9283. [CrossRef]
38. Yang, K.; Fox, J.T. Adsorption of humic acid by acid-modified granular activated carbon and powder activated carbon. *J. Environ. Eng.-ASCE* **2018**, *144*, 04018104. [CrossRef]
39. Maghsoodloo, S.; Noroozi, B.; Haghi, A.K.; Sorial, G.A. Consequence of chitosan treating on the adsorption of humic acid by granular activated carbon. *J. Hazard. Mater.* **2011**, *191*, 380–387. [CrossRef]
40. Lan, T.; Wu, P.; Liu, Z.; Stroet, M.; Liao, J.; Chai, Z.; Mark, A.E.; Liu, N.; Wang, D. Understanding the effect of ph on the solubility and aggregation extent of humic acid in solution by combining simulation and the experiment. *Environ. Sci. Technol.* **2022**, *56*, 917–927. [CrossRef]
41. Wells, M.J.M.; Stretz, H.A. Supramolecular architectures of natural organic matter. *Sci. Total Environ.* **2019**, *671*, 1125–1133. [CrossRef]
42. Huang, Y.F.; Chen, H.L.; Han, G.H. Research on the adsorption of humic acid on pyrite surface. In Proceedings of the Light Metals Symposium at the 147th Annual TMS Meeting and Exhibition, Phoenix, AZ, USA, 11–15 March 2018; pp. 197–202.
43. Lozano, P.; Berge, N.D. Single-walled carbon nanotube behavior in representative mature leachate. *Waste Manage.* **2012**, *32*, 1699–1711. [CrossRef] [PubMed]
44. Han, H.W.; Rafiq, M.K.; Zhou, T.Y.; Xu, R.; Masek, O.; Li, X.K. A critical review of clay-based composites with enhanced adsorption performance for metal and organic pollutants. *J. Hazard. Mater.* **2019**, *369*, 780–796. [CrossRef]
45. Lee, S.; Hur, J. Heterogeneous adsorption behavior of landfill leachate on granular activated carbon revealed by fluorescence excitation emission matrix (EEM)-parallel factor analysis PARAFAC). *Chemosphere* **2016**, *149*, 41–48. [CrossRef]
46. Al-Ghouti, M.A.; Da'ana, D.A. Guidelines for the use and interpretation of adsorption isotherm models: A review. *J. Hazard. Mater.* **2020**, *393*, 122383. [CrossRef]
47. Sun, W.L.; Xia, J.; Li, S.; Sun, F. Effect of natural organic matter (NOM) on Cu (II) adsorption by multi-walled carbon nanotubes: Relationship with nom properties. *Chem. Eng. J.* **2012**, *200*, 627–636. [CrossRef]
48. Wang, M.S.; Liao, L.B.; Zhang, X.L.; Li, Z.H. Adsorption of low concentration humic acid from water by palygorskite. *Appl. Clay Sci.* **2012**, *67–68*, 164–168. [CrossRef]
49. Wang, J.; Yue, D.; Li, M.; Wang, H.; Wang, J.; Wang, C.; Wang, H. Application of carbon nitride nanosheets for adsorption of various humic substances from aqueous solutions. *Chem. Eng. J.* **2023**, *454*, 140296. [CrossRef]
50. Wang, J.; Wang, C.; Shi, A.; Shi, Y.; Yue, D.; Zhang, L.; Wang, J.; Wang, H.; Wang, C.; Cui, D. An innovative approach for landfill leachate treatment based on selective adsorption of humic acids with carbon nitride. *Chem. Eng. J.* **2023**, *461*, 142090. [CrossRef]
51. Karanfil, T.; Kitis, M.; Kilduff, J.E.; Wigton, A. Role of granular activated carbon surface chemistry on the adsorption of organic compounds. 2. Natural organic matter. *Environ. Sci. Technol.* **1999**, *33*, 3225–3233. [CrossRef]
52. Ateia, M.; Apul, O.G.; Shimizu, Y.; Muflihah, A.; Yoshimura, C.; Karanfil, T. Elucidating adsorptive fractions of natural organic matter on carbon nanotubes. *Environ. Sci. Technol.* **2017**, *51*, 7101–7110. [CrossRef] [PubMed]
53. Zhang, Y.R.; Wang, F.; Wang, Y.X. Electrospun cellulose acetate/chitosan fibers for humic acid removal: Construction guided by intermolecular interaction study. *ACS Appl. Polym. Mater.* **2021**, *3*, 5022–5029. [CrossRef]
54. Yang, K.; Xing, B.S. Adsorption of fulvic acid by carbon nanotubes from water. *Environ. Pollut.* **2009**, *157*, 1095–1100. [CrossRef] [PubMed]
55. Lin, J.; Zhan, Y. Adsorption of humic acid from aqueous solution onto unmodified and surfactant-modified chitosan/zeolite composites. *Chem. Eng. J.* **2012**, *200–202*, 202–213. [CrossRef]
56. Ning, Y.; Luo, Z.Q.; Li, Y.L.; Yang, Z.; Liu, D.Q.; Zhang, Y.Y. Alkaline leaching characteristics of uranium from lincang coal: Correlation with the dissolution of coal humic substances. *Fuel* **2021**, *305*, 121507. [CrossRef]
57. Lee, B.-M.; Seo, Y.-S.; Hur, J. Investigation of adsorptive fractionation of humic acid on graphene oxide using fluorescence eem-parafac. *Water Res.* **2015**, *73*, 242–251. [CrossRef]
58. Liu, J.; Cao, J.; Chen, H.; Zhou, D. Adsorptive removal of humic acid from aqueous solution by micro- and mesoporous covalent triazine-based framework. *Colloid. Surface. A* **2015**, *481*, 276–282. [CrossRef]
59. Derakhshani, E.; Naghizadeh, A. Optimization of humic acid removal by adsorption onto bentonite and montmorillonite nanoparticles. *J. Mol. Liq.* **2018**, *259*, 76–81. [CrossRef]
60. Ngah, W.S.W.; Fatinathan, S.; Yosop, N.A. Isotherm and kinetic studies on the adsorption of humic acid onto chitosan-H_2SO_4 beads. *Desalination* **2011**, *272*, 293–300. [CrossRef]
61. Doulia, D.; Leodopoulos, C.; Gimouhopoulos, K.; Rigas, F. Adsorption of humic acid on acid-activated Greek bentonite. *J. Colloid Interface Sci.* **2009**, *340*, 131–141. [CrossRef] [PubMed]
62. Huang, S.Y.; Song, S.; Zhang, R.; Wen, T.; Wang, X.X.; Yu, S.J.; Song, W.C.; Hayat, T.; Alsaedi, A.; Wang, X.K. Construction of Layered Double Hydroxides/Hollow Carbon Microsphere Composites and Its Applications for Mutual Removal of Pb(II) and Humic Acid from Aqueous Solutions. *ACS Sustain. Chem. Eng.* **2017**, *5*, 11268–11279. [CrossRef]

63. Tao, Q.; Xu, Z.Y.; Wang, J.H.; Liu, F.L.; Wan, H.Q.; Zheng, S.R. Adsorption of humic acid to aminopropyl functionalized SBA-15. *Microporous Mesoporous Mater.* **2010**, *131*, 177–185. [CrossRef]
64. Wang, Q.Z.; Chen, X.G.; Liu, N.; Wang, S.X.; Liu, C.S.; Meng, X.H.; Liu, C.G. Protonation constants of chitosan with different molecular weight and degree of deacetylation. *Carbohydr. Polym.* **2006**, *65*, 194–201. [CrossRef]
65. Wang, S.B.; Terdkiatburana, T.; Tade, M.O. Single and co-adsorption of heavy metals and humic acid on fly ash. *Sep. Purif. Technol.* **2008**, *58*, 353–358. [CrossRef]
66. Jayalath, S.; Larsen, S.C.; Grassian, V.H. Surface adsorption of Nordic aquatic fulvic acid on amine-functionalized and non-functionalized mesoporous silica nanoparticles. *Environ. Sci.-Nano* **2018**, *5*, 2162–2171. [CrossRef]
67. Zhang, J.; Gong, J.L.; Zenga, G.M.; Ou, X.M.; Jiang, Y.; Chang, Y.N.; Guo, M.; Zhang, C.; Liu, H.Y. Simultaneous removal of humic acid/fulvic acid and lead from landfill leachate using magnetic graphene oxide. *Appl. Surf. Sci.* **2016**, *370*, 335–350. [CrossRef]

toxics

MDPI

Article

Spatial and Temporal Variations of Heavy Metals' Bioavailability in Soils Regulated by a Combined Material of Calcium Sulfate and Ferric Oxide

Chi Zhang [1,2], Jie Li [3], Yuxia Dai [3], Williamson Gustave [4], Weiwei Zhai [1,3,*], Zhong Zhong [2] and Jianmeng Chen [1]

1 Key Laboratory of Microbial Technology for Industrial Pollution Control of Zhejiang Province, College of Environment, Zhejiang University of Technology, Hangzhou 310058, China
2 Zhejiang Key Laboratory of Environmental Protect Technology, Eco-Environmental Sciences Research & Design Institute of Zhejiang Province, Hangzhou 310007, China
3 Zhejiang Provincial Key Laboratory of Agricultural Resources and Environment, Institute of Soil and Water Resources and Environmental Science, College of Environmental and Resource Sciences, Zhejiang University, Hangzhou 310058, China
4 School of Chemistry, Environmental & Life Sciences, University of the Bahamas, New Providence, Nassau P.O. Box N-4912, Bahamas
* Correspondence: weiweizhai@zjut.edu.cn

Abstract: Heavy metal pollution in soils threatens food safety and human health. Calcium sulfate and ferric oxide are commonly used to immobilize heavy metals in soils. However, the spatial and temporal variations of the heavy metals' bioavailability in soils regulated by a combined material of calcium sulfate and ferric oxide (CSF) remain unclear. In this work, two soil column experiments were conducted to investigate the spatial and temporal variations of CSF immobilized Cd, Pb, and As. In the horizontal soil column, the results showed that CSF's immobilization range for Cd increased over time, and adding CSF in the center of the soil column decreased the concentrations of bioavailable Cd significantly, up to 8 cm away by day 100. The CSF immobilization effect on Pb and As only existed in the center of the soil column. The CSF's immobilization depths for Cd and Pb in the vertical soil column increased over time and extended to 20 cm deep by day 100. However, the CSF's immobilization depths for As only extended to between 5 and 10 cm deep after 100 days of incubation. Overall, the results from this study can serve as a guide to determine the CSF application frequency and spacing distance for the in-situ immobilization of heavy metals in soils.

Keywords: soil; heavy metals; in-situ immobilization; bioavailability; spatial and temporal variations

Citation: Zhang, C.; Li, J.; Dai, Y.; Gustave, W.; Zhai, W.; Zhong, Z.; Chen, J. Spatial and Temporal Variations of Heavy Metals' Bioavailability in Soils Regulated by a Combined Material of Calcium Sulfate and Ferric Oxide. *Toxics* **2023**, *11*, 296. https://doi.org/10.3390/toxics11040296

Academic Editors: Junhao Qin, Peidong Su, Feng Zhu and Lin Ding

Received: 22 February 2023
Revised: 17 March 2023
Accepted: 20 March 2023
Published: 24 March 2023

1. Introduction

Soil heavy metal pollution has increased significantly over the past 50 years due to increased anthropogenic activities associated with rapid urbanization and industrialization. This increase in soil contaminants poses a serious risk to cultivated land quality and food safety globally [1–3]. In China, soil heavy metal pollution is a serious threat and has attracted considerable attention. According to the National Soil Pollution Survey Bulletin issued by the Ministry of Environmental Protection (MEP) and the Ministry of Land and Resources in 2014, 16% of all the arable soils in China exceeded the national standard for soil pollutants (GB15618-1995). Among them, the excessive rates of cadmium (Cd), mercury (Hg), arsenic (As), lead (Pb), and chromium (Cr) were 7.0%, 1.6%, 2.7%, 1.5%, and 1.1%, respectively. Yang et al. conducted an integrated analysis of the concentration of heavy metals in the soil of 1041 agricultural lands in China. They found that heavy metals such as Cd, Pb, and As and their related hazards were particularly serious [2]. These heavy metals in soil can enter the human body through oral and nasal inhalation, dermal contact, and the food chain, ultimately affecting human health [4]. Therefore, the protection and remediation of heavy metal-contaminated soil are essential.

Unlike organic pollutants, heavy metals are non-degradable and persistent in the soil environment. Therefore, once heavy metals contaminate the soil, the contaminants' concentration and ecological toxicity will persist until they have transformed into a less toxic form or are completely removed [5]. Various remediation technologies have been developed, including physical, chemical, and biological remediation techniques, to decrease soil heavy metals' bioavailability and toxicity [6]. For example, phytoremediation is considered an efficient approach to reducing the concentration of heavy metals in soil [7,8]. The replacement technique is used to remediate contaminated soil by completely or partially replacing contaminated soil with clean soil [9]. Among these techniques, in-situ immobilization is a promising soil remediation technology that employs stabilizers to decrease heavy metals' bioavailability. In-situ immobilization has the advantages of practicability, rapid results, and cost-effectiveness [10]. The stabilizer selection is the most crucial criterion for the success of in-situ immobilization. The common stabilizers used for heavy metal immobilization include organic stabilizers, inorganic stabilizers, and organic-inorganic composite stabilizers, such as lime, fly ash, and biochar [11,12]. As cost-effective and environmentally friendly stabilizers, more and more iron-based materials are used for the immobilization of heavy metals, such as zero-valent iron, oxides, iron sulfides, and loaded iron-based materials [13,14], which have a high specific surface area, strong redox capacity, and expand the range of the effective working pH. Yang et al. found that biochar-supported nanoscale zero-valent iron could transform the fraction of unstable heavy metals into a stable form, which substantially decreased the availability of the heavy metals and hence greatly reduced the human health exposure risk [15]. Two iron-based materials, 2-line ferrihydrite and goethite, promote Cd transformation to more stable speciation in contaminated soil [16]. Our previous studies have shown that a combined material of calcium sulfate and ferric oxide (CSF) can effectively decrease the mobility and bioavailability of Cd, Pb, and As in paddy soils [17,18].

The stabilization efficiency was affected by heavy metals and the dosage of stabilizers. For example, Wang et al. showed that 1% and 0.5% biochar had significantly different impacts on Cd in rice roots [12]. The lime, fly ash, and biochar could increase soil pH and decrease Zn, Cd, Cu, and Pb concentrations while increasing the As concentrations [19]. In addition, the immobilization efficiency of heavy metal-contaminated soil is influenced by the stability time. Cui et al. found that adding soil amendments such as apatite and charcoal to contaminated soil can effectively reduce the leaching rates and bioavailability of Cu and Cd in the soil, but both will gradually increase with time [20]. However, there are few studies on the pollution repair process of heavy metal-contaminated soils by composite materials and their stability, timeliness, and effective diffusion range. Due to the convenience of soil column experience, previous scientists usually used soil column experiments to verify heavy metals' temporal and spatial migration characteristics [21–24]. Therefore, CSF as an effective immobilization and remediation material in soils contaminated by Cd, Pb, and As, however, the adequate time and remediation range of CSF needs to be further studied to determine the ideal application frequency and CSF concentration required to reduce the cost and prevent soil hardening.

In this study, two soil column experiments were set up to (1) evaluate the efficiency and persistence of CSF for decreasing the bioavailability of Cd, Pb, and As in paddy soil; and (2) explore the spatial and temporal variation of the bioavailability of Cd, Pb, and As regulated by CSF. Our study aimed to provide new insights for rational and effective remediation procedures for heavy metal-contaminated soils by CSF.

2. Materials and Methods

2.1. Soil and CSF Characterization

Soil samples were collected from the surface layer (that is, from 0 to 20 cm) of a contaminated paddy field near a mining area located in Shangyu, Zhejiang, China (120°87′ E, 30°03′ N). The sampling site has a typical subtropical climate. The basic physical and chemical properties of soil are given in Table 1. The soil pH was 6.48, and the organic matter was 43.33 g kg^{-1}. The Cd, Pb, and As concentrations in the soil were 0.54 mg kg^{-1}, 416.58 mg kg^{-1}, and 94.20 mg kg^{-1}, respectively. The CSF was prepared by mixing calcium sulfate ($CaSO_4{\cdot}2H_2O$) and ferric oxide (Fe_2O_3) at a ratio of 9:1. The $CaSO_4{\cdot}2H_2O$ and Fe_2O_3 were purchased from Zibo Jinshun Chemical Industry Co., Ltd. (Zibo, China) and Sinopharm Chemical Reagent Co., Ltd. (Shanghai, China), respectively. The CSF was ground and passed through a 0.15 mm nylon mesh. The mean particle size was 10.37 μm, and the specific surface area of CSF was 0.65 m^2 g^{-1}. The characteristics of the CSF stabilizer can be found in Table 1.

Table 1. Basic physical and chemical characteristics of the tested soil and CSF.

	pH	Organic Matter (g/kg)	Soil Grain Diameter (%)			Cd (mg/kg)	Pb (mg/kg)	As (mg/kg)
			Sand	Clay	Silt			
Soil	6.48	43.33	10.70	76.40	12.90	0.54	416.58	94.20
Calcium sulfate	8.81	-	-	-	-	nd	nd	5.84
Ferric oxide	3.29	-	-	-	-	nd	1.27	nd

nd: Not detected; (-) = Not measure data.

2.2. Experimental Design

Two different soil column experiments were conducted to explore the spatial variation of Cd, Pb, and As bioavailability regulated by CSF. One was a horizontal soil column, which was used to examine the effect of CSF on the bioavailability and migration of Cd, Pb, and As in the horizontal space (Figure 1a). The main part of the horizontal soil column was 15 cm high and 30 cm in diameter. The other was a vertical soil column, which was used to explore the effect of CSF on the bioavailability and migration of Cd, Pb, and As in the vertical space (Figure 1b). The main part of the vertical soil column was 61 cm high and 10 cm in diameter. A layer of nylon gauze was placed at the bottom of the soil column container, followed by a layer of 3 and 5-cm thick quartz sand in horizontal and vertical soil column containers, respectively. After that, the soil was used to fill in the layers 5 cm apart. The soil was weighed before use to ensure the soil bulk density was the same in each layer. Each soil layer was packed with tamping, especially around the side wall of the soil column container, to prevent the side wall from collapsing. A 10 cm diameter cylindrical gauze bag was loaded in the center for the horizontal soil column experiment. The CSF treatment of the horizontal soil column (CSF-H) was loaded with the original soil mixed with 0.15% CSF in the gauze bag. The control of the horizontal soil column (CK-H) was loaded with the original soil in the gauze bag without CSF. A layer of gauze was laid on the soil layer 38 cm from the bottom for the vertical soil column experiment. The original soil 15 cm thick was used as the control treatment in the vertical soil column (CK-V), and the original soil was mixed with 0.15% CSF of 15 cm thick as CSF in the treatment of the vertical soil column (CSF-V).

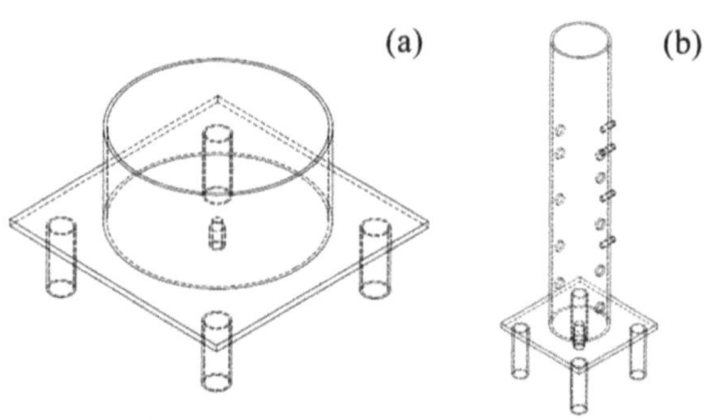

Figure 1. Schematic drawing of the soil columns used in the experiments. (**a**): Horizontal soil column; (**b**): Vertical soil column.

The soil column experiments were placed indoors at 25 °C, and three replicates were set for each treatment. At the beginning of the experiment, deionized water was added to saturate the soil, and then the natural environment was simulated to alternate between dry and wet soil cycles. Based on rainfall at the sampling site and the local fields' water management strategies, we added water to the soil column every three days, with 2000 mL water each time for the horizontal soil column and 250 mL water for the vertical soil column. Soil porewater samples were taken 15, 30, 60, and 100 days after the saturation of the soil column using a Rhizon sampler (2.5 cm × 10 cm, MOM, Rhizon, The Netherlands). The soil porewater samples of the horizontal soil column were taken from the gauze bag in the central soil column (0 cm) and 1, 4, and 8 cm away from the central soil column. The soil porewater samples of the vertical soil column were collected at 0, 5, 10, 20, 30, and 38 cm from the top of the soil column. The collected soil porewater was acidified with 6 M hydrochloric acid to prevent heavy metal precipitation and transformation [25].

2.3. Chemical Analysis of Soil

Soil pH was determined in a 1:2.5 ratio soil/water suspension. The S (SO_4^{2-}) was determined by the barium sulfate turbidimetric method. The bioavailable Cd, Pb, As, and total Fe in soil porewater were determined by inductively coupled plasma mass spectrometry (ICP-MS NEXION300X, PerkinElmer, Inc., Shelton, CT, USA). See the previous study for details [17]. The standard internal method was used to determine the accuracy of the analytical methods. A standard sample of 50 μg L^{-1} was measured after every ten samples as a quality control measure. The recoveries of internal standards for bioavailable heavy metal were within the range of 95.2% and 106.3%, which proved that the detection method was credible.

2.4. Data Analysis

SPSS 19.0 (SPSS, Chicago, IL, USA) software was used for the statistical analysis. The one-way ANOVA followed by the least significant difference (LSD) was used on normally distributed data, and not normally distributed data were compared by the Kruskal–Wallis test. A p-value of less than 0.05 was considered statistically significant. The data in the figures and tables show the average ± standard deviation.

3. Results and Discussion

3.1. Effect of CSF on pH

In the horizontal soil column, the soil column center's pH decreased significantly by 12.2% and 10.2% on days 15 and 100 with the addition of CSF (Table 2). However, the soil pH at different distances from the central soil column (1 cm, 4 cm, and 8 cm) showed no significant change in the CK-H and CSF-H treatments. The soil pH on day 100 was slightly higher than on day 15, which was inconsistent with previous observations [1,26]. This may have occurred because the main components of CSF are $CaSO_4 \cdot 2H_2O$ and Fe_2O_3. Previous studies have shown that the decrease in soil pH with CSF addition is due to increased sulfate (SO_4^{2-}) concentrations [27]. In addition, the soil was not flooded during the incubation, possibly hindering sulfate reduction and thereby limiting proton consumption in the soil.

Table 2. The change of pH in different horizontal soil layers with CSF addition.

Time	Treatment	Distance from the Center of the Soil Column			
		0 cm	1 cm	4 cm	8 cm
15 d	CK-H	5.56 ± 0.30 Aa	5.27 ± 0.30 Aa	5.27 ± 0.30 Aa	5.28 ± 0.28 Aa
	CSF-H	4.88 ± 0.01 Ab	5.00 ± 0.11 Aa	5.01 ± 0.12 Aa	5.00 ± 0.10 Aa
100 d	CK-H	5.75 ± 0.23 Aa	5.58 ± 0.23 Aa	5.58 ± 0.32 Aa	5.57 ± 0.25 Aa
	CSF-H	5.16 ± 0.14 Ab	5.24 ± 0.19 Aa	5.27 ± 0.24 Aa	5.23 ± 0.26 Aa

Note: The capital letters indicate a significant ($p < 0.05$) difference between the different distances for the same treatments, and the lowercase letters indicate a significant difference between CK and CSF treatments for the same distance.

In the vertical soil column, the soil layer of 5 to 38 cm from the topsoil's pH was higher than that of the topsoil on days 15 and 100 in both the CK and CSF treatments (Table 3). Moreover, on the 100th day after adding the CSF, the soil pH 5 cm away from the topsoil was significantly lower than that of the topsoil. The soil pH decreased by 6.3% in treatment 5 cm away from the topsoil. The pH of the CSF treatments showed an increasing trend with the soil depth. It could be that simulated rainfall washes some acid-causing ions into the lower layers, which acidifies the soil.

Table 3. The pH changes in the different vertical soil layers with CSF addition.

Time	Treatment	Distance from the Topsoil					
		0 cm	5 cm	10 cm	20 cm	30 cm	38 cm
15 d	CK-V	5.52 ± 0.10 Ba	6.90 ± 0.08 Aa	6.83 ± 0.29 Aa	6.21 ± 0.94 Aa	6.10 ± 0.98 Aa	6.55 ± 1.19 Aa
	CSF-V	5.46 ± 0.03 Ca	6.96 ± 0.09 ABa	6.70 ± 0.10 Ba	7.12 ± 0.07 Aa	7.02 ± 0.16 Aa	7.16 ± 0.30 Aa
100 d	CK-V	5.24 ± 0.09 Ba	6.66 ± 0.21 Aa	6.46 ± 0.04 Aa	6.68 ± 0.16 Aa	6.27 ± 0.59 Aa	6.34 ± 0.89 Aa
	CSF-V	5.25 ± 0.03 Ca	6.24 ± 0.08 Bb	6.31 ± 0.08 Ba	6.68 ± 0.12 Aa	6.82 ± 0.20 Aa	6.73 ± 0.08 Aa

Note: The capital letters indicate a significant ($p < 0.05$) difference between the different distances for the same treatments, and the lowercase letters indicate a significant difference between CK and CSF treatments for the same distance.

3.2. Effect of CSF on the Concentrations of SO_4^{2-} and Total Fe

The SO_4^{2-} concentrations of the soil column center in the horizontal soil column were significantly higher than that in the soil 1 to 8 cm away from the soil column center (Table 4) in the CSF-H treatment. The SO_4^{2-} concentrations decreased with increasing distance from the soil column center. On day 15, after adding CSF, the SO_4^{2-} concentrations in the center soil and 1 cm away from the soil column center significantly increased by 628.1% and 326.3%, respectively. On day 100, after adding CSF, the SO_4^{2-} concentration in the center of the soil column and the soil 1 cm, 4 cm, and 8 cm away from the soil column center significantly increased by 286.8%, 236.0%, 55.3%, and 86.3%, respectively, compared with the CK treatment. The increase in SO_4^{2-} concentrations may be ascribed to the added $CaSO_4$ through the CSF. Inorganic SO_4^{2-} in the soil solution is highly mobile [28].

Table 4. The concentrations of SO_4^{2-} and total Fe in different horizontal soil layers with CSF addition.

Element	Time	Treatment	Distance from the Center of the Soil Column			
			0 cm	1 cm	4 cm	8 cm
SO_4^{2-} (mg/L)	15 d	CK-H	11.93 ± 2.53 Ab	9.29 ± 1.29 Ab	12.48 ± 3.97 Aa	11.84 ± 3.44 Aa
		CSF-H	86.86 ± 2.95 Aa	39.60 ± 6.83 Ba	16.96 ± 4.47 Ca	12.01 ± 2.44 Ca
	100 d	CK-H	9.37 ± 2.73 Ab	7.87 ± 1.81 Ab	13.10 ± 2.56 Ab	11.11 ± 1.25 Ab
		CSF-H	36.24 ± 3.06 Aa	26.44 ± 1.42 Ba	20.34 ± 3.27 Ca	20.70 ± 3.29 Ca
Fe (μg/L)	15 d	CK-H	15.87 ± 3.87 Ab	15.57 ± 8.60 Aa	23.63 ± 13.61 Aa	21.40 ± 5.17 Aa
		CSF-H	32.17 ± 6.33 Aa	23.60 ± 1.73 BCa	16.63 ± 0.85 Ca	26.80 ± 4.23 ABa
	100 d	CK-H	13.23 ± 8.04 Aa	9.27 ± 6.11 Ab	17.87 ± 4.76 Aa	16.10 ± 1.91 Aa
		CSF-H	23.80 ± 4.16 Aa	23.93 ± 6.84 Aa	20.07 ± 8.70 Aa	18.37 ± 3.51 Aa

Note: The capital letters indicate a significant ($p < 0.05$) difference between the different distances for the same treatments, and the lowercase letters indicate a significant difference between CK and CSF treatments for the same distance.

On day 15, after adding CSF, the total Fe content in the center of the soil column was significantly higher than that in the soil 1 to 4 cm away from the soil column center (Table 4). The total Fe concentrations in the center of the soil column with the addition of CSF significantly increased by 102.7% compared with the CK treatment on day 15, while the total Fe concentrations in the soil 1 to 8 cm away from the center of the soil column were not significantly affected. The total Fe concentrations in the soil 1 cm away from the central soil column with CSF addition were significantly increased by 158.1% after 100 days. However, the CSF addition had no significant influence on the Fe contents in the other soils. In the CK and CSF treatments, the total Fe concentrations decreased on day 100 compared with day 15 at the same distance. The decrease after 100 days may be because by then, Fe oxides and Fe hydroxides in the soil were being reduced and started to dissolve [29,30], thereby increasing the Fe contents in the deeper layers.

In the vertical soil column, the concentrations of SO_4^{2-} in the CK treatment showed no significant changes between the distances from the topsoil on day 15. At the same time, the concentration of SO_4^{2-} in the topsoil significantly increased by 74.4% with the addition of CSF compared with the CK. Moreover, the concentrations of SO_4^{2-} at 5, 10, and 20 cm away from the topsoil significantly decreased by 68.8%, 58.8%, and 45.5%, respectively (Table 5), in the CSF treatment. Compared with the concentration of SO_4^{2-} in the topsoil layer, the SO_4^{2-} concentrations in the soil 5 cm and 10 cm away from the topsoil layer significantly decreased by 65.7% and 45.3%, respectively, in the CK treatment after 100 days. In the CSF treatment, the concentration of SO_4^{2-} in the soil layer from 5, 10, 20, and 30 cm away significantly decreased by 48.5%, 69.0%, 61.5%, and 46.5%, respectively. Previous studies have reported that mineralogical composition, total carbon, particle-size distribution, pH, and the presence of other ions could influence the adsorbed SO_4^{2-} in the soil [31,32]. The concentration of SO_4^{2-} in the soil layer from 5, 10, 20, and 30 cm away significantly decreased, indicating that SO_4^{2-} had been reduced. In addition, SO_4^{2-} could be retained by colloidal Fe oxides or complexed by Fe oxides/hydroxides and sesquihydroxides/sesquioxides [33]. Compared with CK, the concentrations of SO_4^{2-} in the topsoil and at the distance of 5 cm and 38 cm from the topsoil with CSF addition increased significantly by 85.2%, 178.4%, and 95.8%, respectively. In rice paddy soil, SO_4^{2-} is likely retained at depth through anion exchange reactions associated with Fe-oxides and Al-oxides [33].

Table 5. The concentrations of SO_4^{2-} and total Fe in different vertical soil layers with CSF addition.

Element	Time	Treatment	Distance from the Topsoil					
			0 cm	5 cm	10 cm	20 cm	30 cm	38 cm
SO_4^{2-} (mg/L)	15 d	CK-V	1.60 ± 0.42 ABb	1.04 ± 0.08 Ba	1.30 ± 0.20 ABa	1.24 ± 0.47 Ba	1.79 ± 0.42 ABa	2.12 ± 0.70 Aa
		CSF-V	2.79 ± 0.56 Aa	0.87 ± 0.16 Ba	1.15 ± 0.14 Ba	1.52 ± 0.06 Ba	2.05 ± 1.1 ABa	2.00 ± 0.86 ABa
	100 d	CK-V	1.08 ± 0.43 ABb	0.37 ± 0.01 Cb	0.59 ± 0.24 Ca	0.70 ± 0.07 BCa	1.25 ± 0.23 Aa	1.19 ± 0.22 Ab
		CSF-V	2.00 ± 0.34 Aa	1.03 ± 0.11 Ba	0.62 ± 0.14 Ca	0.77 ± 0.22 BCa	1.07 ± 0.12 Ba	2.33 ± 0.21 Aa
Fe (μg/L)	15 d	CK-V	7.58 ± 1.72 Ab	7.41 ± 2.82 Ab	7.61 ± 1.43 Ab	6.53 ± 2.33 Ab	6.80 ± 2.36 Ab	7.49 ± 3.09 Aa
		CSF-V	36.69 ± 8.20 ABa	47.75 ± 3.29 Aa	34.90 ± 10.79 ABa	34.66 ± 7.42 ABa	25.83 ± 11.48 Ba	12.36 ± 4.83 Ba
	100 d	CK-V	8.51 ± 2.38 Ab	9.57 ± 2.50 Ab	9.00 ± 3.49 Aa	9.51 ± 4.18 Aa	6.80 ± 2.14 Ab	11.21 ± 2.63 Ab
		CSF-V	23.67 ± 7.4 Aa	19.62 ± 6.58 Aa	17.60 ± 6.45 Aa	17.77 ± 5.86 Aa	26.87 ± 4.93 Aa	29.03 ± 12.43 Aa

Note: The capital letters indicate a significant ($p < 0.05$) difference between the different distances for the same treatments, and the lowercase letters indicate a significant difference between CK and CSF treatments for the same distance.

Compared with the CK treatment, adding CSF significantly increased the concentrations of total Fe in the topsoil and the soil 5, 10, 20, and 30 cm away from the topsoil by 384.0%, 544.4%, 358.6%, 430.8%, and 279.9%, respectively, on day 15 (Table 5). The concentrations of total Fe in the soil layer 5 cm away from the topsoil had the highest Fe content with CSF addition. Moreover, the total Fe concentrations decreased with the distance from the topsoil (5 to 30cm). However, there was no significant change in total Fe concentrations between the different soil layers after 100 days of incubation in the CSF treatment. The concentrations of total Fe in the topsoil 5, 10, 30, and 38 cm and away from the topsoil significantly increased by 178.1%, 105.0%, 295.1%, and 159.0%, respectively, after 100 days of the CSF treatment. The Fe contents varied with depth and over time, possibly due to the reduction and dissolution of Fe minerals in the soil and to leaching.

3.3. *Effect of CSF on Bioavailable Cd, Pb, and As Concentrations*

The addition of CSF in the soil column center or surface of the soil column affected the bioavailability of Cd, Pb, and As in the surrounding space soil (Figures 2 and 3). In the horizontal soil column, CSF treatment decreased the bioavailable Cd concentrations in the soil column center and soil within 1 cm around the soil column center. Although adding CSF had no significant effect on decreasing Cd bioavailability in all the soils except for the soil column center on day 30, the CSF effective range gradually expanded with incubation time (Figure 2a–d). On day 60, the bioavailable Cd concentrations in the CSF treatment at 0 cm, 1 cm, and 4 cm were lower than that at the corresponding distance of the CK, which was significantly decreased by 11.8%, 13.3%, and 10.5%, respectively. Furthermore, on day 100, adding CSF in the soil column center decreased bioavailable Cd concentrations at 8 cm from the central column. The bioavailable Cd concentrations were significantly decreased by 11.1%, 8.1%, 14.9% and 14.3% at 0, 1, 4, and 8 cm away from the soil column center, respectively. When taken together, these results indicate that CSF's immobilization range for Cd increased over time. The Cd concentration significantly correlated with the SO_4^{2-} contents (Figure 4b). This correlation may be due to the increase in sulfur, leading to Cd precipitation as CdS [34], resulting in a decrease in bioavailable Cd. The effect of CSF on reducing the bioavailability of Pb and As differed from that on Cd. As shown in Figure 2e–l, compared with CK, within 100 days of incubation, the bioavailable Pb concentrations in the soil supplemented with CSF only significantly decreased at the soil column center. The stabilization efficiency of CSF on Pb in the soil column center increased with incubation time, which was 28.5%, 22.0%, 34.9%, and 61.6% on days 15, 30, 60, and 100, respectively. Adding CSF to the soil column center had no significant effect on the bioavailability of Pb in the surrounding soil. Similarly, adding CSF only decreased the bioavailable As at the central soil column within 100 days of incubation. The concentrations of bioavailable As decreased by 5.6%, 6.7%, 13.4%, and 7.4% on days 15, 30, 60, and 100, respectively. Sulfate (SO_4^{2-}) can be reduced to sulfide (S^{2-}) and then immobilize heavy metals, forming a stable sulfide-bound state [35]. The Ksp-CdS (2.6×10^{-29}) is lower than Ksp-PbS (3.4×10^{-28}) and Ksp-As_2S_3 (2.1×10^{-22}) [36]. Therefore, the combination order of Pb, As, and S^{2-} could be limited by Cd, which could explain the farthest CSF's stabilization range for Cd.

In the vertical soil column, the concentrations of bioavailable Cd, Pb, and As of the soil column treated with CSF were lower than that of the control treatment (Figure 3). The bioavailable Cd concentration in the CSF treatment was significantly decreased compared to that of the CK in the topsoil on day 15: the CSF's immobilization depths for Cd increased over time. After 30, 60, and 100 days of incubation, the CSF immobilized Cd at a depth of 5 cm, 10 cm, and 20 cm, respectively. The available Pb in the soil also showed a similar trend to Cd; however, stabilizing was easier. The CSF's immobilization depths for Pb were 10 cm on day 15; however, it took 60 days for Cd to be at this depth.

In contrast, Pb's deepest immobilization depth was 20 cm during the 100 incubation days. The bioavailable Pb concentrations decreased between 36.73% and 91.93%. The concentrations of bioavailable Pb were significantly correlated with total Fe concentration (Figure 5b) with CSF amendment. The greater amount of Pb in the soil than Cd and As would compete with Cd and As to adsorb, making it difficult for Cd and As to combine with CSF. Moreover, the bioavailable As concentrations varied with depths, whereby the concentrations increased and then decreased in all depth layers. The immobilization depths for As were 5 to 10 cm during the 100 incubation days. In addition, regardless of whether CSF was added, the bioavailable As content at 0 cm in the two treatments at the four time points was significantly lower than that at other distances. We speculated that the As content in the topsoil leached into the underlying soil, consequently increasing the bioavailable As contents in the underlying soil over time. Furthermore, the oxic condition of the topsoil may have promoted As precipitation and adsorption of soil mineral oxides [37].

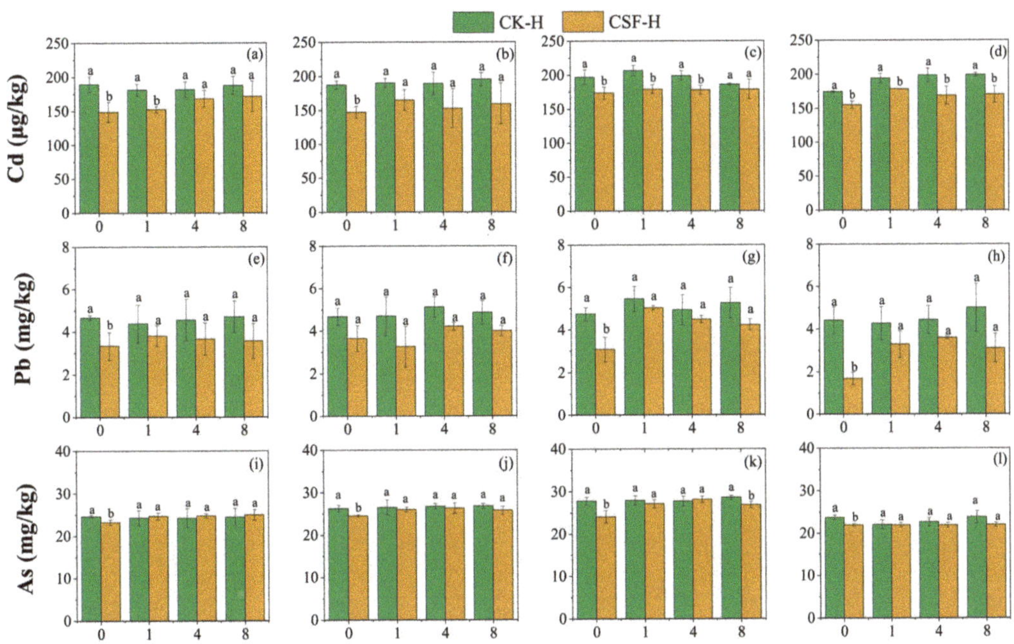

Figure 2. The effect of horizontal migration of CSF on the concentrations of bioavailable Cd, Pb, and As in the surrounding soil, the lowercase letters indicate a significant difference between CK and CSF treatments for the same distance. (**a**,**e**,**i**): Day 15 of incubation; (**b**,**f**,**j**): Day 30 of incubation; (**c**,**g**,**k**): Day 60 of incubation; (**d**,**h**,**l**): Day 100 of incubation.

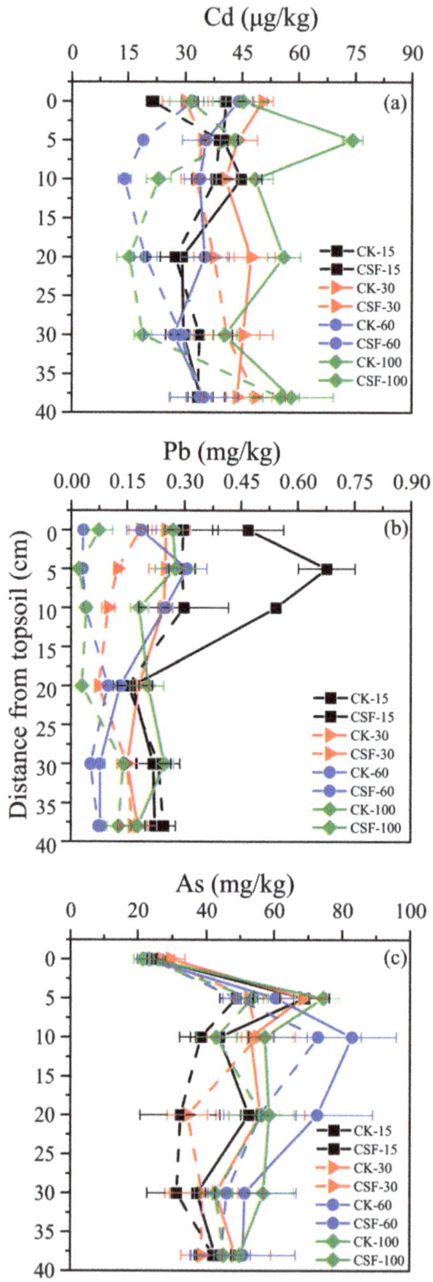

Figure 3. The effect of vertical migration of CSF on the concentrations of bioavailable Cd, Pb, and As in the surrounding soil. (**a**): The concentrations of bioavailableCd in the vertical soil column; (**b**): The concentrations of bioavailable Pb in the vertical soil column; (**c**): The concentrations of bioavailable As in the vertical soil column.

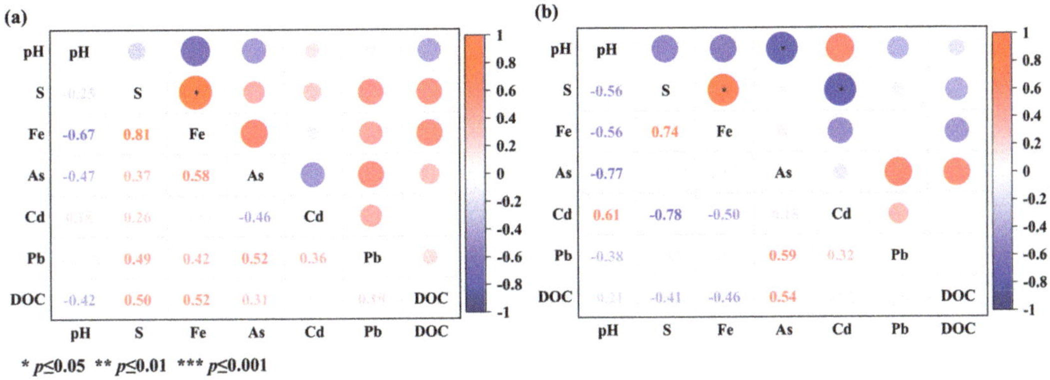

Figure 4. Correlation between physical and chemical properties of soil and bioavailability of Cd, Pb, and As in the horizontal soil column. (**a**): CK; (**b**): CSF.

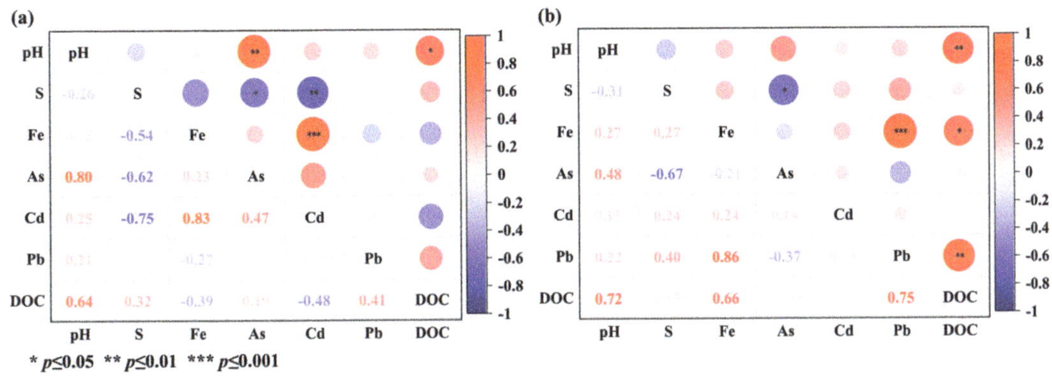

Figure 5. Correlation between physical and chemical properties of soil and bioavailability of Cd, Pb, and As in the vertical soil column. (**a**): CK; (**b**): CSF.

4. Conclusions

This study evaluated the spatial and temporal variations of Cd, Pb, and As bioavailability in paddy soils regulated by CSF. The results showed that CSF decreased the bioavailable Cd, Pb, and As concentrations in the soil column center. The immobilization range and depth for Cd, Pb, and As were different, and the immobilization effect of CSF for Cd was the best. The physical and chemical properties of soil, such as soil particle size fraction and perviousness, may influence the migration of the CSF, which influence the immobilization range and depth for heavy metal. Therefore, the immobilization range and depth for heavy metal by CSF in different soils must be further investigated. This study provides important insights into the application frequency and spacing distance when using CSF to immobilize heavy metals.

Author Contributions: Conceptualization, W.Z. and C.Z.; Methodology, Z.Z.; Software, Y.D. and J.L.; Validation, J.C., W.G. and Z.Z.; Formal Analysis, Y.D.; Investigation, C.Z.; Data Curation, J.L.; Writing—Original Draft Preparation, C.Z. and J.L.; Writing—Review and Editing, J.L., W.G. and W.Z.; Supervision, J.C. and Z.Z. All authors have read and agreed to the published version of the manuscript.

Funding: This research was funded by the National Key R&D Program of China, grant number 2018YFC1800506, the National Natural Science Foundation of China, grant number 41907101, and Talent Cultivation Project of Zhejiang Association for Science and Technology.

Institutional Review Board Statement: Not applicable.

Informed Consent Statement: Not applicable.

Data Availability Statement: Not applicable.

Conflicts of Interest: The authors declare no conflict of interest.

References

1. Liu, J.; Kang, H.; Tao, W.; Li, H.; He, D.; Ma, L.; Tang, H.; Wu, S.; Yang, K.; Li, X. A Spatial Distribution—Principal Component Analysis (SD-PCA) Model to Assess Pollution of Heavy Metals in Soil. *Sci. Total Environ.* **2023**, *859*, 160112. [PubMed]
2. Yang, Q.; Li, Z.; Lu, X.; Duan, Q.; Huang, L.; Bi, J. A Review of Soil Heavy Metal Pollution from Industrial and Agricultural Regions in China: Pollution and Risk Assessment. *Sci. Total Environ.* **2018**, *642*, 690–700.
3. Zhang, J.; Yang, R.; Li, Y.C.; Peng, Y.; Wen, X.; Ni, X. Distribution, Accumulation, and Potential Risks of Heavy Metals in Soil and Tea Leaves from Geologically Different Plantations. *Ecotoxicol. Environ. Saf.* **2020**, *195*, 110475. [PubMed]
4. Zhang, Y.; O'Connor, D.; Xu, W.; Hou, D. Blood Lead Levels among Chinese Children: The Shifting Influence of Industry, Traffic, and e-Waste over Three Decades. *Environ. Int.* **2020**, *135*, 105379. [PubMed]
5. Adimalla, N. Heavy Metals Pollution Assessment and Its Associated Human Health Risk Evaluation of Urban Soils from Indian Cities: A Review. *Environ. Geochem. Health* **2020**, *42*, 173–190. [PubMed]
6. Rajendran, S.; Priya, T.A.K.; Khoo, K.S.; Hoang, T.K.A.; Ng, H.S.; Munawaroh, H.S.H.; Karaman, C.; Orooji, Y.; Show, P.L. A Critical Review on Various Remediation Approaches for Heavy Metal Contaminants Removal from Contaminated Soils. *Chemosphere* **2022**, *287*, 132369.
7. Antonkiewicz, J.; Kowalewska, A.; Mikołajczak, S.; Kołodziej, B.; Bryk, M.; Spychaj-Fabisiak, E.; Koliopoulos, T.; Babula, J. Phytoextraction of Heavy Metals after Application of Bottom Ash and Municipal Sewage Sludge Considering the Risk of Environmental Pollution. *J. Environ. Manag.* **2022**, *306*, 114517.
8. Pouresmaieli, M.; Ataei, M.; Forouzandeh, P.; Azizollahi, P.; Mahmoudifard, M. Recent Progress on Sustainable Phytoremediation of Heavy Metals from Soil. *J. Environ. Chem. Eng.* **2022**, *10*, 108482.
9. Azhar, U.; Ahmad, H.; Shafqat, H.; Babar, M.; Shahzad Munir, H.M.; Sagir, M.; Arif, M.; Hassan, A.; Rachmadona, N.; Rajendran, S.; et al. Remediation Techniques for Elimination of Heavy Metal Pollutants from Soil: A Review. *Environ. Res.* **2022**, *214*, 113918.
10. Fu, T.; Zhang, B.; Gao, X.; Cui, S.; Guan, C.Y.; Zhang, Y.; Zhang, B.; Peng, Y. Recent Progresses, Challenges, and Opportunities of Carbon-Based Materials Applied in Heavy Metal Polluted Soil Remediation. *Sci. Total Environ.* **2023**, *856*, 158810.
11. Li, W.; Ni, P.; Yi, Y. Comparison of Reactive Magnesia, Quick Lime, and Ordinary Portland cement for Stabilization/Solidification of Heavy Metal-Contaminated Soils. *Sci. Total Environ.* **2019**, *671*, 741–753.
12. Wang, Y.Y.; You, L.C.; Lyu, H.H.; Liu, Y.X.; He, L.L.; Hu, Y.D.; Luo, F.C.; Yang, S.M. Role of Biochar–Mineral Composite Amendment on the Immobilization of Heavy Metals for Brassica Chinensis from Naturally Contaminated Soil. *Environ. Technol. Innovation.* **2022**, *28*, 102622.
13. Li, S.; Yang, F.; Zhang, Y.; Lan, Y.; Cheng, K. Performance of Lead Ion Removal by the Three-Dimensional Carbon Foam Supported Nanoscale Zero-Valent Iron Composite. *J. Clean. Prod.* **2021**, *294*, 125350.
14. Diao, Z.H.; Xu, X.R.; Jiang, D.; Kong, L.J.; Sun, Y.X.; Hu, Y.X.; Hao, Q.W.; Chen, H. Bentonite-Supported Nanoscale Zero-Valent Iron/Persulfate System for the Simultaneous Removal of Cr(VI) and Phenol from Aqueous Solutions. *Chem. Eng. J.* **2016**, *302*, 213–222.
15. Yang, D.; Yang, S.; Wang, L.; Xu, J.; Liu, X. Performance of Biochar-Supported Nanoscale Zero-Valent Iron for Cadmium and Arsenic Co-Contaminated Soil Remediation: Insights on Availability, Bioaccumulation and Health Risk. *Environ. Pollut.* **2021**, *290*, 118054.
16. Liu, Q.; Chen, Z.; Chen, Z.; Pan, X.; Luo, J.; Huang, F.; Zhang, X.; Lin, Q. Microbial Community Characteristics of Cadmium Speciation Transformation in Soil after Iron-Based Materials Application. *Appl. Soil Ecol.* **2023**, *183*, 104745.
17. Zhai, W.; Zhao, W.; Yuan, H.; Guo, T.; Hashmi, M.Z.; Liu, X.; Tang, X. Reduced Cd, Pb, and As Accumulation in Rice (*Oryza sativa* L.) by a Combined Amendment of Calcium Sulfate and Ferric Oxide. *Environ. Sci. Pollut. Res.* **2020**, *27*, 1348–1358.
18. Zhai, W.; Dai, Y.; Zhao, W.; Yuan, H.; Qiu, D.; Chen, J.; Gustave, W.; Maguffin, S.C.; Chen, Z.; Liu, X.; et al. Simultaneous Immobilization of the Cadmium, Lead and Arsenic in Paddy Soils Amended with Titanium Gypsum. *Environ. Pollut.* **2020**, *258*, 113790.
19. Tica, D.; Udovic, M.; Lestan, D. Immobilization of Potentially Toxic Metals Using Different Soil Amendments. *Chemosphere* **2011**, *85*, 577–583.
20. Cui, H.; Fan, Y.; Fang, G.; Zhang, H.; Su, B.; Zhou, J. Leachability, Availability and Bioaccessibility of Cu and Cd in a Contaminated Soil Treated with Apatite, Lime and Charcoal: A Five-Year Field Experiment. *Ecotoxicol. Environ. Saf.* **2016**, *134*, 148–155.
21. Li, X.; Bai, S.G.; Xi, B.D.; Yuan, Z.Y.; Wang, Y.Y. Research of Effect of Groundwater Table Fluctuation on Migration Law of Cadmium. *AMR* **2012**, *599*, 455–461.
22. Igloria, R.V.; Hathhorn, W.E.; Yonge, D.R. NOM and Trace Metal Attenuation during Storm-Water Infiltration. *J. Hydrol. Eng.* **1997**, *2*, 120–127.

23. Ekanayake, D.; Loganathan, P.; Johir, M.A.H.; Kandasamy, J.; Vigneswaran, S. Enhanced Removal of Nutrients, Heavy Metals, and PAH from Synthetic Stormwater by Incorporating Different Adsorbents into a Filter Media. *Water Air Soil Pollut.* **2021**, *232*, 96.
24. Xu, Z.; Hu, X.; Ding, Z.; Liu, Y.; Gao, B. Retention of Nano PbO in Saturated Columns and Its Dissolution Kinetics in Soils. *Environ. Sci. Pollut. Res.* **2020**, *27*, 1167–1174.
25. Lomax, C.; Liu, W.; Wu, L.; Xue, K.; Xiong, J.; Zhou, J.; McGrath, S.P.; Meharg, A.A.; Miller, A.J.; Zhao, F. Methylated Arsenic Species in Plants Originate from Soil Microorganisms. *New Phytol.* **2012**, *193*, 665–672.
26. Jin, Y.; Wang, Y.; Li, X.; Luo, T.; Ma, Y.; Wang, B.; Liang, H. Remediation and Its Biological Responses to Cd(II)-Cr(VI)-Pb(II) Multi-Contaminated Soil by Supported Nano Zero-Valent Iron Composites. *Sci. Total Environ.* **2023**, *867*, 161344.
27. Kim, H.S.; Seo, B.H.; Kuppusamy, S.; Lee, Y.B.; Lee, J.H.; Yang, J.E.; Owens, G.; Kim, K.R. A DOC Coagulant, Gypsum Treatment Can Simultaneously Reduce As, Cd and Pb Uptake by Medicinal Plants Grown in Contaminated Soil. *Ecotoxicol. Environ. Saf.* **2018**, *148*, 615–619.
28. Mitchell, M.J.; Lovett, G.; Bailey, S.; Beall, F.; Burns, D.; Buso, D.; Clair, T.A.; Courchesne, F.; Duchesne, L.; Eimers, C.; et al. Comparisons of Watershed Sulfur Budgets in Southeast Canada and Northeast US: New Approaches and Implications. *Biogeochemistry* **2011**, *103*, 181–207.
29. Gao, B.; Chen, Q.; Liu, K.; Li, F.; Fang, L.; Zhu, Z.; Tran, M.T.; Peng, J. Biogeochemical Fe(II) Generators as a New Strategy for Limiting Cd Uptake by Rice and Its Implication for Agricultural Sustainability. *Sci. Total Environ.* **2022**, *820*, 153306.
30. Liu, K.; Li, F.; Pang, Y.; Fang, L.; Hocking, R. Electron Shuttle-Induced Oxidative Transformation of Arsenite on the Surface of Goethite and Underlying Mechanisms. *J. Hazard. Mater.* **2022**, *425*, 127780.
31. Gustafsson, J.P.; Akram, M.; Tiberg, C. Predicting Sulphate Adsorption/Desorption in Forest Soils: Evaluation of an Extended Freundlich Equation. *Chemosphere* **2015**, *119*, 83–89.
32. Takahashi, J.; Higashi, T. Long-Term Changes in Sulfate Concentrations and Soil Acidification of Forested Umbrisols and Andosols of Japan. *Soil Sci.* **2013**, *178*, 69–78.
33. Yang, J.; Liu, Z.; Wan, X.; Zheng, G.; Yang, J.; Zhang, H.; Guo, L.; Wang, X.; Zhou, X.; Guo, Q.; et al. Interaction between Sulfur and Lead in Toxicity, Iron Plaque Formation and Lead Accumulation in Rice Plant. *Ecotoxicol. Environ. Saf.* **2016**, *128*, 206–212. [PubMed]
34. Fulda, B.; Voegelin, A.; Kretzschmar, R. Redox-Controlled Changes in Cadmium Solubility and Solid-Phase Speciation in a Paddy Soil As Affected by Reducible Sulfate and Copper. *Environ. Sci. Technol.* **2013**, *47*, 12775–12783.
35. Zhang, D.; Du, G.; Chen, D.; Shi, G.; Rao, W.; Li, X.; Jiang, Y.; Liu, S.; Wang, D. Effect of Elemental Sulfur and Gypsum Application on the Bioavailability and Redistribution of Cadmium during Rice Growth. *Sci. Total Environ.* **2019**, *657*, 1460–1467. [PubMed]
36. Banfalvi, G. Removal of Insoluble Heavy Metal Sulfides from Water. *Chemosphere* **2006**, *63*, 1231–1234. [PubMed]
37. McGeehan, S.L. Arsenic Sorption and Redox Reactions: Relevance to Transport and Remediation. *J. Environ. Sci. Health Part A Environ. Sci. Eng. Toxicol.* **1996**, *31*, 2319–2336.

toxics

MDPI

Article

Quantitative Source Apportionment of Potentially Toxic Elements in Baoshan Soils Employing Combined Receptor Models

Chunyu Dong [1,2,†], Hao Zhang [1,2,†], Haichan Yang [1,2], Zhaoxia Wei [1], Naiming Zhang [1,2] and Li Bao [1,2,*]

1 Yunnan Agricultural University, Kunming 650201, China
2 Yunnan Laboratory of Improvement of Soil Fertility and Pollution Remediation, Kunming 650201, China
* Correspondence: bbllty@163.com
† These authors contributed equally to this work.

Abstract: Arable soils are crucial for national development and food security; therefore, contamination of agricultural soils from potentially toxic elements (PTEs) is a global concern. In this study, we collected 152 soil samples for evaluation. Considering the contamination factors and using the cumulative index and geostatistical methods, we investigated the contamination levels of PTEs in Baoshan City, China. Using principal component analysis, absolute principal component score-multivariate linear regression, positive matrix factorization, and UNMIX, we analyzed the sources and quantitatively estimated their contributions. The average Cd, As, Pb, Cu, and Zn concentrations were 0.28, 31.42, 47.59, 100.46, and 12.36 mg/kg, respectively. The Cd, Cu, and Zn concentrations exceeded the corresponding background values for Yunnan Province. The combined receptor models showed that natural and agricultural sources contributed primarily to Cd and Cu and As and Pb inputs, accounting for 35.23 and 7.67% pollution, respectively. Industrial and traffic sources contributed primarily to Pb and Zn inputs (47.12%). Anthropogenic activities and natural causes accounted for 64.76 and 35.23% of soil pollution, respectively. Industrial and traffic sources contributed 47.12% to pollution from anthropogenic activities. Accordingly, the control of industrial PTE pollution emissions should be strengthened, and awareness should be raised to protect arable land around roads.

Keywords: cultivated soils; potentially toxic elements; receptor model; source apportionment; soil pollution

check for updates

Citation: Dong, C.; Zhang, H.; Yang, H.; Wei, Z.; Zhang, N.; Bao, L. Quantitative Source Apportionment of Potentially Toxic Elements in Baoshan Soils Employing Combined Receptor Models. *Toxics* **2023**, *11*, 268. https://doi.org/10.3390/toxics11030268

Academic Editors: Junhao Qin, Peidong Su, Feng Zhu and Lin Ding

Received: 17 January 2023
Revised: 9 March 2023
Accepted: 10 March 2023
Published: 14 March 2023

1. Introduction

The expansion of cities and rapid development of modern agriculture have led to increased environmental pollution in cultivated soils. Some of the contaminants of greatest concern are potentially toxic elements (PTEs) [1,2], which not only destroy the quality of cultivated land, but also they indirectly or directly cause damage to human health [3,4]. Therefore, remediation of arable soil PTE contamination has become one of the most urgent problems in environmental science. However, PTEs in soils do not originate from human activities such as mining, metal smelting, pesticide/fertilizer usage, or automobilism [5–7]. Furthermore, several researchers have shown that the mineralization of soil parent material is the main cause of PTE enrichment [8,9].

Yunnan province has a complex geological structure and rich mineral resources. According to a 2020 report on the quality of arable land in the Baoshan region, medium-grade cultivation was performed on 263,500 ha of low-grade cultivated land, which accounted for 79.31% of the cultivated land area [10]. Currently, soil conditions are poor and heavy metal pollution is becoming an increasingly significant problem in the area [11–15]. Moreover, the PTE background value in Baoshan is significantly higher than the national average, making PTE pollution a serious hazard in the region [16,17].

Zhang et al. have found severe PTE enrichment in soils around the Baoshan mining area and have reported the soil to be at risk [18]. Other factors affecting Baoshan soil PTE enrichment include geogenesis and carbonate and basalt parent material differentiation. The Baoshan agricultural environmental protection monitoring station that evaluates the quality of farmland soil indicated that the arable soil in the study area is contaminated to varying degrees by different sources, the most prominent ones being Cd, Pb, As, Cu, and Zn PTEs [19–21]. Importantly, the quality and safety of arable soil determines the quality and safety of agricultural products. Cicchella et al. has recommended paying attention to the nature and sources of contaminants [22]. Still, previous studies have only evaluated the risk of soil contamination in Baoshan and have not distinguished the sources of contamination [23]. Thus, we must understand the sources of PTE pollution in arable land to protect it [24,25].

The methods of pollution source analysis are divided mainly into source identification and source apportionment. Scholars are increasingly using receptor models for source analyses, with the most widely used ones being principal component analysis (PCA), absolute principal component score–multivariate linear regression (APCS-MLR), UNMIX, and positive matrix factorization (PMF). Li et al. used PCA to determine potentially toxic elements in Gansu cropland soils; they found three sources of contamination and evaluated soil data [26]. The same group also used APCS-MLR to analyze the PTE sources in urban farmland. They combined the total hazard index (THI) and total carcinogenic risk (TCR) and found that APCS-MLR analyzed both the pollution sources and pollution status [27]. This model only required the composition of the receptor emission source and did not require the accurate source component spectrum data. Still, the estimated source component spectrum and contribution values were often observed. Chen et al. combined UNMIX with PMF for source analysis of heavy metal data in a suburban area of Kaifeng and found good agreement between the two models for source assignment [28]. The UNMIX model overcomes the drawback of negative source contribution, which does not require prior knowledge of the number of sources and identifies these sources via different source identifiers to infer them. However, it cannot determine the source components, contribution rates, or contribution values. Liu et al. used PMF to determine the pollution sources in a farmland in Wenzhou, Zhejiang Province, and combined it with multivariate statistical methods, which demonstrated the feasibility of PMF models for source analysis [29]. Lv used APCS/MLR and PMF, and both provided three identical sources and were consistent with the mapping of parent material distribution [30]. PMF avoids negative values in the results, thus providing interpretability and clear physical meaning to the obtained source component spectra and source contributions. In addition, PMF does not require the measurement of source component spectra and uses error estimates for each individual data point to deal with missing and imprecise data [31].

Therefore, the main objectives of this study were to determine the concentrations of five soil PTE elements (Cd, Pb, As, Cu, and Zn) and to investigate the sources of PTEs in soil samples from Baoshan cropland. This would allow us to better understand the specific sources of PTEs contamination in Baoshan cropland and to provide a reliable basis for improving pollution in the area. Specifically, our aims were to: (1) investigate the pollution status of five soil PTEs in Baoshan soil; (2) map the distribution of these soil PTEs using GIS; (3) identify the sources of contamination using PCA, APCS-MLR, PMF, and UNMIX models; and (4) compare the contributions of the four receptor models.

2. Materials and Methods

2.1. Study Area

Baoshan is located in the western part of the Yunnan Province of China (98°25′–100°02′ E, 24°08′–25°51′ N). The entire study area has a subtropical plateau climate, with the local annual temperature difference being slight but the daily temperature difference being significant. The average annual temperature is 15.5 °C, the average temperature of the coldest month is 8.2 °C, and the average temperature of the hottest month is 21 °C. The highest precipitation occurs over the western and southern areas. The terrain is high in the northwest and low in the southeast. The city is bordered by Dali, Lincang, the Salween River, and the Dehong Dai and Jingpo Autonomous Prefecture. The city spans three major river systems, namely, the Lancang, Salween, and Longchuan. The region is rich in terms of land, forest, and mineral resources and natural gas reserves; moreover, it is a hub for hydroelectric power generation and has several tourist attraction spots. The types of cultivated land are mainly dryland and a small number of paddy fields. The cultivated land in the study area is shown in Figure 1.

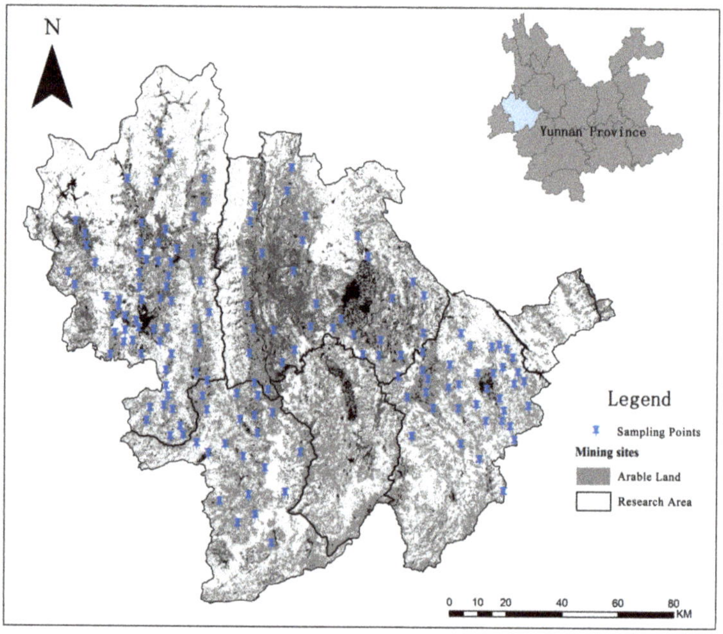

Figure 1. Sampling locations of the study area.

Many experts and scholars investigated the origin of the parent rock mass in Baoshan City and found severe PTE enrichment in soils around the Baoshan mining area. [17,32] The geology of Baoshan City mainly includes carbonate rocks and basalt parent materials; the geological composition of the full study area is shown in Figure 2. In addition, the mining of the rich mineral resources in Baoshan has caused contamination and parent material enrichment of the soil nearby. Moreover, the recent economic development of Baoshan City has driven the expansion of non-ferrous metal smelting, real estate development, animal husbandry, old and outdated production facility construction, and desulfurization, resulting in contaminated farmland.

2.2. Sample Collection and Preparation and Quality Control

We randomly collected 152 topsoil samples (at 0–20 cm depth) from agricultural land in Baoshan in August 2022, using a five-point sampling method with plum blossom. Moreover, 1 kg of soil samples was collected by mixing and dividing the samples into four parts. The sampling points were located by GPS. All the soil samples were stored in plastic bags and transported to the laboratory for air drying and removing plant roots, residues, and visible invaders. Samples were passed through a nylon sieve with a 2 mm pore size to remove soil samples greater than 2 mm. Then, the sieved samples were ground and passed through a nylon sieve with a 0.149 mm pore size (100 mesh), mixed, and prepared for analyses of the soil PTEs, Cd, Pb, Cu, Zn, and As. Table 1 shows the specific methods used for the determination of potentially toxic elements. A flame atomic absorption spectrometer and graphite furnace atomic absorption spectrometer (AA6880, Shimadzu, Kyoto, Japan) and atomic fluorescence spectrophotometer (AFS-230E, Haikou Instruments, Beijing, China) were used for element detection. National standard soil samples (GSS-25) were used for quality control. The test samples were analyzed three times each, and their relative standard deviation was $\leq$5%. Blank samples were tested for the determination of each heavy metal. Heavy metal standard solutions were used for each batch of sample reagents, and the element recovery rate was 90–110%.

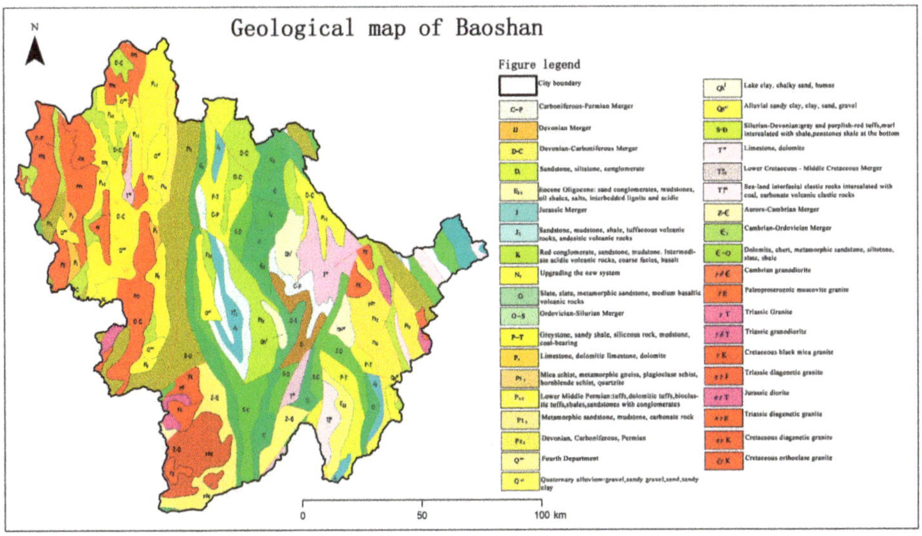

Figure 2. Geological maps of the study area.

Table 1. Detection methods.

Test Items	Detection Method	GB Number	Detection Limits
Copper (Cu) Zinc (Zn)	Furnace atomic absorption spectrophotometry	GB/T17138-1997	0.01 0.5
Cadmium (Cd) lead (Pb)	Graphite furnace atomic absorption spectrophotometry	GB/T17141-1997	0.01 0.2
Arsenic (As)	Atomic fluorescence Spectrophotometer	GB/T22105	0.02

2.3. Pollution Assessment

2.3.1. Pollution Factors

The pollution level of the study area was evaluated by pollution factor evaluation [33–35].

$$PF = \frac{C_i}{B_i} \quad (1)$$

where C_i is the soil PTE concentration (mg/kg), and B_i indicates the corresponding background values of soil PTEs in Yunnan Province (mg/kg) (Cd 0.22, Pb 40.6, Cu 38.38, Zn 89.7, As 18.4). The PF values' environmental indices could be divided into four classes to interpret the pollution levels of soil PTEs (Table 2) [36].

Table 2. PF with different classifications.

PF	Level of Pollution
PF ≤ 1	Low pollution
1 < PF ≤ 3	Moderate pollution
3 < PF ≤ 6	Considerable pollution
PF > 6	Very high pollution

2.3.2. Geoaccumulation Index

The geoaccumulation index (I_{geo}) was developed by Muller (1969) to assess the level of heavy metal and metalloid elements in the sediment by comparing the status of the current concentration with the pre-industrial level. This method assesses heavy metal pollution related to anthropogenic activity, as well as the different rock geology regions, and the natural formation of heavy metal pollution [37].

$$I_{geo} = \log_2 \frac{C_n}{1.5 \times B_n} \tag{2}$$

where C_n is the soil PTE concentration (mg/kg), and B_n is the corresponding background value of soil PTEs in Yunnan Province (mg/kg). The obtained I_{geo} values were classified into seven groups based on Categories (Table 3).

Table 3. Geoaccumulation index with different classifications.

Level	I_{geo}	Level of Pollution
I	$I_{geo} \leq 0$	Not to weakly contaminated
II	$0 < I_{geo} \leq 1$	Weakly to moderately contaminated
III	$1 < I_{geo} \leq 2$	Moderately contaminated
IV	$2 < I_{geo} \leq 3$	Moderately to strongly contaminated
V	$3 < I_{geo} \leq 4$	Strongly contaminated
VI	$4 < I_{geo} \leq 5$	Strongly to extremely contaminated
VII	$I_{geo} \geq 5$	Extremely contaminated

2.4. *Receptor Models*

Based on the receptor model, the contributions of various soil pollution sources were analyzed quantitatively by mathematical method according to the source and concentration of the receptor. Employing PMF, UNMIX, and APCS-MLR, several factors were extracted and identified according to source types, and their contributions were estimated.

2.4.1. APCS-MLR

Principal component analysis is a multivariate statistical analysis method that selects a small number of important variables by linear transformation. Using the linear combination of the original variables after standardization to form the principal component, PCA model can be transformed into several complementary and related comprehensive indexes. The method is used widely in, e.g., demographics, quantitative geography, molecular dynamics simulation, mathematical modeling, and mathematical analysis. The APCS-MLR model was proposed by Thurston [38]. After the data were standardized, the principal component factor was transformed into an absolute principal component score (APCS) by factor analysis. Subsequently, multiple linear regression analyses were performed to determine the content of each selected heavy metal, and the contribution rate of each factor to the pollution source was further calculated.

(1) Standardizing raw data

$$Z_{ij} = \frac{C_{ij} - \overline{C_i}}{\sigma_i} \tag{3}$$

where Z_{ij} is the standardized factor score, and C_{ij} is the soil PTE$_i$ concentration (mg/kg). $\overline{C_i}$ is the average soil PTE$_i$ concentration (mg/kg); σ_i is the standard deviation of soil PTE$_i$ (mg/kg).

(2) Introducing a factor with a concentration of 0

$$Z_{i0} = \frac{0 - \overline{C_i}}{\sigma_i} \tag{4}$$

where APCS for each heavy metal element is obtained using $Z_{ij} - Z_{i0}$. Using the obtained APCS, multiple linear regression analysis is conducted to obtain the regression coefficient:

$$C_i = b_{i0} + \sum\nolimits_{p=1}^{p} (b_{pi} \times APCS_p) \tag{5}$$

where B_{i0} is the constant obtained from multivariate linear regression; b_{pi} is the regression coefficient of source p to the soil PTE; and $b_{pi} \times APCS_p$ is the source contribution to C_i.

2.4.2. PMF

Positive matrix factorization (PMF) analysis, proposed by Paatero et al. [39], is a source analysis method used widely by the United States Environmental Protection Agency (USEPA). The method is often used in sediment, atmospheric, and soil contamination source analyses. The PMF model is least squares through multiple iterations to minimize the objective function Q for obtaining the optimal factor matrix and source profile.

$$Q = \sum_{i=1}^{n} \sum_{j=1}^{m} \left(\frac{x_{ij} - \sum_{k=1}^{p} G_{ik} F_{kj}}{u_{ij}} \right)^2 \rightarrow Q = \sum_{i=1}^{n} \sum\nolimits_{j=1}^{m} \left(\frac{e_{ij}}{u_{ij}} \right)^2 \tag{6}$$

where x_{ij} is the content of the j heavy metal element in the i sample; G_{ik} is the contribution of source k to the i sample; F_{kj} is the content of the j heavy metal element in the k source; u_{ij} is the measured uncertainty (mg/kg); and e_{ij} is the model uncertainty. The sample chemical type uncertainty file is calculated as follows:

$$u_{ij} = \begin{cases} \frac{5}{6} \times MLD, x_{ij} < MLD \\ \sqrt{\left(x_{ij} \times Error\ Fraction\right)^2 + (MLD \times 0.5)^2}\ ,\ x_{ij} > MLD \end{cases} \tag{7}$$

where C_{ij} is the concentration of the j sample chemical type of the i sample, and MDL is the species-specific method detection limit. The error fraction is a percentage of measurement uncertainty.

2.4.3. UNMIX Model

The UNMIX is a receptor model that is based on the pollutant concentration. The major sources and contributions of hand-held pollutants can be obtained directly through extremes. The UNMIX6.0 software be identified to Klstrong in data quality that signal-to-noise (S/n) was more than 2, and the fitting concentration was more than 0.8. Simple software operation, the results do not need their own analyses but cannot evaluate the source component, contribution rate, or contribution value. We used Equation (8), recommended by the USEPA, for performing source apportionment for potentially toxic elements.

$$C_{ij} = \sum_{l=1}^{p} \left(\sum_{k=1}^{p} U_{ik} D_{kl} \right) V_{lj} + \varepsilon_{ij} \tag{8}$$

where U, D, and V are the np diagonal matrix, $p \times p$ diagonal matrix, and $p \times m$ matrix, respectively. ε_{ij} is the error term that contains the variability of C_{ij} and excludes the first major component p.

2.5. Data Treatment with Computer Software

The data was pre-processed using Excel 2010 software. The SPSS® Statistics 23.0 (Armonk, NY, USA) software has performed normal distribution test and statistical analysis of data of potentially toxic elements in soil. The figures were drawn by Origin 2021. The geostatistical analyst tools data exploration tool set analysis was performed by ArcGIS 10.6. The sampling point distribution map and heavy metal pollution spatial distribution feature map were drawn by ArcGIS 10.6. The analysis of PMF was conducted with PMF 5.0.

3. Results and Discussion

3.1. Pollution Characteristics of Soil PTEs

3.1.1. Soil PTE Concentration

Descriptive statistics for Cd, As, Pb, Cu, and Zn in soil PTEs are shown in Table 4. Compared with the background values for Yunnan Province [40], the average concentrations of PTEs Cd (0.28 mg/kg), Cu (47.59 mg/kg), and Zn (100.46 mg/kg) in soils were all higher, and the median values were 1.25, 1.24, and 1.12 times higher, respectively, than the corresponding regional background values. The average concentrations of soil PTEs did not exceed the national soil pollution risk screening values (GB15618-2018), indicating an overall low degree of pollution in the cultivated land of the study area. The box plot in Figure 3 shows the relationship between single soil sample and background values, as well as the risk screening value in Yunnan Province. The Cd content in 25–75% of samples was higher than the background and risk screening values. The contents of the potentially toxic elements As, Pb, Cu, and Zn in 25–75% of the samples were higher than the background values, and the contents of Cu and Zn in 54.6–48.6% of samples were higher than the background values.

Table 4. Descriptive statistics of soil PTEs in the study area (mg/kg). Min = minimum value; Max = maximum value; Med = median value; AM = arithmetic mean; SD = standard deviation; CV = coefficient of variation; Sk = skewness; Ku = kurtosis; k-s test = Kolmogorov–Smirnov test; BG = background values of Yunnan Province (CNEMC, 1990); RSV = risk screening values of the soil PTEs (GB15618-2018).

PTEs	Min	Max	Med	AM	SD	CV (%)	Sk	Ku	k-s Test	BG	RSV
Cd	0.02	1.06	0.19	0.28	0.251	91.3	1.921	2.806	0	0.22	0.3
Pb	15.15	112	28.14	31.42	15.28	49.0	2.168	7.119	0	40.6	120
Cu	5	181.3	44.6	47.59	29.129	61.2	1.575	4.018	0	38.38	100
Zn	15.44	239	87.74	100.46	39.664	39.5	0.493	−0.177	0	89.7	250
As	1.23	54.46	9.39	12.36	10.076	81.5	2.016	4.63	0	18.4	30

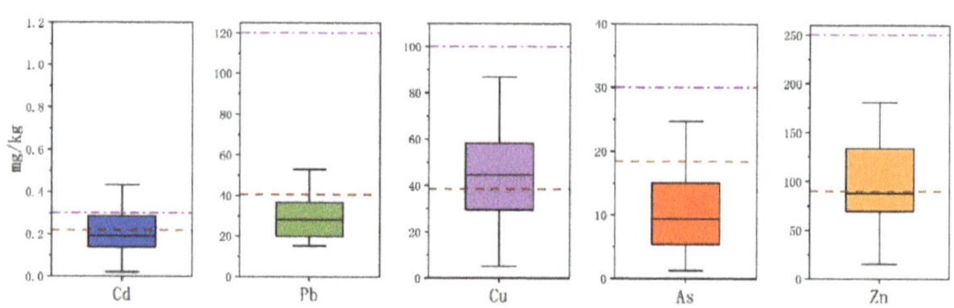

Figure 3. Box plot of soil PTEs in the study area. The red line of the box plot of soil PTE content represents the background value, and the purple line represents the risk screening value.

In 35.53, 21.05, 54.61, 48.68, and 20.39% of the soil samples, the Cd, Pb, Cu, Zn, and As contents, respectively, exceeded the background values. The Cd, Pb, Cu, and As contents exceeded the risk screening values in 21.34, 2.63, 5.26, and 5.26% of the samples, respectively. These findings indicate the contamination of arable land in the study area and thus emphasize the need for further analysis of the contamination status.

The coefficient of variation (CV) reflects the degree of soil disturbance caused by anthropogenic activities. Table 4 shows the high variability of species elements, with variabilities of 0% < CV ≤ 12%, 12% < CV ≤ 31%, and >31% indicating low, moderate, and high variabilities, respectively [41]. The variations of Cd and As were 91.3 and 81.5%, respectively, indicating high degrees of variation and disturbance by anthropogenic activities. These activities include smelting and sugar production and plant fuel burning and sewage discharge causing Cd and As soil pollution. As Baoshan is located in the Cu–Pb–Zn belt, mining of these metals causes severe Cd, As, Cu, Zn, and Pb pollution of the surrounding farmland [42]. However, agricultural activities also lead to increasing Cd and As contents in farmland [43].

Relatively low variations were observed for Zn, Pb, and Cu. Both carbonate and basalt parent materials are developed in the Baoshan area, and soils with basalt parent materials such as granite are rich in Cd, Cu, and Zn. Soils derived from carbonate rocks are rich in Cd, Pb, Cu, and Zn, and their natural genesis results in high Zn and Cu contents and low coefficients of variation [44,45]. However, the Zn and Cu contents of the samples (48.6 and 54.6%, respectively) were higher than the background values. Duan et al. [46] found that improper agricultural practices could lead to an excess of Cu and Zn in soils, and mining could cause Zn, Cu, and Pb pollution. The pollution levels in the overall study area were low, but the soil PTE coefficient of variation was high, which could be owing to the fact that soil pollution is complex and can be caused by a combination of natural and anthropogenic factors. Accordingly, a policy for preventing further risk and a control strategy should be adopted, implying that the source of regional pollution must be identified to facilitate timely prevention and control.

3.1.2. Assessment of Soil PTE Pollution

To further advance our understanding of the extent of soil contamination, the PTEs in the study area were assessed using I_{geo} and PF, as shown in Figure 4. The results showed that the mean I_{geo} median and mean values of PTEs in five soils in the study area were all < 0, indicating non-pollution (Figure 4a); however, the soil PTE, Cd (10.53%), Pb (3.29%), Cu (23.68%), Zn (23.68%), and As (7.24%) samples showed weak to moderate pollution. A small proportion of soil was moderately polluted by PTEs (Cd, 11.8%; Cu, 2.6%). Similarly, as shown in Figure 4b, the average PF values of five PTEs in soils in the study area were all less than 3, indicating that the overall pollution in the study area was moderate. The average PF value of Cd (1.25) was the highest, followed by those of Cu (1.24), Zn (1.12), Pb (0.77), and As (0.67). The PF median values of <1 for Cd and Zn indicate moderate pollution, but the mean values of >1 for Cd and Zn indicate low pollution implying that most cultivated areas were slightly polluted, whereas other areas were weakly to moderately contaminated, with PTE contributions of 23.68, 21.05, 51.97, 48.68, and 20.39% for Cd, Pb, Cu, Zn, and As, respectively. A small part of the soil showed moderately contaminated, and the values for Cd (11.84%) and Cu (2.63%) indicated severe pollution from these elements in the farmland soil.

The relationship between the two pollution indices and soil PTEs is shown in Figure 4c. As the values of the five types of soil PTEs were less than 0, the smaller the PTE, the greater would be its proportion. Therefore, the largest contribution of soil PTEs to I_{geo} was of Zn and Cd, followed by that of Cu, As, and Pb. The greatest contribution of PTEs to PF was of Cd and Cu, followed by that of Zn, Pb, and As. As the results of the various pollution assessment methods differed, neither the degree of soil PTE pollution nor the pollution sources could not be judged directly. Therefore, assessing soil PTE pollution and pollution concentrations must be combined with spatial analyses to gradually explore the source of PTEs for related risk assessment.

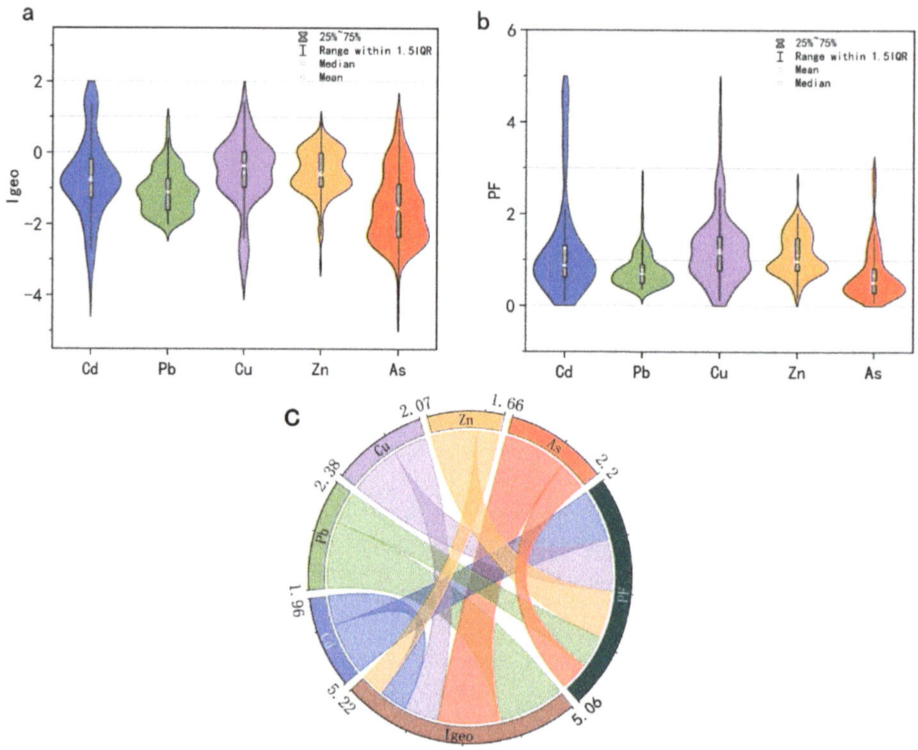

Figure 4. Violin with box plot of soil PTEs in the study area (**a**) I_{geo}, (**b**) PF, and (**c**) the relationship between soil PTEs and environmental indices.

3.2. Spatial Distribution of Soil PTEs

Analyses of the distribution of soil PTEs could provide information on the extent of soil pollution in the study area and improve our understanding of the sources of soil PTEs [47]. The data were interpolated using ArcGIS with the Kriging method, as shown in Figure 5. The distribution of Zn and Cu in soil was partly similar, with the high concentration area located near the northwest of China, where numerous coal mines are situated. Tu et al. [48] found that the development of the coal industry led to Zn and Cu pollution in the surrounding farmland. High concentrations of Cd, Pb, and Cu were found in the northeast of China, with the high Pb values being closer to the South, which are the main concentration areas of cities and towns. Lai et al. [49] indicated that automobile exhaust gas was the main source of Pb pollution. The high Cd and Cu value area is near the North. The investigation showed tungsten and copper deposits in the area, which are mostly located in the high mountain area. As and Zn were mainly concentrated in the southeast, with the spatial distribution characteristics of potentially toxic elements in soils being similar. Hu et al. [50] showed that the use of herbicides, chemical fertilizers, and animal manure could cause As and Zn pollution in soils. Pollution from Cu is distributed widely, and the other four soils (Cd, Zn, Pb, and As) exhibit concentrated distributions of PTEs. Therefore, step-by-step exploration of the sources of PTE contamination was necessary to analyze the source of soil PTE pollution based on a combination of receptor models and geostatistics.

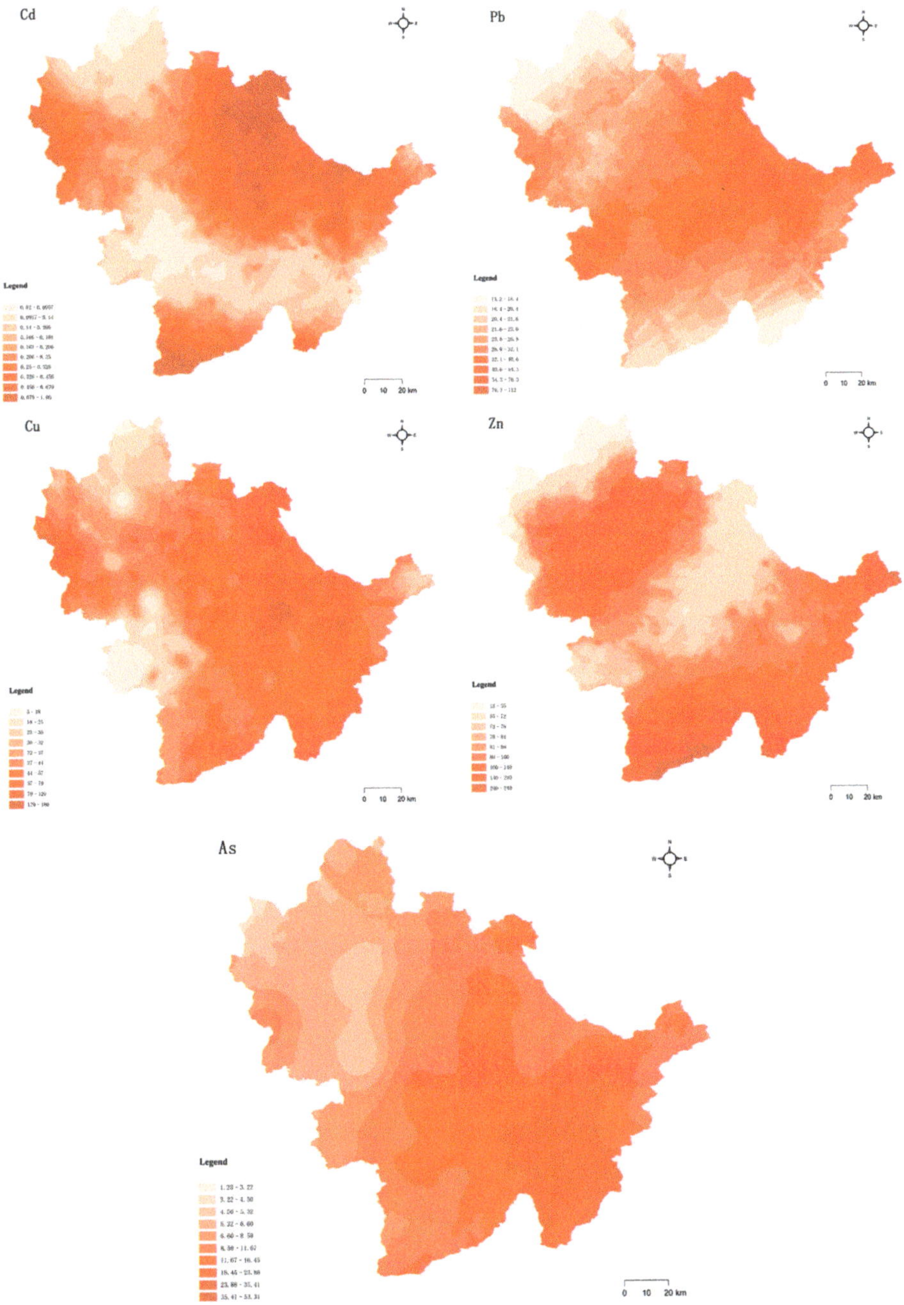

Figure 5. Spatial distribution of soil PTEs in the study area.

3.3. Source Apportionment of Soil PTEs

3.3.1. Multivariate Statistical Analysis

Based on the correlation between PTEs, we used Pearson correlation analysis to determine whether the sources of soil PTEs were consistent [51]. A significant positive correlation between the elements indicates a similar source between the elements; a significant negative correlation between the elements indicates a difference in source between the elements [52]. The results of the correlation coefficient analysis are shown in Figure 6. Positive correlations were found between Pb–Cd, Cd–Cu, Cu–Pb, As–Pb, As–Cu, and As–Pb, and negative correlations were found between Pb–Zn and As–Zn. The Cu–Pb correlation coefficient was 0.589 ($p < 0.01$), indicating that Cu–Pb comes from the same pollution source, possibly industrial or traffic pollution. The following observations were made: (1) Pb is correlated with As, Cu, and Cd, which may be owing to the high proportion of Pb in different pollution sources; (2) the sources of Cd, Pb, Cu, and Zn could be similar; (3) Zn–Pb and As–Zn were correlated negatively with each other, indicating that they could have the same source but in opposite proportions. The source of each element and the correlation between them could be verified further with PCA.

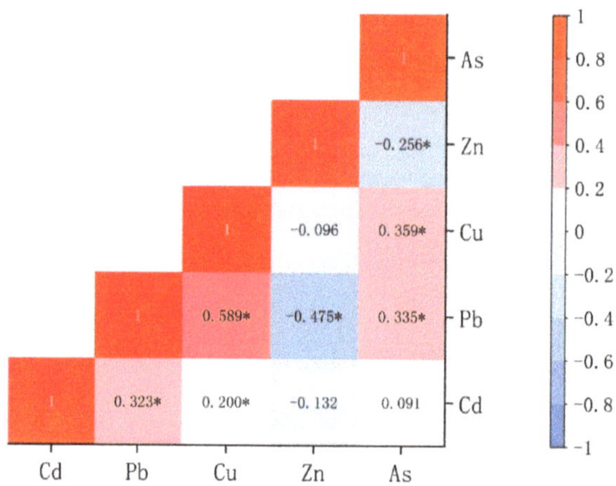

Figure 6. Correlation analysis for soil PTEs. (* indicates significant correlation.)

3.3.2. APCS-MLR Model

We used SPSS Statistics 23.0 to analyze the soil heavy metal concentration. The extraction characteristics of the first two principal components were more than 1; the Kaiser–Meyer–Olkin measure of sampling adequacy and Bartlett values were 0.53, and the Bartlett spherical test p value was 0.00 ($p < 0.05$). These results showed a strong correlation among the potentially toxic elements, which could be analyzed using PCA. The cumulative contribution of variance was 58.25%, which could explain most of the information on the soil PTEs. Potentially toxic elements with a higher factor load under the same principal component have the same origin [53,54]. The first principal component (PC1) was Pb, Zn, and As, the variance contribution was 30.35%, and Zn was the main factor. Although the average values of Pb and As were lower than the background values in Yunnan Province, the samples of Pb (21%) and As (20%) sites exceeded the background values. As shown in Figure 7, the areas with high Pb concentrations were mainly located close to the Longyang District towns and cities, with the As concentrations deriving mainly from fertilizers, fossil fuels, and anthropogenic activities [55]. Investigating the mineral distribution in Baoshan showed the presence of numerous iron and coal mines in the region, the mining of which results in Pb, Zn, As, and Cu pollution [56]. Agricultural activities cause Zn and Cu pollution because of unreasonable fertilizer use, excessive application of conditioning agents, and agricultural machinery activities [57,58]. Therefore, Factor 1 could be considered anthropogenic pollution.

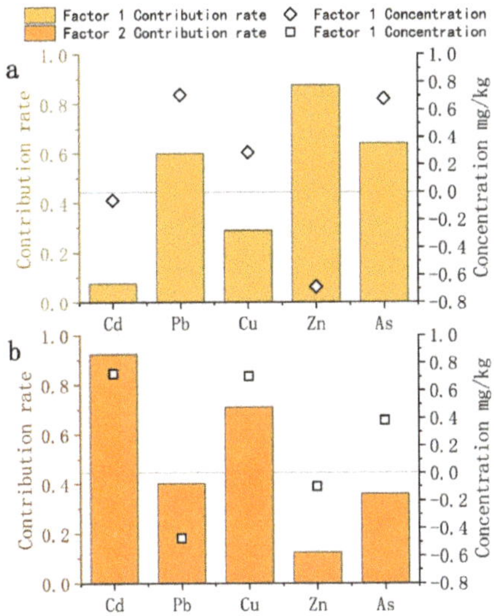

Figure 7. Source profiles and source concentrations of soil potentially toxic elements from PCA. ((**a**) is the source profiles and source concentrations of soil potentially toxic elements for factor 1; (**b**) is the source profiles and source concentrations of soil potentially toxic elements for factor 2).

The second component (PC2), with a higher driving, included Cd and Cu, of which the variance contribution was 27.91%, with Cd being the main factor. Significant correlation was found between Cd and Cu, indicating that both pollutants were likely derived from the same source. Moreover, Baoshan is located on carbonate and basalt parent material, and carbonate parent material can weather Cd, Cu, Pb, Zn, and As. The basalt matrix could be divided into Cd, Cu, and Zn [59], and Cu and Cd could be differentiated from the mixed carbonate matrix [60]. Yunnan Province is characterized by a complex geological structure and rich metal deposits. The main distribution area of Cd and Cu is alpine cultivated land. Therefore, Factor 2 could be artificially induced as a natural cause.

The contribution rates of soil PTEs used to calculate the APCS score and the APCS-MLR receptor model of each soil PTE were obtained by using regression analysis (Figure 7). The main contributing factors were As, Cu, and Pb [61], which could be considered as agricultural pollution, with the main contributing factors being Cd, Zn, and Cu [62]. These results suggested that Factor 2 was a natural cause, and Factor 3 was the main contributing factor of Pb, Cd, and Zn, i.e., industrial and traffic pollution. Our results were consistent with those of [63].

3.3.3. PMF Model

Employing the PMF model, we determined that the signal-to-noise ratio (S/N) of all chemicals was greater than the combined Q value, the resulting source profile, and the scale residuals. The model was run 20 times, resulting in the identification of three sources, as shown in Figure 8. The main contributing factors to Factor 1 were Cd, As, and Cu [64], whereas those to Factor 2 were As and Pb [65]. The results showed that As and Pb pollution could be caused by the application of chemical fertilizers and pesticides in agriculture, i.e., As agricultural pollution. These results were consistent with the analysis results of Han et al. using the PMF model [66]. The main contributing factors of Factor 3 were Zn, Cu, and Pb, which could be considered industrial and traffic pollution. The results were consistent with those of Wang et al. [65].

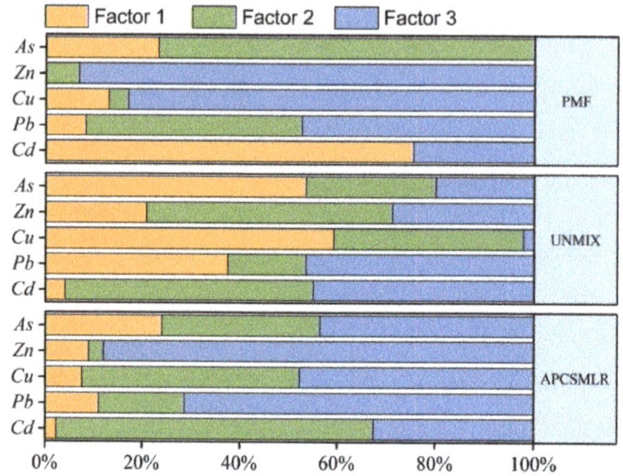

Figure 8. Source contribution of each factor derived from PMF, UNMIX, and APCS/MLR.

3.3.4. UNMIX Model

We used UNMIX 6.0 software to analyze the sample concentration data. The results showed that Min-Rsq was 0.82, i.e., higher than the threshold of 0.8, and S/N was 2.21, greater than the system requirement maximum threshold of 2. Therefore, the approach was considered successful, and three sources could be detected, as shown in Figure 8. Factor 1 was mainly contributed to by Cu, As, and Pb [67], with the assumption being that Factor 1 is agricultural pollution. The main contributing factors to Factor 2 were Cd, Zn, and Cu [68], with Factor 2 considered natural causes. The main contributing factors to Factor 3 were Pb, Cd, and Zn, which could be considered industrial and traffic pollution. These results were consistent with those of Luo et al. [69].

3.4. *Model Evaluation*

The results of the three models were consistent. As shown in Figure 9a, the main sources of pollution derived from industrial activities and traffic, followed by natural causes and agricultural activity. Clearly, the pollution originated from human activity. Combining the results from the three receptor models enabled us to compile the information more comprehensively. As shown in Figure 9b, the main source of pollution in the cultivated land was disturbance by humans, which severely affected the quality of the land. Combined with Figures 7–9, the four models identified three common source categories other than PCA, namely, natural sources (35.23%), agricultural sources (17.67%), and industrial and transportation sources (47.12%). The difference between the model results was that UNMIX yielded a low resolution of industrial and transportation sources and a high resolution of natural and agricultural sources, whereas APCS-MLR and PMF and resolution of the same contribution rate. In addition to UNMIX, the APCS-MLR and PMF models revealed that the agricultural source was dominated mainly by As, whereas the natural source was dominated primarily by Cd. We employed the UNMIX model to analyze the Cu with the highest contribution rate to As, with As being considered the main factor in agricultural pollution. The results of the other three models accounted for a substantial proportion of Zn and Pb in the analysis of industrial and traffic pollution. The potential variables could not be estimated by PCA, and the negative values limited the results of the analysis. Although APCS-MLR evolved from PCA, negative numbers remained in the calculation process. In contrast, UNMIX is a simple model that does not require setting the number of pollution sources and well explains the species concentrations [70]. Employing PMF, we could estimate the error at each site. The reasonable treatment of missing and imprecise data is used widely in soil and air pollution analyses; however, the results are affected by uncertainties of the data and model structure [29]. Three source apportionment models could be used to quantitatively identify pollution, with the contribution rate being the only difference [71,72]. Accordingly, an error that could be caused by employing a single model could be counteracted by the comparative analysis of several receptor models [73,74].

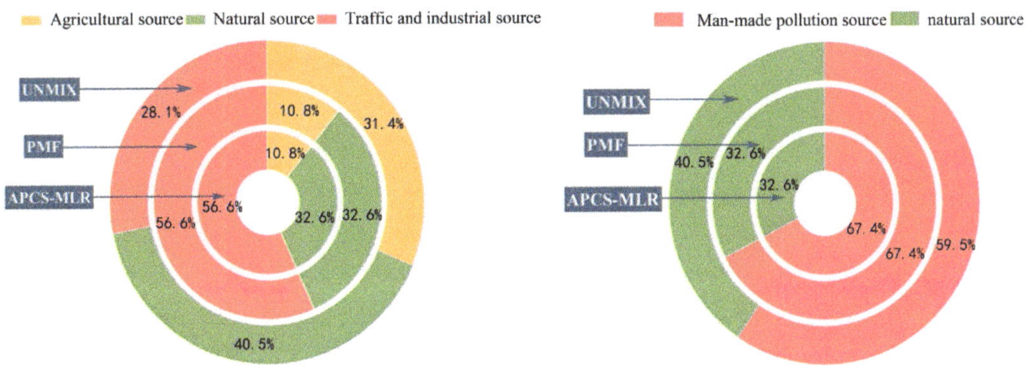

Figure 9. Estimated average source contribution (%).

4. Conclusions

The pollution characteristics and source contributions of PTEs in cultivated soils in Baoshan were investigated, and three receptor models were used to analyze and compare the pollution sources in the study area. The concentrations of Cd, Cu, and Zn in soil were higher than the corresponding background values in Yunnan province. According to the I_{geo} and PF values, the cultivated soil was polluted to different degrees. In particular, pollution from Cd and Cu was severe. The analytical results of the APCS-MLR, PMF, and UNMIX models indicated that As and Pb pollution was caused by agricultural activities, with As pollution deriving mostly from the improper use of agricultural chemicals, together with Cu and Zn pollution. Moreover, the use of agricultural machinery in the region led to Pb pollution. The natural source is mainly caused by the elements Cd and Cu. As the study site is a carbonate-weathering area, Cd is the most significant controlling factor. Automobile exhaust emissions were the main source of Pb and Zn pollution. The emissions of Cu and Cd were the main source of Zn pollution. Overall, 64.76% of the pollution in Baoshan derived from agriculture, of which 47.12% of the pollution derived from industrial transport. Accordingly, this local study should pay more attention to the pollution by industrial production, and more attention is required to mitigate pollution caused by industrial production and traffic emissions.

Author Contributions: Conceptualization, C.D. and L.B.; methodology, C.D.; software, H.Z.; validation, C.D. and H.Z.; formal analysis, H.Y.; investigation, H.Y. and Z.W.; resources, N.Z.; data curation, L.B.; writing—original draft preparation, C.D. and H.Z.; writing—review and editing, C.D. and L.B.; visualization, H.Y. and Z.W.; supervision, L.B.; project administration, N.Z.; funding acquisition, N.Z. All authors have read and agreed to the published version of the manuscript.

Funding: This research was funded by the Studies on the transformation, translocation regularity, and control mechanisms of cadmium for high geological background and anthropogenic pollution soils in Yunnan Province, grant number U200220167 and Key Technologies of green production of agricultural products quality safety control, grant number 202002AE320005.

Institutional Review Board Statement: Not applicable.

Informed Consent Statement: Informed consent was obtained from all subjects involved in the study.

Data Availability Statement: Not applicable.

Acknowledgments: This research was supported by the Studies on the transformation, translocation regularity, and control mechanisms of cadmium for high geological background and anthropogenic pollution soils in Yunnan Province (U200220167), Key Technologies of green production of agricultural products quality safety control (202002AE320005), and the Transport and accumulation of cadmium and gene expression in low-cumulative cadmium in soil-corn system (201801YB00003).

Conflicts of Interest: The authors declare no conflict of interest.

References

1. Barsby, A.; McKinley, J.M.; Ofterdinger, U.; Young, M.; Cave, M.R.; Wragg, J. Bioaccessibility of trace elements in soils in Northern Ireland. *Sci. Total Environ.* **2012**, *433*, 398–417. [CrossRef]
2. Ahado, S.K.; Nwaogu, C.; Sarkodie, V.Y.O.; Borůvka, L. Modeling and Assessing the Spatial and Vertical Distributions of Potentially Toxic Elements in Soil and How the Concentrations Differ. *Toxics* **2021**, *9*, 181. [CrossRef]
3. Hou, D.; O'Connor, D.; Nathanail, P.; Tian, L.; Ma, Y. Integrated GIS and multivariate statistical analysis for regional scale assessment of heavy metal soil contamination: A critical review. *Environ. Pollut.* **2017**, *231*, 1188–1200. [CrossRef] [PubMed]
4. Jin, Y.; O'Connor, D.; Sik Ok, Y.; Tsang, D.C.W.; Liu, A.; Hou, D. Assessment of sources of heavy metals in soil and dust at children's playgrounds in Beijing using GIS and multivariate statistical analysis. *Environ. Int.* **2019**, *124*, 320–328. [CrossRef]
5. Jadoon, S.; Muhammad, S.; Hilal, Z.; Ali, M.; Khan, S.; Khattak, N. Spatial distribution of potentially toxic elements in urban soils of Abbottabad city, (N Pakistan): Evaluation for potential risk. *Microchem. J.* **2020**, *153*, 104489. [CrossRef]
6. Moghtaderi, T.; Shakeri, A.; Rodríguez-Seijo, A. Potentially Toxic Element Content in Arid Agricultural Soils in South Iran. *Agronomy* **2020**, *10*, 564. [CrossRef]
7. Wang, X.; Wang, L.; Zhang, Q.; Liang, T.; Li, J.; Bruun, H.H.C.; Sabry, M.S.; Vasileios, A.; Nanthi, B.; Jörg, R. Integrated assessment of the impact of land use types on soil pollution by potentially toxic elements and the associated ecological and human health risk. *Environ. Pollut.* **2022**, *299*, 118911. [CrossRef] [PubMed]
8. Zuzolo, D.; Cicchella, D.; Lima, A.; Guagliardi, I.; Cerino, P.; Pizzolante, A.; Thiombane, M.; De Vivo, B.; Albanese, S. Potentially toxic elements in soils of Campania region (Southern Italy): Combining raw and compositional data. *J. Geochem. Explor.* **2020**, *213*, 106524. [CrossRef]
9. Nogueira, T.A.R.; Abreu-Junior, C.H.; Alleoni, L.R.F.; He, Z.; Soares, M.R.; Vieira, C.d.S.; Lessa, L.G.F.; Capra, G.F. Background concentrations and quality reference values for some potentially toxic elements in soils of São Paulo State, Brazil. *J. Environ. Manag.* **2018**, *221*, 10–19. [CrossRef] [PubMed]
10. Available online: http://www.baoshan.gov.cn/info/egovinfo/1001/zfxxgkpt/zfxxgkptzn-content/01525525-X-/2021-122900 1.htm (accessed on 12 December 2022.).
11. Li, Q.; Hu, Q.; Zhang, C.; Jin, Z. Effects of Pb, Cd, Zn, and Cu on Soil Enzyme Activity and Soil Properties Related to Agricultural Land-Use Practices in Karst Area Contaminated by Pb-Zn Tailings. *Pol. J. Environ. Stud.* **2018**, *27*, 2623–2632. [CrossRef]
12. Jaradat, Q.M.; Massadeh, A.M.; Momani, K.A.; Al Saleem, M.A. The Spatial Distribution of Pb, Cd, Zn, and Cu in Agricultural Roadside Soils. *SoilSediment Contam.* **2010**, *19*, 58–71. [CrossRef]
13. Zhou, R.; Liu, X.; Luo, L.; Zhou, Y.; Wei, J.; Chen, A.; Tang, L.; Wu, H.; Deng, Y.; Zhang, F.; et al. Remediation of Cu, Pb, Zn and Cd-contaminated agricultural soil using a combined red mud and compost amendment. *Int. Bioeterioration Biodegrad.* **2017**, *118*, 73–81. [CrossRef]
14. Liu, Y.; Liu, D.; Zhang, W.; Chen, X.; Zhao, Q.; Chen, X.; Zou, C. Health risk assessment of heavy metals (Zn, Cu, Cd, Pb, As and Cr) in wheat grain receiving repeated Zn fertilizers. *Environ. Pollut.* **2020**, *257*, 113581. [CrossRef]
15. Liao, Z.; Chen, Y.; Ma, J.; Islam, M.S.; Weng, L.; Li, Y. Cd, Cu, and Zn Accumulations Caused by Long-Term Fertilization in Greenhouse Soils and Their Potential Risk Assessment. *Int. J. Environ. Res. Public Health* **2019**, *16*, 2805. [CrossRef]
16. Zhan, J.; Li, X.; Christie, P.; Wu, L. A review of soil potentially toxic element contamination in typical karst regions in southwest China. *Curr. Opin. Environ. Sci. Health* **2021**, *23*, 100284. [CrossRef]
17. Zhang, L.; Yang, Z.; Peng, M.; Cheng, X. Contamination Levels and the Ecological and Human Health Risks of Potentially Toxic Elements (PTEs) in Soil of Baoshan Area, Southwest China. *Appl. Sci.* **2022**, *12*, 1693. [CrossRef]
18. Zhang, C.; Wang, Z.; Liu, L.; Liu, Y. Source Analysis of Soil Heavy Metals in Agriccultural Land the Ming Area Based on APCS-MLR Receptor Model and Geostatistical Method. *Environ. Sci.* **2022**, 1–14. [CrossRef]
19. Khorasanipour, M.; Karimabadi, F.; Espahbodi, M.; Ebrahimnejad, M. The effect of flotation desulfurization on the trace element geochemistry of Sarcheshmeh mine tailings, SE of Iran: Recycling and the environmental opportunities. *Environ. Earth Sci.* **2021**, *80*, 420. [CrossRef]
20. Liu, Y.; Zhang, W.; Yang, W.; Bai, Z.; Zhao, X. Chemical Compositions of PM2.5 Emitted from Diesel Trucks and Construction Equipment. *Aerosol Sci. Eng.* **2018**, *2*, 51–60. [CrossRef]
21. Rozanski, S.; Castejon, J.; McGahan, D. Child risk assessment of selected metal(loid)s from urban soils using in vitro UBM procedure. *Ecol. Indic.* **2021**, *127*, 107726. [CrossRef]
22. Cicchella, D.; Zuzolo, D.; Albanese, S.; Fedele, L.; Di Tota, I.; Guagliardi, I.; Thiombane, M.; De Vivo, B.; Lima, A. Urban soil contamination in Salerno (Italy): Concentrations and patterns of major, minor, trace and ultra-trace elements in soils. *J. Geochem. Explor.* **2020**, *213*, 106519. [CrossRef]

23. Li, J.; Wu, J.; Jiang, J.; Teng, Y.; He, L.; Song, L. Review on Source Apportionment of Soil Pollutants in Recent Ten Years. *Soil Sci.* **2018**, *49*, 232–242. [CrossRef]
24. Wu, J.; Long, J.; Liu, L.; Li, J.; Liao, H.; Zhang, M.; Zhao, C.; Wu, Q. Risk Assessment and Source Identification of Toxic Metals in the Agricultural Soil around a Pb/Zn Mining and Smelting Area in Southwest China. *Int. Environ. Res. Public Health* **2018**, *15*, 1838. [CrossRef] [PubMed]
25. Negahban, S.; Mokarram, M. Potential Ecological Risk Assessment of Ni, Cu, Zn, Cd, and Pb in Roadside Soils. *Earth Space Sci.* **2021**, *8*, e2020EA001120. [CrossRef]
26. Li, Y.; Gou, X.; Wang, G.; Zhang, Q.; Su, Q.; Xiao, G. Heavy metal contamination and source in arid agricultural soil in central Gansu Province, China. *Environ. Sci.* **2008**, *20*, 607–612. [CrossRef]
27. Li, N.; Li, Y.; Wang, G.; Zhang, H.; Zhang, X.; Wen, J.; Cheng, X. The sources risk assessment combined with APCS/MLR model for potentially toxic elements in farmland of a first-tier city, China. *Environ. Sci. Pollut. Res.* **2022**, *29*, 50717–50726. [CrossRef] [PubMed]
28. Chen, Z.; Ding, Y.; Jiang, X.; Duan, H.; Ruan, X.; Li, Z.; Li, Y. Combination of UNMIX, PMF model and Pb-Zn-Cu isotopic compositions for quantitative source apportionment of heavy metals in suburban agricultural soils. *Ecotoxicol. Environ. Saf.* **2022**, *234*, 113369. [CrossRef] [PubMed]
29. Liu, H.; Anwar, S.; Fang, L.; Chen, L.; Xu, W.; Xiao, L.; Zhong, B.; Liu, D. Source Apportionment of Agricultural Soil Heavy Metals Based on PMF Model and Multivariate Statistical Analysis. *Environ. Forensics* **2022**, 1–9. [CrossRef]
30. Lv, J. Multivariate receptor models and robust geostatistics to estimate source apportionment of heavy metals in soils. *Environ. Pollut.* **2018**, *244*, 72–83. [CrossRef]
31. Wang, J.; Yang, J.; Chen, T. Source appointment of potentially toxic elements (PTEs) at an abandoned realgar mine: Combination of multivariate statistical analysis and three common receptor models. *Chemosphere* **2022**, *307*, 135923. [CrossRef]
32. Zhang, L.; McKinley, J.; Cooper, M.; Peng, M.; Wang, Q.; Song, Y.; Cheng, H. A regional soil and river sediment geochemical study in Baoshan area, Yunnan province, southwest China. *Geochem. Explor.* **2020**, *217*, 106557. [CrossRef]
33. Lei, M.; Li, K.; Guo, G.; Ju, T. Source-specific health risks apportionment of soil potential toxicity elements combining multiple receptor models with Monte Carlo simulation. *Sci. Total Environ.* **2022**, *817*, 152899. [CrossRef] [PubMed]
34. Candeias, C.; Avila, P.; da Silva, E.; Teixeira, J. Integrated approach to assess the environmental impact of mining activities: Estimation of the spatial distribution of soil contamination (Panasqueira mining area, Central Portugal). *Environ. Monit. Assess.* **2015**, *187*, 135. [CrossRef]
35. Turan, D.; Kocahakimoglu, C.; Kavcar, P.; Gaygısız, H.; Atatanir, L.; Turgut, C.; Sofuoglu, S. The use of olive tree (*Olea europaea* L.) leaves as a bioindicator for environmental pollution in the Province of Aydın, Turkey. *Env. Sci. Pollut. Res.* **2011**, *18*, 355–364. [CrossRef]
36. Rehman, I.; Ishaq, M.; Ali, L.; Khan, S.; Ahmad, I.; Din, I.; Ullah, H. Enrichment, spatial distribution of potential ecological and human health risk assessment via toxic metals in soil and surface water ingestion in the vicinity of Sewakht mines, district Chitral, Northern Pakistan. *Ecotoxicol. Environ. Saf.* **2018**, *154*, 127–136. [CrossRef] [PubMed]
37. Mukhopadhyay, S.; Chakraborty, S.; Bhadoria, P.B.S.; Li, B.; Weindorf, D. Assessment of heavy metal and soil organic carbon by portable X-ray fluorescence spectrometry and NixProTM sensor in landfill soils of India. *Geoderma Reg.* **2020**, *20*, e00249. [CrossRef]
38. Thurston, G.; Spengler, J. A quantitative assessment of source contributions to inhalable particulate matter pollution in metropolitan Boston. *Atmos. Environ.* **1985**, *19*, 9–25. [CrossRef]
39. Paatero, P.; Tapper, U. Positive matrix factorization: A non-negative factor model with optimal utilization of error estimates of data values. *Environmetrics* **1994**, *5*, 111–126. [CrossRef]
40. CNEMC. *Chinese Soil Element Background Value*; Environmental Science Press: Beijing, China, 1990.
41. Gholizadeh, A.; Borůvka, L.; Saberioon, M.M.; Kozák, J.; Vašát, R.; Němeček, K. Comparing Different Data Preprocessing Methods for Monitoring Soil Heavy Metals Based on Soil Spectral Features. *Soil and Water Research.* **2015**, *10*, 218–227. [CrossRef]
42. Xing, R.; Wu, Z.; Du, G. Risk assessment and source analysis soil heavy metal pollution in Xuanzhou District, Anhui Province. *East China Geol.* **2022**, *43*, 336–344. [CrossRef]
43. Gao, Y.; Li, F.; Mao, L.; Gu, B.; Peng, C.; Yang, Q.; Lu, L.; Chen, X.; Zhang, D.; Tao, H. Potential Loss of Toxic Elements from Slope Arable Soil Erosion into Watershed in Southwest China: Effect of Spatial Distribution and Land-Uses. *Minerals* **2021**, *11*, 1422. [CrossRef]
44. Xiao, J.; Chen, W.; Wang, L.; Zhang, X.; Wen, Y.; Bostick, B.; Wen, Y.; He, X.; Zhang, L.; Zhuo, X.; et al. New strategy for exploring the accumulation of heavy metals in soils derived from different parent materials in the karst region of southwestern China. *Geoderma* **2022**, *417*, 115806. [CrossRef]
45. Cabral Pinto, M.; Silva, M.; Ferreira da Silva, E.; Dinis, P.; Rocha, F. Transfer processes of potentially toxic elements (PTE) from rocks to soils and the origin of PTE in soils: A case study on the island of Santiago (Cape Verde). *Geochem. Explor.* **2017**, *183*, 140–151. [CrossRef]
46. Duan, B.; Qiang, F. Comparison of the Potential Ecological and Human Health Risks of Heavy Metals from Sewage Sludge and Livestock Manure for Agricultural Use. *Toxics* **2021**, *9*, 145. [CrossRef] [PubMed]
47. Francos, N.; Gholizadeh, A.; Ben Dora, E. Spatial distribution of lead (Pb) in soil: A case study in a contaminated area of the Czech Republic. *Geomat. Nat. Hazards Risk* **2022**, *12*, 610–620. [CrossRef]

48. Tu, C.; Yang, K.; He, C.; Zhang, L.; Li, B.; Wei, Z.; Jian, X.; Yang, M. Sources and Risk Assessment of Heavy Metals in Small Watersheds in Typical Coal Mining Areas of Eastern Yunnan. *East China Geol.* 2022, pp. 1–16. Available online: http://kns.cnki.net/kcms/detail/11.1167.P.20220330.1029.002.html (accessed on 16 January 2023).
49. Lai, S.; Cao, R.; Tan, G. Study on accumulation characteristics of heavy metal elements in topsoil of southern Longhai City, Fujian Province. *East China Geol.* **2021**, *42*, 29–36. [CrossRef]
50. Hu, Y.; He, K.; Sun, Z.; Chen, G.; Cheng, H. Quantitative source apportionment of heavy metal(loid)s in the agricultural soils of an industrializing region and associated model uncertainty. *Hazard. Mater.* **2020**, *391*, 122244. [CrossRef]
51. Gong, C.; Wang, S.; Wang, D.; Lu, H.; Dong, H.; Liu, J.; Yan, B.; Wang, L. Ecological and human health risk assessment of heavy metal(loid)s in agricultural soil in hotbed chives hometown of Tangchang, Southwest China. *Sci. Rep.* **2022**, *12*, 8563. [CrossRef]
52. Singh, H.; Pandey, R.; Singh, S.K.; Shukla, D.N. Assessment of heavy metal contamination in the sediment of the River Ghaghara, a major tributary of the River Ganga in Northern India. *Appl. Water Sci.* **2017**, *7*, 4133–4149. [CrossRef]
53. Jin, G.; Fang, W.; Shafi, M.; Wu, D.; Li, Y.; Zhong, B.; Ma, J.; Liu, D. Source apportionment of heavy metals in farmland soil with application of APCS-MLR model: A pilot study for restoration of farmland in Shaoxing City Zhejiang, China. *Ecotoxicol. Environ. Saf.* **2019**, *184*, 109495. [CrossRef]
54. Samsudin, M.; Azid, A.; Khalit, S.; Saudi, A.; Zaudi, M. River water quality assessment using APCS-MLR and statistical process control in Johor River Basin, Malaysia. *Adv. Appl. Sci.* **2017**, *4*, 84–97. [CrossRef]
55. Jiang, H.; Cai, L.; Wen, H.; Hu, G.; Chen, L.; Luo, J. An integrated approach to quantifying ecological and human health risks from different sources of soil heavy metals. *Total Environ.* **2020**, *701*, 134466. [CrossRef] [PubMed]
56. Chen, X.; Lei, M.; Zhang, S.; Zhang, D.; Guo, G.; Zhao, X. Apportionment and Spatial Pattern Analysis of Soil Heavy Metal Pollution Sources Related to Industries of Concern in a County in Southwestern China. *Environ. Res. Public Health* **2022**, *19*, 7421. [CrossRef]
57. Li, C.; Zhang, C.; Yu, T.; Liu, X.; Xia, X.; Hou, Q.; Yang, Y.; Yang, Z.; Wang, L. Annual net input fluxes of cadmium in paddy soils in karst and non-karst areas of Guangxi, China. *Geochem. Explor.* **2022**, *241*, 107072. [CrossRef]
58. Wu, J.; Chen, Y.; Ma, J.; Cao, J.; Jiang, Y. Sustainable Strategies for the Agricultural Development of Shaanxi Province Based on the Risk Assessment of Heavy Metal Pollution. *Foods* **2022**, *11*, 1409. [CrossRef]
59. Wu, H.; Wang, J.; Guo, J.; Hu, X.; Bao, H.; Chen, J. Record of heavy metals in Huguangyan Maar Lake sediments: Response to anthropogenic atmospheric pollution in Southern China. *Sci. Total Environ.* **2022**, *831*, 154829. [CrossRef] [PubMed]
60. Xiao, G.; Chen, J.; Bai, B.; Li, Y.; Zhu, N. Content Characteristics and Risk Assessment of Heavy Metals in Soil of Typical High Geological Background Areas, Yunnan Province. *Geol. Explor.* **2021**, *57*, 1077–1086.
61. Zhang, C.; Zou, X.; Yang, H.; Liang, J.; Zhu, T. Bioaccumulation and Risk Assessment of Potentially Toxic Elements in Soil-Rice System in Karst Area, Southwest China. *Front. Environ. Sci.* **2022**, *10*, 866427. [CrossRef]
62. Li, Y.; Xu, Z.; Ren, H.; Wang, D.; Wang, J.; Wu, Z.; Cai, P. Spatial Distribution and Source Apportionment of Heavy Metals in the Topsoil of Weifang City, East China. *Front. Environ. Sci.* **2022**, *10*, 893938. [CrossRef]
63. Wang, J.; Wu, H.; Wei, W.; Xu, C.; Tan, X.; Wen, Y.; Lin, A. Health risk assessment of heavy metal(loid)s in the farmland of megalopolis in China by using APCS-MLR and PMF receptor models: Taking Huairou District of Beijing as an example. *Total Environ.* **2022**, *835*, 155313. [CrossRef]
64. Chen, Z.; Xu, J.; Duan, R.; Lu, S.; Hou, Z.; Yang, F.; Peng, M.; Zong, Q.; Shi, Z.; Yu, L. Ecological Health Risk Assessment and Source Identification of Heavy Metals in Surface Soil Based on a High Geochemical Background: A Case Study in Southwest China. *Toxics* **2022**, *10*, 282. [CrossRef] [PubMed]
65. Wang, S.; Cai, L.; Wen, H.; Luo, J.; Wang, Q.; Liu, X. Spatial distribution and source apportionment of heavy metals in soil from a typical county-level city of Guangdong Province, China. *Sci. Total Environ.* **2019**, *655*, 92–101. [CrossRef]
66. Han, L.; Xu, X. Quantitative Evaluation of Human Health Risk of Heavy Meyals in Soil Based on Postitive Matrix Factorization Model and Geo-statistics. *Environ. Sci.* **2020**, *41*, 5114–5124. [CrossRef]
67. Atafar, Z.; Mesdaghinia, A.; Nouri, J.; Homaee, M.; Yunesian, M.; Ahmadimoghaddam, M.; Mahvi, A. Effect of fertilizer application on soil heavy metal concentration. Environ Monit Assess. *Environ. Monit. Assess.* **2010**, *160*, 83. [CrossRef]
68. Zhang, F.; Peng, M.; Wang, H.; Ma, H.; Xu, R.; Cheng, X.; Hou, Z.; Chen, Z.; Li, K.; Cheng, H. Ecological risk assessment of heavy metals at township scale in thehigh background of heavy metals, southwestern China. *Environ. Sci.* **2020**, *41*, 4197–4209. [CrossRef]
69. Luo, X.; Wu, C.; Lin, Y.; Li, W.; Deng, M.; Tan, J.; Xue, S. Soil heavy metal pollution from Pb/Zn smelting regions in China and the remediation potential of biomineralization. *Environ. Sci.* **2022**, *125*, 662–677. [CrossRef] [PubMed]
70. Li, Y.; Kuang, H.; Hu, C.; Ge, G. Source Apportionment of Heavy Metal Pollution in Agricultural Soils around the Poyang Lake Region Using UNMIX Model. *Sustainability* **2021**, *13*, 5272. [CrossRef]
71. Zhang, S.; Wang, L.; Zhang, W.; Wang, L.; Shi, X.; Lu, X. Pollution assessment and source apportionment of trace metals in urban topsoil of Xi' a City in Northwest China. *Arch. Environ. Contam. Toxicol.* **2019**, *77*, 575–586. [CrossRef]
72. Guan, Q.; Zhao, R.; Pan, N.; Wang, F.; Yang, Y.; Luo, H. Source apportionment of heavy metals in farmland soil of Wuwei, China: Comparison of three receptor models. *Clean. Prod.* **2019**, *237*, 117792. [CrossRef]

73. Yang, B.; Zhou, L.; Xue, N.; Li, F.; Li, Y.; Vogt, R.; Cong, X.; Yan, Y.; Liu, B. Source apportionment of polycyclic aromatic hydrocarbons in soils of Huanghuai Plain, China: Comparison of three receptor models. *Sci. Total Environ.* **2013**, *443*, 31–39. [CrossRef]
74. Jin, Z.; Lv, J. Integrated receptor models and multivariate geostatistical simulation for source apportionment of potentially toxic elements in soils. *CATENA* **2020**, *194*, 104638. [CrossRef]

toxics

MDPI

Article

Isolation and Identification of Mercury-Tolerant Bacteria LBA119 from Molybdenum-Lead Mining Soils and Their Removal of Hg^{2+}

Hanyue Yao [1], Hui Wang [1,*], Jiangtao Ji [2,*], Aobo Tan [1], Yang Song [2] and Zhi Chen [3]

1 School of Chemical Engineering and Pharmacy, Henan University of Science and Technology, Luoyang 471023, China
2 College of Agricultural Equipment Engineering, Henan University of Science and Technology, Luoyang 471023, China
3 Department of Building, Civil and Environmental, Concordia University, Montreal, QC H3G 1M8, Canada
* Correspondence: wanghui_peony@163.com (H.W.); jjt0907@163.com (J.J.)

Abstract: Aims: To screen heavy metal-tolerant strains from heavy metal-contaminated soil in mining areas and determine the tolerance of the strains to different heavy metals and their removal rates through experiments. Methods: Mercury-resistant strain LBA119 was isolated from mercury-contaminated soil samples in Luanchuan County, Henan Province, China. The strain was identified by Gram staining, physiological and biochemical tests, and 16S rDNA sequences. The LBA119 strain showed good resistance and removal rates to heavy metals such as Pb^{2+}, Hg^{2+}, Mn^{2+}, Zn^{2+}, and Cd^{2+} using tolerance tests under optimal growth conditions. The mercury-resistant strain LBA119 was applied to mercury-contaminated soil to determine the ability of the strain to remove mercury from the soil compared to mercury-contaminated soil without bacterial biomass. Results: Mercury-resistant strain LBA119 is a Gram-positive bacterium that appears as a short rod under scanning electron microscopy, with a single bacterium measuring approximately 0.8×1.3 μm. The strain was identified as a *Bacillus* by Gram staining, physiological and biochemical tests, and 16S rDNA sequence analysis. The strain was highly resistant to mercury, with a minimum inhibitory concentration (MIC) of 32 mg/L for mercury. Under a 10 mg/L mercury environment, the optimal inoculation amount, pH, temperature, and salt concentration of the LBA119 strain were 2%, 7, 30 °C, and 20 g/L, respectively. In the 10 mg/L Hg^{2+} LB medium, the total removal rate, volatilization rate, and adsorption rate at 36 h were 97.32%, 89.08%, and 8.24%, respectively. According to tolerance tests, the strain showed good resistance to Pb^{2+}, Mn^{2+}, Zn^{2+}, Cd^{2+}, and other heavy metals. When the initial mercury concentration was 50 mg/L and 100 mg/L, compared with the mercury-contaminated soil that contained an LB medium without bacterial biomass, LBA119 inoculation increased 15.54–37.67% after 30 days of culture. Conclusion: This strain shows high bioremediation potential for mercury-contaminated soil.

Keywords: *Bacillus*; 16S rDNA; heavy metal mercury; soil

check for updates

Citation: Yao, H.; Wang, H.; Ji, J.; Tan, A.; Song, Y.; Chen, Z. Isolation and Identification of Mercury-Tolerant Bacteria LBA119 from Molybdenum-Lead Mining Soils and Their Removal of Hg^{2+}. *Toxics* **2023**, *11*, 261. https://doi.org/10.3390/toxics11030261

Academic Editors: Junhao Qin, Peidong Su, Feng Zhu and Lin Ding

Received: 14 February 2023
Revised: 4 March 2023
Accepted: 10 March 2023
Published: 12 March 2023

1. Introduction

Mercury pollution in water bodies due to human activities is mainly caused by wastewater from industries that produce items such as chlor-alkali, plastics, batteries, electronics, and used medical devices [1]. It is estimated that approximately 3.2×10^4 hm of agricultural land in China is contaminated with mercury. Mercury is highly toxic and can cause severe damage to human cells and limit normal cellular function when it enters the human body [2]. Mercury is distributed in the atmosphere, soil, and water bodies; both soluble and insoluble mercury compounds are partially volatilized into the atmosphere, and the transport transformation of mercury occurs between land, water, and air. A general rule of thumb is that, compared to inorganic mercury, organic mercury is more volatile [1], and methylmercury is the most volatile. Through microorganisms, inorganic mercury can be converted into more toxic methylmercury, dimethylmercury [3], etc.

Numerous studies have examined the possibility of removing mercury with microbial organisms. For example, Yang Wen screened a mercury-resistant strain in a mercury-rich soil upstream of Miyun Reservoir in Beijing, and this strain was identified as *Pseudomonas aeruginosa* [4]. When mercury concentrations were 10 mg/L for 24 h under optimum conditions, the rates of mercury removal were over 90%. Zheng Yan et al. reported a highly resistant strain of mercury, *Pseudomonas aeruginosa*, which can rapidly volatilize more than 68% of Hg^0 within 8 h, and Fan Taotao et al. isolated three strains of mercury-resistant and mercury-reducing bacteria from mercury-contaminated water in a chemical plant [5], and these strains exhibit a mercury removal rate of more than 90% under optimal conditions. The mercury-tolerant strains in this study were selected from agricultural fields that were contaminated with heavy metals near molybdenum mines in Luanchuan County, and the rate of mercury removal was approximately 96% at mercury concentrations of 2–12 mg/L, a higher percentage than that previously reported in the literature.

Heavy metal-contaminated soil located around mining areas is a valuable source for isolating strains tolerant to heavy metals [6]. A strain with mercury tolerance was isolated from agricultural fields in the molybdenum mining area of Luanchuan County, Luoyang. Its biological properties and volatile mercury energy were initially investigated. The possibility of using this strain to remediate mercury-contaminated environments was evaluated to provide a reference for future microbial remediation of mercury-contaminated soils in mining areas.

2. Materials and Methods

2.1. Soil Sample Source

In this experiment, mercury-contaminated soils were collected from agricultural fields in Luanchuan County, Luoyang City, Henan Province (33° N, 111° E) using the five-point sampling method, and the samples were primarily from the protoplasmic soil layer within the top 20 cm of the soil. The mercury content of the soil was determined by atomic fluorescence with a detection limit of 0.002 mg/kg.

2.2. Screening Mercury-Resistant Strains

Soil samples were collected from the heavy metal-contaminated soil in the mining area; 10 g of soil was placed in a 250 mL beaker, 90 mL of sterile water was added to create a bacterial suspension and the suspension was magnetically stirred for 30 min and left to stand. Then, 1 mL of supernatant was inoculated with a pipette gun in an LB medium containing 10 mg/L Hg^{2+} in a 30 °C; the samples were shaken at 150 r/min in an incubator and the samples were incubated for 2 h for 10× series (10^{-2}, 10^{-3}, 10^{-4} and 10^{-5}) gradient dilutions [1]. A dilution of the sample was spread on a solid LB medium containing 20 mg/L Hg^{2+} and incubated in a biochemical incubator at 37 °C for 48 h. After the colonies grew on the medium, the concentration of mercuric chloride in the LB solid medium was increased continuously. A high concentration of mercury-tolerant strains was screened by picking and culturing individual colonies on an LB solid medium containing a high mercury concentration.

The selected mercury-resistant strains were cultured in line on LB solid medium without mercury, and the strains were repeatedly purified and screened. The morphological characteristics of the colonies on the medium were observed and recorded. Inoculations of the mercury-resistant strain were placed into LB liquid medium and were incubated at 30 °C and 150 rpm in a constant temperature shaker until the solution became cloudy. The solution was preserved using glycerol. With the aid of a pipette on an ultraclean table, 0.2 mL of glycerol was pipetted into a 1.5 mL centrifuge tube, and 0.2 mL of the bacterial seed solution was mixed with gentle shaking. The 1.5 mL centrifuge tube containing the bacterial solution was placed in a sealed bag and placed in a refrigerator at −40 °C for use in subsequent experiments.

2.3. *Strain Identification*

The bacteria were identified by colony morphology and biochemical tests. Biochemical tests included Gram stain [7], the starch hydrolysis test, gelatin hydrolysis test, oil hydrolysis test, indole test, sugar fermentation test, methyl red test, volt-pop (VP) test, H_2S test, and the strain motility test.

2.4. *Molecular Biology Identification*

Phylogenetic trees are used to represent the affinities between various organisms and to infer the evolutionary history of species through the study of biological sequences, mainly through DNA sequences, protein sequences, protein structures, etc. Phylogenetic trees are constructed. The phylogenetic tree in this study was constructed based on the 16S rDNA of LBA119 bacteria and was compared with the validated 16S rDNA in the NCBI library.

Genomic DNA of the strain was extracted by the boiling template method, and the 16S rDNA gene [8] was amplified by polymerase chain reaction (PCR) using bacterial universal primers 27F (5′-AGAGTTTGATCMTGGCTCAG-3′) and 1492R (5′-ACGGCTACCTTGTTACGA-3′). The temperature program was to preheat at 95 °C for 5 min, denaturation at 95 °C for 1 min, 55 °C annealing for 1 min, and 72 °C extension for 2 min, repeated for 30 cycles, and finally 72 °C extension for 10 min. The composition of the PCR reaction system is shown in Table 1. The PCR [9] products were sent to Shanghai Bioengineering Co., Ltd (Sangon Biotech (Shanghai) Co., Ltd., Shanghai, China). for sequencing, and the sequencing results were compared with the GenBank database of the National Center for Biotechnology Information (NCBI) to identify the genus of the strain. The phylogenetic tree was constructed by the neighbor-joining method (distance-based approach) of the MEGA X software (Version 11.0.13).

Table 1. Components of PCR reaction.

Reaction System	Volume (μL)	Reaction System	Volume (μL)
ddH_2O	9.5	Primers 1492R	1
DNA Template	1	2 × Taq DNA Polymerase	12.5
Primers 27F	1	General System	25

2.5. *Exploration of Optimal Growth Conditions for Bacteria Strains*

To investigate the effect of different media on the growth of experimental strains [10], 50 mL of each of eight liquid media, including an LB medium, beef paste peptone medium, high type 1 medium, PDA medium, Choi medium, sand medium, glycerol medium, and PSA medium, was prepared in a 250 mL conical flask, and a standard liquid medium containing 10 mg/L Hg^{2+} and no Hg^{2+} was prepared. Inoculation was performed after the medium was adjusted to pH 7.0, sterilized at 121 °C for 30 min, and cooled (fresh bacterial seed solution to medium = 2:100). The culture was incubated for 36 h at 37 °C and 150 r/min in a constant temperature shaking incubator. The cultured bacterial solution was centrifuged at 8000 rpm for 10 min, the supernatant was discarded, and the centrifuge tube was placed in a vacuum drying oven and dried at 70 °C to a constant weight. Each group was divided into three replicates. The dry weight of the strain was calculated as follows:

$$m = (M_1 - m_1) - (M_2 - m_2)$$

where M_1 is the initial mass of the experimental group, M_2 is the initial mass of the control group, m_1 is the mass of the experimental group after drying, and m_2 is the mass of the control group after drying.

Under the optimum conditions, LB liquid medium containing 10 mg/L Hg^{2+} and LB liquid medium without Hg^{2+} were used. The following range of inoculums was used for the test strains: 1%, 2%, 3%, 4%, 5%, 6%, 7%, and 8%, and the OD_{600nm} values were determined after 36 h of incubation at 150 rpm and 30 °C.

The pH of the liquid LB medium was adjusted to 4, 5, 6, 7, 8, 9, and 10 with hydrochloric acid (HCl) or sodium hydroxide (NaOH), and the OD_{600nm} values were measured after inoculating fresh bacterial solutions into LB liquid medium and incubating the solutions at 37 °C for 36 h at 150 rpm to obtain the proper pH range for strain growth.

Based on the same medium preparation method, pH, and inoculation method, the OD600 nm values were determined after 36 h of incubation [11] at 20 °C, 25, 30, 35 °C, and 40 °C in a constant temperature shaking incubator. Under the above optimal conditions, the bacteria were inoculated in liquid LB [12] media containing 0, 0.5%, 1%, 2%, 4%, 6%, and 8% sodium chloride, and the $OD_{600\ nm}$ values were measured after incubation to obtain the optimal salt concentration. Each group of experiments was repeated three times.

2.6. Determination of the Tolerance of Strains to Mercury

Under optimal growth conditions, LB liquid medium was prepared with Hg^{2+} concentrations of 0, 1, 2, 4, 8, 16, 32, 64, and 128 mg/L, sterilized at 121 °C for 30 min, inoculated with 2% fresh bacterial broth into LB liquid medium, and incubated with shaking for 120 h. To determine the minimum inhibitory concentration of heavy metal mercury on the experimental strains, the optical density ($OD_{600nm)}$ values of each medium were measured regularly. The minimum inhibitory concentration (MIC) of Hg^{2+} in the experimental strains was determined at regular intervals by measuring the $OD_{600\ nm}$ values of each medium for 120 h.

According to the minimum inhibitory concentration (MIC) of Hg^{2+} on the strains, 2% fresh bacterial solution was inoculated into the liquid LB medium with 5, 10, and 20 mg/L Hg^{2+}; an LB medium without mercury was used as a blank control, and the optical density was measured regularly in a UV spectrophotometer to study the changes in the growth pattern of the strains under mercury stress.

2.7. Determination of Mercury Volatilization and Adsorption Capacity of the Strain

As part of the experiment, three groups were established. The positive control group contained no mercury, the negative control group was not inoculated with the strain. The experimental group was inoculated with 2% fresh bacterial broth in a liquid LB medium containing 10 mg/L Hg^{2+} and under optimal growth conditions, and three parallel treatments were conducted for each group. The samples were removed at regular intervals and centrifuged at 8000 rpm for 10 min. The process of digesting the supernatant and bacterium was performed as follows. A total of 1 mL supernatant from the centrifuged culture solution was mixed with 10 mL of HCl in a 50 mL Teflon crucible, the sample was heated continuously on an electric furnace, and the sample was digested until the volume was reduced to 3 mL. A total of 5 mL nitric acid was added to the solution, and the solution was heated for 1 h. The steps were repeated until the digestion was complete and the white fumes were exhausted. Then, the digestion solution was transferred to a 50 mL volumetric flask after it had cooled to room temperature and was shaken well with a fixed volume.

A cold atomic absorption spectrophotometer was used to measure the mercury concentration in the supernatant and the mercury content in the bacteria. The efficiency of total mercury removal was the initial mercury content of the culture medium minus the mercury content of the supernatant, divided by the initial mercury content of the culture medium; the efficiency of mercury adsorption by the bacteria was determined by the mercury content of the bacteria divided by the initial mercury content of the culture medium.

The mercury concentration of the supernatant and the mercury content of the bacteria were measured using a cold atomic absorption spectrophotometer. The efficiency of total mercury removal is the initial mercury content of the culture medium minus the mercury content of the supernatant, divided by the initial mercury content of the culture medium, and the efficiency of mercury adsorption by the bacteria is the mercury content of the bacteria divided by the initial mercury content of the culture medium. The total removal rate of mercury minus the adsorption efficiency of the bacteria on mercury is the volatilization efficiency of mercury [13]. The heavy metal concentrations in the digestion solution and the blank control were measured three times. The calculation was performed as follows:

$$q_1 = \frac{(a_1 - b_1)}{a_1}$$

$$q_2 = \frac{c_1}{a_1}$$

$$q_3 = q_1 - q_2$$

where a_1 is the initial mercury content of the culture solution (mg/L); b_1 is the mercury content of the supernatant (mg/L); c_1 is the mercury content of the bacteria (mg/L); q_1 is the total removal rate (%); q_2 is the adsorption efficiency (%); and q_3 is the volatilization efficiency (%).

2.8. Transformation Infrared Spectroscopy

To investigate the response mechanism of the LBA119 strain to mercury stress, Fourier transform infrared (FTIR) spectroscopy [11] was used to analyze the changes in the adsorption functional groups of bacteria before and after treatment with the heavy metal Hg^{2+}. The bacteria were incubated in an LB medium with mercury (10 mg/L) and without mercury for 36 h. The cultures were centrifuged at 8000 rpm for 10 min, the supernatant was discarded, and the organisms were washed three times with phosphate buffer solution. The strains were freeze-dried under a vacuum, and the dried samples were thoroughly mixed with potassium bromide (KBr) powder in an agate mortar (bacterial mass: KBr mass = 1:100), pressed into thin slices using a solid press, and measured using a Fourier infrared spectrometer to record the infrared spectra in the region of 4000–400 cm^{-1}. The obtained spectral data were processed using Omnic 9.0 software (OMNIC 8.2.0.387).

2.9. Testing the Strains for Resistance to Different Types of Heavy Metal Ions

LB liquid medium containing single heavy metal ions (Mn^{2+}, Zn^{2+}, Pb^{2+}, Cd^{2+}, and Cr^{6+}) was prepared and inoculated with 2% fresh bacterial broth and incubated on a shaker at 30 °C for approximately 36 h. An LB medium with different concentration gradients (1, 100, 500, 1000, 1500, 2000, 3000, 4000, and 5000 mg/L) of heavy metals was first set to determine the crude tolerance of the strain to heavy metal concentrations [14]. The concentration interval was then narrowed to determine the maximum concentration of heavy metals that could be tolerated. Bacterial concentrations were determined using the 600 nm optical density method (OD_{600nm}) to determine the minimum inhibitory concentration (MIC). Five heavy metal resistance assays were performed independently [15], with three parallel groups for each group.

2.10. Study on the Bioremediation Effect of Strains of Bacteria on Mercury-Contaminated Soil

In this experiment, mercury-contaminated soil was collected in Luanchuan County, Henan Province, from farmland in the molybdenum mining area. The soil was air-dried and sieved, and then a specific quantity of $HgCl_2$ solution was added to it. The soil samples were prepared with mercury concentrations of 5, 50, and 100 mg/L, stirred well, and left to stand for several days before being stirred again; then, the mercury soil was aged for 60 days.

2.11. Strain for Remediation of Mercury-Contaminated Soil

A logarithmic growth stage was achieved by incubating the bacteria. An adjustment of pH 4.6 was made to the seed solution used for inoculating LBA119 bacteria. A blank LB medium was added to the control group at the same concentration. After inoculation, the soil and inoculum were gently mixed in a shaker at 30 rpm for 20 min to conduct an exploratory experiment on the biological removal of mercury from the soil [16].

The study involved several experimental groups, which were prepared as follows. (1) Restoration group: 100 g of mercury-aged soil samples, 95 mL of distilled water, and 5 mL of bacterial seed solution were added and shaken well to ensure that all the soil was suspended in the liquid. Sterile distilled water was replenished quantitatively every day to maintain the water content of the soil, and the new bacterial solution was added every 7 days, with an additional amount of 5 mL each time. (2) Control group: 100 g of aged soil samples containing mercury, 95 mL of distilled water, and 5 mL of the LB medium were added and shaken well, and 5 mL of fresh medium was added every 7 days.

In each experimental group, three replicates were conducted. Incubation for 7, 15, and 30 days was conducted in an incubator (temperature 30 °C, 8 h of light time and 16 h of darkness (simulated mine light time)). Each group was provided 0.5 g of air-dried soil that was ground and digested. Using cold atomic fluorescence spectrometry [17], the mercury content was determined, and the mercury removal efficiency was calculated.

Mercury remediation efficiency was calculated by subtracting the initial mercury content of the soil from the mercury content at the time of sampling. The results are the mean ± standard deviation of three independent replicates. The efficiency of the strains for remediating mercury in the soil was calculated as follows:

$$q(\%) = \frac{a - b}{a} \times 100\%$$

where q is the removal efficiency (%) of mercury in soil by the strain; a is the initial mercury content of soil; and b is the mercury content of soil at different sampling times.

3. Results

3.1. Determination of Soil Mercury Content

According to the results of the atomic fluorescence spectrometry analysis, the mean mercury concentration in the soil was 10.46 mg/kg, which is considerably higher than the background value for Chinese soil in the environment.

3.2. Screening and Purification of Mercury-Resistant Strains

The bacterial suspension was diluted in LB solid plates containing 20 mg/L Hg^{2+} and was incubated at 37 °C for 48 h in a biochemical incubator to obtain mercury-resistant colonies. The mercury-resistant strains with a positive growth status were selected from the plates, numbered LBA119, and enriched after multiple scribing to separate individual colonies. The enriched strains were preserved in glycerol. The LBA119 colonies demonstrated the morphological characteristics shown in Figure 1a after incubation at 37 °C for 24 h under constant temperature and inversion. Specifically, the colony size was 0.5–1 mm; the bacteria exhibited a smooth edge, round shape, smooth surface, consistent uniformity, no halo, a creamy white color with a light yellow. The bacteria were not sticky, easy to pick, highly flat and produced a rotten egg smell. By Gram staining, these bacteria were observed as Gram-positive bacteria under the biological microscope (Figure 1b). Scanning electron microscopy was used to observe the morphology of the bacterium: the LBA119 bacteria were fixed in 1% glutaraldehyde solution for 20 min and then centrifuged, rinsed and centrifuged several times using phosphate buffer, then dehydrated (3–4 times) with increasing concentrations of ethanol (30%, 50%, 70%, 80%, 90%, 100%) step by step, and finally the ethanol solution containing bacteria was dropped on clean polished aluminum sheets, dried, sprayed with gold, and microscopic images of bacterial cells were observed with scanning electron microscopy. The size of individual organisms was approximately 0.8 × 1.3 μm [18] under electron microscopy (Figure 1c).

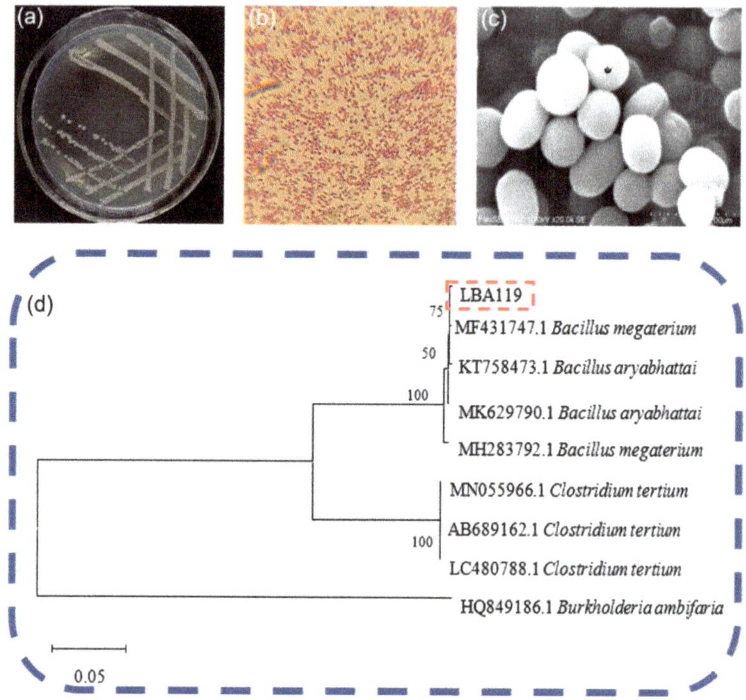

Figure 1. (**a**) LBA119 morphological growth diagram; (**b**) LBA119 Gram stain; (**c**) LBA119 electron microscope scan; (**d**) LBA119 phylogenetic tree.

3.3. Identification of Mercury-Resistant Strains

The bacterial isolates were identified according to biochemical characteristics and the 16S rDNA gene [19]. Referring to Table 2, the experimental results showed that strain LBA119 was positive for starch hydrolysis, gelatin hydrolysis, glucose fermentation, lactose fermentation, indole assay, methyl red, citrate utilization assay, and motility assay. It was negative for oil hydrolysis, VP, and H_2S. The phylogenetic tree of the strain is shown in Figure 1d, and the 16S rDNA gene sequence comparison showed that the strain was 99% similar to strain LBA119 and strain KT758473.1 *Bacillus* aryabhattai. The strain LBA119 was identified as *Bacillus* [20].

Table 2. Physiological and biochemical characteristics of the LBA119 strain.

Identification Test	Test Results	Identification Test	Test Results
Starch hydrolysis	+	Methyl red test	+
Oil and grease hydrolysis	−	V-P	−
Glucose fermentation	+	Lactose fermentation	+
Indole test	+	H_2S	−
Citrate utilization test	+	Motility experiments	+
Gelatin hydrolysis	+	Gram stain	+

Note: "+" indicates positive; "−" indicates negative.

3.4. Optimizing the Optimal Growth Conditions for Mercury-Tolerant Strains

A comparison of the growth of LBA119 in different media is shown in Figure 2a. The experimental results showed that the magnitude of growth of the LBA119 strain in different media (10 mg/L Hg^{2+}) was LB Culture (LB) > Potato Dextrose Agar (PDA) > Czapek Dox Medium (Czapek's Medium)> Beef extract peptone medium (BPD) > Glycerol Medium

(Glycerol) > Sabouraud Dextrose Broth Medium (SDB) > Gauze's Synthetic Broth Medium (GAUZE's Medium)> Potato Saccharose Agar (PSA).

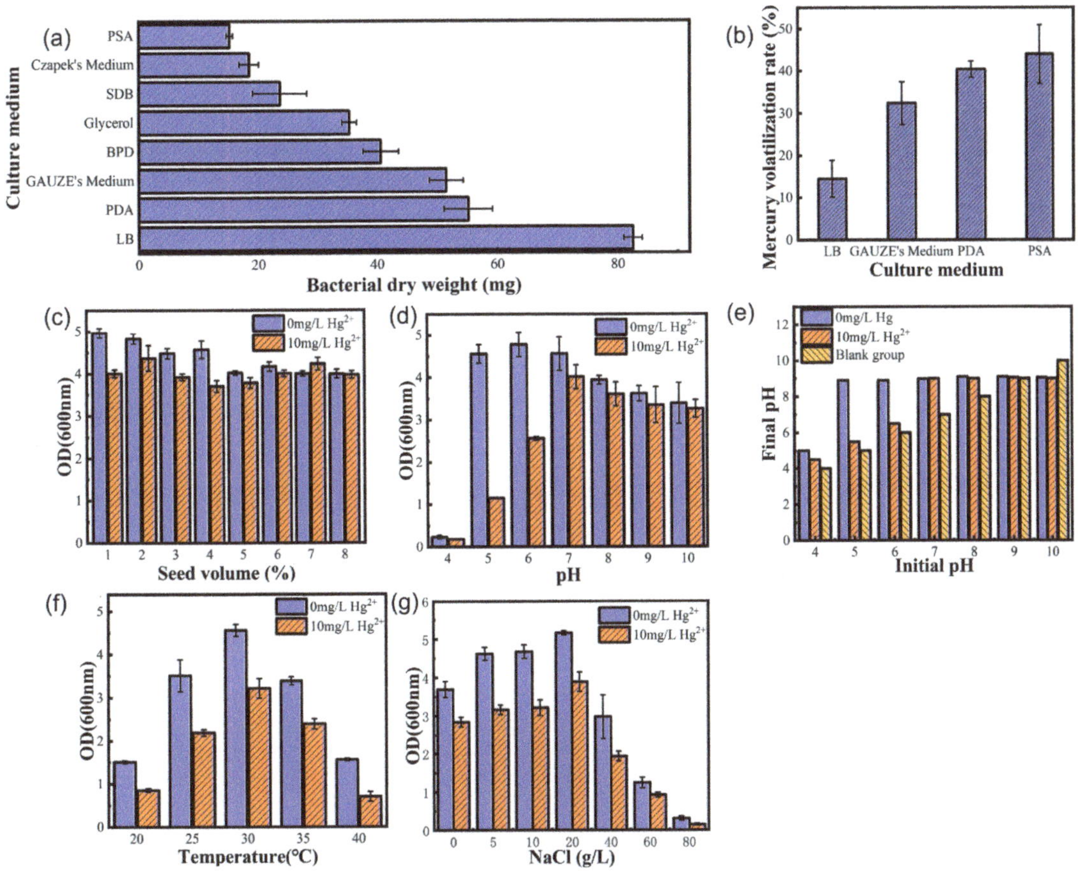

Figure 2. (**a**) Growth of the LBA119 strain in different media and (**b**) volatilization of mercury in different culture media; (**c**) effect of different strain inocula on the growth status of the LBA119 strain; (**d**) effect of pH on the growth of the LBA119 strain and (**e**) effect of the LBA119 strain on environmental pH; (**f**) effect of temperature on the growth of the LBA119 strain; (**g**) effect of salt concentration on the growth of the LBA119 strain.

The volatilization of mercury in different media is shown in Figure 2b. The experimental results showed that, after 48 h of shaking incubation, the magnitude of mercury volatilization in the medium was in the following order: LB > GAUZE's Medium > PDA > PSA.

The effects of different strain inocula on the growth status of the LBA119 strain are shown in Figure 2c. The biomass of this strain reached a maximum at 1% inoculum when incubated for 36 h without mercury. In the mercury-containing case, the inoculum level had less effect on the growth status of the strain, and the maximum biomass was reached at 2% inoculum, with a small increase in the growth of the strain when the inoculum level was increased to 7%. In the subsequent experiments with mercury addition, the optimal inoculum for this strain should be 2%.

The effect of the initial pH of the medium on the growth of the LBA119 strain is shown in Figure 2d. The experimental results showed that the LBA119 strain was adapted to weakly acidic, neutral, and alkaline environments but not to acidic solid environments, and its optimum growth pH was 6.0 in a mercury-free environment; its biomass (OD_{600nm}) in a 10 mg/L Hg^{2+} environment was reduced compared with that in a mercury-free state, and its optimum growth pH was 7.0.

The effect of the LBA119 strain on the pH of the medium is shown in Figure 2e. When the initial pH value was 5.0–10.0, the pH value of the medium was stable in an 8.9–9.1 range after 36 h of incubation, which indicates that the microbial metabolic activity of the mercury-resistant strain changes the pH value of the environmental medium, and the LBA119 strain causes the pH value of the medium increase (or decrease) to be approximately 9.

As shown in Figure 2f, temperature causes a significant effect on the growth of the strain. The LBA119 strain exhibits a good growth rate from 25 to 35 °C, and its optimal growth temperature is 30 °C in mercury-containing and nonmercury-containing environments. Growth decreases significantly when the temperature falls below 20 °C and reaches 40 °C, so 30 °C is the optimal growth temperature for this strain.

A comparison of the growth of the LBA119 strain with varying salt concentrations is shown in Figure 2g. Based on the experimental results, without mercury, the LBA119 strain can grow in an LB medium containing 0–60 g/L NaCl; its biomass increases as the salt concentration increases, and its biomass reaches its maximum biomass at a salt concentration of 20 g/L. In an environment with 10 mg/L Hg^{2+}, the strain can grow in 0–60 g/L NaCl, and its optimal salt concentration is 20 g/L. As a result, in mercury-containing environments, 20 g/L salt is the most appropriate salt level for the LBA119 strain.

3.5. *Effect of Mercury Concentration on the Growth Status of the Strain*

The minimum inhibitory mercury concentration of strain LBA119 is shown in Figure 3a. The growth of strain LBA119 decreased with increasing heavy metal mercury concentration, and the growth of this strain in the LB liquid medium with 1, 2, 4, 8, 16, and 32 mg/L Hg^{2+} decreased by 11.24%, 8.30%, 11.75%, 25.48%, 78.39%, and 96.57%, respectively. The growth of the LBA119 strain almost stopped at Hg^{2+} concentrations greater than 32 mg/L. Therefore, the minimum inhibitory concentration (MIC) of Hg^{2+} for the LBA119 strain was 32 mg/L.

The growth curves of the LBA119 strain are presented under different mercury concentrations in Figure 3b. The growth pattern of LBA119 in the blank control was 0–30 h for the logarithmic growth period, and the OD_{600nm} started to decline after 36 h for the decay period. The OD_{600nm} of LBA119 at a low concentration (5 mg/L) of mercury decreased by 14.82%; when the mercury content was 15 mg/L, the growth of the LBA119 strain showed a delayed period (24 h), and its OD_{600nm} maximum value appeared at 36 h, which decreased by 44.81% compared with the control, but the overall growth curve shape was similar to the control growth curve. At a mercury concentration of 20 mg/L, the growth was severely inhibited, and its OD_{600nm} maximum value decreased by 96.09% compared with the control.

3.6. *Mercury Volatilization and Adsorption Efficiency of the Strains*

The volatilization and adsorption efficiency of the LBA119 strain with different mercury concentrations is shown in Figure 3c. At mercury concentrations of 2–12 mg/L, the mercury removal rate by this strain reached more than 90% and up to 96.89%. However, as the mercury concentration increased, the mercury removal rate decreased rapidly (from 65.6% at 16 mg/L to less than 32.28% at 24 mg/L).

Figure 3d shows the results of inoculating fresh bacterial solution in an LB liquid medium containing 10 mg/L Hg^{2+} and investigating the effect of incubation time on the mercury removal efficiency of the LBA119 strain. The experimental results indicated that the efficiency of mercury removal by the strain increased as time progressed, and the rate of mercury volatilization was significantly greater than the rate of mercury adsorption. The total removal rate of the LBA119 strain increased significantly after 12 h and reached a maximum at 36 h. The rates of total removal, volatilization, and adsorption were 97.32%, 89.08%, and 8.24%, respectively.

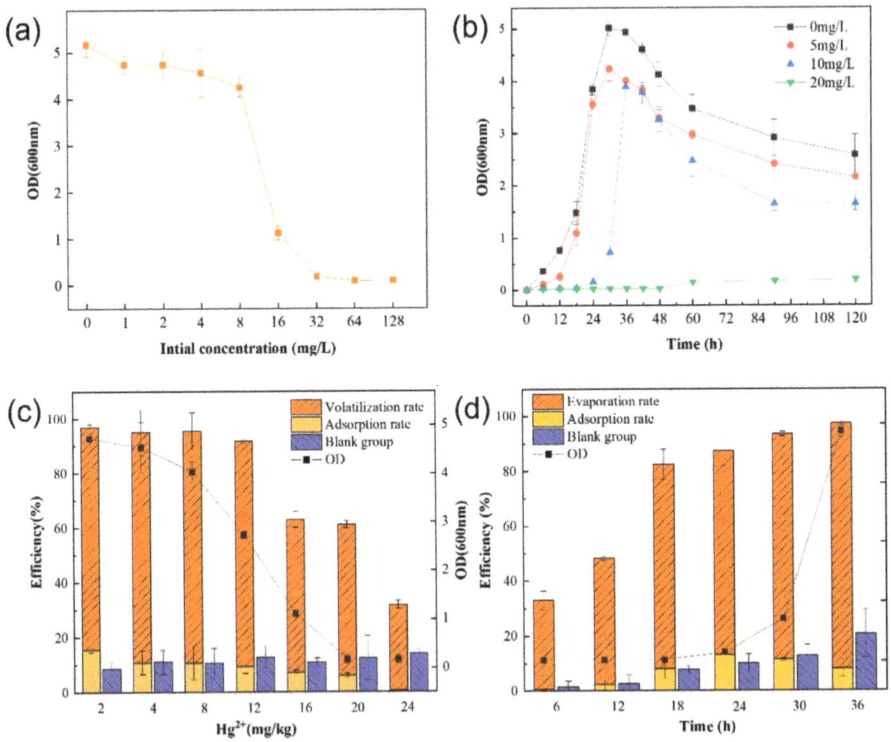

Figure 3. (**a**) Minimum inhibitory mercury concentration of the LBA119 strain and (**b**) growth curves at different mercury concentrations; (**c**) removal efficiency of LBA119 for different mercury concentrations; (**d**) effect of time on the efficiency of LBA119 for mercury removal.

3.7. Fourier Transform Infrared Spectroscopy Studies-Adsorption Mechanism

To investigate the mechanism by which strain LBA119 responds to Hg stress, changes in the adsorption functional groups of bacteria before and after treatment with the heavy metal Hg^{2+} were analyzed by Fourier transform infrared (FTIR) spectroscopy [21]. Figure 4a showed that the strain showed a significant increase in absorption peaks at 3300.82, 2927.26, 1651.04, 1536.86, 1398.39, 1238.97, 1079.39, and 558.78 cm^{-1} when the environmental Hg^{2+} concentration was 10 mg/L. The spectral peak at 3300.82 cm^{-1} is the stretching vibration of the OH from water bound by the stretching of the NH group from CH in the cell wall; the spectrum peak at 2927.26 cm^{-1} is CH3 of fatty acid; the peaks at 1651.04 cm^{-1} and 1536.86 cm^{-1} are the characteristic spectral bands of the protein; and the intense absorption peak at 1651.04 cm^{-1} is mainly caused by the stretching vibration of C=O. The strong absorption peak at 1536.86 cm^{-1} is primarily driven by the bending beat of NH, which are from the amide I (C=O) and amide II (NH) bands of the protein and peptide, respectively. The results indicate that the proteins amide I and amide II of the bacterial cell wall are actively involved in the adsorption process of heavy metal Hg^{2+}. The peak at 1398.39 cm^{-1} is the stretching vibration of COOH. The peak at 1238.97 cm^{-1} is the stretching vibration of P-O and C-S and the superimposed absorption peak of C-O and C-OH. The peak at 1079.39 cm^{-1} is the stretching vibrational peaks of C-O-C and C-O-P, and 558.78 cm^{-1} is the vibrational absorption peaks of M-O and O-M-O (M is the metal ion), in which the absorption peaks are enhanced. Furthermore, the functional group may provide lone pairs of electrons to ligand-bind to Hg^{2+}, indicating that, for the LBA119 strain under heavy metal Hg stress, Hg-O and O-Hg-O uptake levels were increased.

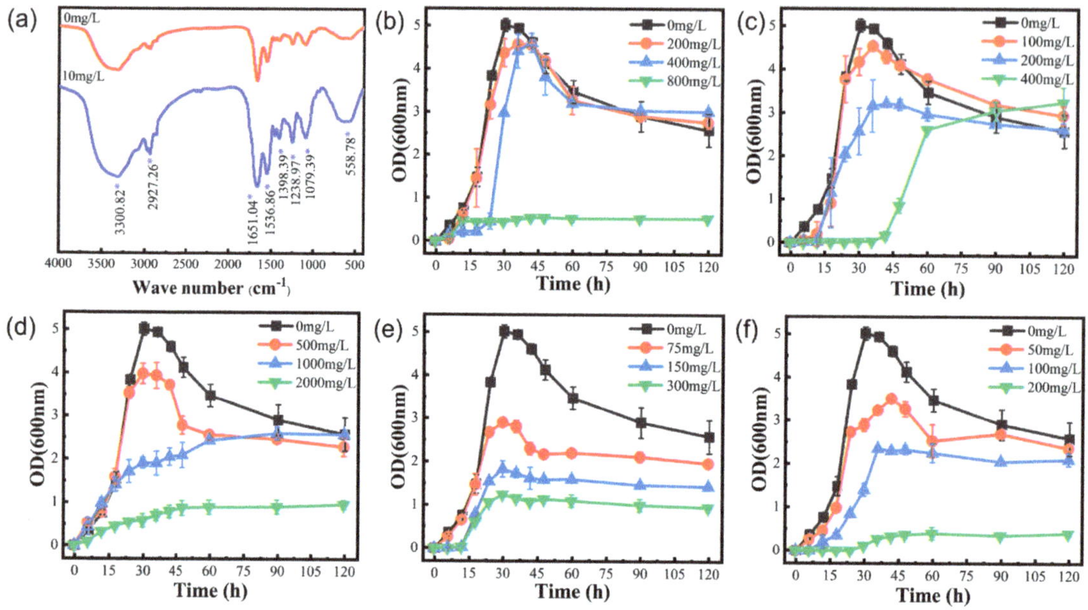

Figure 4. (**a**) FTIR spectra of the LBA 119 strain (* indicates peak increase). (**b**) Growth curves of LBA119 at different Mn^{2+} concentrations. (**c**) Growth curves of LBA119 at different Zn^{2+} concentrations. (**d**) Growth curves of LBA119 at different Pd^{2+} concentrations. (**e**) Growth curves of LBA119 at different Cd^{2+} concentrations. (**f**) Growth curves of LBA119 at different Cr^{6+} concentrations.

3.8. Effectiveness of Removing Different Heavy Metals

The results of the maximum tolerance concentrations of the LBA119 strain to a single heavy metal, as shown in Table 3, indicate that the LBA119 strain is highly resistant to heavy metal ions other than Hg^{2+}. The strain showed extremely high resistance to Pb^{2+}, Mn^{2+}, Zn^{2+}, Cd^{2+}, and Cr^{6+} and weak resistance to Co^{2+} and Ag^{+}. The tolerance concentration of LBA119 bacteria to heavy metal ions was $Pb^{2+} > Mn^{2+} > Zn^{2+} > Cd^{2+} > Cr^{6+} > Hg^{2+}$. At low concentrations, heavy metal ions did not significantly affect the growth of the 119 strain. Consequently, this strain is heavy metal-resistant with remediation potential in heavy metal-contaminated environments.

Table 3. Tolerance of strain LBA119 to single heavy metals.

Heavy Metal Ions	MTC_G (mg/L)	MTC_Y (mg/L)
$Mn^{2+}(MnCl_2{\cdot}4H_2O)$	500	800
$Zn^{2+}(ZnSO_4{\cdot}7H_2O)$	180	400
$Pb^{2+}(PbC_4H_6O_4{\cdot}3H_2O)$	600	2000
$Cd^{2+}(CdCl_2{\cdot}5/2H_2O)$	300	300
$Cr^{6+}(K_2Cr_2O_7)$	200	200

The effects of other heavy metals on the growth curve of LBA119 are shown in Figure 4. The experimental results indicated that the growth of the strain was severely inhibited at the maximum tolerated concentration of heavy metal ions. The OD_{600nm} of the LBA119 strain reached its maximum at 30 h in the absence of heavy metals; the maximum peaks of Mn^{2+} (200 mg/L), Zn^{2+} (100–200 mg/L), Pb^{2+} (500 mg/L) and Cd^{2+} (75–300 mg/L) all occurred at 36 h. Under Zn^{2+} (400 mg/L) and Cr^{6+} (200 mg/L) environments, the lag period at the beginning of the growth of the LBA119 strain was prolonged, but the shape of the overall growth curve was similar to that of the control. Compared to that of the control, the OD_{600nm} maximum values of this strain were reduced by 89.10%, 35.19%, 75.26%, 82.15%, and 92.02% at the highest concentrations of Mn^{2+}, Zn^{2+}, Pb^{2+}, Cd^{2+}, and Cr^{6+}, respectively, indicating that the growth of the LBA119 strain was substantially inhibited at the maximum tolerated concentration.

3.9. Remediation Effect of the LBA119 Strain on Mercury-Contaminated Soil

The effects of performing bioremediation with mercury-contaminated soil using the LBA119 strain was analyzed, as presented in Figure 5b. When the initial mercury concentration in the soil was 5 mg/L, both the remediation group inoculated with LBA119 and the control group inoculated with the LB medium without bacterial biomass reached a 100% removal rate after 7 days [22]. When the baseline Hg concentration was 50 mg/L, the total Hg removal rate was significantly increased by inoculation with LBA119, from 37.23% to 74.90% at 30 days, compared to that of the Hg-contaminated soil with an LB medium without bacterial biomass. When the initial mercury concentrations were 50 and 100 mg/L, the total mercury removal rate of the remediation group was higher than that of the control group. This value increased from 37.23% to 74.9% and from 46.81% to 62.35% after 30 days, respectively.

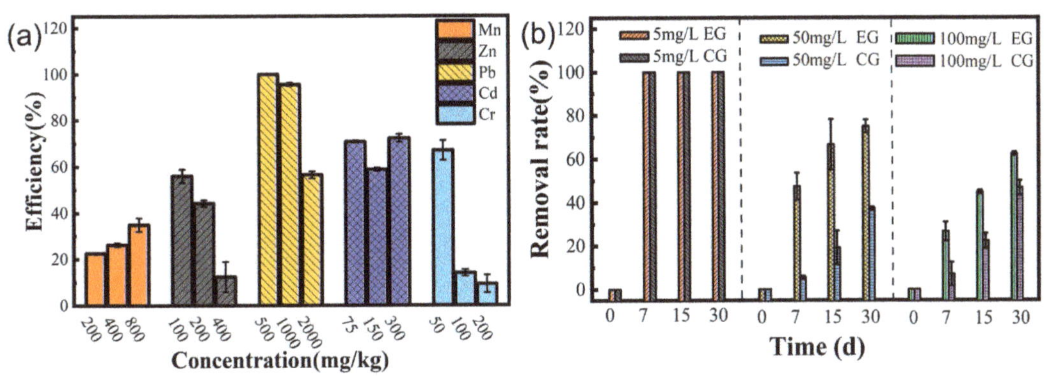

Figure 5. (**a**) Heavy metal removal rate of the LBA119 strain; (**b**) Bioremediation effect of the LBA119 strain on mercury-contaminated soil.

4. Discussion

Four factors affect the growth and reproduction of the strain: optimum medium, optimum temperature, optimum pH, and optimum inoculum. Firstly, the mercury experiments in different media showed (Figure 2b) that the volatility of mercury was higher in PSA, PDA, and GAUZE's medium, probably because the media contained reducing sugars that could reduce Hg^{2+} and volatilize after reducing Hg^{2+} to Hg^0. Therefore, these media are not suitable for studies related to the study of strain tolerance to Hg. Taner Sar et al. [23] studied the effect of inoculum level on ethanol production. We found that a low inoculum level resulted in a prolonged delay period. In contrast, a high inoculum level resulted in violent competition for survival among microorganisms and the death of a significant number of microorganisms. As shown in Figure 2d, too high or too low pH environments resulted in significant differences in strain growth, which is because extreme pH not only affects the activity and stability of enzymes in bacteria but also reduces microbial resistance to high temperatures and nutrient uptake, and also affects the ability of bacteria to produce fatty acids, which is consistent with the extremophile green alga Coccomyxa melkonianii at different pH values, which is consistent with the lipid production and morphological behavior of the extremophile green alga [24]. Since microorganisms do not possess thermoregulatory mechanisms, external temperature directly affects the activity of microbial intracellular enzymes, which are reduced at low temperatures. The metabolic rate and the ability of bacteria to synthesize substances decrease at low temperatures, leading to a decrease in reproductive capacity (Figure 2f), in agreement with the view of Rosa Margesin et al. [25], who studied the effect of different temperatures on a series of growth parameters of three bacteria. As shown in Figure 2c, in the mercury-containing environment, the strains showed easy protein secretion and expression under optimal growth conditions compared to the control, which would increase tolerance to inorganic mercury. As a result, we determined that the optimal growth conditions for this strain in a mercury-containing environment were an LB medium with 2% inoculum, 30 °C, and an initial pH of 7.0.

The logarithmic growth period of strain LBA119 was continuously delayed with increasing mercury concentration (Figure 3b). The growth and removal of mercury by this strain decreased with increasing mercury concentration (Figure 3a,c). However, mercury removal continuously increased with time (Figure 3d). Mercuric chloride is an inorganic mercury agent; the higher the mercury concentration, the more toxic it is. High mercury concentrations inhibit bacterial strains' growth and their bacterial sulfhydrylase activity, causing impairment of bacterial metabolism. At mercury concentrations up to 32 mg/L, the LBA119 strain stopped growing because mercury ion reductase was inhibited at high mercury concentrations, the same as the mercury tolerance of Halomonas zincidurans [26].

As shown in Figure 4a, by comparing the IR spectra of strain LBA119 in the LB medium containing mercury (10 mg/L) and that without mercury, it was found that OH, -CH3, C=O in the amide I band, NH in the amide II band, P-O, C-S C-O, C-OH, -COOH, C-O-C, C-O-P, M-O, and O-M-O groups were enhanced. Amide I and amide II are characteristic absorption peaks of proteins. From these peaks, it can be assumed that proteins are involved in the adsorption process to the heavy metal Hg^{2+}. The positively charged mercury ions combined with protein molecules to form protein salts, and irreversible precipitation of proteins (heavy metal ions precipitated proteins) occurred. The relationship between protein content and heavy metals indicated that specific proteins were critical for binding heavy metals [27]. The presence of excessive mercury concentrations may inhibit strain growth or even result in strain death.

The main mechanisms of bacterial resistance to heavy metals are biosorption, extracellular precipitation, biotransformation, bioaccumulation, and efflux. Table 2 and Figure 4b–f show the extensive resistance of strain LBA119 to other heavy metals, with severe inhibition of growth at the maximum tolerated concentration of heavy metal ions, in agreement with the previously discussed results.

Biosorption depends on many factors, such as surface adsorption, ion exchange, ligand, chelation, and micro-precipitation, as well as the type of functional groups of the bacterial biomass. As shown in Figure 5a, the strain showed some removal of different heavy metals, with the most effective removal of Pb. According to Hoyle-Gardner, J. et al. [28], at neutral pH, the electrostatic attraction between Pb ions and receptors on the bacterial cell wall might facilitate the adsorption of Pb by the strain, and Pb precipitates at weak acidic pH 2 and alkaline pH above 7, in the absence of bacteria, which is consistent with our results. The LBA119 strain adjusted the environmental pH (Figure 1e), and the mechanism of lead removal mainly relied on adsorption and the adjustment of environmental pH to precipitate lead ions.

Microbial soil remediation technology mainly relies on adsorption and transformation to remediate the soil and also has the characteristics of low investment, low pollution, and environmental friendliness. In the Orbetello Lagoon, Italy, Milva Pepi et al. [11] have isolated mercury-tolerant bacteria that can volatilize both organic and inorganic mercury into elemental mercury and contribute to mercury removal from sediment leachate as in the removal of mercury by the strain LBA119 we studied. Thus, LBA119 is resistant to mercury and can volatilize the ionic form of mercury metal (Hg^{2+}) into its elemental form (Hg^0). As shown in Figure 5b, inoculation of the LBA119 strain in high-concentration mercury-contaminated soil could improve the mercury removal rate compared with the control group, thus reducing the heavy metal mercury content in the soil. In conclusion, strain LBA119 has potential applications in the bioremediation of mercury-contaminated soils.

5. Conclusions

Heavy metal-contaminated soils from agricultural fields in mining areas may be an essential source for the isolation of many potentially heavy metal-tolerant bacteria. The mercury-tolerant bacterial strain LBA119 isolated from Luanchuan, Luoyang: (1) was identified as *Bacillus* by Gram staining, electron microscopy analysis, physiological and biochemical tests, and 16S rDNA identification; (2) the optimal growth conditions of the strain in the absence and presence of mercury were explored, and the growth, as well as the enzymatic activity of the strain, could be promoted under optimal conditions; (3) the strain was tested for minimum mercury inhibition concentration under optimum growth conditions, showing that the strain is highly tolerant of mercury and is capable of surviving in mercury-rich environments; (4) the bacteria isolated from the mine were found to remove mercury primarily by volatilization supplemented by adsorption, which can promote the volatilization of mercury in the soil and reduce the mercury content in the soil, emphasizing the significance of the strain to the area; (5) the bioremediation effect of isolated bacteria on mercury ranged from 62 to 97.36%, indicating that the strain is a promising candidate to be used for the bioremediation of heavy metals (especially Hg). This strain could be ideal for the microbial remediation of heavy metal-contaminated soil in mining areas.

Author Contributions: H.Y. and H.W. contributed to the study's conception and design. H.Y. and H.W. performed material preparation, data collection, and analysis. J.J. and A.T. were responsible for the drawing and layout of graphics. Y.S. and Z.C. were in responsible for the grammatical revision of the paper. The first draft of the manuscript was written by H.Y. and all authors commented on previous versions of the manuscript. All authors have read and agreed to the published version of the manuscript.

Funding: This work was supported by the National Natural Science Foundation of China [No.41471256]; Project for the Training of Young Backbone Teachers of Higher Education Institutions in Henan Province (2018GGJS047); Henan Province Science and Technology Research Project (192102110050).

Institutional Review Board Statement: Not applicable.

Informed Consent Statement: Not applicable.

Data Availability Statement: The authors declare that the data supporting the findings of this study are available within the article and its additional information files.

Conflicts of Interest: The authors declare no conflict of interest.

References

1. Naguib, M.M.; Khairalla, A.S.; El-Gendy, A.O.; Elkhatib, W.F. Isolation and characterization of mercury-resistant bacteria from wastewater sources in Egypt. *Can. J. Microbiol.* **2019**, *65*, 308–321. [CrossRef] [PubMed]
2. Priyadarshanee, M.; Chatterjee, S.; Rath, S.; Dash, H.R.; Das, S. Cellular and genetic mechanism of bacterial mercury resistance and their role in biogeochemistry and bioremediation. *J. Hazard. Mater.* **2022**, *423*, 126985. [CrossRef]
3. Møller, A.K.; Barkay, T.; Al-Soud, W.A.; Sørensen, S.J.; Skov, H.; Kroer, N. Diversity and characterization of mercury-resistant bacteria in snow, freshwater and sea-ice brine from the High Arctic. *FEMS Microbiol. Ecol.* **2011**, *75*, 390–401. [CrossRef] [PubMed]
4. Yang, W.; Chen, X.; Zhu, B.; Zhu, Y. Isolation of mercury-resistant strains and its characteristic for mercury-removed. *Chin. J. Environ. Eng.* **2017**, *11*, 602–607.
5. Fan, T.T.; Zhao, M.M.; Xue, L.G.; Wang, S.M.; Emaneghemi, B.; Xiao-Yan, H.E.; Wang, X.J.H.A.S. Research Status of Anti-mercury Bacteria and Bioremediation Mechanism. *Hunan Agric. Sci.* **2019**, *02*, 115–119.
6. Hao, L.; Zhang, B.; Feng, C.; Zhang, Z.; Lei, Z.; Shimizu, K. Human health risk of vanadium in farmland soils near various vanadium ore mining areas and bioremediation assessment. *Chemosphere* **2021**, *263*, 128246. [CrossRef] [PubMed]
7. Oliveira, R.; Wu, L.; Luo, Y.; Rajkumar, M.; Rocha, I.; Freitas, H. Inoculatlon with metal-mobilizing plant-growth-promoting rhizobacterium *Bacillus* sp. SC2b and its role in rhizoremediation. *J. Toxicol. Environ. Health Part A* **2015**, *78*, 931–944.
8. Pushkar, B.; Sevak, P.; Sounderajan, S. Assessment of the bioremediation efficacy of the mercury resistant bacterium isolated from the Mithi River. *Water Supply* **2018**, *19*, 191–199. [CrossRef]
9. Aatif, A.; Arslan, S.; Mushtaq, A.S.; Zakia, L.; Stanley, J.O. Expression and Purification of Transmembrane Protein MerE from Mercury-Resistant *Bacillus cereus*. *J. Microbiol. Biotechnol.* **2019**, *29*, 274–282.
10. Bravo, G.; Vega-Celedón, P.; Gentina, J.C.; Seeger, M. Effects of Mercury II on Cupriavidus metallidurans Strain MSR33 during Mercury Bioremediation under Aerobic and Anaerobic Conditions. *Processes* **2020**, *8*, 893. [CrossRef]

11. Pepi, M.; Gaggi, C.; Bernardini, E.; Focardi, S.; Lobianco, A.; Ruta, M.; Focardi, S.E. Mercury-resistant bacterial strains *Pseudomonas* and *Psychrobacter* spp. isolated from sediments of Orbetello Lagoon (Italy) and their possible use in bioremediation processes. *Int. Biodeterior. Biodegrad.* **2011**, *65*, 85–91. [CrossRef]
12. Abu-Dieyeh, M.H.; Alduroobi, H.M.; Al-Ghouti, M.A. Potential of mercury-tolerant bacteria for bio-uptake of mercury leached from discarded fluorescent lamps. *J. Environ. Manag.* **2019**, *237*, 217–227. [CrossRef] [PubMed]
13. Ghosh, S.; Sadhukhan, P.C.; Ghosh, D.K.; Chaudhuri, J.; Mandal, A. Volatilization of Mercury by Resting Mercury-Resistant Bacterial Cells. *Bull. Environ. Contam. Toxicol.* **1996**, *56*, 259–264. [CrossRef] [PubMed]
14. Ruiz-Díez, B.; Quiñones, M.A.; Fajardo, S.; López, M.A.; Higueras, P.; Fernández-Pascual, M. Mercury-resistant rhizobial bacteria isolated from nodules of leguminous plants growing in high Hg-contaminated soils. *Appl. Microbiol. Biotechnol.* **2012**, *96*, 543–554. [CrossRef]
15. Zhang, W.; Chen, L.; Liu, D. Characterization of a marine-isolated mercury-resistant *Pseudomonas* putida strain SP1 and its potential application in marine mercury reduction. *Appl. Microbiol. Biotechnol.* **2012**, *93*, 1305–1314. [CrossRef]
16. Koçberber, N.; Dönmez, G. Chromium(VI) bioaccumulation capacities of adapted mixed cultures isolated from industrial saline wastewaters. *Bioresour. Technol.* **2007**, *98*, 2178–2183. [CrossRef]
17. Ji, H.; Zhang, Y.; Bararunyeretse, P.; Li, H. Characterization of microbial communities of soils from gold mine tailings and identification of mercury-resistant strain. *Ecotoxicol. Environ. Saf.* **2018**, *165*, 182–193. [CrossRef]
18. Li, J.; Zheng, T.; Yuan, D.; Gao, C.; Liu, C. Probing the single and combined toxicity of PFOS and Cr(VI) to soil bacteria and the interaction mechanisms. *Chemosphere* **2020**, *249*, 126039. [CrossRef]
19. Niane, B.; Devarajan, N.; Poté, J.; Moritz, R. Quantification and characterization of mercury resistant bacteria in sediments contaminated by artisanal small-scale gold mining activities, Kedougou region, Senegal. *J. Geochem. Explor.* **2019**, *205*, 106353. [CrossRef]
20. Cardona, G.I.; Escobar, M.C.; Acosta-González, A.; Marín, P.; Marqués, S. Highly mercury-resistant strains from different Colombian Amazon ecosystems affected by artisanal gold mining activities. *Appl. Microbiol. Biotechnol.* **2022**, *106*, 2775–2793. [CrossRef]
21. Pattnaik, S.; Dash, D.; Mohapatra, S.; Pattnaik, M.; Marandi, A.K.; Das, S.; Samantaray, D.P. Corrigendum to "Improvement of rice plant productivity by native Cr(VI) reducing and plant growth promoting soil bacteria Enterobacter cloacae". *Chemosphere* **2020**, *238*, 125011. [CrossRef]
22. Chang, J.; Shi, Y.; Si, G.; Yang, Q.; Dong, J.; Chen, J. The bioremediation potentials and mercury(II)-resistant mechanisms of a novel fungus *Penicillium* spp. DC-F11 isolated from contaminated soil. *J. Hazard. Mater.* **2020**, *396*, 122638. [CrossRef]
23. Sar, T.; Stark, B.C.; Akbas, M.Y. Bioethanol production from whey powder by immobilized E. coli expressing Vitreoscilla hemoglobin: Optimization of sugar concentration and inoculum size. *Biofuels-UK* **2019**, *12*, 1103–1108. [CrossRef]
24. Soru, S.; Malavasi, V.; Caboni, P.; Concas, A.; Cao, G. Behavior of the extremophile green alga *Coccomyxa melkonianii* SCCA 048 in terms of lipids production and morphology at different pH values. *Extremophiles* **2019**, *23*, 79–89. [CrossRef] [PubMed]
25. Margesin, R. Effect of temperature on growth parameters of psychrophilic bacteria and yeasts. *Extrem. Life Under Extrem. Cond.* **2009**, *13*, 257–262. [CrossRef] [PubMed]
26. Zhou, P.; Huo, Y.Y.; Xu, L.; Wu, Y.H.; Meng, F.X.; Wang, C.S.; Xu, X.W. Investigation of mercury tolerance in *Chromohalobacter israelensis* DSM 6768(T) and *Halomonas zincidurans* B6(T) by comparative genomics with *Halomonas xinjiangensis* TRM 0175(T). *Mar. Genom.* **2015**, *19*, 15–16. [CrossRef]
27. Kim, S.U.; Cheong, Y.H.; Seo, D.C.; Hur, J.S.; Heo, J.S.; Cho, J.S. Characterisation of heavy metal tolerance and biosorption capacity of bacterium strain CPB4 (*Bacillus* spp.). *Water Sci. Technol.* **2007**, *55*, 105–111. [CrossRef]
28. Hoyle-Gardner, J.; Jones, W.; Badisa, V.L.D.; Mwashote, B.; Ibeanusi, V.; Gaines, T.; Tucker, L. Lead metal biosorption and isotherms studies by metal-resistant *Bacillus* strain MRS-2 bacterium. *J. Basic Microbiol.* **2021**, *61*, 697–708. [CrossRef] [PubMed]

toxics

Article

Microbial Removal of Petroleum Hydrocarbons from Contaminated Soil under Arsenic Stress

Qu Su [1,2], Jiang Yu [1,2,3], Kaiqin Fang [2,4], Panyue Dong [1,2], Zheyong Li [1,2], Wuzhu Zhang [2,4], Manxia Liu [1,2], Luojing Xiang [1,2,*] and Junxiong Cai [1,2,*]

1 Hubei Provincial Academy of Eco-EnvironmentSal Science, Wuhan 430070, China
2 State Environmental Protection Key Laboratory of Soil Health and Green Remediation, Wuhan 430070, China
3 Institute of Advanced Studies, China University of Geosciences, Wuhan 430079, China
4 Huazhong Agricultural University, Wuhan 430070, China
* Correspondence: xlj19651717@163.com (L.X.); cjxhbhjkx@163.com (J.C.)

Abstract: The contamination of soils with petroleum and its derivatives is a longstanding, widespread, and worsening environmental issue. However, efforts to remediate petroleum hydrocarbon-polluted soils often neglect or overlook the interference of heavy metals that often co-contaminate these soils and occur in petroleum itself. Here, we identified *Acinetobacter* baumannii strain JYZ-03 according to its Gram staining, oxidase reaction, biochemical tests, and FAME and *16S* rDNA gene sequence analyses and determined that it has the ability to degrade petroleum hydrocarbons. It was isolated from soil contaminated by both heavy metals and petroleum hydrocarbons. Strain JYZ-03 utilized diesel oil, long-chain n-alkanes, branched alkanes, and polycyclic aromatic hydrocarbons (PAHs) as its sole carbon sources. It degraded 93.29% of the diesel oil burden in 7 days. It also had a high tolerance to heavy metal stress caused by arsenic (As). Its petroleum hydrocarbon degradation efficiency remained constant over the 0–300 mg/L As(V) range. Its optimal growth conditions were pH 7.0 and 25–30 °C, respectively, and its growth was not inhibited even by 3.0% (*w*/*v*) NaCl. Strain JYZ-03 effectively bioremediates petroleum hydrocarbon-contaminated soil in the presence of As stress. Therefore, strain JYZ-03 may be of high value in petroleum- and heavy-metal-contaminated site bioremediation.

Keywords: bioremediation; heavy metal stress; petroleum hydrocarbon removal; soil pollution

Citation: Su, Q.; Yu, J.; Fang, K.; Dong, P.; Li, Z.; Zhang, W.; Liu, M.; Xiang, L.; Cai, J. Microbial Removal of Petroleum Hydrocarbons from Contaminated Soil under Arsenic Stress. *Toxics* **2023**, *11*, 143. https://doi.org/10.3390/toxics11020143

Academic Editors: Junhao Qin, Peidong Su, Feng Zhu and Lin Ding

Received: 9 November 2022
Revised: 20 January 2023
Accepted: 30 January 2023
Published: 1 February 2023

1. Introduction

Alkanes, aromatic compounds, nitrogen, and sulfur-oxygen-containing compounds are the major constituents of total petroleum hydrocarbons (TPHs). The aromatic fraction comprises compounds with benzene rings and includes polycyclic aromatic hydrocarbons (PAHs). These latter contain multiple fused aromatic rings and are listed as priority pollutants, as they are carcinogenic, mutagenic, toxic, and environmentally recalcitrant [1,2]. The development of land that has been polluted by oil refineries and other chemical industries is highly problematic. Prior research on the disposal or reclamation of these contaminated soils has ignored the interference of heavy metals. However, many oil-polluted sites are co-contaminated with heavy metals [3]. In earlier studies, petroleum hydrocarbons and heavy metals were detected in contaminated soils near gas stations, automobile repair workshops, and power stations. In fact, crude oil already contains heavy metals. Osuji [4] performed a qualitative analysis of the ash content in crude oil from domestic and foreign oilfields and detected Zn, Cu, Pb, Cd, Ni, Mn, Co, and 27 other metals. These elements and hydrocarbons constitute petroleum components and may interact with each other [5]. Heavy metal pollution may also be introduced to soils via the drilling fluid additives and low-quality barite used during extraction and mining [6]. Common hazardous metals such as Zn, Pb, Cu, Cd, Ni, Hg, Ba, and Cr were detected in at least trace amounts in all drilling waste tested [7]. Giller [8] and Olaniran [9] reported that heavy metals have different toxic effects on microorganisms and the environment as they vary in terms of bonding and mobility. Each heavy metal species can occur in the form of colloids or soluble complexes differing in toxicity and substrate competition. The physicochemical properties of heavy metals and petroleum hydrocarbons in the soil influence the environmental behavior of the pollutants and affect their transmembrane transport and biodegradation.

Heavy metals have complex effects on the soil environment. The efficiency of microbial petroleum hydrocarbon biodegradation tends to decrease with increasing heavy metal concentration [10–12]. Few studies have investigated microbial co-removal of petroleum hydrocarbon and heavy metals. Nevertheless, *Pseudarthrobacter phenanthrenivorans*, *Sphingomonas paucimobilis*, *Citrobacter freundii*, *Stenotrophomonas maltophilia*, and other species induce defense systems in response to heavy metal stress [13–15]. Oriomah [11] reported that *Achromobacter xylosoxidans* separately tolerated Cu (II) and waste oil. When both contaminants coexisted, however, the ability of *A. xylosoxidans* to degrade waste oil decreased from 40% to 5% after the Cu (II) concentration increased to 200 mg/L. Baltrons [16] discovered that inhibition of microbial degradation of 3–4-ring PAHs increased with heavy metal concentration. Heavy metals inhibit bacterial petroleum hydrocarbon biodegradation by damaging cells and/or reducing their viability [17]. Heavy metals also affect the functional properties of biologically active substances such as cellular enzymes and proteins. However, certain enzymes and organic acids may intracellularly and extracellularly reduce and/or complex heavy metals [18,19]. As these defense responses may occupy or even monopolize the normal metabolic pathways of bacterial cells, they may also bypass or deviate from petroleum hydrocarbon metabolism [20,21]. Heavy metals can also affect bacterial metabolic pathways by inhibiting petroleum hydrocarbon-degrading enzymes. Liu [22] found that exposure to 100–300 mg/L Pb (II) altered the spatial conformation of *Bacillus malacitensis* catechol 2,3-dioxygenase. The catalytic active site of the enzyme was blocked, reactive oxygen species (ROS) were induced, and the enzyme protein was damaged. Earlier studies have largely ignored or neglected the impact of heavy metals on petroleum-contaminated sites. Bioremediation research has focused almost exclusively on biological oil removal without regarding the potential interference of heavy metals in this process. For these reasons, the mechanisms of petroleum hydrocarbon removal by soil microorganisms under heavy metal stress must be elucidated.

2. Materials and Methods

2.1. Sample Collection

A contaminated soil sample was collected from the soil surface (0–20 cm) of a plot near a gas station in Wuhan, Hubei Province, China. The soil sample was placed in cloth bags, transferred to a laboratory, and stored at −20 °C until the subsequent analyses. Certain soil samples were stored at 4 °C until microbial screening and petroleum hydrocarbon content determination. The remaining soil was air-dried and passed through a 2 mm mesh sieve to remove large particles. Selected physicochemical properties of the soil were then determined (Table 1).

Table 1. Basic soil physicochemical properties.

Soil Property	Value
pH	8.24
Soil organic matter (SOM) (mg/kg)	78,650
Total N (mg/kg)	3600
Total P (mg/kg)	760
As (mg/kg)	8.6
Total petroleum hydrocarbons (TPH) (mg/kg)	318.88

2.2. Media

Hydrocarbon-degrading bacteria were isolated in mineral salt medium (MSM) consisting of 1.0 g NH_4NO_3, 1.0 g NaCl, 0.2 g $MgSO_4\ 7H_2O$, 7.5 g K_2HPO_4, 2.0 g KH_2PO_4, and 1% (*w*/*v*) diesel oil in 1 L pure water at pH 7.0−7.2.

The Luria-Bertani (LB) medium consisted of 10 g NaCl, 10 g tryptone, 5 g yeast extract, and 1 L pure water at pH 7.2–7.4. The beef extract-peptone medium consisted of 5 g NaCl, 10 g peptone, 3 g beef extract, and 1 L pure water at pH 7.0–7.2.

2.3. Enrichment and Isolation of Petroleum Hydrocarbon-Degrading Bacteria

A total of 5 g of contaminated soil sample was blended with 20 sterilized glass beads, and the mixture was added to 95 mL sterilized ultrapure water, shocked, and incubated at 150 rpm and 30 °C for 4 h. The suspension stood 10 min, and the bacteria-laden supernatant was collected. The first enrichment was performed by inoculating 5 mL soil suspension into 45 mL MSM containing 100 mg/L diesel oil as the sole carbon source. The suspension was cultured on a shaker table for 7 days and transferred to fresh MSM every 7 days. Bacterial growth was monitored daily by the optical density at 600 nm (OD_{600}). When the medium became turbid or discolored, it was diluted with a concentration gradient, coated with beef extract peptone medium, and incubated at 30 °C for 1–2 days. Single colonies were purified to isolate those derived from a single strain. All isolates were as stored at −80 °C in the form of liquid cultures containing 20% sterilized glycerol.

2.4. Petroleum Hydrocarbon Degradation Assay

The petroleum hydrocarbon-degrading bacteria were cultured in LB medium until the middle of the logarithmic growth phase and centrifuged in a 50 mL centrifuge tube at 8000× *g* rpm for 5 min. The supernatant was discarded, and the bacteria were suspended in sterile physiological saline. The preceding operations were repeated thrice. The OD_{600} of the final bacterial suspension was adjusted to 1.2, and it was inoculated into fresh MSM with different environmental factors. The efficiency of strain JYZ-03 at degrading the various components in diesel oil was investigated. The effect of varying the NaCl concentration (0.1, 0.5, 1.0, 2.0, or 3.0% as follows: add 5.85 g, 27.9 g, 55.8 g, 111.6 g, or 167.4 g NaCl to 100 mL MSM, respectively), the initial seeding dose (1.0, 2.0, 3.0, 4.0, or 5.0%), the pH (5.0, 6.0, 7.0, 8.0, or 9.0), and the temperature (20, 25, 30, 35, or 40 °C) of the culture medium were also investigated.

All biodegradation experiments were performed in MSM containing diesel oil as the sole carbon and energy source. The flasks were incubated at 150 rpm for 7 days, and bacterial growth was monitored daily by measuring the increases in culture OD_{600}. After 7 days of incubation, any remaining petroleum hydrocarbon was extracted with n-hexane (1:1 *v*/*v*). One milliliter organic phase was passed through an organic filter membrane and then analyzed by gas chromatography (GC) in flame ionization detection (FID) mode. The gas chromatograph used in the test was a GC2010-Plus model with a quartz capillary column type SH-Rtx-1 (30 m × 0.25 mm × 0.25 μm). The operating parameters were as follows: split ratio of 33.81:1; air flow rate: 400 mL/min; hydrogen flow rate: 40 mL/min; carrier gas (nitrogen) flow rate: 3.0 mL/min; makeup gas flow rate: 30 mL/min; injection volume: 1 μL; injection port temperature: 300 °C; FID detector temperature: 325 °C; temperature program: increase to 230 °C at a rate of 40 °C/min, increase to 320 °C at a rate of 20 °C/min, and hold for 20 min.

The degradation rate was estimated by calculating the GC profile of the petroleum hydrocarbon substrate. The strain cell growth was evaluated by measuring the increase in culture OD_{600}.

The degradation rate was calculated as follows:

$$\text{Degradation}(\%) = \frac{a - \text{b}}{a} \times 100 \quad (1)$$

where *a* is the mass of the petroleum hydrocarbon in the control and b is the mass of the petroleum hydrocarbon remaining after treatment.

2.5. Carbon Source Utilization

The strain JYZ-03 was also tested for its ability to grow on and degrade 500 mg/L long-chain alkanes (n-hexadecane, n-triacontane, n-dotriacontane, n-tetratriacontane, n-hexatriacontane, and n-octatriacontane) and 50 mg/L polycyclic aromatic hydrocarbons (PAHs; naphthalene and phenanthrene) in MSM as sole carbon source, respectively. All chemicals were of analytical grade and obtained from Sigma-Aldrich Corp., St. Louis, MO, USA. The foregoing extraction and degradation efficiency measurement methods in 2.4 were applied to the long-chain alkanes. The PAHs were extracted with 1:1 (v/v) acetone:n-hexane. Bacterial growth was monitored by measuring the increase in culture OD_{600}.

2.6. Biochemical Test

The shape, Gram staining, spore formation, and colony morphology of the bacteria growing on a solid LB medium were examined by transmission electron microscopy (TEM). Biochemical reactions, contact enzyme, methyl red (M-R), and emulsification properties were also determined.

2.7. Effects of Heavy Metal Stress on Bacterial Growth and Petroleum Hydrocarbon Degradation

A JYZ-03 suspension (OD_{600} = 0.8–1.0; 2% v/v) was added to LB medium containing 0, 20, 50, 100, 150, 200, 250, 300 mg/L As(V) or Pb^{2+} to explore the inhibitory effect of heavy metals on strain JYZ-03 growth. The flasks were incubated at 30 °C, and 150 rpm for 12 h, and the OD_{600} were measured and cultured under the same conditions for 7 days to investigate the effects of various As(V) and Pb^{2+} concentrations in MSM on the petroleum hydrocarbon degradation efficiency of strain JYZ-03. Petroleum hydrocarbons were added at a 1:1 ratio of diesel volume to culture medium volume.

2.8. Identification of the Selected Bacterial Isolates

The genomic DNA of the bacterial isolates was extracted by standard methods, and 16S rRNA gene sequencing was performed by the dideoxy sanger method. The 16S rRNA gene was amplified with universal bacterial primers in a thermal cycler. The reaction was conducted in a total volume of 25 mL. The PCR program was 10 min at 95 °C, 35 cycles of 30 s at 94 °C, 30 s at 52–65 °C, and 90 s at 72 °C. The sequences were compared against available nucleotide sequences deposited in the National Center for Biotechnology Information (NCBI) database (https://www.ncbi.nlm.nih.gov/nuccore (accessed on 7 September 2022)) by using the Basic Local Alignment Search Tool (BLAST; https://blast.ncbi.nlm.nih.gov/Blast.cgi (accessed on 7 September 2022)). Phylogenetic trees were plotted by the neighbor-joining (NJ) method using MEGA X software (https://www.megasoftware.net/dload_win_gui (accessed on 8 September 2022)). The sequences of the selected strains were submitted to the National Center for Biotechnology Information (NCBI) GenBank database (https://www.ncbi.nlm.nih.gov/genbank/ (accessed on 8 September 2022)) and assigned accession numbers.

2.9. Evaluation of the Efficiency of TPH from Soil by JYZ03

The actual soil collected was air-dried, ground through 2 mm sieve and mixed with diesel fuel, and aged for 10 d away from light to make the petroleum hydrocarbons distributed evenly in the soil. The bacteria were incubated in LB medium with strain JYZ-03 for 24 h, centrifuged at 8000× *g* rpm for 5 min, washed with sterile water 3 times, and adjusted the bacterial suspension OD_{600} = 0.9. Three groups were designed, petroleum hydrocarbon single contamination group, petroleum hydrocarbon-arsenic composite group, and control group; each group was set up in three parallel.

Petroleum hydrocarbon single pollution group: the initial concentration of petroleum hydrocarbon was 4103.90 mg/kg, and 10 mL of bacterial suspension of degradation bacteria was inserted. Petroleum hydrocarbon-arsenic composite group: the initial concentration of petroleum hydrocarbon was 4322.30 mg/kg, and Na_2HAsO_4 solution and 10 mL of bacterial suspension were added; the inoculum was about 5×10^7 CFU/g soil. The concentration of arsenic was 201.65 mg/kg. Sterilized pure water was added to the control group, and the initial content of petroleum hydrocarbons in the control was 4303.2 mg/kg. pH of the soil in the three groups was 6.8.

The three groups were incubated in a biochemical incubator at 30 °C, and the percentage of water content was maintained at 10% by adding sterilized pure water at regular intervals. The remaining petroleum hydrocarbon content in the soil was measured every 5~7 days.

3. Results and Discussion

3.1. Metabolic and Taxonomic Characteristics of the Isolates

Five strains JYZ(01-05) were isolated from soil samples contaminated with petroleum hydrocarbons near a gas station in Wuhan, Hubei Province, China. The objective was to determine whether these strains could form a clear zone on MSM containing diesel as the sole carbon source. After 5 days of culture, the degradation efficiency of JYZ-03 was 64.92%, and the degradation efficiency of the other 4 strains was lower than 50%. Unlike other strains isolated in this experiment, JYZ-03 retained the capacity to degrade petroleum hydrocarbons even under arsenic (As) stress. Gram staining, contact enzyme and amylase reactions, methyl red tests, strain emulsification performance evaluations, and SEM analysis, Strain JYZ-03 was a kind of Gram-negative rod-shaped bacterium (Figure 1), and 16S rDNA gene sequence analysis verified that strain JYZ-03 belonged to *Acinetobacter* sp. (Figure 2). Prior studies showed that *Acinetobacter* sp. are ubiquitous in the environment and have been isolated from oil-polluted samples in which they utilized crude oil as their sole carbon source. *Acinetobacter* sp. is the main bacterial taxon implicated in petroleum hydrocarbon biodegradation [23,24].

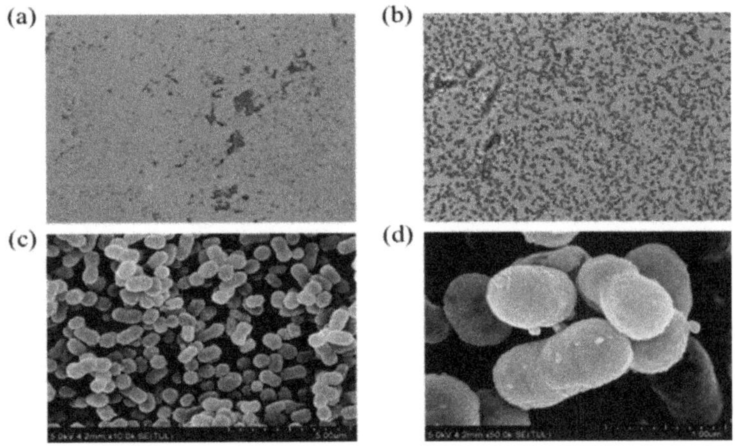

Figure 1. Photographs of cell of strain JYZ-03. (**a**,**b**) Gram stain photo of JYZ-03 (×20 microscopic view; ×100 microscopic view); (**c**,**d**) Scanning electron microscope photo of JYZ-03 (5.00 μm SEM; 1.00 μm SEM).

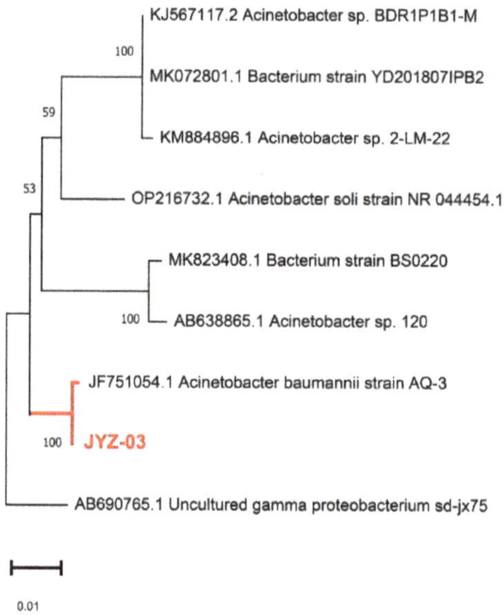

Figure 2. Phylogenetic tree of soil bacterial strains capable of degrading total petroleum hydrocarbons.

3.2. Diesel Oil Degradation Assay

Diesel oil is a complex mixture of medium- and long-chain n-alkanes (C11–C25). The bacterial strain alters the alkane composition of the diesel oil it degrades, as shown in Table 2. The chromatogram of the residual diesel oil in MSM revealed the alkanes n-undecanone (C11) to n-pentadecane (C25). The characteristic peak intensity of each n-alkane was significantly reduced compared with the control. We also measured the degradation efficiency of each n-alkane component in the diesel oil at the end of the culture. The most n-alkane degradation efficiency of strain JYZ-03 in oil-containing MSM was in the range of 75–100%, and the n-alkanes degradation efficiency was 84.05%, as shown in Table 2. Hence, strain JYZ-03 effectively degraded the major n-alkanes in diesel oil. It is generally more difficult to biodegrade long-carbon chains than short-carbon chain alkanes [23] We found that the degradation efficiency of n-undecane was 48.78%, which was lower than those of other medium- and long-carbon chain alkanes. Undecane is an intermediate in several serial oxidations involving the hydroxylation, dehydrogenation, and decarboxylation of other medium- and long-carbon chain alkanes. The biodegradation efficiency of n-alkanes depends upon their chain length, physicochemical properties, and the characteristics of the bacterial strains using them as carbon and energy sources.

Table 2. Residual n-alkanes concentrations of 1% diesel oil in MSM medium for 7 days.

N-Alkanes	Initial Concentrations (mg·L^{-1})	Final Concentrations (mg·L^{-1})	Degradation Rate (%)
C11	11.94 ± 0.80	6.12 ± 0.01	48.78 ± 0.97
C12	113.70 ± 5.87	28.82 ± 0.3.6	74.65 ± 6.05
C13	294.27 ± 18.76	36.75 ± 0.82	87.51 ± 4.36
C14	644.28 ± 43.60	120.80 ± 3.03	81.25 ± 6.95
C15	794.74 ± 54.45	128.35 ± 2.78	83.85 ± 5.10
C16	596.48 ± 40.25	93.47 ± 1.78	84.33 ± 4.42
C17	355.69 ± 25.30	52.64 ± 0.89	85.20 ± 3.41
C18	360.10 ± 12.02	57.51 ± 0.61	84.03 ± 5.09
C19	346.46 ± 22.48	48.95 ± 0.86	85.87 ± 3.95
C20	235.60 ± 18.70	37.84 ± 1.44	83.94 ± 7.68

Table 2. Cont.

N-Alkanes	Initial Concentrations (mg·L^{-1})	Final Concentrations (mg·L^{-1})	Degradation Rate (%)
C21	202.03 ± 14.08	21.64 ± 0.54	89.29 ± 3.85
C22	114.92 ± 9.14	12.43 ± 0.33	89.18 ± 3.59
C23	47.27 ± 3.43	10.07 ± 0.29	78.70 ± 8.59
C24	26.43 ± 2.30	1.68 ± 0.21	93.65 ± 8.98
C25	13.14 ± 0.80	0.00 ± 0.00	100.00 ± 0.00
Average	4157.00 ± 246.00	663.04 ± 12.57	84.05 ± 5.11

3.3. Carbon Source Utilization

The principal components of diesel oil are C11–C25 normal alkanes. However, they do not represent all petroleum hydrocarbon pollutants. Branched alkanes and polycyclic aromatic hydrocarbons (PAHs) are also important components of petroleum hydrocarbons. Thus, more than 10 carbon sources were measured and tested as potential carbon substrates for strain JYZ-03, including long-chain and branched alkanes as well as polycyclic aromatic hydrocarbons (PAHs). The foregoing chemical classes are the major components of petroleum hydrocarbons and are ubiquitous in contaminated soil. Long-chain alkanes are often more toxic than short-chain alkanes. For this reason, it is equally important to study the microbial degradation efficiency of long-chain normal alkanes. Table 3 shows that strain JYZ-03 reached the highest degradation efficiency in long-chain (C26–C38) n-alkanes in the culture medium with C32.

Several strains of *Acinetobacter* sp. may preferentially degrade C10–C30 alkanes, as these bacteria harbor genes encoding the hydrocarbon-degrading enzymes n-alkane dioxygenase and n-alkane hydroxylase [24,25]. These bacterial enzymes aerobically biodegrade alkanes.

Strain JYZ-03 degrades long-chain alkanes with far lower efficiency than medium-chain alkanes.

Table 3. Bacterial degradation efficiency of various carbon sources.

Carbon Sources	Degradation Rate (%)
C26	17.56 ± 0.04
C28	18.56 ± 1.77
C30	9.94 ± 2.72
C32	25.88 ± 3.44
C34	20.24 ± 1.32
C36	11.16 ± 1.42
C38	3.11 ± 1.90
Pristane	63.18 ± 9.06
Phytane	70.52 ± 8.58
Naphthalene	36.62 ± 2.78
Phenanthrene	26.71 ± 4.46

In many cases, the toxicity of branched alkanes and PAHs is higher, and their degradation is more difficult than those of linear alkanes. Branched alkanes have complex structures and are sterically hindered. Therefore, it is challenging to identify the enzymes that are capable of degrading them. Branched alkane biodegradation is complex and difficult [22]. The branched alkanes in diesel oil include mainly pristane (2,6,10,14-tetramethylpentadecane) and phytane (2,6,10,14-tetramethylhexadecane). We selected various branched alkanes and PAHs as sole carbon sources to study the ability of strain JYZ-03 to degrade them. Table 3 shows that the pristane and phytane degradation efficiencies of strain JYZ-03 were 63.18% and 70.52%, respectively. Thus, the bacterium could effectively decompose them. PAHs are stable, have unique molecular structures, and their toxicity increases with the number of benzene rings. Microbial action is primarily responsible for removing PAHS from the environment [26]. Naphthalene (NAP) and phenanthrene (PHE) have two and three benzene rings, respectively, and were used as the sole carbon sources for strain JYZ-03 here to determine its capacity to degrade this class of substances. Table 3 shows that in MSM containing 50 mg/L naphthalene or 50 mg/L phenanthrene, the degradation efficiencies were 36.62% and 26.71%, respectively. Strain JYZ-03 could effectively degrade PAHs with two and three benzene rings and use branched alkanes and PAHs as its sole carbon and energy sources. The carbon source utilization experiment partially elucidated the degradation ability of strain JYZ-03 to a different component of petroleum hydrocarbon.

3.4. *Effects of Environmental Factors on Petroleum Hydrocarbon Removal by JYZ-03*

Several petroleum hydrocarbon degradation tests were conducted to test the effect of different environmental factors such as pH, temperature, salinity, and inoculum size on petroleum hydrocarbon degradation by the JYZ-03 strain. The results are presented in Figure 3. Figure 3A shows that pH directly affected bacterial growth, metabolism, and enzyme activity. Since pH directly affects the growth and metabolism of bacteria and the activity of related degrading enzymes, the pH of the culture medium is one of the key factors affecting the degradation of petroleum hydrocarbons. With the increase in pH value, the degradation efficiency generally increased first and then decreased. When pH was 7, the optimum growth pH of bacteria was obtained, and the degradation efficiency of petroleum hydrocarbon was 81.18%. When pH was 8, the degradation efficiency of petroleum hydrocarbons decreased slightly to 70.51%, indicating that the strain grew faster in neutral and weakly alkaline environments and had a higher degradation efficiency of petroleum hydrocarbons, which may be caused by the long-term maintenance of neutral pH of the medium during the acclimation stage. However, in the weak acidic environment with pH 5–6, the growth rate of bacteria showed a delayed period, and the bacterial density was significantly lower than that in the neutral and weak alkaline environment when it reached the stable stage, which may be related to the sudden change of pH in the medium during acclimation and activation. Degradation efficiency first rose and then fell with increasing pH. Thus, culture medium pH was a key factor affecting petroleum hydrocarbon degradation [27–29].

Temperature also plays an important role in petroleum hydrocarbon degradation as it affects both diesel physicochemistry and microbial metabolism. The effects of various culture temperatures on strain JYZ-03 growth rate and degradation efficiency are shown in Figure 3B. Microbial petroleum hydrocarbon degradation efficiency first increased and then decreased with increasing temperature. The optimal culture temperature range for strain JYZ-03 was 25–30 °C. The bacterium had a low tolerance for higher temperatures. *Acinetobacter* are strictly aerobic organisms and use oxygen as the terminal electron acceptor. All strains can grow well between 20 and 30 °C. Most strains have an optimum of 33–35 °C, although some fail to grow at 37 °C. It is recommended that an incubation temperature of 30 °C is used, and in some cases, the use of a lower temperature may be advisable in addition to 30 °C. Verma [30] reported similar results for bacterial degradation of oily sludge.

Salinity directly affects osmotic pressure in bacterial cells. The effects of different NaCl concentrations on strain JYZ-03 growth and degradation efficiency are shown in Figure 3C. Salinity somewhat affected degradation efficiency, possibly because there were other inorganic salts in the culture medium and the degrading bacteria had only limited salinity tolerance [31–33]. Zhang et al. also reported that although the salt tolerance of the microbial consortia was not high, the excellent hydrocarbon removal capacity could make the consortia a potential candidate for the bioremediation of petroleum hydrocarbon-contaminated saline–alkaline fields [34,35].

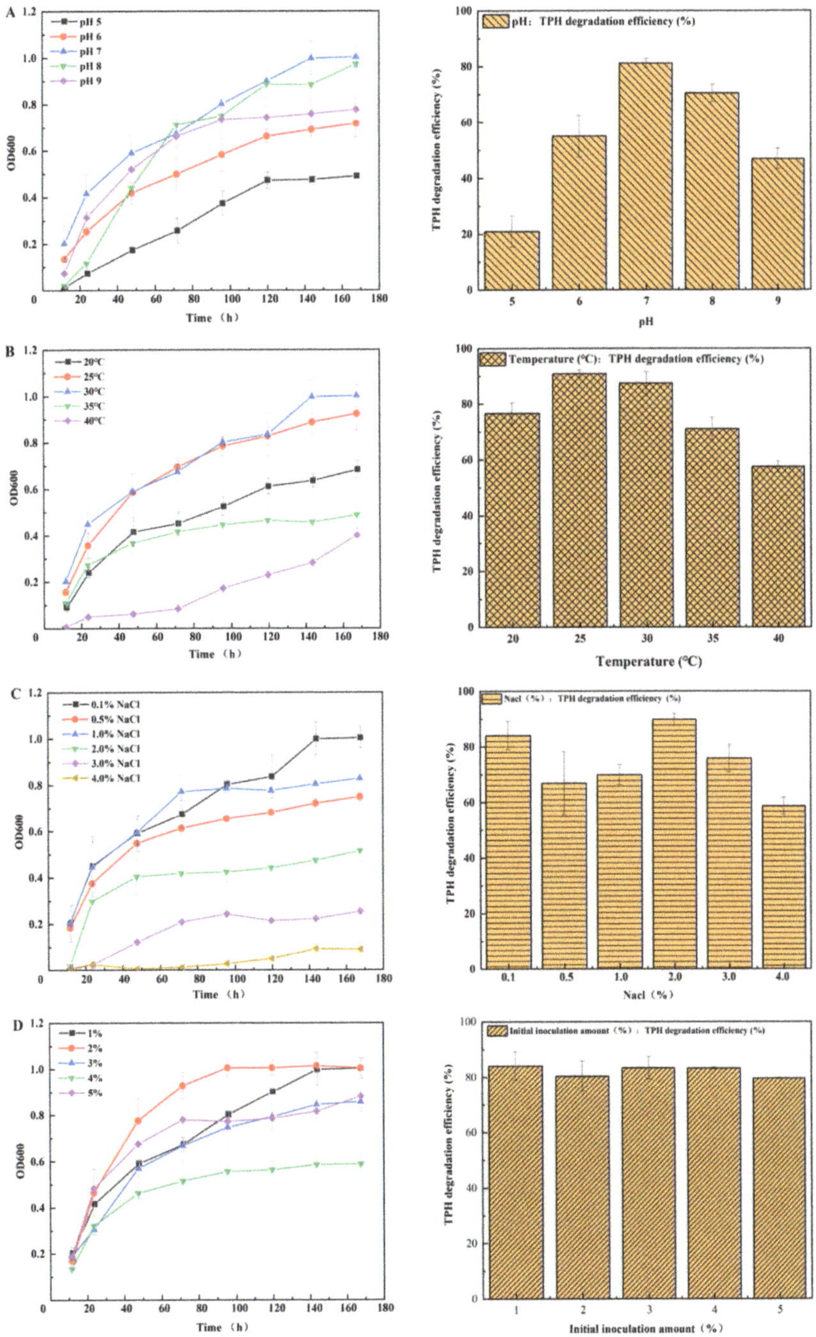

Figure 3. Effect of environmental factors on diesel oil removal by strain JYZ-03. (**A**) pH; (**B**) incubation temperature; (**C**) NaCl concentration (%); (**D**) inoculum concentration. The experiments of "TPH degradation efficiency" were performed in MSM medium, and the experiments of the "Turbidity" (OD_{600}) were performed in LB medium.

Strain JYZ-03 at initial inoculum concentrations of 1–5% removed to 80% of the diesel oil within 7 days (Figure 3D). The removal of hydrocarbon mainly depends on the capabilities of the microorganisms [36–39]. Changes in inoculum concentration had no apparent effect on the petroleum hydrocarbon degradation efficiency of strain JYZ-03; that is, very high inoculum concentrations do not necessarily improve petroleum hydrocarbon degradation efficiency. However, the bacterial growth rate has been indeed affected by increasing inoculum size, as the highest turbidity was achieved with 1% and 2%. The higher the inoculum size with the same nutrient content implies the faster the nutrient depletion, so turbidity reaching larger inoculum sizes was lower (Figure 3D, right).

3.5. Effects of Heavy Metal Stress on Petroleum Hydrocarbon Removal by JYZ-03

Heavy metals are toxic to microorganisms. To determine the toxicity of each heavy metal to strain JYZ-03, we first assessed the effects of As and Pb^{2+} on the growth of strain JYZ-03 by adding various heavy metal concentrations to the LB medium. Figure 4A shows that As(V) markedly and uniquely affected strain JYZ-03 growth. The JYZ-03 biomass remained high even at elevated As(V) concentrations. Hence, strain JYZ-03 has strong tolerance for As(V). The effects of Pb^{2+} on strain JYZ-03 are shown in Figure 4C. At medium-range Pb^{2+} concentrations (0–200 mg/L), the OD_{600} was in the range of 1.61–1.72 after 12 h culture, and the growth of strain JYZ-03 was only slightly affected. Thus, strain JYZ-03 has a certain degree of Pb^{2+} tolerance, at least at Pb^{2+} concentrations tested.

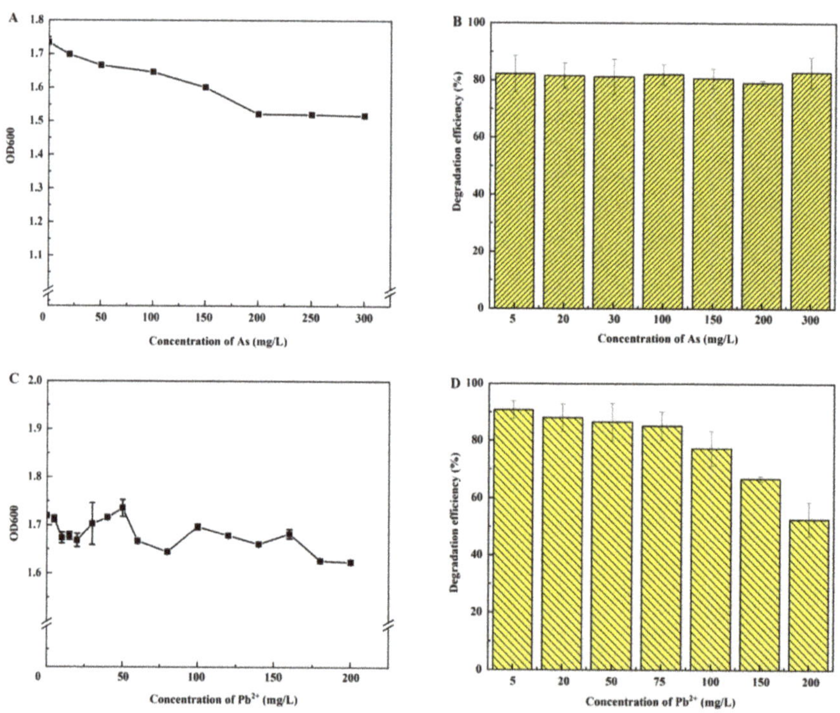

Figure 4. Effects of different heavy metal concentrations on strain JYZ-03 growth and petroleum hydrocarbon degradation. (**A**) Effects of As(V) on strain JYZ-03 growth; (**B**) petroleum hydrocarbon degradation by strain JYZ-03 at various As concentrations; (**C**) effects of Pb^{2+} on strain JYZ-03 growth; (**D**) petroleum hydrocarbon degradation by strain JYZ-03 at various Pb^{2+} concentrations.

Different heavy metal concentrations were added to MSM to establish their effects on the efficiency of petroleum hydrocarbon degradation by strain JYZ-03. Figure 4B shows that even at 300 mg/L As(V), the petroleum hydrocarbon degradation efficiency of strain JYZ-03 was more than 80%. Furthermore, the petroleum hydrocarbon degradation efficiency of strain JYZ-03 did not apparently decrease with increasing As(V) concentration. Therefore, strain JYZ-03 could potentially be bioremediate sites co-contaminated with As and petroleum hydrocarbons.

In MSM containing Pb^{2+}, the petroleum hydrocarbon degradation efficiency of strain JYZ-03 significantly decreased with increasing Pb^{2+} concentration (Figure 4D). At a Pb^{2+} concentration of 200 mg/L, the petroleum hydrocarbon degradation efficiency was only 52.65%. Thus, high Pb^{2+} concentrations strongly inhibit petroleum hydrocarbon degradation by strain JYZ-03.

3.6. Evaluation of the Efficiency of TPH Removal from Soil by JYZ03

In order to evaluate the degradation capability of the JYZ-03 strain in real polluted soils both in the presence or in the absence of heavy metals, this strain was inoculated in samples of hydrocarbon-polluted soils prepared as described in Section 2.9 of Material and Methods. The results are presented in Figure 5. Figure 5 shows that there was no significant difference in the removal of TPHs from the soil by JYZ03 in the compound-polluted soil of petroleum hydrocarbon with As(V) and single petroleum hydrocarbon-polluted soil. The degradation efficiency increased gradually within 10 days after the biological enhancement treatment, and then, the content of petroleum hydrocarbons in the soil tended to become stable, possibly because during the degradation process, the pollutants are adsorbed by the soil colloidal components, the carbon source required by microorganisms decreases, and the strains gradually decay or are outcompeted by indigenous microorganisms. Therefore, the degradation rate of petroleum hydrocarbons is far lower than that in the initial stage of remediation, as well as reported by Palanisamy [40].

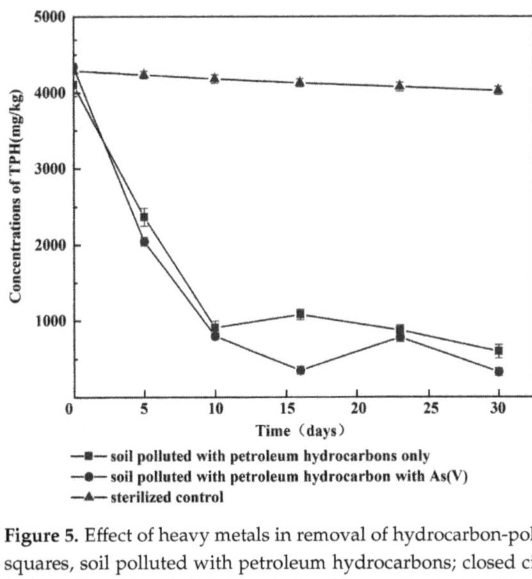

Figure 5. Effect of heavy metals in removal of hydrocarbon-polluted soils by JYZ-03 strain. Closed squares, soil polluted with petroleum hydrocarbons; closed circles, soil polluted with petroleum hydrocarbons plus As(V); closed triangles, sterilized soil polluted without inoculation.

4. Conclusions

Heavy metals occur in various forms in petroleum hydrocarbon-contaminated environments. However, certain bacterial strains that utilize petroleum hydrocarbons as carbon and energy sources are also resistant to heavy metals and are, therefore, suitable for application in bioremediation technology. We detected and isolated *Acinetobacter* baumannii strain JYZ-03 from soils contaminated with both petroleum hydrocarbons and heavy metals and established that it could effectively remove the former and had a high tolerance for the latter, including As(V). The petroleum hydrocarbon degradation efficiency of strain JYZ-03 varied with the molecular structure of the contaminant, including carbon chain length, degree of branching, and number of carbon rings present. The efficiencies with which strain JYZ-03 degraded the medium carbon chain length (C11–C25) and long carbon chain length (C26–C38) n-alkanes in diesel oil were 75–100% and 3.11–17.55%, respectively. It degraded branched alkanes and phytane by 63.18% and 70.52%, respectively, and the PAHs naphthalene and phenanthrene by 36.62% and 26.71%, respectively. It also degraded >76.97% of the diesel oil under physicochemical conditions optimal for its growth and performance, namely, pH 7, 25–30 °C, and 0.1–1% salinity. Petroleum hydrocarbon degradation by strain JYZ-03 first increased and then decreased with increasing pH. Strain JYZ-03 grew best in neutral and weakly alkaline environments. Its petroleum hydrocarbon degradation efficiency first increased and then decreased with increasing temperature. Strain JYZ-03 grew best at 25–30 °C. It had high salt tolerance, and even 3.0% (w/v) NaCl did not inhibit its growth. Within a certain range, the initial inoculum size did not markedly affect petroleum hydrocarbon degradation. The present study demonstrated that strain JYZ-03 effectively bioremediates petroleum hydrocarbon-contaminated soil even in the presence of As stress.

Author Contributions: Conceptualization, L.X. and J.C.; methodology, Q.S. and K.F.; formal analysis, Q.S.; investigation, Q.S. and K.F.; resources, Q.S.; data curation, Q.S. and W.Z.; writing—original draft preparation, Q.S. and P.D.; writing—review and editing, J.Y., Z.L. and M.L.; funding acquisition, L.X. and J.C. All authors have read and agreed to the published version of the manuscript.

Funding: This research was funded by the National Key Research and Development Program of China, grant number 2019YFC1804003.

Institutional Review Board Statement: Not applicable.

Informed Consent Statement: Not applicable.

Data Availability Statement: Not applicable.

Acknowledgments: This work was jointly supported by the National Key Research and Development Program of China (No. 2019YFC1804003), the Environmental Protection Scientific Research Project of Hubei Province, and the Knowledge Innovation Program of Wuhan-Basic Research.

Conflicts of Interest: The authors declare no conflict of interest.

References

1. Czarny, J.; Staninska-Pięta, J.; Piotrowska-Cyplik, A.; Juzwa, W.; Wolniewicz, A.; Marecik, R.; Ławniczak, Ł.; Chrzanowski, Ł. *Acinetobacter* sp. as the key player in diesel oil degrading community exposed to PAHs and heavy metals. *J. Hazard. Mater.* **2020**, *383*, 121–168. [CrossRef]
2. Khan, M.A.I.; Biswas, B.; Smith, E.; Naidu, R.; Megharaj, M. Toxicity assessment of fresh and weathered petroleum hydrocarbons in contaminated soil—A review. *Chemosphere* **2018**, *212*, 755–767. [CrossRef]
3. Adeniyi, A.; Afolabi, J.A. Determination of total petroleum hydrocarbons and heavy metals in soils within the vicinity of facilities handling refined petroleum products in Lagos metropolis. *Environ. Int.* **2002**, *28*, 79–82. [CrossRef]
4. Osuji, C.; Onojake, C.M. Trace heavy metals associated with crude oil: A case study of ebocha-8 oil-spill-polluted site in Niger Delta, Nigeria. *Chem. Biodivers.* **2004**, *1*, 1708–1715. [CrossRef]
5. Verla, A.; Opara, A.I.; Enyoh, C.E.; Ngozi, V.E.; Natheniel, C.O.; Kingsley, O.U.; Chizoruo, I.F.; Amaka, A.P. Petroleum hydrocarbons and heavy metals risk of consuming fish species from Oguta Lake, Imo State, Nigeria. *J. Chem. Health Risks* **2021**, *1*, 1–15.
6. Ebunu, I.A.; Bamidele, L.B.; Peter, A.O.; Adelana, R.A.; Azikiwe, P.O. Characterization of barite ores from selected locations in Nigeria for drilling fluid formulation. *Sci. Afr.* **2021**, *14*, e01057.
7. Lakmun, C.; Nithiya, A.; Sathiabama, T.T.; Shreeshivadasan, C. Drilling fluids: Presence of hazardous BTEXs and crystalline silica. *Environ. Manag. Tech.* **2020**, *8*, 1029–1035.
8. Giller, K.E.; Witter, E.; McGrath, S.P. Toxicity of heavy metals to microorganisms and microbial processes in agricultural soils: A review. *Soil Biol. Biochem.* **1998**, *30*, 1389–1414. [CrossRef]
9. Olaniran, A.O.; Balgobind, A.; Pillay, B. Bioavailability of heavy metals in soil: Impact on microbial biodegradation of organic compounds and possible improvement strategies. *Int. J. Mol. Sci.* **2013**, *14*, 10197–10228. [CrossRef]

10. Thavamani, P.; Megharag, M.; Naidu, R. Bioremediation of high molecular weight polyaromatic hydrocarbons co-contaminated with metals in liquid and soil slurries by metal tolerant PAHs degrading bacterial consortium. *Biodegradation* **2012**, *23*, 823–835. [CrossRef]
11. Oriomah, C.; Adelowo, O.O.; Adekanmbi, A.O. Bacteria from spent engine-oil-contaminated soils possess dual tolerance to hydrocarbon and heavy metals, and degrade spent oil in the presence of copper, lead, zinc and combinations thereof. *Ann. Microbiol.* **2015**, *65*, 207–215. [CrossRef]
12. Pimda, W.; Bunnag, S. Impact of inorganic nutrients and heavy metals present as co-contaminants on biodegradation of petroleum hydrocarbons by *Phormidium ambiguum* Strain TISTR 8296. *Water Air Soil Pollut.* **2017**, *228*, 1–13. [CrossRef]
13. Ibarrolaza, A.; Coppotelli, B.M.; Delpanno, M.T.; Donati, E.R.; Morelli, I.S. Application of the knowledge-based approach to strain selection for a bioaugmentation process of phenanthrene-and Cr (VI)-contaminated soil. *J. Appl. Microbiol.* **2011**, *111*, 26–35. [CrossRef]
14. Sorkhoh, N.A.; Ali, N.; Alawadhi, H.; Dashti, N.; Al-Mailem, D.; Eliyas, M.; Radwan, S. Phytoremediation of mercury in pristine and crude oil contaminated soils: Contributions of rhizobacteria and their host plants to mercury removal. *Ecotoxicol. Environ. Saf.* **2010**, *73*, 1998–2003. [CrossRef] [PubMed]
15. Chen, S.; Yin, H.; Ye, J.; Peng, H.; Liu, Z.; Dang, Z.; Chang, J. Influence of co-existed benzo[a]pyrene and copper on the cellular characteristics of Stenotrophomonas maltophilia during biodegradation and transformation. *Bioresour. Technol.* **2014**, *158*, 181–187. [CrossRef]
16. Baltrons, O.; Lopez-Mesas, M.; Vilaseca, M.; Gutiérrez-Bouzán, C.; Derf, F.L.; Portet-Koltalo, F.; Palet, C. Influence of a mixture of metals on PAHs biodegradation processes in soils. *Sci. Total Environ.* **2018**, *628–629*, 150–158. [CrossRef]
17. Biswas, B.; Sarkar, B.; Mandal, A.; Naidu, R. Heavy metal-immobilizing organoclay facilitates polycyclic aromatic hydrocarbon biodegradation in mixed-contaminated soil. *J. Hazard. Mater.* **2015**, *298*, 129–137. [CrossRef]
18. Tchounwou, P.B.; Yedjou, C.G.; Patlolla, A.K.; Sutton, D.J. Heavy metal toxicity and the environment. *Exp. Suppl.* **2012**, *101*, 133–164.
19. Briffa, J.; Sinagra, E.; Blundell, R. Heavy metal pollution in the environment and their toxicological effects on humans. *Heliyon* **2020**, *6*, 2405–8440. [CrossRef]
20. Glahn, F.; Schmidt-Heck, W.; Zellmer, S.; Guthke, R.; Wiese, J.; Golka, K.; Hergenröder, R.; Degen, G.H.; Lehmann, T.; Hermes, M.; et al. Cadmium, cobalt and lead cause stress response, cell cycle deregulation and increased steroid as well as xenobiotic metabolism in primary normal human bronchial epithelial cells which is coordinated by at least nine transcription factors. *Arch. Toxicol.* **2008**, *82*, 513–524. [CrossRef]
21. Salam, L.B.; Obayori, S.O.; Nwaokorie, F.O.; Suleiman, A.; Mustapha, R. Metagenomic insights into effects of spent engine oil perturbation on the microbial community composition and function in a tropical agricultural soil. *Environ. Sci. Pollut. Res.* **2017**, *24*, 7139–7159. [CrossRef] [PubMed]
22. Liu, S.H.; Zeng, G.M.; Niu, Q.Y.; Gong, J.-L.; Hu, X.-J.; Lu, L.-H.; Zhou, Y.-Y.; Hu, X.; Chen, M.; Yan, M. Effect of Pb (II) on phenanthrene degradation by new isolated *Bacillus* sp. P1. *RSC Adv.* **2015**, *5*, 55812–55818. [CrossRef]
23. Kaur, D.; Singh, R.P.; Gupta, S. Screening and characterization of next-generation biofuels producing bacterial strains. *Curr. Microbiol.* **2022**, *79*, 85.
24. Maeng, J.H.; Sakai, Y.; Ishige, T.; Tani, Y.; Kato, N. Diversity of dioxygenases that catalyze the first step of oxidation of long-chain n-alkanes in *Acinetobacter* sp M-1. *FEMS Microbiol. Lett.* **1996**, *141*, 177–182. [CrossRef]
25. Tani, A.; Tsuchimochi, S.; Nakabeppu, Y.; Nakajo, M. Bone and Tl-201 scintigraphy in a case of hereditary multiple exostoses. *Clin. Nucl. Med.* **2001**, *26*, 1028–1031. [CrossRef] [PubMed]
26. Piubeli, F.; Dos Santos, L.G.; Fernandez, E.N.; Silva, F.H.D.; Durrant, L.R.; Grossman, M.J. The emergence of different functionally equivalent PAH degrading microbial communities from a single soil in liquid PAH enrichment cultures and soil microcosms receiving PAHs with and without bioaugmentation. *Pol. J. Microbiol.* **2018**, *67*, 365–375. [CrossRef]
27. Whang, L.M.; Liu, P.; Ma, C.-C.; Cheng, S.-S. Application of rhamnolipid and surfactin for enhanced diesel biodegradation—Effects of pH and ammonium addition. *J. Hazard. Mater.* **2009**, *164*, 1045–1050. [CrossRef]
28. Xia, W.; Li, J.; Xia, Y.; Song, Z.; Zhou, J. Optimization of diesel oil biodegradation in seawater using statistical experimental methodology. *Water Sci. Technol.* **2012**, *66*, 1301–1309. [CrossRef]
29. Sathishkumar, M.; Binupriya, A.R.; Baik, S.-H.; Yun, S.-E. Biodegradation of crude oil by individual bacterial strains and a mixed bacterial consortium isolated from hydrocarbon contaminated areas. *Clean Soil Air Water* **2008**, *36*, 92–96. [CrossRef]
30. Verma, S.; Bhargava, R.; Pruthi, V. Oily sludge degradation by bacteria from Ankleshwar, India. *Int. Biodeterior. Biodegrad.* **2006**, *57*, 207–213. [CrossRef]
31. Pointin, X.; Tang, J.C.; Li, D.S.; Zhang, Q.M. Effect of salinity on the bioremediation of petroleum hydrocarbons in a saline-alkaline soil. *Lett. Appl. Microbiol.* **2012**, *55*, 210–217.
32. Gurav, R.; Lyu, H.; Ma, J.; Tang, J.; Liu, Q.; Zhang, H. Degradation of nalkanes and PAHs from the heavy crude oil using salt-tolerant bacterial consortia and analysis of their catabolic genes. *Environ. Sci. Pollut. Res.* **2017**, *24*, 11392–11403. [CrossRef] [PubMed]
33. Mnif, S.; Sayadi, S.; Chamkha, M. Biodegradative potential and characterization of a novel aromatic-degrading bacterium isolated from a geothermal oil field under saline and thermophilic conditions. *Int. Biodeterior. Biodegrad.* **2014**, *86*, 258–264. [CrossRef]

34. Al Farraj, D.A.; Alkufeidy, R.M.; Alkubaisi, N.A.; Alshammari, M.K. Polynuclear aromatic anthracene biodegradation by psychrophilic *Sphingomonas* sp., cultivated with Tween-80. *Chemosphere* **2021**, *263*, 128115. [CrossRef] [PubMed]
35. Chandran, P.; Das, N. Degradation of diesel oil by immobilized Candida tropicalis and biofilm formed on gravels. *Biodegradation* **2011**, *22*, 1181–1189. [CrossRef]
36. Greenwood, P.F.; Wibrow, S.; George, S.J.; Tibbett, M. Sequential hydro-carbon biodegradation in a soil from arid coastal Australia treated with oil under laboratory controlled conditions. *Org. Geochem.* **2008**, *39*, 1336–1346. [CrossRef]
37. Han, G.Q.; Wang, B.; Xu, W.H.; Chen, G.Q.; Wang, H.X.; Zhang, H.B.; Zhang, X.J.; Xiong, Z.T. Effects of heavy metal combined pollution soil microbial indicators and soil enzymatic activity. *J. Soil Water Conserv.* **2010**, *24*, 238–242.
38. Luo, Q.; Zhang, J.-G.; Shen, X.-R.; Sui, X.; Fan, Z.-Q. Characterization of a novel diesel oil-degrading *Pseudomonas* sp. strain F4. *Fresenius Environ. Bull.* **2013**, *22*, 689–697.
39. Kaczynska, G.; Borowik, A.; Wyszkowska, J. Soil dehydrogenases as an indicator of contamination of the environment with petroleum products. *Water Air Soil Pollut.* **2015**, *226*, 372. [CrossRef]
40. Palanisamy, N.; Ramya, J.; Kumar, S.; Vasanthi, N.S.; Chandran, P.; Khan, S. Diesel biodegradation capacities of indigenous bacterial species isolated from diesel contaminated soil. *J. Environ. Health Sci. Eng.* **2015**, *12*, 142. [CrossRef]

toxics

MDPI

Article

Adsorption Characteristics of Indigenous Chromium-Resistant *Aspergillus niger* Strain Isolated from Red Soil for Remediation of Toxic Chromium in Red Soil Environments

Jiwei Xu [1,2], Lumeng Li [1], Huabin Wang [1], Zhanyuan Gao [1], Chuanshu Wang [1,2], Rong Sun [1], Yong Zhang [1,2], Wumei Xu [1], Xiying Hou [1] and Rui Xu [1,2,*]

[1] School of Energy and Environment Science, Yunnan Normal University, Kunming 650500, China
[2] Provincial Key Laboratory of Rural Energy Engineering, Kunming 650500, China
* Correspondence: ecowatch_xr@163.com; Tel.: +86-13888083229

Abstract: The microbial treatment of soil has great potential to reduce chromium pollution. Here, an indigenous chromium-resistant *Aspergillus niger* strain (A1) was isolated and screened from heavily chromium-contaminated red soil in Yunnan Province, China using a traditional isolation method and a selective culture experiment. The molecular identification of A1 was achieved using 18S rRNA sequencing. The tolerance of the strain to toxic chromium was evaluated through pure laboratory culture. The adsorption effect and mechanism of A1 on chromium in red soil were further studied. The study concluded that A1 exhibited strong activity with exposure to 500 mg·L^{-1} Cr^{6+}. Chromium adsorption by *A. niger* occurred mainly through intracellular metabolism, surface complexations with EPS, and chemical reduction with -C=C-, -OXuH, NH_2, and -C=0. The optimized results showed that A1 had the best Cr^{6+} removal effect at pH 4, 40 °C, and a 60 h culture time. Compared with the inoculating of exogenous microbial agents, after inoculating A1 into the chromium-contaminated red soil, Cr^{6+} content was significantly reduced, and the high-toxicity chromium state (water-soluble and exchange states) decreased, whereas the low-toxicity chromium state (precipitation and residue states) increased. The results of red soil ITS also showed that the inoculation of indigenous microorganisms can better colonize the red soil. This study proves the feasibility of the application of indigenous *A. niger* to address red soil chromium pollution and provides a new idea and theoretical support for red soil remediation.

Keywords: heavy metal; mechanism; adsorption kinetics; microbial remediation; state transformation

Citation: Xu, J.; Li, L.; Wang, H.; Gao, Z.; Wang, C.; Sun, R.; Zhang, Y.; Xu, W.; Hou, X.; Xu, R. Adsorption Characteristics of Indigenous Chromium-Resistant *Aspergillus niger* Strain Isolated from Red Soil for Remediation of Toxic Chromium in Red Soil Environments. *Toxics* **2023**, *11*, 31. https://doi.org/10.3390/toxics11010031

Academic Editors: Rafael Clemente and Myung Chae Jung

Received: 11 December 2022
Accepted: 26 December 2022
Published: 29 December 2022

1. Introduction

With rapid industrial development and agricultural modernization, heavy metals in soil have become serious agricultural pollutants and ecological risk factors [1,2]. Even trace concentrations of heavy metal can be toxic to organisms [3], and after being absorbed by plants, heavy metals can be accumulated up the food chain, increasing foods' toxicity risks [4]. A soil environment is more complex than a solution environment, and it is difficult to determine toxicity directly via the valence of heavy metal elements. At the same time, heavy metal pollutants are difficult to extract from soil using conventional methods and cannot be degraded, except through transformation between different forms [5]. Thus, the deactivation of heavy metals into low-toxicity forms with low bioavailability is an important method to reduce the harm of heavy metals in soil.

Red soil is a common soil type that is mainly distributed in Africa, Asia, Oceania, South America, and tropical and subtropical areas of North America. Red soil formation occurs primarily due to the interaction of vigorous bioaccumulation, bioclimate factors, desiliconization, iron rich aluminization, and weathering processes [6]. Therefore, red soil is usually acidic to strongly acidic and rich in exchangeable acids [7]. Heavy metals (such as chromium and nickel) are prone to forming water-soluble and exchangeable states with extremely high mobility and toxicity in acidic environments, thereby increasing their ecological risks [8]. Yunnan Province in China is a concentrated area of red soil. The cultivated area of red soil is 49.36 million mu, accounting for about 60% of the cultivated area of the province and 73% of the land area. Yunnan is characterized by acidic red soil that is rich in iron and aluminum [9] and suitable for growing various crops; however, the high background value of heavy metals and the forms of heavy metals easily absorbed by plants under acidic conditions constitute potential safety hazards. Common heavy metals include chromium, cadmium, arsenic, and lead [10]. A survey showed that the enrichment concentrations of As and Cr in Yunnan red soil were higher than the background value and average concentration of national standard soil [11]. Additionally, red soil pollution has concealment and hysteresis. Air pollution and water pollution are generally relatively straight and can be perceived through the senses, and red soil pollution is often determined via soil sample analysis, crop testing, and even research on the impact of human and animal health. Harm caused by the generation of pollution from red soil usually takes longer, meaning red soil pollution is cumulative [12]. Compared with ordinary soil, it is more difficult for pollutants to migrate, diffuse, and dilute in soil. Chromium is one of the most toxic heavy metals, with serious carcinogenic and teratogenic properties [13]. Chromium in soil has different oxidation states: −2, −1, 0, +1, +2, +3, +4, +5, and +6 [14]. The valence states +3 and +6 are the main chromium forms in soil [15]. Cr^{3+} has high stability and low toxicity, is difficult to absorb by living organisms, is easily precipitated in soil environments [16], and mainly exists in precipitated and residual states. Cr^{6+}, a common oxidized form of chromium oxyanions, is highly toxic and highly soluble in water [17]. Cr^{6+} has a strong migration ability in soil and is the main form of the water-soluble and exchange states of chromium. Chromium in cultivated soil can be absorbed by crops, entering the human food chain, and pollutes water bodies through surface runoff and leaching, further propagating the contamination. Therefore, when treating soil chromium pollution, attention should be paid to adsorption methods and drivers that can transform chromium to a low-toxicity valence state, which reduces the toxicity and migration ability of chromium in soil.

At present, chromium pollution in soil is mainly treated using conventional physical or chemical methods, super-enriched plants, or microbial methods [18]. However, these traditional methods can be costly, impractical, and ineffective [19]; some even produce toxic sludge, adding more harmful chemicals to the environment [20]. Super-enriched plants can effectively extract heavy metal elements from soil; however, the long recovery time and accompanying land resource occupation problems limit their practical application [21]. Microbial remediation has emerged in recent years as a method for passivating or adsorbing heavy metals using natural and low-cost biomass resources, such as fungi, bacteria, algae, and moss. Compared with other methods, microbial remediation has low operating costs, high removal efficiency, and no secondary pollution [22]. Microbial remediation technology is the use of soil or water in the natural flora or the genetic engineering of bacteria, in the right environment, using microbial adsorption, precipitation, or redox, or through microbial respiration or metabolic pathways using toxic compounds as growth energy, so as to achieve or reduce the concentration of pollutant repair technology [23]. According to the current literature, there are two main mechanisms of microbial remediation of soil. First, EPS containing carboxyl (R-COO-), phosphate ($R\text{-}HPO_4^-$), sulfhydryl (R-SH), amine ($R\text{-}NH_3^+$), phenol ($R\text{-}C_6H_4OH$), and hydroxyl (R-OH) functional groups improves its bioavailability and uses the bonding of EPS surface functional groups with heavy metals to improve its biosorption capacity [24]. Another mechanism is microbial oxidation and reduction. The environmental toxicity of different valence states of heavy metals varies greatly [25]. Microorganisms convert toxic heavy metals into non-toxic and relatively unavailable forms through redox reactions. This is of great significance in the remediation process of heavy metal-contaminated soil, especially in environments that other remediation technologies cannot reach. [26]. Considering the physical and chemical properties of acidic red soil and its worldwide distribution, screening indigenous chromium-resistant strains from red soil could help treat heavy metal pollution in red soil environments. Additionally, exogenous microorganisms are often difficult to colonize in a soil environment, resulting in unstable remediation effects. For example, Jorquera found that the exogenous microorganisms Bacillus-like phosphobacteria (BLP) were difficult to keep alive and colonize in soil without bio-fertilizer [27]. Ghiglione found that the wild strain colonizes maize rhizosphere soil more easily than the exogenous strain [28]. Therefore, this study screened indigenous chromium-resistant strains from red soil and proved the feasibility of activating indigenous resistant microorganisms to reduce chromium toxicity in red soil. At the same time, the mechanism of chromium passivation was studied to lay a foundation for subsequent method improvement.

In this study, indigenous chromium-resistant *Aspergillus niger* strains were isolated and purified from red soil with severe chromium pollution, and strains with strong chromium adsorption capacity were screened. The effects of different operating variables on chromium adsorption capacity were studied by optimizing the conditions. The mechanism by which *A. niger* adsorbs chromium from red soil was studied by analyzing the changes in its surface functional groups before and after chromium adsorption and the energy spectrum of *A. niger*-adsorbed chromium. The ability of indigenous *Aspergillus niger* to reduce chromium availability in red soil was also evaluated by studying the morphological changes of chromium in red soil after inoculation with *Aspergillus niger*. Finally, we used metagenomic technology to study the effects of inoculation with indigenous *Aspergillus niger* on red soil microflora and evaluated the practicability and ecological risks of this method. Our findings support the feasibility of using indigenous strains to reduce chromium pollution in red soil, thereby improving the application of microbial remediation.

2. Materials and Methods

2.1. Culture Medium and Reagents

SDAY medium [29] was used for the isolation and preservation of microorganisms from red soil. The reagents diphenylcarbazide, potassium dichromate ($K_2Cr_2O_7$), glucose, agar, peptone, and ethanol were analytically pure. Fresh ddH_2O was used for the experimental work.

2.2. Preparation of Chromium-Contaminated Red Soil

The red soil used in this experiment was collected from farmland in the Chenggong District, Kunming City, Yunnan Province, China. After natural air drying, the soil sample was screened twice through 20-mesh screens. The organic matter content in the red soil was 5.4%, pH was 4.5, and no chromium was detected. The soil used for strain screening was weighed, 1 $g \cdot kg^{-1}$ Cr^{6+} was added, and the soil was then allowed to rest for 60 d. The chromium content of the red soil was 922.34 $mg \cdot kg^{-1}$ when the strains were tested and screened. Sterile soil was prepared through sterilization at high temperature, followed by the addition of 400 $mg \cdot kg^{-1}$ of Cr^{6+}.

2.3. Isolation of Chromium-Resistant Strains

Ten grams of soil was weighed for strain screening, and placed into a 250 mL conical flask, and 90 mL of ddH_2O was added to it. The mixture was then shaken at 100 $r \cdot min^{-1}$ for 15–30 min and mixed thoroughly to obtain a soil suspension. Under sterile conditions, 1 mL of the soil suspension was removed and 9 mL of ddH_2O was added. This mixture was mixed well; 1 mL of the mixture was removed and 9 mL of ddH_2O was added, and this was repeated five times to obtain 10^{-2}, 10^{-3}, 10^{-4}, and 10^{-5} dilutions. Then, 0.1 mL of each dilution of each concentration was taken, spread on SDAY plates, and incubated at 37 °C for 5 d. During the culturing period, colonies with evident differences were selected, scribed, and purified on the improved SDAY plate. After several repetitions, the strain was preserved in test tube slant medium at 4 °C.

2.4. Selection of Strains with Chromium Adsorption Capacity

The fungi obtained via screening were numbered A1–A7, and the bacteria were numbered B1–B3. The screened chromium-resistant strains were cultured in a liquid medium containing 50 mg·L^{-1} or 500 mg·L^{-1} Cr^{6+} for 10 d. The Cr^{6+} concentration in the culture medium was measured at the end of the culturing period, and the Cr^{6+} adsorption rates of different strains were calculated. The strain with the strongest adsorption capacity was screened.

2.5. Identification and Phylogenetic Analysis of Strains

The screened fungi were identified via 18S rRNA sequencing. DNA was extracted according to standard procedures and analyzed via agarose gel electrophoresis. Fungal ITS1 and ITS4 fragments were amplified through polymerase chain reaction (PCR). A 25 µL PCR reaction system containing 5 µL strain DNA (50 mg·L^{-1}) with ITS1 and ITS4 primers, 12.5 µL Taq Master Mix, and 7.5 µL RNase-Free Water was used and mixed well, and then, 5 µL mineral oil seal was applied. After denaturation (94 °C, 4 min), PCR was performed for 35 cycles under the following thermal conditions: (1) denaturation (94 °C, 45 s), (2) annealing (55 °C, 35 s), and (3) extension (72 °C, 1 min). The amplified products were sequenced after the specified PCR cycle. Fungal sequences were compared with other species sequences using the NCBI blast tool. Phylogenetic analysis between the amplified product and its relatives was carried out using MEGA11.

2.6. A. niger Colony Characteristics under Chromium Stress

The screened *A. niger* was inoculated on the modified SDAY plate containing Cr^{6+} concentrations of 0, 50, 100, 200, and 500 mg·L^{-1} using single-spore isolation, with three biological replicates. The diameters of single colonies under different Cr^{6+} concentrations were measured at 24 h intervals. *A. niger* was placed into a liquid medium at 30 °C and 150 r·min^{-1} for 24 h. Then, the liquid medium was inoculated into the liquid medium with Cr^{6+} concentrations of 0 (control (CK)), 100, and 500 mg·L^{-1} at a rate of 2%. The spore yield was measured every 3 h. After the experiment, *A. niger* bodies (in 500 mg·L^{-1} Cr^{6+}) were filtered and collected. Fourier transform infrared spectroscopy (FTIR) was used to analyze changes in the surface functional groups of *A. niger* before and after contact with Cr^{6+}. The energy spectrum of *A. niger* adsorbed chromium was analyzed via X-ray photoelectron spectroscopy (XPS).

2.7. Optimization Studies

After inoculating *A. niger* in liquid medium at 30 °C and 150 r·min^{-1} for 24 h, washing with sterile water, and filtering twice with four layers of gauze, a 6.7·10^{6}/mL spore suspension was prepared for later use. For the optimization study, the following experiments both added 5 mL spore suspension (6.7·10^{6}/mL) and were carried out in Erlenmeyer flasks under different conditions: pH (4–9), culture temperature (10–50 °C), contact time (18–72 h), and initial Cr^{6+} concentration (10–150 mg·L^{-1}). The concentration of unabsorbed Cr was calculated as follows:

$$Cr^{6+}\ removal\ rate = (c_0 - c_e)/c_0 \times 100\%, \tag{1}$$

$$Cr^{6+}\ ratio\ in\ solution = c_e/c_t \times 100\% \tag{2}$$

where c_o, c_e, and c_t are the initial and final Cr^{6+} concentrations (mg·L^{-1}) and total Cr concentrations (mg·L^{-1}) in solution, respectively.

The unabsorbed chromium was examined using atomic absorption spectroscopy. The concentration of Cr^{6+} was examined using the diphenylcarbazide spectrophotometric method.

2.8. Kinetic and Isotherm Validation

The relationship between the Cr^{6+} distributions in solution and *A. niger* was studied via adsorption kinetic analysis. Adsorption characteristic types were verified using pseudo-first-order models and pseudo-second-order models. The adsorption process was verified using models including Langmuir models, Freundlich models, and intra-particle diffusion models. Using the model-fitting tool in Origin 2021, the isotherm parameters and correlation coefficient R^2 were calculated. Kinetic analysis was used to explain the mechanism of Cr^{6+} removal by *A. niger*. Compared with the Langmuir isotherm, the Freundlich isotherm described multilayer and heterogeneous adsorption. Therefore, the adsorption type of chromium ions by *A. niger* could be determined by comparing the fitting degree of the following five equations:

Pseudo-first-order:

$$ln(q_e - q) = lnq_e - K_1 t, \tag{3}$$

where qe (mg/g) is the equilibrium adsorption capacity; qt (mg/g) is the adsorption capacity at t time; and K_1 is the pseudo-first-order constant (h^{-1}).

Pseudo-second-order:

$$\frac{1}{q_e - q} = \frac{1}{q_e} + K_2 t, \tag{4}$$

where K_2 is the pseudo-second-order constant (g/mg/h).

Langmuir model:

$$q_e = q_{max} \times \frac{K_L C_e}{1 + K_L C_e}, \tag{5}$$

where q_{max} is the Langmuir monolayer capacity (mg/g) and K_L is the Langmuir constant (L/mg).

Freundlich model:

$$q_e = K_F C_e^{1/n}, \tag{6}$$

where n is the Freundlich exponent and K_F is the Freundlich constant ((mg/g)/(L/mg)$^{(1/n)}$).

Intra-particle diffusion model (Weber-Morris equation) (Martins et al., 2015):

$$q_t = K_i t^{1/2} + C_i, \tag{7}$$

where K_i ($mg \cdot g^{-1} \cdot h^{-1/2}$) is the intra-particle diffusion rate of the adsorption process at any stage and C_i is the intercept describing the effects of the boundary layer thickness.

2.9. Effects of A. niger Inoculation Amount on Chromium State Transformation

A total of 200 g of the chromium-containing soil was weighed into a 250 mL beaker and inoculated with 3 mL and 10 mL (M3 and M10, respectively) of *A. niger* culture solution ($6.7 \cdot 10^6$/mL spore suspension) that had been cultivated for 48 h; each treatment was repeated three times. During the experiment, sterile water was added to maintain soil moisture at ~60%. Taking sterile soil as a blank control (CK) and sterile soil inoculated with 3 mL commercial microbial agent ($6.7 \cdot 10^6$/mL spore suspension) as a control (B-CK), 10 g soil samples were taken at 0, 5, 10, 20, and 30 d after inoculation.

After freeze-drying, crushing, and screening, 5 g of the collected soil samples were placed in a 250 mL beaker; then, 50 mL alkaline extraction solution (Na_2CO_3-NaOH mixed solution, pH > 11.5) and 400 mg $MgCl_2$ and 0.5 mL K_2HPO_4-KH_2PO_4 buffer solution were added. After sealing with polyethylene film, it was stirred and heated to 95 °C for 60 min. After cooling and filtration, pH was adjusted to 7.5 ± 0.5 with HNO_3, and the concentration of Cr^{6+} was determined via atomic absorption spectrophotometry.

The content of different forms of Cr in soil was determined using the sequential extraction method. Firstly, the 5 g red soil was fully soaked, and the concentration of water-soluble chromium in the leached solution was detected after extraction and filtration. Then, for the exchange, precipitation (Fe-Mn combined), and organic-bound states of chromium, 1 $mol \cdot L^{-1}$ NH_4Ac, 2 $mol \cdot L^{-1}$ HCl, and 5% H_2O_2-2 $mol \cdot L^{-1}$ HCl were used as the extraction agents, respectively. Finally, the residual state was determined after the 0.2 g residual solid was fully digested by the mixed acid (1.19 g/mL HCl: 1.42 g/mL HNO_3: 1.49 g/mL HF = 3:6:2) in a Microwave Digestion System. The above steps were completed in an artificial climate chamber at 25 $\pm$ 0.1 °C. The ratio of each extraction agent to soil was 5:1. The single extraction time was 4 h (oscillating for 2 h, standing for 2 h). The unabsorbed chromium was examined using atomic absorption spectroscopy; the concentration of Cr^{6+} was examined using the diphenylcarbazide spectrophotometric method.

2.10. Analysis of Red Soil Fungus Diversity after A. niger Inoculation

Total DNA was extracted from the red soil treated with CK and M10 on the 30th day. PCR amplification was performed, and the products were purified, quantified, and homogenized to form a sequencing library. The library was sequenced using Illumina Novaseq 6000 after passing quality inspection, and the original image data files obtained were transformed into Sequenced Reads via Base Calling analysis. Sequenced Reads were filtered by low-quality filtering length; then, OTUs/ASVs were obtained via clustering and denoising. Species classification was performed according to the sequence composition of the feature, and a Wayne diagram was drawn.

3. Results

3.1. Isolation and Identification of Chromium-Resistant Strains

Ten strains with high activity were screened from chromium-contaminated soil. The results obtained after screening with liquid medium containing Cr^{6+} are provided in Table 1. A1 strains showed stronger activity than the other strains under different chromium concentrations. When the Cr^{6+} concentration was 477.25 mg·L^{-1}, the A1 adsorption rate was 44.31%; when the Cr^{6+} concentration was 48.9 mg·L^{-1}, the A1 adsorption rate was as high as 65.75%. The sequencing results of *A. niger* are shown in Figure 1a. The phylogenetic analysis results are presented in Figure 1b.

```
>A1
GTGCGGGTCCTTTGGGCCCAACCTCCCATCCGTGTCTATTGTACCCTGTTGCTTCGGCGGGCCCGC
CGCTTGTCGGCCGCCGGGGGGGCGCCTCTGCCCCCCGGGCCCGTGCCCGCCGGAGACCCCAACA
CGAACACTGTCTGAAAGCGTGCAGTCTGAGTTGATTGAATGCAATCAGTTAAAACTTTCAACAAT
GGATCTCTTGGTTCCGGCATCGATGAAGAACGCAGCGAAATGCGATAACTAATGTGAATTGCAGA
ATTCAGTGAATCATCGAGTCTTTGAACGCACATTGCGCCCCCTGGTATTCCGGGGGGCATGCCTGT
CCGAGCGTCATTGCTGCCCTCAAGCCCGGCTTGTGTGTTGGGTCGCCGTCCCCCTCTCCGGGGGG
ACGGGCCCGAAAGGCAGCGGCGGCACCGCGTCCGATCCTCGAGCGTATGGGGCTTTGTCACATG
CTCTGTAGGATTGGCCGGCGCCTGCCGACGTTTTCCAACCATTCTTTCCAGGTTGACCTCGGATCA
GGTAGGGATACCCGCTGAACTTAAGCATATCATAAAAGCGGAGGAAATCATTACCGAGTGCGG
```

(a)

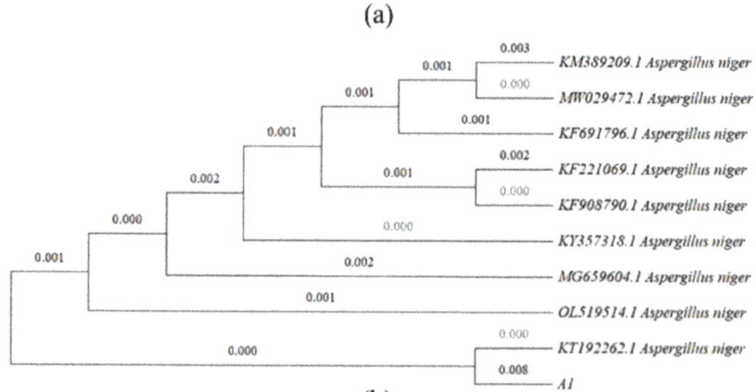

(b)

Figure 1. (**a**) Sequencing result of A1. (**b**) Phylogenetic tree of A1.

Table 1. Screening of chromium tolerance of *A. niger* strains.

Strain	Initial Cr^{6+} Concentration: 50 $mg \cdot L^{-1}$			Initial Cr^{6+} Concentration: 500 $mg \cdot L^{-1}$		
	Cr^{6+} Concentration after Culture (mg/L)	Adsorption Capacity (mg/L)	Adsorption Rate (%)	Cr^{6+} Concentration after Culture (mg/L)	Adsorption Capacity (mg/L)	Adsorption Rate (%)
CK	48.9	—	—	477.25	—	—
A1	10.93	37.97 ± 0.73 h	77.64	265.8	211.46 ± 0.94 f	44.31
A2	25.51	23.39 ± 0.44 g	47.83	353.05	124.29 ± 0.26 e	26.02
A3	34.05	14.85 ± 0 f	30.36	446.75	30.28 ± 0.18 d	6.39
A4	37.42	11.48 ± 0.11 d	23.47	453.5	23.81 ± 0.12 c	4.98
A5	43.17	5.73 ± 0.29 b	11.71	477.22	—	—
A6	36.31	12.59 ± 0.12 e	25.74	457	20.82 ± 0.26 b	4.24
A7	44.99	3.91 ± 0.09 a	7.99	477.3	—	—
B1	36.57	12.33 ± 0.08 e	25.21	460	17.35 ± 0.88 a	3.61
B1	34.15	14.75 ± 0.11 f	30.16	447	30.29 ± 0.11 d	6.33
B3	40.65	8.25 ± 0.13 c	16.87	458.85	18.50 ± 0.35 a	3.86

Note: Data are mean values ± SE. Different lowercase letters in the same column indicate significant differences determined by Duncan's multiple range test ($p < 0.05$). CK—control.

3.2. *A. niger* Colony Characteristics under Chromium Stress

The *A. niger* community growth rates under stress from different chromium concentrations are shown in Table 2. In the first 24 h of culturing, the effect of chromium stress on *A. niger* colony growth was not evident, whereas it was significantly inhibited after 48 h. When the culturing time was 96 h, the colony diameter in the CK group reached 70.13 mm; the difference between treatments was significant ($p < 0.05$). When Cr^{6+} concentration was 500 $mg \cdot L^{-1}$, the diameter of a single colony was 41.03 mm. Although high Cr^{6+} concentrations evidently inhibited *A. niger* growth, strain A1 still showed strong activity.

Table 2. Single colony diameters of *A. niger* strain A1 during exposure to different Cr^{6+} concentrations.

Initial Chromium Concentration (mg/L)	Single Colony Diameter (mm)			
	24 h	48 h	72 h	96 h
CK	4.50 ± 0.29 b	17.16 ± 0.17 d	30.08 ± 0.08 e	70.13 ± 0.08 e
50	3.83 ± 0.16 ab	16.16 ± 0.16 c	27.25 ± 0.14 d	65.18 ± 0.09 d
100	3.91 ± 0.08 ab	15.25 ± 0.14 b	24.13 ± 0.08 c	58.56 ± 0.29 c
200	4.50 ± 0.22 a	15.00 ± 0.14 b	22.21 ± 0.06 b	50.80 ± 0.60 b
500	3.66 ± 0.22 a	14.00 ± 0.00 a	19.51 ± 0.13 a	41.03 ± 1.18 a

Note: Data are mean values ± SE. Different lowercase letters in the same column indicate significant differences determined by Duncan's multiple range test ($p < 0.05$). CK—control.

Figure 2a shows the effects of different Cr^{6+} concentrations on *A. niger* sporulation characteristics. When chromium concentration was low, sporulation ability did not show significant change in the first 24 h of culturing; however, in the 500 $mg \cdot L^{-1}$ Cr^{6+} treatment, sporulation ability was inhibited since the beginning of culturing. The experimental group exposed to 500 $mg \cdot L^{-1}$ Cr^{6+} entered the stable spore growth period at 36 h, and other treatments reached equilibrium after 39 h. During the whole process, A1 showed strong tolerance to both low and high Cr^{6+} concentrations.

Figure 2b shows the peaks in the Cr 2p spectra of *A. niger* after contact with 50 $mg \cdot L^{-1}$ Cr^{6+}. The peaks were located at 576.9 eV and 580.0 eV, indicating that Cr^{6+} and Cr^{3+} coexisted on the surface of *A. niger*. The results of peak separation showed that the two strong peaks were formed through the combination of five peaks. The peaks at binding energies of 587.5 eV and 577.9 eV corresponded to the existence of Cr^{6+}, and the peak area accounted for 45.44%. The peaks at binding energies of 586.5 eV, 577.0 eV, and 576.2 eV corresponded to the existence of Cr^{3+}, and the peak area accounted for 54.56%. XPS analysis directly proved that the *A. niger* membrane system had a fixation effect on Cr^{6+}, and further indicated that *A. niger* has the ability to reduce Cr^{6+} to Cr^{3+}. After 2d contact, more than half of the Cr^{6+} fixed by the cell surface was reduced to Cr^{3+}.

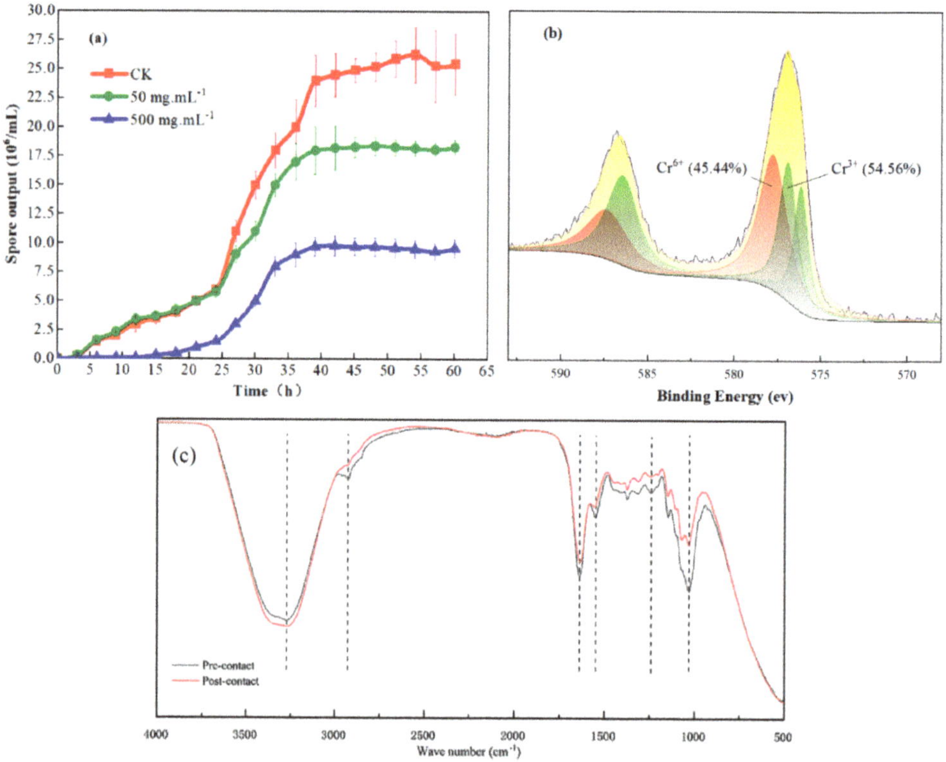

Figure 2. (**a**) Sporulation curves of *Aspergillus niger* strain A1 in chromium solution. CK—control. (**b**) X-ray photoelectron spectroscopy patterns of *A. niger* after 24 h of exposure to Cr^{6+}. (**c**) Fourier transform infrared spectroscopy analysis of *A. niger* pre-contact or post-contact with Cr^{6+}.

The changes in functional groups on the surface of *A. niger* between post-contact and pre-contact with 50 mg·L^{-1} Cr^{6+} ions were determined via FTIR at wavelengths of 400–4000 cm^{-1} (Figure 2c). The peak difference in the two curves indicates the change in functional groups on the surface of *A. niger*. The peak near 3400 cm^{-1} was the stretching vibration of –OH. The peak at 3270 cm^{-1} shows the presence of the hydroxyl group and the stretching vibration of—NH_2. The peak between 2930, 1680–1040 cm^{-1} shows the stretching vibration of -CH_2, C=O, and C-O, respectively. The peak intensity of post-contact was higher than that of pre-contact, indicating that hydroxyl groups were barely involved in the chromium adsorption process. After contact, the peak at 2930 cm^{-1} disappeared, implying that the reaction consumed the methylene group (>CH_2) from the surface of *A. niger*. The loss of peaks at 1640 cm^{-1} and 1550 cm^{-1} suggest that the glycoproteins on *A. niger*'s surface may have been involved in the chromium adsorption process. The peaks at 1000–1500 cm^{-1} were characteristic of carboxyl and amine groups. Consistent with the result of the XPS, FTIR analysis also revealed that the *A. niger* cell surface had great potential for Cr^{6+} adsorption. A variety of reductive groups on the membrane system are involved in the fixation and reduction of Cr^{6+}.

3.3. Optimization of Adsorption Conditions

When the culture temperature was 30 °C and the initial Cr^{6+} concentration was 50 mg·L^{-1}, the Cr^{6+} removal rate decreased continuously and the Cr^{6+} ratio in solution increased continuously as pH increased from 4 to 9 (Figure 3a). The adsorption rate of Cr^{6+} and the reduction rate of Cr^{3+} at pH 4 were the highest (reaching 86% and 35%), and the percentage of Cr fixed by membrane and intracellular adsorption in *A. niger* reached 51%. According to the experimental results, *A. niger*(A1) had a better removal effect on Cr^{6+} in an acidic environment; at the same time, A1 can effectively promote the transformation of Cr^{6+} to low-toxicity Cr^{3+}.

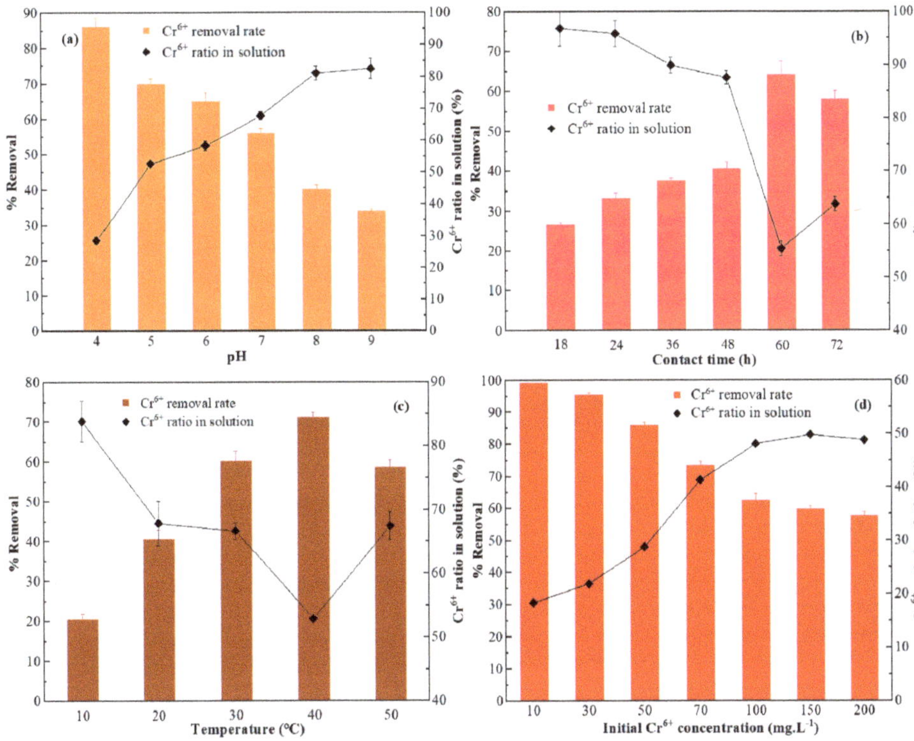

Figure 3. Influences of (**a**) pH, (**b**) contact time, (**c**) temperature, and (**d**) initial Cr^{6+} concentration on Cr^{6+} removal rate and Cr^{6+} ratio in the remaining solution.

When the culture temperature was 30 °C, pH was 7, and the initial Cr^{6+} concentration was 50 mg·L^{-1}, the adsorption effect of strain A1 on Cr^{6+} increased with the increase in culture time within 0–60 h. (Figure 3b). The adsorption effect was the largest when the contact time was 60 h, at which time, the Cr^{6+} removal rate peaked at 64% and the Cr^{6+} ratio in solution reduced to 36%, and subsequently, the Cr^{6+} removal rate began to decline. At 60 h, *A. niger* adsorbed and fixed about 35% Cr, and the intracellular adsorption basically reached dynamic equilibrium.

When pH was 7 and the initial Cr^{6+} concentration was 50 mg·L^{-1}, the Cr^{6+} removal rate increased significantly when the temperature was increased from 10 °C to 40 °C (Figure 3c). When the temperature was 40 °C, the Cr^{6+} removal rate and adsorption rate of *A. niger* reached maximum values of 71% and 45.2%, respectively, and the Cr^{6+} ratio in solution reduced to 29%; subsequently, the Cr^{6+} removal rate began to decline as the temperature continued to increase. The optimum temperature range for Cr^{6+} adsorption by strain A1 was 30–40 °C. An increase in temperature within the suitable range can increase the Cr^{6+} removal capacity and the adsorption capacity of cells, but temperatures that are too high or too low will damage the cell structure or inhibit the activity of cells, resulting in a decrease in adsorption capacity.

When the culture temperature was 30 °C, pH was 4, and the initial Cr^{6+} concentration was 10 mg·L^{-1}, the Cr^{6+} removal rate and adsorption rate of *A. niger* reached 99.1% and 95.1%, respectively, i.e., the lower the concentration, the higher the adsorption rate (Figure 3d). When Cr^{6+} concentration was 200 mg·L^{-1}, the Cr^{6+} removal rate and adsorption rate of *A. niger* were 57.5% and 12.9%, respectively. Without a carbon source, the higher the initial Cr^{6+} concentration was, the more Cr^{6+} was converted into Cr^{3+} by *A. niger*. At the same time, the adsorption rate of *A. niger* to Cr decreased with the increase in the initial Cr^{6+} concentration in the solution, but there was no significant difference in the total adsorption amount. Therefore, with a thicker solution, the biomass of *A. niger* should be increased to ensure the adsorption effect.

3.4. Adsorption Kinetics and Isotherm Analyses of A. niger in Chromium Solution

To further determine the chromium adsorption capability, Langmuir (Equation (5)) and Freundlich (Equation (6)) isotherms were used to study the equilibrium data of adsorption at different temperatures (Figure 4a).

All values of the two isotherm constants are given in Table 3. The results indicate that the Langmuir isotherm model was most suitable for Cr^{6+} ion elimination onto *A. niger*. The Langmuir isotherm had a higher correlation coefficient (R^2) and a better fitting effect. This indicated that the adsorption process of Cr^{6+} in water by *A. niger* was mainly monolayer adsorption. The Langmuir monolayer capacity ($q_e$max) for Cr^{6+} at 40 °C was 117.71 mg·g^{-1} when the initial Cr^{6+} concentration was 50 mg·L^{-1}.

Table 3. Fitting constants of two isotherm models for different adsorbents.

Temperature (°C)	Langmuir Model			Freundlich Model		
	$q_e$max(mg/L)	K_L(L/mg)	R^2	K_F((mg/g)/(L/mg)$^{(1/n)}$)	n	R^2
20	95.13	0.0085	0.9929	1.7261	1.3288	0.9810
30	104.80	0.0083	0.9882	1.9242	1.3129	0.9736
40	117.71	0.0076	0.9879	1.9676	1.2878	0.9757

The experimental data for the kinetic experiments were fitted with two adsorption kinetics models, i.e., pseudo-first-order and pseudo-second-order models, to explain the adsorption process between *A. niger* and Cr^{6+} (Figure 4b). The calculated kinetic parameters and R^2 values are given in Table 4. Under different initial Cr^{6+} concentrations, the pseudo-second-order model had a higher correlation coefficient (R^2) and a better fitting effect. The calculated values ($q_e$cal) of the pseudo-second-order model were 38.353 mg/g and 57.282 mg/g when the initial Cr^{6+} concentrations were 50 mg·L^{-1} and 100 mg·L^{-1}, respectively. These results revealed that adsorption kinetics could be described by pseudo-second-order kinetic models, suggesting that the process controlling the adsorption rate was complicated and may have been physico-chemical composite adsorption dominated by chemisorption.

Table 4. Fitting parameter values of the two kinetic models.

Initial Cr^{6+}	$q_e$exp	Pseudo-First-Order Model			Pseudo-Second-Order Model		
Concentration (mg/L)	(mg/g)	$q_e$cal (mg/g)	K_1 (h^{-1})	R^2	$q_e$cal (mg/g)	K_2 (g/mg/h)	R^2
25	21.2	19.83	0.0258	0.9927	22.941	0.00165	0.9974
50	35.1	32.25	0.0188	0.9825	38.353	0.00057	0.9934
100	54.0	49.45	0.0212	0.9745	57.282	0.00046	0.9892

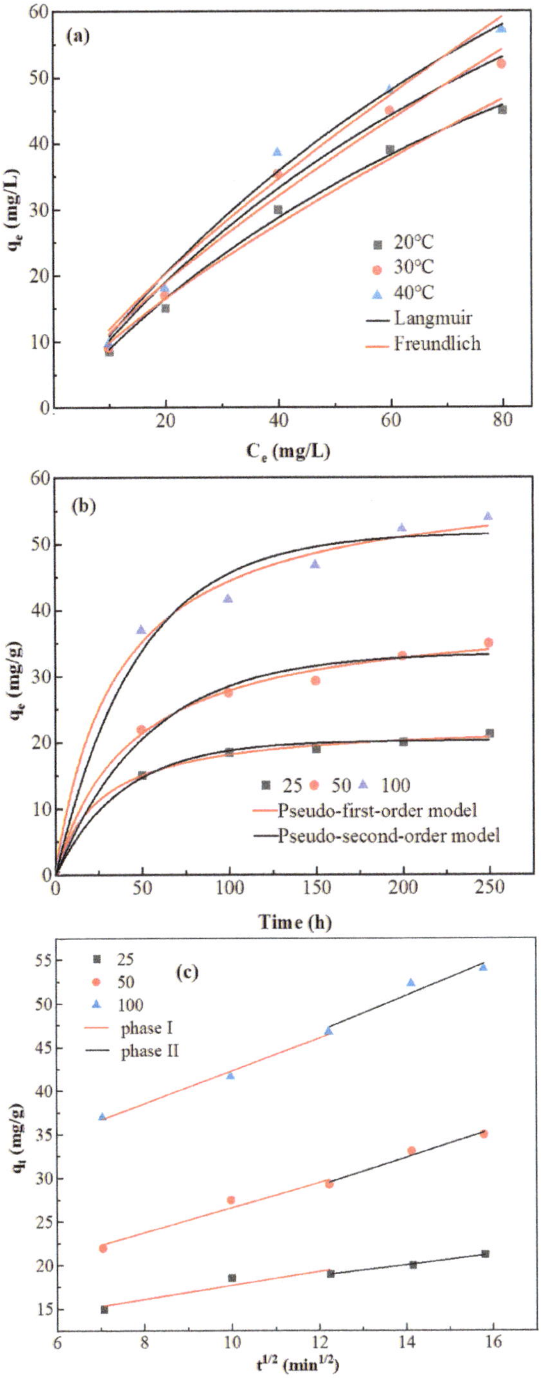

Figure 4. (**a**) Adsorption isotherm study for *Aspergillus niger*. (**b**) Kinetic analysis for *Aspergillus niger*. (**c**) The intra-particle diffusion model of *Aspergillus niger*.

In the intra-particle diffusion model, the straight lines did not pass through the origin (Figure 4c), indicating that intra-particle diffusion was not the only mechanism controlling the adsorption rate. The intra-particle diffusion model could fit two adsorption stages with different initial chromium concentrations. Two different adsorption rate constants for stepwise adsorption were obtained (Table 5). The results suggested that the diffusion rate of Cr^{6+} into *A. niger* was very slow and was restricted by other factors.

Table 5. Fitting constant values of intra-particle diffusion model.

Initial Cr^{6+}	Intra-Particle Diffusion Model			
Concentration (mg/L)	K_{i1} (g/mg/min1/2)	R^2	K_{i2} (g/mg/min1/2)	R^2
25	0.79366	0.78666	0.61527	0.98419
50	1.43346	0.9141	1.60844	0.95323
100	1.87709	0.98207	2.04	0.86863

3.5. Remediation of Chromium Pollution in Red Soil by A. niger (Strain A1)

The change in soil chromium content in each treatment was continuously measured for 30 d (Figure 5a). Compared with CK, *A. niger* inoculation significantly promoted reductions in chromium content in red soil. Compared with B-CK, improving the biomass of indigenous *A. niger* achieved a better removal effect of Cr^{6+} than inoculation with the same amount of exogenous microbial agent. After only 2 d of treatment, the Cr^{6+} content in M3 and M10 soils had decreased to 180 mg·kg^{-1} and 175 mg·kg^{-1}, respectively. The results showed that the amount of *A. niger* positively correlated with the remediation effect of chromium pollution in red soil. After 30 d of microbial remediation, the chromium concentration in red soil decreased from 300 mg·kg^{-1} to < 50 mg·kg^{-1}, demonstrating a strong remediation effect.

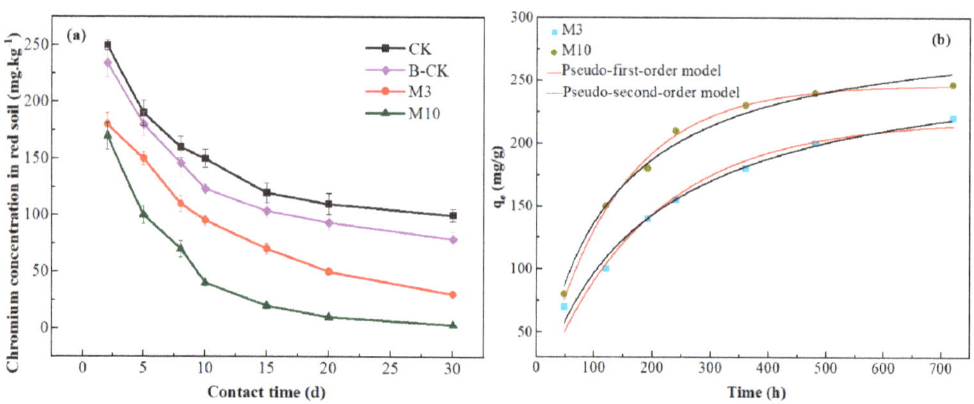

Figure 5. (**a**) Cr^{6+} concentration in red soil with increased contact time. CK—control; B-CK—commercial microbial agent. (**b**) Kinetic analysis for Cr^{6+}–*Aspergillus niger*.

The chromium adsorption curves of *A. niger* were calculated using two kinetic models (Figure 5b), and the calculated kinetic parameters and R^2 values are given in Table 6. The pseudo-second-order model had a higher correlation coefficient (R^2) and a better fitting effect in M3. However, the pseudo-first-order model had a higher correlation coefficient (R^2) and a better fitting effect in M10, and the theoretically calculated value ($q_e$cal) of the pseudo-first-order model was closer than the experimental results ($q_e$exp), but lower. The $q_e$cal calculated by the pseudo-second-order model was much higher than the $q_e$exp. The results showed that the removal process of Cr^{6+} may be a physico-chemical composite process dominated by physical adsorption.

Table 6. Fitting parameter values of the two kinetic models.

Process Mode	$q_e$exp (mg/g)	Pseudo-First-Order Model			Pseudo-Second-Order Model		
		$q_e$cal (mg/g)	K_1 (h^{-1})	R^2	$q_e$cal (mg/g)	K^2 (g/mg/h)	R^2
M3	220	208.14	0.00539	0.96344	259.66	0.0002	0.98527
M10	247	242.03	0.0076	0.99297	284.04	0.00028	0.98461

The intra-particle diffusion model was used to fit the data of the two experimental groups, and two adsorption stages were obtained. The adsorption rate constants for stepwise adsorption were obtained (Table 7). The straight lines did not pass through the origin, indicating that the intra-particle diffusion model was not the only mechanism controlling the adsorption rate. In the first stage, the internal diffusion rates of M3 and M10 were similar, whereas the internal diffusion rate of M3 was higher in the second stage.

Table 7. Fitting constant values of the intra-particle diffusion model.

Process Mode	Intra-Particle Diffusion Model			
	K_{i1} (g/mg/min1/2)	R^2	K_{i2} (g/mg/min1/2)	R^2
M3	12.30598	0.9819	5.7385	0.96802
M10	12. 87473	0.94838	3.17912	0.84897

After the screened strain A1 was inoculated in the red soil with severe chromium pollution (300 mg·kg^{-1} chromium), samples were taken every 5 d, and the morphological distributions of chromium in the red soil under various treatments were continuously measured for 30 d using a sequential extraction method (Figure 6a). The forms of chromium in soil were divided into: water-soluble, exchange, precipitation (Fe-Mn combined), organic-bound, and residue states. Among them, the water-soluble and exchange states have strong migration ability, are highly toxic, and are easily absorbed by organisms, mainly Cr^{6+}; meanwhile, the precipitation, organic-bound, and residue states are more stable, their direct use by organisms is more difficult, and most of them are Cr^{3+}. At the beginning of the experiment, water-soluble and exchange states were the main components of the three treatment groups. In M3 treatment, highly toxic forms accounted for >60%, while in M10 treatment, they accounted for >65%. The transformation rate of chromium from high-toxicity to low-toxicity forms in M3 and M10 was significantly increased after inoculation with *A. niger*; the transformation rate positively correlated with the inoculation amount. After only 5 d of treatment, the proportion of the highly toxic form in M3 and M10 was reduced to <50%, which was lower than that in CK. After 30 d of treatment, the proportion of the highly toxic form in CK was still nearly 30%, whereas it was <10% in M3 treatment and <5% in M10 treatment. The contents of various forms of chromium in red soil on the 30th day are shown in Figure 6b. After 30 d of treatment, the concentration of chromium in a residual state in CK increased to 170 mg·kg^{-1}. The concentration of chromium in a residual state increased by 16%, and the concentration in the precipitation state increased by 10% after 30 d of treatment. However, in M3 and M10 inoculated with *A. niger*, the concentrations of the residual and precipitation states increased significantly; the concentrations of the residual state increased to 180 mg·kg^{-1} and 158 mg·kg^{-1} after 30 d, while those of the precipitation state increased to 80 mg·kg^{-1} and 150 mg·kg^{-1}, and those of the organic-bound state increased to 60 mg·kg^{-1} and 80 mg·kg^{-1}, respectively. During the experimental period, by increasing indigenous *A. niger* biomass, the Cr in red soil was transformed from a highly toxic state to a less toxic state; toxic Cr^{6+} was largely removed and Cr migration capacity greatly decreased.

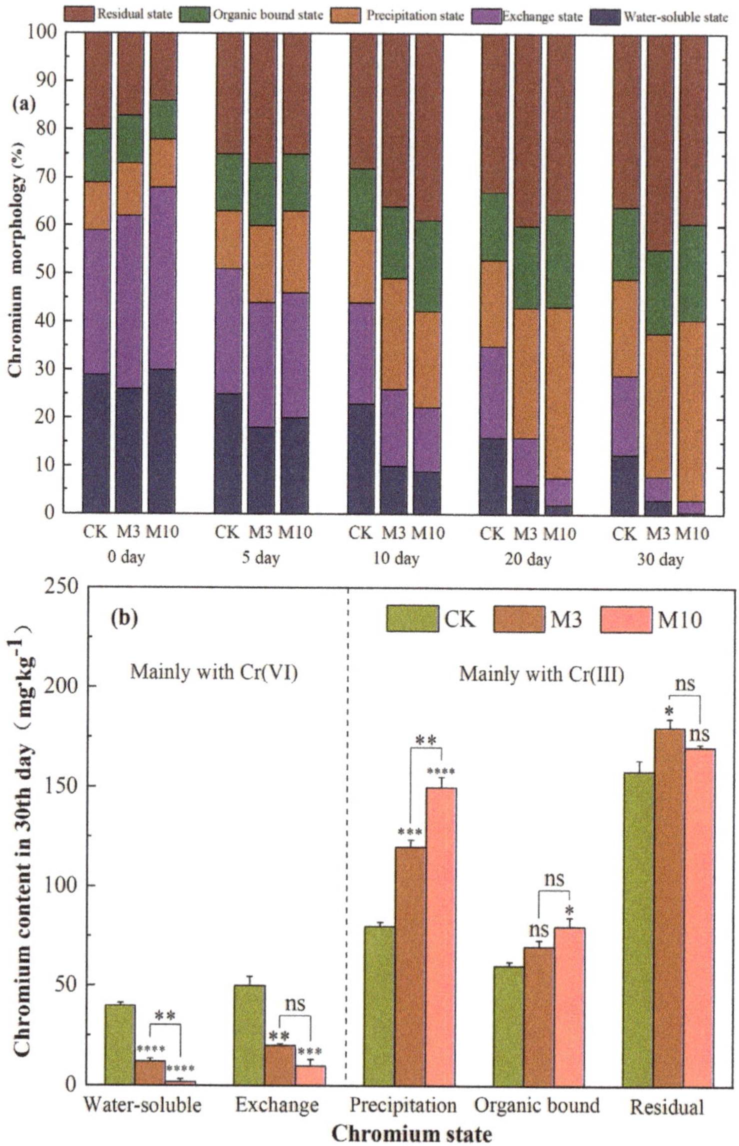

Figure 6. (**a**) *Aspergillus niger* drives the transformation of chromium form in red soil. (**b**) Chromium distribution in soil after *A. niger* inoculation. CK—control. (****, $p < 0.0001$; ***, $0.0001 < p < 0.001$; **, $0.001 < p < 0.01$; *, $0.01 < p < 0.05$; ns, no significant difference).

High-throughput sequencing was used to analyze soil microbial diversity in M10 and CK treatments after 30 d of *A. niger* inoculation. The results showed that the inoculation of *A. niger* significantly increased the abundance of *A. niger* in red soil with severe chromium pollution (Figure 7a). At the same time, after *A. niger* was inoculated, the total amount of fungi in the red soil did not change much, but the species changed greatly (Figure 7b).

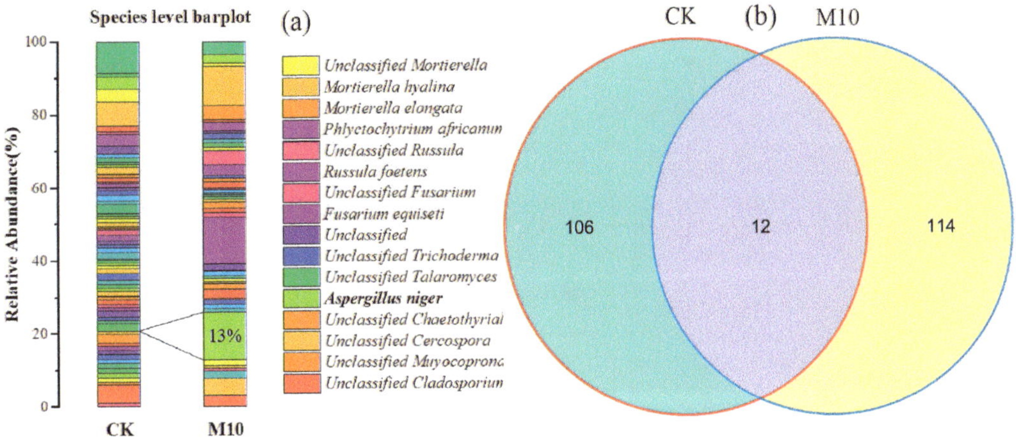

Figure 7. (**a**) Variation of soil fungi relative abundance at species level after 30 d of *A. niger* inoculation. (**b**) OTUs—Venn diagram analysis.

4. Discussion

4.1. Significance of Screening Indigenous Chromium-Tolerant Microorganisms to Treat Heavy Metal Pollution in Red Soil

In recent years, red soil has been seriously degraded, increasing ecological problems [30], and the application of conventional physical and chemical methods has had limited success. In contrast, microbial remediation is environmentally friendly and pollution-free, and highly suitable for remediating heavy metal pollution in red soil. Owing to the bioclimate in red soil, colonization by exogenous microorganisms is difficult after inoculation. This is a considerable problem while using microorganisms to remediate heavy metal pollution in red soil, which directly affects the remediation [31]. Thus, we isolated and screened indigenous chromium-resistant *A. niger* from chromium-polluted red soil using heavy metal chromium as an ecological filtration factor. After sufficient theoretical verification, *A. niger* screened in red soil was used for the treatment of chromium pollution in it, and encouraging results were obtained. The inoculation of indigenous *A. niger* could not only protect the microbial diversity of red soil, but also significantly increase the abundance of *A. niger* and enhance the ability of Cr passivation. These results proved the feasibility of activating indigenous microorganisms to deactivate soil heavy metals; at the same time, the problem of the unstable effect of exogenous microorganisms was also solved in this way.

In this study, the colony characteristics of strain A1 screened from red soil under Cr stress were studied, which proved that strain A1 could tolerate high concentrations of Cr^{6+} and had the capacity to adsorb Cr^{6+}. On this basis, the effects of temperature, pH, initial Cr^{6+}concentration and contact time on Cr^{6+}adsorption by strain A1 were discussed in terms of condition optimization. It is proven that Indigenous *A. niger* exhibited the highest adsorption capacity under acidic conditions and was highly adaptable to medium and high ambient temperatures. The selected indigenous *A. niger* demonstrated strong potential for the remediation of chromium pollution in red soil in warm climates.

4.2. Potential of Indigenous A. niger to Reduce Chromium Pollution in Red Soil Environments and Its Adsorption Mechanism

The main functions of microorganisms for alleviating heavy metal pollution can be categorized as: reducing intake, increasing excretion, and metabolic detoxification. The use of gene editing to knock out specific channel protein genes can produce microorganisms that are hyperaccumulators of specific heavy metals [32]. Strengthening the metabolism and detoxification ability of microorganisms can effectively improve their heavy metal adsorption abilities, and the stress caused by heavy metals can induce them to produce many functional proteins [33]. These proteins have strong binding ability for cadmium, copper, mercury, chromium, and other metals, and can transform water-soluble heavy metals into less toxic precipitated and organic-bound states, thereby reducing metal bioavailability in cells [34]. The *A. niger* isolated and purified in the present study has previously been used to remove heavy metals from sewage [35]. Recent studies show that chitosan or other extracellular polymeric substances (EPS) on the cell wall of *A. niger* can cooperate with various functional groups to complete the extracellular adsorption of heavy metal ions [36–38]. In this study, FTIR analysis showed that many reducing groups on the surface of *A. niger* were involved in the fixation of Cr^{6+}, such as -C=C-, -C=0, -C-O-C, and -NH_3. In order to explore the binding form of chromium on *A. niger*'s surface, we compared the differences between extraction with 8000 rpm centrifugation and filtration, and the results showed that the chromium content of centrifugal extraction was slightly lower than that of filtration extraction. Combined with XPS analysis, this shows that *A. niger*'s membrane system can fix Cr^{6+} with a relatively firm structure (surface complexation) and reduce it to Cr^{3+} in large quantities, except that a small amount of Chromium is attracted by ion exchange. In the subsequent supplementary experiment, the filtered *A. niger* bodies were broken to compare the amount of chromium fixed by the membrane system and absorbed by the cytoplasm. The results showed that the main way for *A. niger* to absorb chromium is through intracellular adsorption, and the surface adsorption amount is positively correlated with the biomass of *A. niger*; moreover, there was little difference in the fixation capacity per unit membrane area. There is also an upper limit for intracellular adsorption. When the intracellular chromium content is too high, *A. niger* can reduce the intake of chromium by closing the transport channel, which is also reflected in the kinetic analysis of this study. In this study, three kinetic models (including the intra-particle diffusion model) were used to elucidate the adsorption process of *A. niger* in solution and soil. The fitting results of the intra-particle diffusion model show that chromium could not diffuse through *A. niger* membranes and required endocytosis or active transport to enter the cell; in the early stage of adsorption, it was mainly the surface adsorption of membrane system, and the internal diffusion rate was very slow. In the middle stage of adsorption, the surface adsorption reached a dynamic balance and the internal diffusion effect was enhanced; additionally, detoxification by *A. niger* occurred through intracellular metabolism, and it stored chromium in the cells. At the late stage of adsorption, there was no internal diffusion effect, the absorption rate of chromium by *A. niger* was equal to the discharge rate or *A. niger* stopped absorbing chromium, and the adsorption effect reached a dynamic balance. Based on the other analysis, we found that the increase in Cr^{6+} concentration could improve the chromium adsorption rate of *A. niger*; however, the increase in the adsorption rate was limited by the combined action of the strong driving force of high concentration and binding sites on the surface of *A. niger*. These conclusions improve on the theory proposed by Chhikara [39].

In the red soil environment, a low dose of *A. niger* inoculation was more consistent with the intra-particle diffusion model. This is most likely because the complex soil environment can passivate chromium in addition to *A. niger*; therefore, when *A. niger* concentration was low, the adsorption process was not dominated by *A. niger* only. Compared with the microbiological agents purchased on the market, the application of the same amount of *A. niger* achieved significantly better results. Commercial microbial agents had a good effect within 7–10 d after inoculating, and then, the effect weakened. This phenomenon may be related to the difficulty in the colonization of exogenous microorganisms or the inadaptability of exogenous microorganisms to the acidic red soil environment. However, by increasing the biomass of *A. niger*, the passivation effect of hexavalent chromium in red soil was significantly improved, and the passivation rate was positively correlated with the addition amount.

As *A. niger* has a certain enrichment effect of chromium from the environment, in follow-up research, it could be transformed into a chromium hyperaccumulator through gene elimination and destruction of the channel protein synthesis gene that pumps out chromium. In addition, the results of the optimization study indicated that the optimal pH for chromium pollution remediation using *A. niger* was consistent with the main pH distribution of red soil, and that *A. niger* could drive the transformation of water-soluble and exchangeable states of chromium to precipitation and residue states while absorbing soil chromium. Moreover, its addition directly affected the rate of transformation of chromium state and greatly reduced the toxicity of chromium in red soil. In this study, the potential for indigenous *A. niger* to be used in treating heavy metal pollution in red soil was verified in many aspects. At the same time, this study proved that the activation of indigenous microorganisms was more effective than adding exogenous microorganisms, and the mechanism of chromium adsorption by *A. niger* was studied. Based on these conclusions, we can further put forward various improvement measures to improve the effectiveness of applications of indigenous *A. niger*.

4.3. Contributions and Limitations of Current Research and Future Development Directions

The method used in this study can not only be used for the remediation of chromium pollution in red soil but also for the in situ remediation of other soil types and heavy metal elements by screening native strains with high tolerance. Furthermore, similar research can be applied to the current treatment of organic pollutants and soil microplastics. After screening the functional strains for pollutants, pollutant degradation can be achieved by activating or externally supplementing indigenous microorganisms. Organic pollutants can form non-toxic metabolites after being degraded by microorganisms and return to the carbon and nitrogen cycles [40]. Therefore, the activation of indigenous microorganisms is more suitable for the treatment of organic pollutants than for heavy metal pollution, considering the recovery of pollutants from the red soil.

This study showed that the use of indigenous *A. niger* to remediate soil chromium pollution has high feasibility and low cost, is convenient, has a stable effect, and produces no secondary pollution. However, microbial remediation, as with traditional remediation methods, faces the problem that the treated chromium remains in the soil. There is a risk that chromium may be re-released under certain circumstances after being adsorbed into cells or transformed into a less toxic form. On this basis, through the immobilized *A. niger* method, further research on technology for the recovery of bioremediation can effectively solve these problems. In addition, the soil's microenvironmental stability is a dynamic equilibrium. In soil, plants, microorganisms, and other organisms can maintain the relative stability of their internal chemical composition by adjusting the concentration of chemical elements and the measurement ratio of different chemical elements in the body to cope with the influence of external conditions such as soil nutrient limitation on their own growth. This mechanism is called homeostasis [41]. After the addition of exogenous strains, the biomass would be greatly increased in a short time; however, the soil bacterial colony structure would eventually be stabilized. The present study found that the excessive addition of *A. niger* significantly improved the treatment effect in a short period. However, the influence of *A. niger* on adsorption gradually decreased with the stabilization of soil bacterial colony structure. Therefore, there may be a potential upper limit of bioremediation, which may be related to the colony characteristics and niche competitiveness of the selected microorganisms. Further research is necessary to solve this problem, and to improve bioremediation via strain improvement or gene editing.

4.4. Virulence Test

Aspergillus niger, which was screened, is one of the most common molds in daily life. It is widely distributed in soil and air, and is also widely used in the treatment of heavy metal pollution, making it valuable for use in the actual environment. However, virulence tests still need to be carried out before using *A. niger* as a fungal agent to control heavy metal pollution in red soil. For this reason, we inoculated *A. niger* into corn and cabbage fields, and continued to observe the effects of *A. niger* inoculation on the growth of corn and cabbage for 2 months. The observation results are shown in Figure 8.

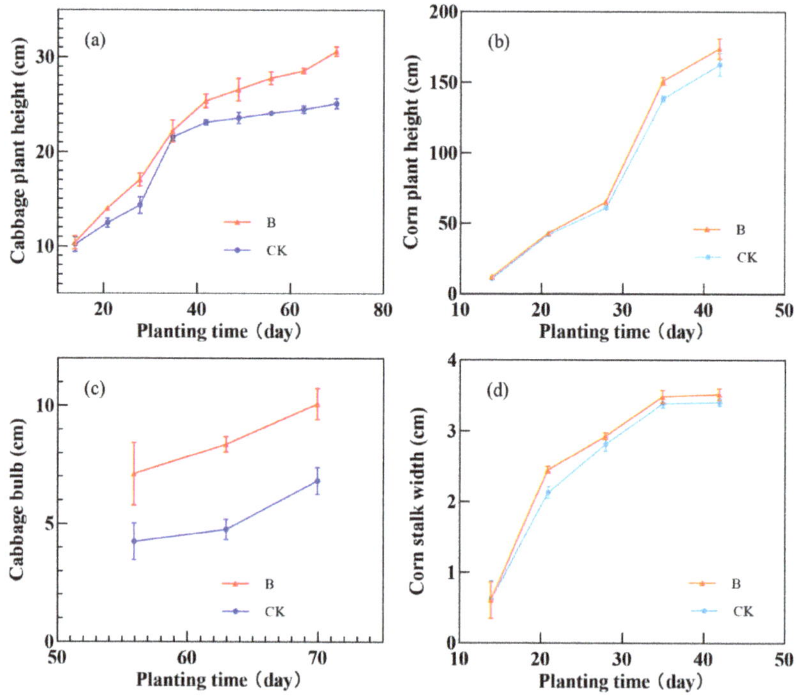

Figure 8. (**a**) Height of cabbage plant; (**b**) height of corn plant; (**c**) bulb of cabbage plant; (**d**) width of corn plant.

Cabbage and corn are the representative crops of Cruciferae and Gramineae, respectively. The observation results showed that the inoculation of *A. niger* had no significant adverse effect on crop growth. Therefore, the use of indigenous *A. niger* to control chromium pollution in red soil is a safe and green new scheme.

5. Conclusions

An indigenous chromium-resistant *A. niger* strain was screened from red soil. The strain showed high activity with exposure to 500 mg·L^{-1} of Cr^{6+}. *Aspergillus niger* had a strong and rapid adsorption effect on Cr^{6+} ions under acidic conditions at 20–50 °C. Our investigations pertaining to its adsorption mechanism showed that *A. niger* mainly adsorbed chromium through intracellular adsorption, surface complexation, and ion exchange. Extracellular adsorption is accomplished through the cooperation of EPS and various functional groups on the cell surface. Complete adsorption on the surface was followed by intracellular metabolic detoxification, and it was evident that this effect would be continuously improved with an increase in biomass after colonization by *A. niger*. A large amount of chromium was enriched into *A. niger* cells because it was restricted from freely crossing the surface of *A. niger* cells, consequently reducing the available chromium content in soil. Moreover, the inoculation of indigenous *A. niger* can greatly improve the dominance of *A. niger* in red soil, drive the transformation of a large amount of chromium into a less toxic form, and greatly reduce the content of Cr^{6+}. In conclusion, the proposed method of application of indigenous *A. niger* to treat chromium pollution in red soil is a feasible strategy, and the effect is better than using traditional exogenous microbial agents.

Author Contributions: R.X.: conceptualization, methodology; J.X.: data curation, writing—original draft; L.L.: visualization; H.W.: project administration; Z.G. and C.W.: software; R.S.: resources; Y.Z.: writing—review and editing; W.X.: supervision; X.H.: investigation. All authors have read and agreed to the published version of the manuscript.

Funding: This research was funded by the Provincial Natural Science Foundation of Yunnan and the Yunnan Provincial Young Talent Program, grant number QNBJ2019248 and the Basic Research Program Foundation of Yunnan Province, grant number 202201AS070020 and the China National Innovative Training Program for college students, grant number202210681030 and the University Student Scientific Research Training Fund Project of Yunnan Normal University KX2021036.

Institutional Review Board Statement: Not applicable.

Informed Consent Statement: Not applicable.

Data Availability Statement: Not applicable.

Conflicts of Interest: The authors declare no conflict of interest.

Abbreviations

ddH_2O—double distilled water; FTIR—Fourier transform infra-red spectroscopy; PCR—polymerase chain reaction; XPS—X-ray photoelectron spectroscopy; EPS— extracellular polymeric substances; SDYA—a medium used to isolate and culture fungi, Sabouraud Dextrose Agar with Yeast Extract.

References

1. Hu, G.; Bi, S.; Xu, G.; Zhang, Y.; Mei, X.; Li, A. Distribution and assessment of heavy metals off the Changjiang River mouth and adjacent area during the past century and the relationship of the heavy metals with anthropogenic activity. *Mar. Pollut. Bull.* **2015**, *96*, 434–440. [CrossRef]
2. Yao, Y.; Li, J.; He, C.; Hu, X.; Yin, L.; Zhang, Y.; Zhang, J.; Huang, H.; Yang, S.; He, H.; et al. Distribution Characteristics and Relevance of Heavy Metals in Soils and Colloids Around a Mining Area in Nanjing, China. *Bull. Environ. Contam. Toxicol.* **2021**, *107*, 996–1003. [CrossRef]
3. Jeevanantham, S.; Saravanan, A.; Hemavathy, R.; Kumar, P.S.; Yaashikaa, P.; Yuvaraj, D. Removal of toxic pollutants from water environment by phytoremediation: A survey on application and future prospects. *Environ. Technol. Innov.* **2019**, *13*, 264–276. [CrossRef]
4. Jobby, R.; Jha, P.; Yadav, A.K.; Desai, N. Biosorption and biotransformation of hexavalent chromium [Cr(VI)]: A comprehensive review. *Chemosphere* **2018**, *207*, 255–266. [CrossRef]
5. Yuan, X.; Xiong, T.; Wang, H.; Wu, Z.; Jiang, L.; Zeng, G.; Li, Y. Immobilization of heavy metals in two contaminated soils using a modified magnesium silicate stabilizer. *Environ. Sci. Pollut. Res.* **2018**, *25*, 32562–32571. [CrossRef]
6. Gu, C.; Wilson, S.G.; Margenot, A.J. Lithological and bioclimatic impacts on soil phosphatase activities in California temperate forests. *Soil Biol. Biochem.* **2019**, *141*, 107633. [CrossRef]
7. Zhao, K.; Cai, Z.; Wang, B.; Wen, S.; Zhou, X.; Sun, N. Changes in pH and exchangeable acidity at depths of red soils derived from 4 parent materials under 3 vegetations. *Sci. Agric. Sin.* **2015**, *48*, 4818–4826. [CrossRef]
8. Fan, X.; Ding, S.; Chen, M.; Gao, S.; Fu, Z.; Gong, M.; Wang, Y.; Zhang, C. Mobility of chromium in sediments dominated by macrophytes and cyanobacteria in different zones of Lake Taihu. *Sci. Total Environ.* **2019**, *666*, 994–1002. [CrossRef]
9. Yin, L.; Liao, Y.; Liu, L.; Song, Y.; He, G. Screening identification and tolerance characteristics of Fe/Al/Cu tolerant bacteria in acidic groundwater in red soil area. *Fresenius Environ. Bull.* **2022**, *31*, 2109–2116.
10. Garg, U.K.; Kaur, M.; Garg, V.; Sud, D. Removal of hexavalent chromium from aqueous solution by agricultural waste biomass. *J. Hazard. Mater.* **2007**, *140*, 60–68. [CrossRef]
11. Liang, B.; Han, G.; Zeng, J.; Qu, R.; Liu, M.; Liu, J. Source and Risk Assessment of Trace Metals in Red Soils from Yunnan Province, Southwest China. *CLEAN Soil Air Water* **2021**, *49*, 2000288. [CrossRef]
12. Wang, X.-F.; Deng, C.-B.; Sunahara, G.; Yin, J.; Xu, G.-P.; Zhu, K.-X. Risk Assessments of Heavy Metals to Children Following Non-dietary Exposures and Sugarcane Consumption in a Rural Area in Southern China. *Expo. Health* **2018**, *12*, 1–8. [CrossRef]
13. Pradhan, D.; Sukla, L.B.; Sawyer, M.; Rahman, P.K. Recent bioreduction of hexavalent chromium in wastewater treatment: A review. *J. Ind. Eng. Chem.* **2017**, *55*, 1–20. [CrossRef]
14. Malaviya, P.; Singh, A. Bioremediation of chromium solutions and chromium containing wastewaters. *Crit. Rev. Microbiol.* **2014**, *42*, 607–633. [CrossRef] [PubMed]
15. Kalidhasan, S.; Kumar, A.S.K.; Rajesh, V.; Rajesh, N. The journey traversed in the remediation of hexavalent chromium and the road ahead toward greener alternatives—A perspective. *Co-ord. Chem. Rev.* **2016**, *317*, 157–166. [CrossRef]

16. Chen, S.; Zhang, X.; Liu, J.; You, S.; Zhang, H.; Lin, H. Reduction of Cr (VI) to Cr (III): A detoxification mechanism in hyperaccumulator leersia hexandra swartz. *Fresenius Environ. Bull.* **2016**, *25*, 959–968.
17. Saravanan, A.; Kumar, P.S.; Preetha, B. Optimization of process parameters for the removal of chromium(VI) and nickel(II) from aqueous solutions by mixed biosorbents (custard apple seeds and *Aspergillus niger*) using response surface methodology. *Desalination Water Treat.* **2015**, *57*, 14530–14543. [CrossRef]
18. Chang, X.; Li, M.; Liu, Q.; Liu, Q.; Yao, J. Adsorption–reduction of chromium(vi) from aqueous solution by phenol–formaldehyde resin microspheres. *RSC Adv.* **2016**, *6*, 46879–46888. [CrossRef]
19. Bao, Z.; Feng, H.; Tu, W.; Li, L.; Li, Q. Method and mechanism of chromium removal from soil: A systematic review. *Environ. Sci. Pollut. Res.* **2022**, *29*, 35501–35517. [CrossRef]
20. Elahi, A.; Arooj, I.; Bukhari, D.A.; Rehman, A. Successive use of microorganisms to remove chromium from wastewater. *Appl. Microbiol. Biotechnol.* **2020**, *104*, 3729–3743. [CrossRef]
21. Wang, R.; Shafi, M.; Ma, J.; Zhong, B.; Guo, J.; Hu, X.; Xu, W.; Yang, Y.; Ruan, Z.; Wang, Y.; et al. Effect of amendments on contaminated soil of multiple heavy metals and accumulation of heavy metals in plants. *Environ. Sci. Pollut. Res.* **2018**, *25*, 28695–28704. [CrossRef] [PubMed]
22. Lin, Z.; Li, J.; Luan, Y.; Dai, W. Application of algae for heavy metal adsorption: A 20-year meta-analysis. *Ecotoxicol. Environ. Saf.* **2019**, *190*, 110089. [CrossRef] [PubMed]
23. Kavamura, V.N.; Esposito, E. Biotechnological strategies applied to the decontamination of soils polluted with heavy metals. *Biotechnol. Adv.* **2010**, *28*, 61–69. [CrossRef]
24. da Costa Waite, C.C.; da Silva, G.O.A.; Bitencourt, J.A.P.; Sabadini-Santos, E.; Crapez, M.A.C. Copper and lead removal from aqueous solutions by bacterial consortia acting as biosorbents. *Marine Pollut. Bull.* **2016**, *109*, 386–392. [CrossRef]
25. Li, P.; Tao, H.-C. Cell surface engineering of microorganisms towards adsorption of heavy metals. *Crit. Rev. Microbiol.* **2013**, *41*, 140–149. [CrossRef] [PubMed]
26. Wiatrowski, H.A.; Barkay, T. Monitoring of microbial metal transformations in the environment. *Curr. Opin. Biotechnol.* **2005**, *16*, 261–268. [CrossRef]
27. Jorquera, M.A.; de la Luz Mora, M. Bacillus-like phosphobacteria in agronomic volcanic soils from Chile. In Proceedings of the 19th World Congress of Soil Science: Soil Solutions for a Changing World, Brisbane, Australia, 1–6 August 2010; pp. 1–6.
28. Ghiglione, J.-F.; Gourbiere, F.; Potier, P.; Philippot, L.; Lensi, R.; Bouterige, S.; Robert, R.; Bouchara, J.P.; Marot-Leblond, A.; Molinero, V.; et al. Role of Respiratory Nitrate Reductase in Ability of *Pseudomonas fluorescens* YT101 To Colonize the Rhizosphere of Maize. *Appl. Environ. Microbiol.* **2000**, *66*, 3277–3282. [CrossRef]
29. Pachamuthu, P.; Kamble, S.T. In Vivo Study on Combined Toxicity of Metarhizium anisopliae (Deuteromycotina: Hyphomycetes) Strain ESC-1 with Sublethal Doses of Chlorpyrifos, Propetamphos, and Cyfluthrin Against German Cockroach (Dictyoptera: Blattellidae). *J. Econ. Entomol.* **2000**, *93*, 60–70. [CrossRef]
30. Huang, G.; Zhao, Q. Initial exploration of red soil ecology. *Acta Ecol. Sin.* **2014**, *34*, 5173–5181. [CrossRef]
31. Ma, L.; Wang, X.; Tao, J.; Feng, X.; Zou, K.; Xiao, Y.; Liang, Y.; Yin, H.; Liu, X. Bioleaching of the mixed oxide-sulfide copper ore by artificial indigenous and exogenous microbial community. *Hydrometallurgy* **2017**, *169*, 41–46. [CrossRef]
32. Pérez-Palacios, P.; Funes-Pinter, I.; Agostini, E.; Talano, M.A.; Ibáñez, S.G.; Humphry, M.; Edwards, K.; Rodríguez-Llorente, I.D.; Caviedes, M.A.; Pajuelo, E. Targeting Acr3 from *Ensifer medicae* to the plasma membrane or to the tonoplast of tobacco hairy roots allows arsenic extrusion or improved accumulation. Effect of *acr3* expression on the root transcriptome. *Metallomics* **2019**, *11*, 1864–1886. [CrossRef] [PubMed]
33. Cui, L.; Fan, L.; Li, Z.; Wang, J.; Chen, R.; Zhang, Y.; Cheng, J.; Wu, X.; Li, J.; Yin, H.; et al. Characterization of extracellular polymeric substances from Synechocystis sp. PCC6803 under Cd (II), Pb (II) and Cr (VI) stress. *J. Environ. Chem. Eng.* **2021**, *9*, 105347. [CrossRef]
34. Cavet, J.; Borrelly, G.P.; Robinson, N.J. Zn, Cu and Co in cyanobacteria: Selective control of metal availability. *FEMS Microbiol. Rev.* **2003**, *27*, 165–181. [CrossRef]
35. Guimarães-Soares, L.; Felícia, H.; Bebianno, M.J.; Cássio, F. Metal-binding proteins and peptides in the aquatic fungi Fontanospora fusiramosa and Flagellospora curta exposed to severe metal stress. *Sci. Total Environ.* **2006**, *372*, 148–156. [CrossRef]
36. Gu, Y.; Xu, W.; Liu, Y.; Zeng, G.; Huang, J.; Tan, X.; Jian, H.; Hu, X.; Li, F.; Wang, D. Mechanism of Cr(VI) reduction by Aspergillus niger: Enzymatic characteristic, oxidative stress response, and reduction product. *Environ. Sci. Pollut. Res.* **2014**, *22*, 6271–6279. [CrossRef]
37. Inyang, M.I.; Gao, B.; Yao, Y.; Xue, Y.; Zimmerman, A.; Mosa, A.; Pullammanappallil, P.; Ok, Y.S.; Cao, X. A review of biochar as a low-cost adsorbent for aqueous heavy metal removal. *Crit. Rev. Environ. Sci. Technol.* **2015**, *46*, 406–433. [CrossRef]
38. Chaudhary, P.; Beniwal, V.; Kaur, R.; Kumar, R.; Kumar, A.; Chhokar, V. Efficacy of *Aspergillus fumigatus* MCC 1175 for Bioremediation of Tannery Wastewater. *CLEAN Soil Air Water* **2019**, *47*, 1900131. [CrossRef]
39. Chhikara, S.; Hooda, A.; Rana, L.; Dhankhar, R. Chromium (VI) biosorption by immobilized Aspergillus niger in continuous flow system with special reference to FTIR analysis. *J. Environ. Biol.* **2010**, *31*, 561–566. [CrossRef]

40. Wang, H.; Zhang, L.; Hu, C.; Wang, X.; Lyu, L.; Sheng, G. Enhanced degradation of organic pollutants over Cu-doped $LaAlO_3$ perovskite through heterogeneous Fenton-like reactions. *Chem. Eng. J.* **2018**, *332*, 572–581. [CrossRef]
41. Persson, J.; Fink, P.; Goto, A.; Hood, J.M.; Jonas, J.; Kato, S. To be or not to be what you eat: Regulation of stoichiometric homeostasis among autotrophs and heterotrophs. *Oikos* **2010**, *119*, 741–751. [CrossRef]

toxics

Article

Novel Insights into the Influence of Soil Microstructure Characteristics on the Migration and Residue of Light Non-Aqueous Phase Liquid

Xiaodong Li [1,2], Qian Zhang [1], Xueli Zhang [2], Jialun Shen [2], Zongquan Sun [2], Fujun Ma [2], Bin Wu [1,*] and Qingbao Gu [2,*]

1 Technical Centre for Soil, Agriculture and Rural Ecology and Environment, Ministry of Ecology and Environment, Beijing 100012, China
2 State Key Laboratory of Environmental Criteria and Risk Assessment, Chinese Research Academy of Environmental Sciences, Beijing 100012, China
* Correspondence: wubin@craes.org.cn (B.W.); guqb@craes.org.cn (Q.G.); Tel.: +86-10-84757917 (B.W.); +86-10-84915187 (Q.G.); Fax: +86-10-84757904 (B.W.); +86-10-84932813 (Q.G.)

Abstract: Understanding the influence of soil microstructure on light non-aqueous phase liquids (LNAPLs) behavior is critical for predicting the formation of residual LNAPLs under spill condition. However, the roles of soil particle and pore on LNAPLs migration and residue remains unclear. Here, the experiment simulated an LNAPLs (diesel) spill that was performed in fourteen types of soils, and the key factors affecting diesel behavior are revealed. There were significant differences between fourteen types of soils, with regard to the soil particle, soil pore, and diesel migration and residue. After 72 h of leakage, the migration distance of diesel ranged from 3.42 cm to 8.82 cm in the soils. Except for sandy soil, diesel was mainly distributed in the 0–3 cm soil layer, and the residual amounts were 7.85–26.66 g/kg. It was further confirmed from microstructure that the consistency of soil particle and volume of soil macropores (0.05–7.5 μm) are important for diesel residue in the 0–1 cm soil layer and migration distance. The large soil particles corresponding to 90% of volume fraction and volume of soil mesopores (<0.05 μm) are key factors affecting diesel residue in the 1–3 cm soil layer. The result helps to further comprehend the formation mechanism of residual LNAPLs in the soil.

Keywords: diesel; soil particle and pore; statistical analysis; key factor

check for updates

Citation: Li, X.; Zhang, Q.; Zhang, X.; Shen, J.; Sun, Z.; Ma, F.; Wu, B.; Gu, Q. Novel Insights into the Influence of Soil Microstructure Characteristics on the Migration and Residue of Light Non-Aqueous Phase Liquid. *Toxics* **2023**, *11*, 16. https://doi.org/10.3390/toxics11010016

Academic Editors: Junhao Qin, Peidong Su, Feng Zhu and Lin Ding

Received: 2 December 2022
Revised: 20 December 2022
Accepted: 21 December 2022
Published: 24 December 2022

1. Introduction

Chemical spill caused by tank bottom corrosion, pipeline damage, or facility explosion has increased the frequency and harm degree of soil pollution incidents, with the spill of petroleum products especially being the most frequent [1–3]. Petroleum products (e.g., diesel) are water insoluble or sparingly water soluble, which belongs to light non-aqueous phase liquids (LNAPLs) [4]. After the spill of petroleum products, it will migrate downward at the soil surface under the action of gravity. During the migration process, petroleum products could accumulate at different locations underground, and form residual states such as accumulation areas or lenses of NAPLs, resulting in the long-term pollution of soil and groundwater environment [5,6].

The formation process of residual LNAPLs is affected by the characteristics of LNAPLs, environmental conditions, and soil structure (e.g., soil particle size, soil pore size, and spatial heterogeneity) [7–10]. Under spill conditions of specific LNAPLs, soil structure is the main factor to control the formation of residual LNAPLs [7,11]. Therefore, the quantitative characterization of soil structure is conducive to understanding the influence degree and mechanism of different types of soils on the formation of residual LNAPLs. Due to different environmental factors, soil properties vary greatly in different regions, which resulted in the disparity of soil structure characteristics in China [12–14]. However, to the best of our knowledge, there are no relevant studies on the systematic quantitative analysis of soil structure characteristics in the large-scale region of China.

In recent years, a few studies have explored the impacts of soil structure characteristics on the formation process of residual LNAPLs under spill condition [15–18]. Saripalli et al. characterized the relationship between the soil pore size and spatial distribution of residual LNAPLs by the interfacial tracers technique and morphological method [16]. Soil macropores and mesopores were the key factors affecting the migration rate of fluid, and the number of soil micropores controlled the residual amount of fluid in the soil [15]. With the increase in specific pore surface area, the migration rate of LNAPLs in porous medium tended to decrease [10,19]. It has been reported that although two types of soils had the same soil porosity and organic matter content (OM), soil with high clay content possessed more micropores, resulting in a higher residual amount of LNAPLs in the soil [20]. Moreover, soil properties are largely determined by soil weathering, which thus affected soil classification order [21]. For example, the Vertisols are highly fertile due to their high clay content, but most of the water content is inaccessible to plants. Entisols are the most recently developed soils abounding illites and kaolinitic minerals or well-crystalline oxides. Alfisols have a high content of Al and Fe oxides, good structure, and typical slightly acidic activity [22]. Golia et al. [23] reported that high regression coefficients were observed between soil pH and potentially toxic elements concentrations in Alfisols, reflecting that soil pH is the dominant characteristic influencing potentially toxic elements. In Vertisols, clay content proved to be the most important parameter influencing potentially toxic elements concentrations levels. Therefore, the soil classification order-dependent diversity can be assumed to lead to the different formation processes of LNAPLs in soils. However, most previous studies have been restricted to the impact of a single soil structure characteristic on the formation process (migration and residue) of residual LNAPLs in the soil [24]. Due to the lack of statistical quantitative analysis, the key factors of soil structure affecting migration and residue of LNAPLs under spill condition remained unclear.

In this study, fourteen types of soils in China were selected from surface soils of different geographic regions, and further identified the distribution characteristics of soil particles and pore size using several characterization methods. Diesel is chosen as the typical petroleum pollutant due to being widely used in society. Then, the simulated accidental spill was performed in fourteen types of soils to investigate the migration distance and vertical distribution of diesel. Based on soil structure characteristics in the large-scale region of China, the multivariate linear regression of the migration distance and residual amount of diesel with soil structure characteristics are established through the statistical product service solutions (SPSS) statistical analysis, and revealed the sensitive factors which affect the migration distance and residual amount of diesel. The result will be a benefit to further comprehending the formation process and mechanism of residual LNAPLs in the soil.

2. Materials and Methods

2.1. Soil Samples

Fourteen types of soils were collected from surface soils (0–20 cm) distributed in different geographic reigns across north to south China: Heilongjiang and Jilin in northeast China; Hebei in North China; Henan and Hunan in central China; Zhejiang, Jiangsu, Jiangxi, and Shandong in east China; Shaanxi and Qinghai in northwest China; Hainan in south China; Guizhou and Chongqing in southwest China. All soil samples were naturally air-dried indoors, passed through a 2 mm sieve, and stored in plastic containers at room temperature. Basic physiochemical properties including pH, OM, and cation exchange capacity (CEC) of soil samples were analyzed by routine methods [25,26]. The soil pH was measured in a 1:2.5 (w/v) soil/water suspension after 1 h equilibration using a SevenCompact™ S210 pH Meter (Mettler-Toledo). The soil OM was measured by the potassium dichromate volumetric method. The ammonium acetate method was used to measure the CEC. Soil particle size distribution was determined by a laser particle size analyzer. Soil pore size distribution was characterized by a mercury piezometric method.

The Characteristic parameters included Dv10, Dv30, Dv50, Dv60, Dv90, Cu, and Cc [27–29]. Dv10, Dv30, Dv50, Dv60, and Dv90 represent the soil particles corresponding to 10%, 30%, 50%, 60%, and 90% of volume fraction, respectively. The nonuniform coefficient (Cu = Dv60/Dv10) is defined as the uniformity of soil particle size distribution. The curvature coefficient ($Cc = (Dv30)^2/Dv10 \cdot Dv60$) reflects the overall shape of soil particle size distribution curve. The consistency and diameter distance are the characteristic parameters used to describe the deviation and width of soil particle size distribution curve, respectively.

2.2. Experimental Device

A one-dimensional soil column was used to simulate the migration of diesel in the soil under spill condition. The soil column (238.5 cm^3) was obtained by filling a cylindrical glass tube with a length of 25.0 cm and a 4.5 cm inner diameter, which was lined with quartz flakes in the lower part and connected to the atmosphere in the upper part. Specifically, 1 cm-thick soil was weighed in batches according to actual soil bulk density (a total of 15 cm soil column), then slowly poured and compacted into a cylindrical glass tube. The soil column is composed of 2 cm coarse quartz sand (20–40 mesh), 15 cm test soil, and 1 cm fine quartz sand (40–70 mesh) from top to bottom. Coarse quartz sand could prevent the loss of diesel due to splashing from the spill process, and also make diesel evenly distributed on the soil surface.

2.3. Experimental Procedures

To simulate the instantaneous diesel spill under no rainfall condition, 10 mL of diesel was dumped above the soil column at one time. Then, the migration distance (S) of diesel front was observed and recorded at regular intervals to analyze the dynamic process of diesel migration. The 5.0 g soil sample was withdrawn using a five-point sampling method in each soil layer and mixed to guarantee the homogeneity of the samples. The mixed soil samples were collected and determined the residual amount of diesel (*Y*). In the pre-experiment, three types of soil-filled columns (Jiangxi red earth, Heilongjiang black soil, and Shaanxi loessal soil) were selected, and the samples were taken at 1 h, 5 h, 12 h, 1 d, 2 d, 3 d, and 5 d after the spill occurred, respectively. By analyzing the variation of diesel residue in the soil, the steady state of residual diesel was achieved in 3 days. The migration experiment was conducted at a constant temperature of (20 $\pm$ 2) °C and in triplicate.

2.4. Analytical Methods

Take 2.00 $\pm$ 0.01 g of contaminated soil and put it into a polytetrafluoroethylene centrifuge tube. Add 5 mL of dichloromethane for ultrasonic extraction for 30 min, and then centrifuge at 3000 rpm for 15 min. The method for determining the concentration of diesel refers to the previous studies [26]. A DB-5 capillary column was used, and the carrier gas was helium (purity, 99.999%). The flow rate was 1.5 mL/min. The oven temperature of GC was programmed from 50 °C (2 min) to 250 °C at 8 °C/min (3 min). The temperatures of the injector, ion source, and transfer line were set to 250 °C, 230 °C, and 280 °C, respectively.

3. Results and Discussion

3.1. Basic Physiochemical Properties and Structure Characteristics of Soils

3.1.1. Basic Physiochemical Properties of Soils

Fourteen types of soils were collected from different geographic regions in China with a wide distribution range. Soil classification of the studied pedons according to the U.S. Soil Taxonomy is conducted and soils were classified into 4 orders, including Mollisols (Heilongjiang and Jilin black soil), Alfisols (Shaanxi loessal soil, Hebei and Henan fluvo-aquic soil, and Hunan red earth), Entisols (Shandong fluvo-aquic soil, Qinghai gray desert soil, and Chongqing purple soil), and Ultisols [30]. The basic physiochemical properties were quite different in various regions (Table 1). Guizhou yellow soil is a strongly acidic soil with the lowest pH value (3.96), Shaanxi loessal soil has the highest pH value of 8.76, and most of the soils are weakly alkaline. The organic matter content of Jilin black soil was the highest at 47.90 g/kg, and the cation exchange capacity of Hainan sandy soil was the lowest (1.92 cmol/kg). The soil bulk density in fourteen geographic regions was between 1.295 g/cm^3 (Jilin black soil) and 1.873 g/cm^3 (Hainan sandy soil). The soil clay content in Heilongjiang was the highest at 43.3%. This great variation in the basic physiochemical properties of soils was attributed to the corresponding differences in terrain, climate, vegetation, and soil age [31]. According to soil particle size distribution [32], fourteen types of soils were divided into sandy soil, sandy loam soil, sandy clay loam soil, clay loamy soil, and sandy clay soil.

Table 1. Physiochemical properties of fourteen types of soils.

Classification (Location)	pH	OM g/kg	CEC cmol/kg	Soil Bulk Density g/cm^3	Sand %	Silt %	Clay %	Soil Texture
Gray desert soil (Qinghai)	7.90	23.10	5.81	1.3537	73.38	7.36	19.26	sandy clay loam soil
Yellow soil (Guizhou)	3.96	29.00	12.00	1.3537	82.03	2.80	15.17	sandy clay loam soil
Loessal soil (Shaanxi)	8.76	1.35	2.57	1.492	79.07	4.55	16.38	sandy clay loam soil
Fluvo-aquic soil (Hebei)	7.96	11.50	12.20	1.4669	60.47	17.02	22.51	sandy clay loam soil
Red earth (Jiangxi)	7.77	14.60	10.50	1.6052	63.41	13.08	23.51	sandy clay loam soil
Paddy soil (Zhejiang)	5.06	23.40	9.24	1.4040	59.99	26.62	13.39	clay loamy soil
Purple soil (Chongqing)	7.44	8.38	18.20	1.6219	66.69	5.65	27.66	clay loamy soil
Black soil (Jilin)	6.88	31.30	26.00	1.2951	33.06	26.71	40.23	clay loamy soil
Black soil (Heilongjiang)	5.55	47.90	34.90	1.3998	40.60	16.10	43.30	clay loamy soil
Red earth (Hunan)	7.69	7.46	21.20	1.5423	40.81	18.19	41.00	clay loamy soil
Fluvo-aquic soil (Henan)	7.98	13.60	9.35	1.5507	85.40	3.86	10.74	sandy loam soil
Fluvo-aquic soil (Shandong)	7.81	11.00	14.3	1.3789	74.84	11.76	13.40	sandy loam soil
Sandy soil (Hainan)	5.65	3.26	1.92	1.8734	94.30	0.80	4.90	sandy soil
Paddy soil (Jiangsu)	7.78	20.20	15.70	1.4878	64.34	10.33	25.33	sandy clay soil

3.1.2. Characteristic Parameters of Soil Particle Size Distribution

Soil particle size distribution is one of the factors affecting the migration distance and residual amount of pollutants. As shown in Table 2, the characteristic parameters of soil particle size distribution varied greatly. The average particle size (D[3,2] and D[4,3]) of Hainan sandy soil was the largest (15.53 μm and 303.27 μm), while it was lower in Guizhou yellow soil and Hunan red earth. In all, the average particle size of fourteen types of soils followed by sandy soil > sandy clay loam soil > clay loamy soil. The diameter distance and consistency of Zhejiang paddy soil were 15.38 and 3.97, respectively, indicating that the soil particle size of paddy soil in China showed great heterogeneity [33]. Notably, there was an autocorrelation between characteristic parameters of soil particle size distribution (Table S1), so it is necessary to conduct a multicollinearity diagnosis of characteristic parameters of soil particle size distribution (Table S2). The SPSS software was used for principal component analysis on characteristic parameters to obtain the main components containing all information on soil particle size distribution (Table S3). The two principal component factors were obtained (F_1 and F_2), and their characteristic roots were determined to be 8.987 and 2.213, respectively. The variance contribution rates to each variable are 74.893% and 18.445%, respectively, which can explain 93.337% of soil particle size distribution.

$$F_1 = 0.1395Z_1 + 0.1264Z_2 + 0.1111Z_3 + 0.1069Z_4 + 0.1107Z_5 + 0.1115Z_6 + 0.1004Z_7 + 0.1023Z_8 + 0.1017Z_9 - 0.0662Z_{10} + 0.0939Z_{11} + 0.0777Z_{12} \quad (1)$$

$$F_2 = 0.1416Z_1 + 0.0786Z_2 + 0.0129Z_3 - 0.007Z_4 + 0.0121Z_5 + 0.0173Z_6 - 0.0305Z_7 - 0.0133Z_8 - 0.0129Z_9 + 0.0612Z_{10} + 0.5167Z_{11} + 0.4945Z_{12} \quad (2)$$

Table 2. Characteristic parameters of soil particle size distribution of fourteen types of soils.

Classification (Location)	Diameter Distance	Consistency	SSA m^2/g	D[3,2] μm	D[4,3] μm	Dv10 μm	Dv30 μm	Dv50 μm	Dv60 μm	Dv90 μm	Cc	Cu
Gray desert soil (Qinghai)	4.06	2.08	1.03	5.85	45.61	2.36	8.54	18.27	25.37	76.54	10.76	1.22
Yellow soil (Guizhou)	4.62	1.40	2.32	2.59	9.21	0.98	2.48	4.96	6.98	23.92	7.16	0.90
Loessal soil (Shaanxi)	2.17	0.67	0.85	7.09	43.03	2.87	20.74	39.52	48.37	88.66	16.83	3.09
Fluvo-aquic soil (Hebei)	3.66	1.17	1.15	5.22	24.27	2.03	6.05	15.74	23.48	59.54	11.60	0.77
Red earth (Jiangxi)	4.73	2.54	1.67	3.60	25.65	1.36	4.09	8.66	12.31	42.37	9.06	1.00
Paddy soil (Zhejiang)	15.38	3.97	1.35	4.45	54.38	1.61	5.65	12.42	18.65	192.62	11.61	1.06
Purple soil (Chongqing)	5.71	1.80	1.66	3.62	19.00	1.34	3.75	8.62	12.94	50.59	9.69	0.81
Black soil (Jilin)	5.95	1.97	1.53	3.92	17.58	1.55	3.71	7.90	12.49	48.56	8.04	0.71
Black soil (Heilongjiang)	3.17	1.06	1.40	4.27	13.87	1.72	4.69	9.14	12.29	33.43	7.14	1.04
Red earth (Hunan)	5.27	1.69	1.91	3.14	12.49	1.27	3.07	5.78	8.03	31.73	6.31	0.92
Fluvo-aquic soil (Henan)	3.03	1.00	0.99	6.09	39.25	2.24	10.84	28.76	38.16	89.23	17.06	1.38
Fluvo-aquic soil (Shandong)	4.07	1.35	1.00	6.02	35.80	2.28	7.10	21.40	37.33	89.29	16.36	0.59
Sandy soil (Hainan)	2.16	0.65	0.39	15.53	303.27	10.13	160.31	281.98	342.58	619.32	33.81	7.41
Paddy soil (Jiangsu)	4.58	1.48	1.35	4.45	18.69	1.83	4.85	9.69	13.57	46.24	7.41	0.95

Note: SSA represents the specific surface area; D[3,2] represents the surface area average particle size; D[4,3] represents the volume average particle size.

3.1.3. Characteristic Parameters of Soil Pore Size Distribution

The curves of soil pore size distribution were obtained according to the measurement results of mercury piezometric method [34,35]. As shown in Figure 1, the pore volume percentages of fourteen types of soils decreased gradually with the increase in pore size. When the pore size ranged from 0.5 μm to 30.0 μm, there was an obvious decreasing trend in the curves of soil pore size distribution, which indicated that the soil pore size was mainly concentrated at 0.5–30.0 μm (Table S4). According to the curves of soil pore size distribution, the pore volume of different soil pore sizes (>75.0, 30.0–75.0, 7.5–30.0, 0.5–7.5, 0.05–0.5, and <0.05 μm) were calculated (Table S5). The pore volumes of soil pore size (7.5–30.0 μm) of Henan fluvo-aquic soil, Hainan sandy soil, Shaanxi loessal soil, Chongqing purple soil, Shandong fluvo-aquic soil, Heilongjiang black soil, Hebei fluvo-aquic soil, and Jilin black soil were mainly distributed from 0.09 cm^3/g to 0.17 cm^3/g. The pore volumes of Qinghai brown soil and Zhejiang paddy soil in the range of 0.5–7.5 μm were 0.0965 cm^3/g and 0.0956 cm^3/g, respectively. Most of the soil pore size of Jiangsu paddy soil, Jiangxi red earth, Hunan red earth, and Guizhou yellow soil was lower than 7.5 μm, and the pore volumes were 0.0243, 0.052, 0.0462, and 0.0317 cm^3/g, respectively. This phenomenon was due to the fine-grained soils with high content that could easily enter the interior of the mesopores, which reduced the connectivity of soil pores and formed fine pores [36].

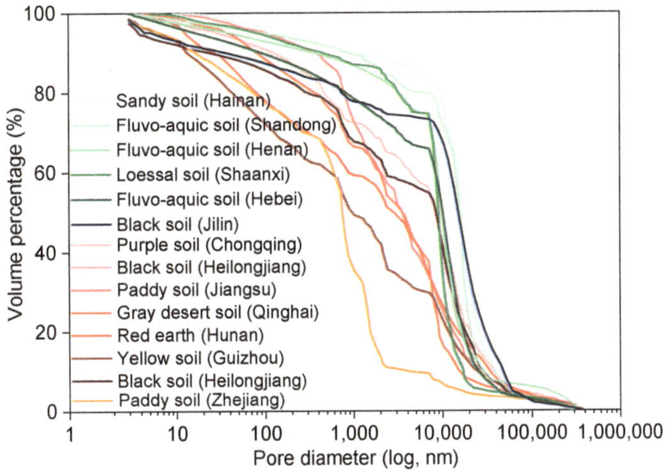

Figure 1. Soil pore size distribution curves of fourteen types of soils.

3.2. Migration Distance of Diesel in Fourteen Types of Soils

With the leakage of 10 mL diesel on fourteen types of soil surfaces, the migration process of diesel front was fast initially and then slowed down to reach a steady state in the soil (Figure 2). The migration rate of diesel before 0.5 h was about 1 cm/h, then gradually trended to 0. The previous study reported that NAPLs are strongly driven by gravity at the initial stage of leakage [37]. Compared with the migration distance of diesel in fourteen types of soils, the migration process of diesel could be divided into three categories (I: 3.0 cm < S < 5.0 cm, II: 5.0 cm < S < 7.0 cm, III: S > 8.0 cm). For example, the migration distance of diesel in seven types of soils including gray desert soil (Qinghai), yellow soil (Guizhou), red earth (Jiangxi and Hunan), paddy soil (Zhejiang and Jiangsu), and purple soil (Chongqing) ranged 3.0 cm to 5.0 cm.

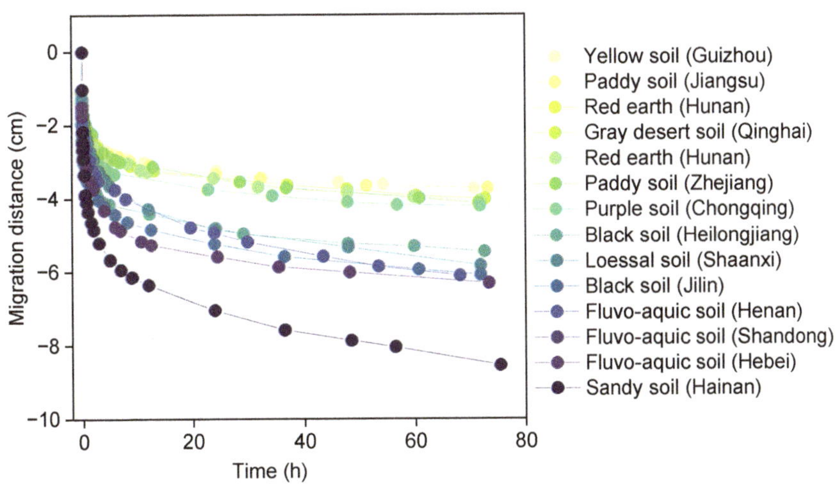

Figure 2. Migration kinetics of diesel in fourteen types of soils.

Combined with the migration data of diesel, the third-order exponential equation (Equation (3)) was used to fit the dynamic process of diesel migration in fourteen types of soils [38].

$$S = Y_e - Ae^{(-t/a)} - Be^{(-t/b)} - Ce^{(-t/c)} \quad (3)$$

where S is the migration distance of diesel front in the soil column, cm; Y_e represents the theoretical migration distance when diesel formed steady state in the soil column, cm; and other values are equation fitting parameters.

The migration kinetics could be satisfactorily described with the third-order exponential equation ($R^2 > 0.99$), and the equilibrium migration distance ranged from 3.42 cm (Guizhou yellow soil) to 8.82 cm (Hainan sandy soil) (Table S6). The equilibrium migration distance of diesel in fourteen types of soils is as follows: sandy soil > sandy loam soil > sandy clay loam soil > clay loamy soil > sandy clay soil, which further confirmed that soil structure characteristics are key factors that lead to the migration distance of diesel under spill condition [38].

3.3. Residual Amount of Diesel in Fourteen Types of Soils

Figure 3 showed the variation of the residual amount of diesel after migrating 72 h in fourteen types of soils. The residual amount of diesel in seven types of soils gradually increased first and then decreased with the increase in soil depth (Hunan red earth, Shandong fluvo-aquic soil, Shaanxi loessal soil, Guizhou yellow soil, Henan fluvo-aquic soil, Hebei fluvo-aquic soil, and Jilin black soil) (Figure 3a). For example, the maximum residual amount of diesel in Hunan red earth, Shandong fluvo-aquic soil, Shaanxi loessal soil, and Guizhou yellow soil were observed at 1.5 cm, the values were 7.8, 7.4, 7.2, and 8.2 g/kg, respectively. A similar phenomenon was reported by previous studies, which was attributed that the migration rate gradually decreased with the increase in soil depth [38,39]. Meanwhile, the high viscosity of diesel is also an important factor for its interception by the upper layer of soil [40]. As shown in Figure 3b, the residual amount of diesel in gray desert soil (Qinghai), purple soil (Chongqing), paddy soil (Zhejiang and Jiangsu), black soil (Heilongjiang), and red earth (Hunan) decreased with the increase in soil depth, even the residual amount of diesel decreased rapidly with the soil depth exceeding 2.5 cm. Due to the larger particle size and porosity, the residual amount of diesel is uniformly distributed in Hainan sandy soil. Since the shortest migration distance of diesel in fourteen types of soils was 3.45 cm, only the residual amount of diesel in the 0-3 cm soil layer were focused on in this study. In the 0–1 cm soil layer, the residual amount of diesel (Y_1) ranged from 2.8 g/kg (Hainan sandy soil) to 7.7 g/kg (Zhejiang paddy soil). In the 1–2 cm soil layer, the residual amount of diesel (Y_2) in nine types of soils is more than 6.0 g/kg, and the average concentration was 1.2 times that of diesel in the 0–1 cm soil layer. In summary, the residual amount of diesel in fourteen types of soils was as follows: sandy clay loam soil > clay loamy soil > sandy clay soil > sandy loam soil > sandy soil.

3.4. Statistical Analysis of Soil Characteristics Affecting Diesel Migration and Residue

Basic physiochemical properties and structure characteristics of soil could affect the behavior of NAPLs in the soil [11,41]. Herein, the basic physiochemical properties and structure characteristics of soil were selected as the independent variables of multivariate statistical analysis, and the main controlling factors affecting the migration distance and residual amount of diesel were analyzed by principal component regression and multiple linear regression analysis.

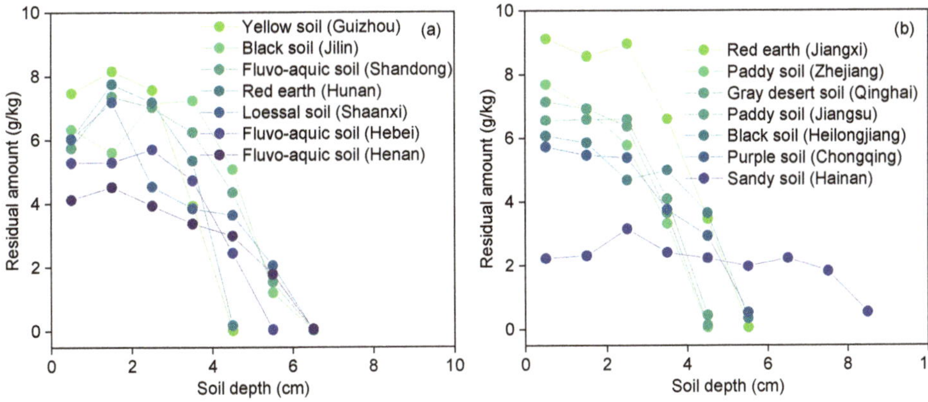

Figure 3. Residual amount of diesel in fourteen types of soils after 72 h.

3.4.1. Effect of Basic Physiochemical Properties of Soil

Correlation analysis was conducted between the migration distance and residue amount of diesel in the 0–3 cm soil layer and the basic physicochemical properties of soil (Figure 4). Residual amounts of diesel in the 0–1 cm and 1–2 cm soil layers have a negative correlation with soil bulk density ($R^2 > 0.55$, $p < 0.05$). In comparison, there is a poor correlation between migration distance or residue amount of diesel in the 2–3 cm (Y_3) soil layer and other basic physiochemical properties (e.g., organic matter content and cation exchange capacity). This result indicated that during the spill of diesel, the chemical properties of soil have little effect, while soil structure characteristics have an obvious effect on the migration distance and residue amount of diesel in unsaturated soil. The phenomenon was inconsistent with the current results that soil organic matter is a key factor in determining the adsorption of organic pollutants by soil [42,43], which was attributed to the fact that diesel with a fast infiltration rate is hard to fully interact with the soil [44]. Han et al. [20] found that although two types of soils with the same soil porosity and organic matter content, soil with high clay content possessed more micropores, resulting in the higher residual amount of NAPL in the soil.

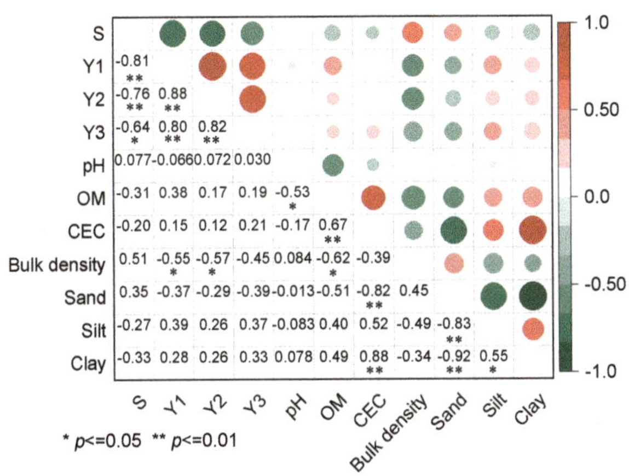

Figure 4. Correlation analysis between physiochemical properties, migration distance, and residual amount of diesel in the 0–3 cm soil layer.

3.4.2. Effect of Soil Particle Size Distribution

As shown in Table S7, there was a negative correlation between migration distance and characteristic parameters of soil particle size distribution (except for specific surface area, diameter distance, and consistency of soil particle), and the residual amount of diesel in the 0-3 cm soil layer is negatively correlated with specific surface area of soil particle ($R^2 > 0.627$, $p < 0.01$). Due to the multicollinearity between the characteristic parameters of soil particle size distribution (Table S2), we extract the principal components (F_1 and F_2) and then carry out a stepwise regression analysis with migration distance and residual amount of diesel. The regression equations are $S = 5.201 + 1.074F_1 - 0.574F_2$ ($R^2 = 0.722$, $p < 0.01$), $Y_1 = 5.962 - 0.842F_1 + 0.659F_2$ ($R^2 = 0.671$, $p < 0.01$), $Y_2 = 6.355 - 1.014F_1$ ($R^2 = 0.493$, $p < 0.01$), and $Y_3 = 5.901 - 0.96F_1$ ($R^2 = 0.433$, $p < 0.01$), respectively.

The regression coefficients of standardized independent variables in the regression equation are listed in Table S8. It can be seen that in the regression equation of migration distance, the absolute value of the consistency coefficient is the largest (0.2004), indicating that the consistency is the main influencing factor on the migration distance of diesel in fourteen types of soils; similarly, the consistency has also the greatest influence on the residual amount of diesel in the 0–1 cm soil layer. The consistency is defined as the distribution uniformity of soil particle size distribution [45]. The larger the consistency value was, the stronger heterogeneity of soil pore size distribution and retardation to diesel migration, which reduced the migration ability of diesel [46], further increased the residual amount of diesel in the topsoil. In the regression equation of residual amounts of diesel in the 1–2 cm and 2–3 cm soil layers, the absolute value of Dv90 coefficient was 0.1415 and 0.1339, respectively, indicating that Dv90 has the greatest impact on the residual amounts of diesel in the 1–2 cm and 2–3 cm soil layer. Dv90 referred the soil particle size corresponding to 90% of the volume fraction [47]. The specific surface area of soil particle decreased as Dv90 value increased, which is adverse to the adsorption capacity of soil to diesel [48]. Moreover, the more soil pores formed by soil large particles could provide channels for diesel migration, resulting in the reduction in the residual amounts of diesel in 1–2 cm and 2–3 cm soil layers.

3.4.3. Effect of Soil Pore Size Distribution

Figure 5 showed the correlation between migration distance and residual amount of diesel in the 0-3 cm soil layer and soil pore size distribution. Migration distance of diesel was positively correlated with the pore volume of soil pores at 30.0–75.0 μm ($R^2 = 0.55$) and 7.5–30.0 μm ($R^2 = 0.62$), but negative correlated with the pore volume of soil pores at 7.5–0.5 μm ($R^2 = 0.82$), 0.5–7.5 μm ($R^2 = 0.68$), and smaller than 0.05 μm ($R^2 = 0.61$). It is known that the large pores contributed to reducing the capillary pressure that the NAPL need to overcome to go through these pores and thus migrate through the soil [10,49]. Similar results have been reported that soil macropores (>60 μm diameter) and mesopores (15 μm < diameter < 60 μm) was the main factor affecting the migration of fluid in the porous medium [15]. The correlation coefficient between the residual amount of diesel in the 0–1 cm soil layer and the pore volume of soil macropores (0.05–7.5 μm) was higher than 0.60, and the residual amount of diesel in the 1–3 cm soil layer was higher than the correlation with the pore volume of soil mesopores (<0.05 μm). The results indicated that soil macropores volume are considered to be important for the diesel residue on the soil surface, while the diesel residue under topsoil was affected by soil mesopores volume smaller than 0.05 μm significantly. The higher capillary force in the soil, which results from a small pore body and narrow throats, caused more diesel to be trapped in the pores [24].

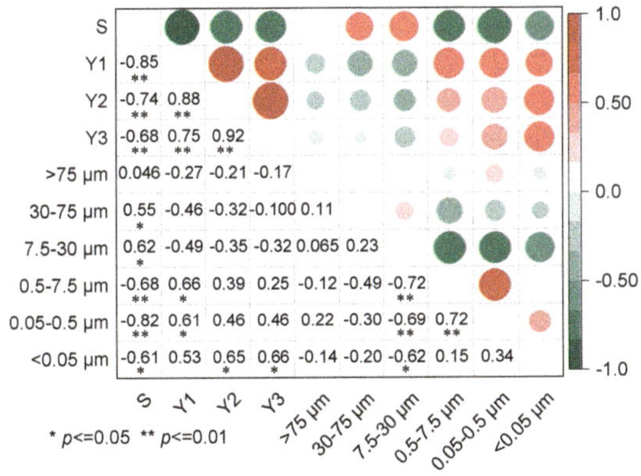

Figure 5. Correlation analysis between soil pore size distribution, migration distance, and residual amount of diesel in the 0–3 cm soil layer.

4. Conclusions

In this study, the experiment simulated a diesel spill that was performed in fourteen types of soils distributed in the large-scale region of China to investigate the migration distance and vertical distribution of diesel. Based on statistical analysis, the key factors affecting diesel behavior are revealed in terms of soil particle and pore size distribution. There were significant differences between fourteen types of soils, with regard to the soil particle, soil pore, and diesel migration and residue. After 72 h of leakage, the migration distance of diesel ranged from 3.42 cm to 8.82 cm in fourteen types of soils. Except for sandy soil, diesel is mainly distributed in the 0–3 cm soil layer, and the residual amount were 7.85–26.66 g/kg, in the order of sandy clay loam soil > clay loamy soil > sandy clay soil > sandy loam soil > sandy soil. Further confirmed from macro-level that soil bulk density controlled the residue amount of diesel in the topsoil (0–2 cm). Soil microstructure including the consistency of soil particle and volume of soil macropores (0.05–7.5 μm) are considered to be important for the residual amount in the 0–1 cm soil layer and migration distance of diesel. The large soil particles corresponding to 90% of volume fraction and volume of soil mesopores (<0.05 μm) are key factors affecting the residual amount of diesel in the 1–3 cm soil layer. This study would provide support for investigating the formation process and mechanism of residual LNAPLs in the soil.

Supplementary Materials: The following supporting information can be downloaded at: https://www.mdpi.com/article/10.3390/toxics11010016/s1, Table S1: Correlation analysis between parameters of soil particle size distribution; Table S2: Multicollinearity diagnostic between soil particle size distribution parameters; Table S3: Eigenvalues and proportion of factor analysis; Table S4: Volume percentage of different soil pore tested by mercury intrusion porosimetry; Table S5: Volume of different soil pore tested by mercury intrusion porosimetry; Table S6: Fitted results of diesel transport kinetics by third-order exponential function; Table S7: Correlation analysis between parameters of soil particle size distribution, residual concentration and migration depth; Table S8: Principal component regression coefficients and constants.

Author Contributions: Conceptualization, X.L. and B.W.; methodology, Q.Z.; validation, X.L., B.W. and Q.G.; formal analysis, X.L. and B.W.; investigation, X.Z.; resources, X.Z.; data curation, J.S. and Z.S.; writing—original draft preparation, X.L.; writing—review and editing, B.W.; visualization, Q.Z.; supervision, F.M.; project administration, X.L.; funding acquisition, X.L. and Q.G. All authors have read and agreed to the published version of the manuscript.

Funding: This research was funded by China Postdoctoral Science Foundation, grant number 2022M722997, National Key Research and Development Program of China, grant number 2019YFC1803800. The APC was funded by China Postdoctoral Science Foundation.

Institutional Review Board Statement: Not applicable.

Informed Consent Statement: Not applicable.

Data Availability Statement: Not applicable.

Acknowledgments: We would like to thank China Postdoctoral Science Foundation (2022M722997) and National Key Research and Development Program of China (2019YFC1803800) for their financial support.

Conflicts of Interest: The authors declare no conflict of interest.

References

1. Kaimi, E.; Mukaidani, T.; Miyoshi, S.; Tamaki, M. Ryegrass enhancement of biodegradation in diesel-contaminated soil. *Environ. Exp. Bot.* **2006**, *55*, 110–119. [CrossRef]
2. Gupta, D.; Sarker, B.; Thadikaran, K.; John, V.; Maldarelli, C.; John, G. Sacrificial amphiphiles: Eco-friendly chemical herders as oil spill mitigation chemicals. *Sci. Adv.* **2015**, *1*, e1400265. [CrossRef] [PubMed]
3. Erofeevskya, L.A.; Aleksandrov, A.R. Liquidation of oil and petroleum products spills based on use of NA and CA-differences of zeolite. *IOP Conf. Ser. Earth Environ. Sci.* **2020**, *459*, 052011. [CrossRef]
4. Wang, S.; Wang, D.; Yu, Z.C.; Dong, X.G.; Liu, S.M.; Cui, H.M.; Sun, B. Advance in research on petroleum biodegradability in soil. *Environ. Sci. Proc. Imp.* **2020**, *23*, 9–27. [CrossRef]
5. Lari, K.S.; Rayner, J.L.; Davis, G.B. Towards optimizing LNAPL remediation. *Water Resour. Res.* **2019**, *55*, 923–936. [CrossRef]
6. Garg, S.; Newell, C.J.; Kulkarni, P.R.; King, D.C.; Adamson, D.T.; Renno, M.I.; Sale, T. Overview of natural source zone depletion: Processes, controlling factors, and composition change. *Ground Water Monit. Remediat.* **2017**, *37*, 62–81. [CrossRef]
7. Luo, L.; Zhang, S.Z.; Ma, Y.B. Evaluation of impacts of soil fractions on phenanthrene sorption. *Chemosphere* **2008**, *72*, 891–896. [CrossRef]
8. Yoon, H.; Werth, C.J.; Valocchi, A.J.; Oostrom, M. Impact of nonaqueous phase liquid (NAPL) source zone architecture on mass removal mechanisms in strongly layered heterogeneous porous media during soil vapor extraction. *J. Contam. Hydrol.* **2008**, *100*, 58–71. [CrossRef]
9. Chen, B.L.; Huang, W.H. Effects of compositional heterogeneity and nanoporosity of raw and treated biomass-generated soot on adsorption and absorption of organic contaminants. *Environ. Pollut.* **2011**, *159*, 550–556. [CrossRef]
10. Su, K.N.; Rahman, N.A.; Bob, M.M. Observation of light non-aqueous phase liquid migration in aggregated soil using image analysis. *Transp. Porous Med.* **2012**, *92*, 83–100.
11. Rahman, N.A.; Foong, L.K.; Nazir, R.; Lewis, R.W. Vibration effect influence upon non-aqueous phase liquid migration in double-porosity soil. *Geol. Croat.* **2018**, *71*, 163–171. [CrossRef]
12. Liang, Y.T.; Nostrand, J.D.V.; Deng, Y.; He, Z.L.; Wu, L.Y.; Zhang, X.; Li, G.H.; Zhou, J.Z. Functional gene diversity of soil microbial communities from five oil-contaminated fields in China. *ISME J.* **2011**, *5*, 403–413. [CrossRef] [PubMed]
13. Liang, Y.T.; Zhang, X.; Wang, J.; Li, G.H. Spatial variations of hydrocarbon contamination and soil properties in oil exploring fields across China. *J. Hazard. Mater.* **2012**, *241–242*, 371–378. [CrossRef]
14. Tang, S.; She, D.; Wang, H. Effect of salinity on soil structure and soil hydraulic characteristics. *Can. J. Soil Sci.* **2021**, *101*, 62–73. [CrossRef]
15. Sasal, M.C.; Andriulo, A.E.; Taboada, M.A. Soil porosity characteristics and water movement under zero tillage in silty soils in Argentinian Pampas. *Soil. Tillage Res.* **2006**, *87*, 9–18. [CrossRef]
16. Saripalli, K.P.; Rao, P.; Annable, M. Determination of specific NAPL–water interfacial areas of residual NAPLs in porous media using the interfacial tracers technique. *J. Contam. Hydrol.* **1998**, *30*, 375–391. [CrossRef]
17. Arshadi, M.; Gesho, M.; Qin, T.; Goual, L.; Piri, M. Impact of mineralogy and wettability on pore-scale displacement of NAPLs in heterogeneous porous media. *J. Contam. Hydrol.* **2020**, *230*, 103599. [CrossRef]
18. Liao, S.C.; Saleeba, Z.; Bryant, J.D.; Abriola, L.M.; Pennell, K.D. Influence of aqueous film forming foams on the solubility and mobilization of non-aqueous phase liquid contaminants in quartz sands. *Water Res.* **2021**, *195*, 116975. [CrossRef]
19. Montero, E. Rényi dimensions analysis of soil particle-size distributions. *Ecol. Model.* **2005**, *182*, 305–315. [CrossRef]
20. Han, C.M.; Zhang, H.; Gu, Q.B.; Guo, G.L.; Li, Y.; Li, F.S. Toluene sorption behavior on soil organic matter and its composition using three typical soils in China. *Environ. Earth Sci.* **2013**, *68*, 741–747. [CrossRef]
21. Alloway, B.J. *Heavy Metals in Soils: Trace Metals and Metalloids in Soils and Their Bioavailability*, 3rd ed.; Alloway, B.J., Ed.; Blackie Academic and Professional: London, UK, 2013.
22. Kabata-Pendias, A.; Pendias, H. *Trace Elements in Soils and Plants*, 3rd ed.; CRC Press: London, UK, 2001.
23. Golia, E.E.; Tsiropoulos, G.N.; Füleky, G.; Floras, S.; Vleioras, S. Pollution assessment of potentially toxic elements in soils of different taxonomy orders in central Greece. *Environ. Monit. Assess.* **2019**, *191*, 106. [CrossRef] [PubMed]
24. Kim, M.; Kim, K.Y.; Lim, J.H.; Chan, Y.K.; Kim, S.G.; Han, G.; Han, W.S.; Park, E. Pore-scale investigation of dynamic immiscible displacement in layered media using synchrotron X-ray microtomography. *Environ. Sci. Technol.* **2021**, *56*, 282–292. [CrossRef] [PubMed]
25. Lu, R. *Analytical Methods of Soil and Agricultural Chemistry*; China Agricultural Science and Technology Press: Beijing, China, 1999; pp. 107–240.

26. Li, X.D.; Wu, B.; Zhang, Q.; Liu, Y.Q.; Wang, J.Q.; Xu, D.P.; Li, F.S.; Ma, F.J.; Gu, Q.B. Effects of soil properties on the remediation of diesel-contaminated soil by Triton X-100-aided washing. *Environ. Sci. Pollut. Res.* **2020**, *27*, 23323–23330. [CrossRef] [PubMed]
27. Giménez, D.; Perfect, E.; Rawls, W.J.; Pachepsky, Y. Fractal models for predicting soil hydraulic properties: A review. *Eng. Geol.* **1997**, *48*, 161–183. [CrossRef]
28. Gan, D.Q.; Yang, X.; Zhang, Y.P. Experimental analysis on permeability characteristics of iron tailings. *Math. Probl. Eng.* **2019**, *2019*, 6539846. [CrossRef]
29. Smedt, F.D.; Wierenga, P.J. Solute transport through soil with nonuniform water content. *Soil Sci. Soc. Am. J.* **1978**, *42*, 7–10. [CrossRef]
30. Shi, X.Z.; Yu, D.S.; Warner, E.D.; Sun, W.X.; Petersen, G.W.; Gong, Z.T.; Lin, H. Cross-reference system for translating between genetic soil classification of china and soil taxonomy. *Soil Sci. Soc. Am. J.* **2005**, *70*, 78–83. [CrossRef]
31. Xia, J.; Ren, R.; Chen, Y. Multifractal characteristics of soil particle distribution under different vegetation types in the Yellow River Delta chenier of China. *Geoderma* **2020**, *368*, 114311. [CrossRef]
32. Shamkhi, M.S.; Albadry, H.J. Soil texture distribution for east wasit province, Iraq. *IOP Conf. Ser. Earth Environ. Sci.* **2022**, *961*, 012073. [CrossRef]
33. Wang, H.J.; Liu, Q.H.; Shi, X.Z.; Yu, D.S.; Zhao, Y.C. Carbon storage and spatial distribution patterns of paddy soils in China. *Front. Agric. China* **2007**, *1*, 149–154. [CrossRef]
34. Lima, R.; Rolim, M.M.; Toledo, M.; Tormena, C.A.; Silva, A.; Silva, I.; Pedrosa, E. Texture and degree of compactness effect on the pore size distribution in weathered tropical soils. *Soil Till. Res.* **2022**, *215*, 105215. [CrossRef]
35. Ren, W.G.; Zhou, H.W.; Zhong, J.C.; Xue, D.J.; Wang, C.S.; Liu, Z.L. A multi-scale fractal approach for coal permeability estimation via MIP and NMR methods. *Energies* **2022**, *15*, 2807. [CrossRef]
36. Wang, B.Y.; Liu, B.; Sun, G.X.; Bai, L.H.; Chi, Y.; Liu, Q.; Liu, M. Evaluation of the shale oil reservoir and the oil enrichment model for the first member of the lucaogou formation, western jimusaer depression, junggar basin, NW China. *ACS Omega* **2021**, *6*, 12081–12098. [CrossRef] [PubMed]
37. Ryzhik, V. Spreading of a NAPL lens in a double-porosity medium. *Computat. Geosci.* **2007**, *11*, 1–8. [CrossRef]
38. Wang, Y.Q.; Shao, M.A. Infiltration characteristics of non-aqueous phase liquids in undisturbed loessal soil cores. *J. Environ. Sci.* **2009**, *21*, 1424–1431. [CrossRef]
39. Li, Y.X.; Zheng, X.L.; Ma, Y.F. Experimental study on the infiltration of diesel through unsaturated zone of soil. *Environ. Pollut. Control.* **2011**, *33*, 27–31.
40. Zhang, L.; Zhang, L.Z.; Xu, Z.M.; Guo, X.Y.; Xu, C.M.; Zhao, S.Q. Viscosity mixing rule and viscosity-temperature relationship estimation for oil sand bitumen vacuum residue and fractions. *Energy Fuels* **2019**, *33*, 206–214. [CrossRef]
41. Simantiraki, F.; Aivalioti, M.; Gidarakos, E. Implementation of an image analysis technique to determine LNAPL infiltration and distribution in unsaturated porous media. *Desalination* **2009**, *248*, 705–715. [CrossRef]
42. Kim, P.G.; Tarafdar, A.; Kwon, J.H. Effects of soil pH on the sorption capacity of soil organic matter for polycyclic aromatic hydrocarbons in unsaturated soils. *Pedosphere* **2022**, *in press*. [CrossRef]
43. Potts, J.; Jones, D.L.; Macdonald, A.; Ma, Q.X.; Cross, P. Acetamiprid fate in a sandy loam with contrasting soil organic matter contents: A comparison of the degradation, sorption and leaching of commercial neonicotinoid formulations. *Sci. Total Environ.* **2022**, *842*, 156711. [CrossRef]
44. Kodešová, R.; Vignozzi, N.; Rohošková, M.; Hájková, T.; Kočárek, M.; Pagliai, M.; Kozák, J.; Šimůnek, J. Impact of varying soil structure on transport processes in different diagnostic horizons of three soil types. *J. Contam. Hydrol.* **2009**, *104*, 107–125. [CrossRef] [PubMed]
45. Chen, C.H.; Li, X.L.; Zhang, L.; Ying, Y.; Jiang, L.Q.; Che, S.L. Influence of particle size distribution on the consistency of high permeability MnZn Ferrite. Mater. *Sci. Forum* **2014**, *787*, 357–361. [CrossRef]
46. Zhang, Y.J.; Zhu, S.Q.; Xiao, R.; Wang, J.; Li, F.S. Vertical transport of polycyclic aromatic hydrocarbons in different particle-size fractions of sandy soils. *Environ. Geol.* **2008**, *53*, 1165–1172. [CrossRef]
47. Yang, S.C.; Liu, J.W.; Xu, L.F.; Zhang, M.Y.; Jeng, D.S. A new approach to explore the surface profile of clay soil using white light interferometry. *Sensors* **2020**, *20*, 3009. [CrossRef]
48. Tang, L.; Gudda, F.O.; Wu, C.X.; Ling, W.; El-Ramady, H.; Mosa, A.; Wang, J. Contributions of partition and adsorption to polycyclic aromatic hydrocarbons sorption by fractionated soil at different particle sizes. *Chemosphere* **2022**, *301*, 134715. [CrossRef] [PubMed]
49. Alazaiza, M.Y.D.; Ngien, S.K.; Bob, M.M.; Kamaruddin, S.A.; Ishak, W.M.F. Influence of macro-pores on DNAPL migration in double-porosity soil using light transmission visualization method. *Transp. Porous Med.* **2017**, *117*, 103–123. [CrossRef]

toxics

Review

Recent Advances and Future Prospects on the Tailing Covering Technology for Oxidation Prevention of Sulfide Tailings

Meiyan Si [1], Yunjian Chen [2], Chen Li [1], Yichao Lin [1], Jianhong Huang [1,*], Feng Zhu [3,*], Senlin Tian [1] and Qun Zhao [1]

1 Faculty of Environmental Science and Engineering, Kunming University of Science and Technology, Kunming 650504, China
2 Yunnan Geological Engineering Second Survey Institute Co., Ltd., Kunming 650213, China
3 Faculty of Metallurgy and Environment, Central South University, Changsha 410083, China
* Correspondence: huangjianhong78@163.com (J.H.); zhufeng1990@csu.edu.cn (F.Z.)

Abstract: Acid mine drainage, produced from sulfur-containing mine waste exposed to air, water, and bacteria, is considered as a serious environmental pollutant because of its extremely low pH and excessive heavy metals. In order to solve the ecological environment problems caused by the acid mine drainage, treatment methods such as neutralization, adsorption, passivation, bio-inhibition, and physical coverage have been developed. Nevertheless, these methods are terminal treatment methods, which are unable to prevent the generation of acid mine drainage at the source. Recently, it is noteworthy that the tailing covering technology is particularly emphasized, owing to its superior source control capability. By reducing the contact with air, water, and bacteria, the oxidation of sulfide tailings is significantly reduced, thus avoiding the production of acid mine drainage. To date, massive research has been studied and parts of technologies have been applied, but the review on the principles, processes, and applications of these technologies are still lacking. Thus, the present review aims to increase the knowledge related to the most relevant application of tailing covering technology with the following aspects: (i) the background, concepts, and performance of tailing covering technology; (ii) the applicable conditions for each tailings coverage system and their advantages and limitations; and (iii) the future perspective of this technology.

Keywords: acid mine drainage; tailing covering technology; sulfide tailings; oxidation prevention

Citation: Si, M.; Chen, Y.; Li, C.; Lin, Y.; Huang, J.; Zhu, F.; Tian, S.; Zhao, Q. Recent Advances and Future Prospects on the Tailing Covering Technology for Oxidation Prevention of Sulfide Tailings. *Toxics* **2023**, *11*, 11. https://doi.org/10.3390/toxics11010011

Academic Editor: Tiziana Missana

Received: 15 November 2022
Revised: 15 December 2022
Accepted: 20 December 2022
Published: 22 December 2022

1. Introduction

Continuous exploitation of mineral resources plays a great role in industrial development, but also produces massive mine solid waste. As reported, the annual discharge of mine tailings in the world has exceeded 10 billion tons [1]. The accumulation of such a large amount of tailings is a great hidden danger to the surrounding environment. Acid mine drainage (AMD), produced by the oxidation of sulfide tailings (Figure 1) [1], is considered to be the most challenging pollutant to treat because of its extremely low pH, high heavy metals content, and high sulfate concentration. Furthermore, AMD contains large quantities of toxic substances, such as cyanides, hazardous heavy metals (e.g., Cu, Zn, Cd, Mn, Pb, Cr, Ni, and Fe), and toxic metalloid (e.g., As and Se) [2,3]. Once the AMD leaks, it can pose a long-term and large-scale threat to the environment around the mining area, especially the surface water, groundwater, and soil, thereby affecting the health of the residents and the biodiversity of ecosystems [3–10]. Therefore, preventing and controlling AMD pollution are two of the top priorities of today's global mining industry.

To solve the problem of AMD pollution, massive technologies have been proposed and applied, such as neutralization [11], sulfide precipitation [12], electrocoagulation [13], wetlands construction [14], and microbial treatment [15]. Although these methods have exhibited great performance in AMD treatment, they are all limited to taking effect after AMD has been generated and spread. Inevitably, ecological risks exist in the runoff, collection, and treatment of AMD [5]. Instead, source control to avoid the generation of AMD is a better approach, and tailing covering technology has emerged [16]. Based on the principle that the AMD is produced by the interaction of sulfur-containing tailings with water, oxygen and microbial components, various technical branches have been proposed for applicability to different mining environments and natural conditions. Through the unremitting efforts of research, so far, tailing covering technology has proved to have great advantages in AMD source control. However, the review of these technologies is still lacking. Herein, we believe that a comprehensive review of tailing covering technology and the evaluation of single technology could be key steps towards the exploration of new materials for AMD covering technology in the future.

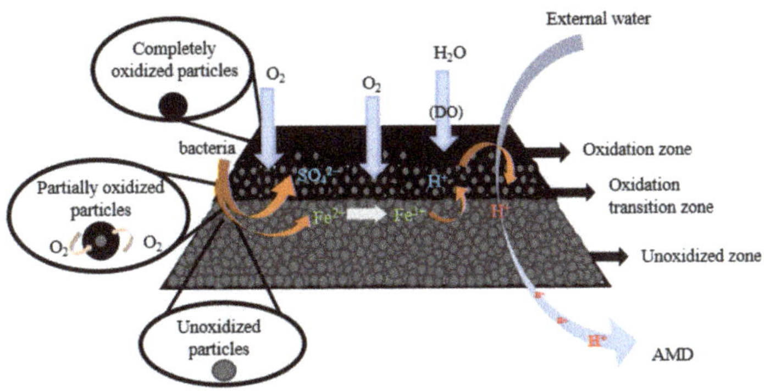

Figure 1. Oxidation of sulfide minerals and formation of AMD.

In this review, we systematically reviewed a large number of foreign studies and studied how the covering system was implemented and worked and mapped out the previous studies and current trends of tailings covering technologies. The background, concepts, and performance of tailing covering technology are comprehensively discussed. The applicable conditions for each tailings coverage system and their advantages and limitations are systematically reviewed and compared. Furthermore, the future perspective outlook for advances in this technology is proposed.

2. Tailings Cover Technology

Water and oxygen are the main reactants to produce AMD, and the inhibition methods should consider controlling the availability of one or two components. Therefore, two strategies are adopted: (i) preventing oxygen from entering the tailings pile and thus reducing the rate of sulfide oxidation; and (ii) isolating the infiltration of external water and thus weakening the role of dissolved oxygen. According to different coverage principles, tailing cover systems can be divided into dry covers, wet covers, and organic covers, which are expatiated in the following sections.

2.1. Wet Covers

Wet covers forms a fully saturated water cover to prevent tailing oxidation caused by the advection transport and diffusion of oxygen [17], thus reducing the rate of tailing oxidation.

2.1.1. Flooded Cover

When sulfur-containing tailings are placed below the water surface, the low solubility (25 °C, 8.6 mg·L^{-1}) and low effective diffusion coefficient ($D_e = 2 \times 10^{-9}$ m^2·s^{-1}) of oxygen in water can be utilized to create a natural anoxic environment, which can theoretically permanently inhibit the oxidation activity of tailings (Figure 2). Garcilaso et al. reported that the flooded cover is an efficient treatment scheme [18]. The natural isolation of water made the flux of O_2 through the water layer to the tailings reduce by 10,000 times, and

D_e decreases from 3.86×10^{-6} to 8×10^{-12} $m^2 \cdot s^{-1}$ [19,20]. Romano et al. [21] used a couples oxygen diffusion and sulfide-mineral oxidation (PYROX) model to simulate the oxidation of sulfide minerals over a 100-year period, with sulfate as the target product. Results showed that the oxidation rate of flooded cover system decreased by about 99.1% compared with open tailings.

Unfortunately, the immersed sulfide tailings are not completely inert, and the oxidation reaction continued to occur at the tailing–water interface slowly. An investigation on a historical tailing pond (Savage River mine, located in northwest Tasmania, Australia) found that there was an oxidation zone of 0.05 m under the water surface of 1.5 m, and the acidic pH values were measured of the tailings pore water, and the heavy metals increased (Ni: 76–123 $mg \cdot kg^{-1}$, Zn: 45–54 $mg \cdot kg^{-1}$, Cu: 91–679 $mg \cdot kg^{-1}$) [22]. Under the action of wind and waves of shallow water overburden layer, turbulence and tailing re-suspension occurred at the air–water interface, which are important reasons for the oxidation of underwater tailings [23]. In addition, the increase in interstitial water acidity (pH, from 9.5 to 8.5) and SO_4^{2-} production just under the interface also proves that the tailings gradually oxidized [24]. Research shows that the presence of a transitional oxidation front at approximately 0.3–0.6 m across the sub-aerial zone, with interlayer oxidation tailings containing pyrite enriched in Cu, Co and Zn, is observed. Fortunately, the low sulfur alteration intensity (SAI < 2/10) makes the combined organic and water cover effectively limiting, and the risk for AMD production was low [22] To tackle this problem, Mustafa et al. [25] weakened the influence of wind by optimizing the overlying depth, thus controlling the sediment re-suspension within an acceptable range.

Flooded cover is effective as an oxygen barrier, but it needs high economic expenditure, such as water supplement, dam construction, and maintenance. More importantly, it is not suitable for arid and semi-arid areas where annual evaporation is greater than precipitation, and water supplement will greatly increase the maintenance cost in the later stage [26]. In addition, the stability of dam is a consideration, especially in earthquake-prone areas.

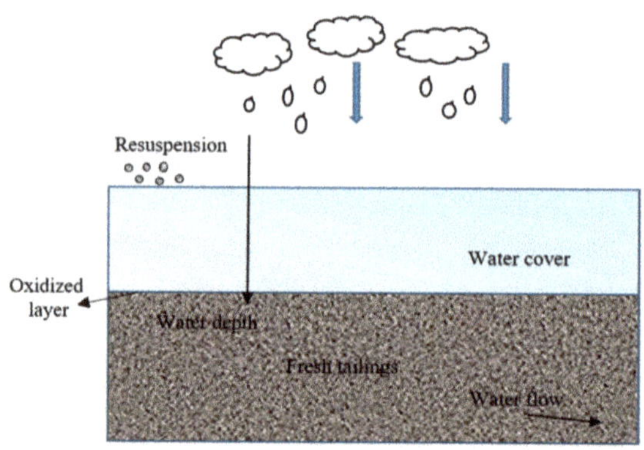

Figure 2. Schematic diagram of flooded cover system [22]. (Note: the thick arrow represents rainfall; the long arrow indicates the depth of water cover; the arrow on the left indicates the extension of the oxidized layer; the arrow on right indicates the runoff direction of water in the fresh tailing.) This picture was adapted from [22] with permission from publisher Elsevier in May 2016.

2.1.2. Cover with Capillary Barrier Effects (CCBE)

CCBE is a typical multilayer cover mode, which consists of at least three layers of cover [27] (Figure 3). A capillary break layer (CBL) made of coarse-grained material is located at the bottom and serves as a support. The middle part is a moisture-retaining layer (MRL) made of fine particles, which has low saturated hydraulic conductivity (Ksat > 10^{-5} $cm \cdot s^{-1}$) and high water retention capacity to limit oxygen migration [28]. The surface layer is an evaporation protective layer (EPL) made of coarse particles, which plays a role in limiting evaporation and promoting horizontal drainage. When two kinds of particle layers with different particle sizes contact, the vertical flow between the two layers is often limited due to the difference in unsaturated hydraulic properties, resulting in a capillary barrier effect [29,30], which maintains high saturation in the layer with finer particle sizes (SR > 85%). Therefore, the substitution of various cheap materials (industrial waste, mining waste, natural materials, etc.) has become the research hotspot of CCBE.

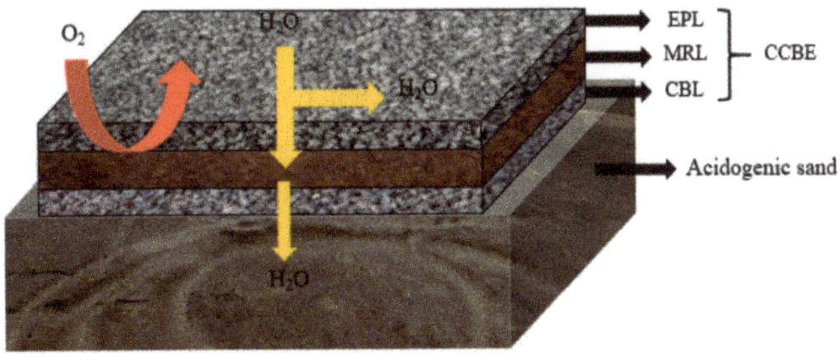

Figure 3. Schematic diagram of cover with capillary barrier effects.

A large number of findings have confirmed the effectiveness of CCBE in limiting oxygen diffusion (about 99%) and AMD production [31–33]. Dagenais et al. [34] monitored the CCBE performance data for four consecutive years and found that the MRL remained saturated for many years, with oxygen flux as low as 10 $g \cdot m^{-2} \cdot a^{-1}$. Larochelle et al. [35] used acid-producing waste rock as CBL to conduct laboratory column experiments. They found that the saturation of MRL remained around 85–90%, and the pH remained close to neutral during the whole test period. A similar result was obtained by Molson et al. [33], wherein the O_2 flux was only related to the porosity of the material when the MRL layer was close to saturation, and its thickness had relatively little influence. For example, in five pilot-scale field tests (four sites were constructed with CCBE over the tailings, a site was uncovered) of the Norebec-Manitou mine site, Canada, the finite volume model MIN3P was used to simulate the reduction of AMD discharge by using the cover with CCBE [33]. The monitored data after 1200 days showed that the pH remained neutral (about 6.5~7), the concentration of $SO_4{}^{2-}$ and Fe^{2+} decreased to 1700 $mg \cdot L^{-1}$ and 8 $mg \cdot L^{-1}$, respectively, and Cu^{2+} concentrations decreased to 0.001 $mg \cdot L^{-1}$. Overall, 1–7 orders of magnitude are reduced compared with the uncovered.

Generally, CCBE can effectively inhibit the generation of AMD. However, in the area where evaporation is greater than precipitation, it is easy to cause excessive desaturation of MRL layer and gradually lose the function as oxygen barrier. In addition, the CCBE has a disadvantage on long-term stability. According to the report of Pabst et al. [30], the inhibition efficiency of sulfur-containing tailings decreased significantly after being covered for 10 years. More importantly, CCBE is not being an economic option. The covering materials required for its construction are usually shipped from other locations because on-site alternatives such as non-acid-producing tailings are not readily available, which greatly increases construction costs [36]. Therefore, CCBE is not the optimal coverage option in terms of efficiency and economy.

2.1.3. Elevated Water Table (EWT)

The EWT method mainly relies on raising the groundwater level to keep the depth at the edge height of the capillary effect layer to increase the saturation of the tailings [37–39]. Hence, controlling the height of the water level is the key to the effectiveness of this technology, and it is also the main difference from flooded cover (Figure 4). In saturated or nearly saturated porous media, the effective diffusion coefficient of oxygen is very low, resulting in a lower oxygen flux that reduces the oxidation rate of sulfide minerals [40]. Therefore, this technology is usually coupled with a monolayer cover method. Bussière et al. [41]

combined EWT with monolayer cover technique to significantly inhibit sulfide mineral oxidation. Related studies have confirmed this conclusion. For example, the combination of coarse-grained covering materials and EWT could increase penetration and limit water loss caused by evaporation [40,42]. The combination with fine-grained covering materials could promote moisture retention and prevent evapotranspirative demand, and no significant tailing oxidation is observed ($<10^{-4}$ kg·(m^2·a^{-1})$^{-1}$) [43,44]. Some studies have shown that there is a correlation between the depth of groundwater level and the rate of oxygen consumption [45]. Ouangrawa et al. [37] demonstrated that oxygen diffusion and oxidation of sulfide minerals can be prevented by keeping the water table depth less than the air entry value (AVE) of tailings and keeping the tailings highly saturated (Sw $\geq$ 90%). In the simulated profiles experiment of SO_4^{2-} and Fe(III), within column 1, peak SO_4^{2-} and Fe(III) concentrations reached about 12,000 mg·L^{-1} and 2000 mg·L^{-1}, respectively. However, as the EWT increased, within column 6, SO_4^{2-} and Fe(III) concentrations were significantly lower, at 3300 mg·L^{-1} and 10 mg·L^{-1}, respectively. Furthermore, the observed discharge pH became more neutral (pH~7–8) [38]. The decrease in SO_4^{2-} and Fe(III) concentrations were limited by gypsum and ferrihydrite precipitation. Similar to flooded cover and CCBE technologies, EWT technique adopts the low effective diffusion coefficient of O_2 in saturated and near-saturated media [19,40]. The transmission of O_2 is the main factor limiting the oxidation of sulfide minerals.

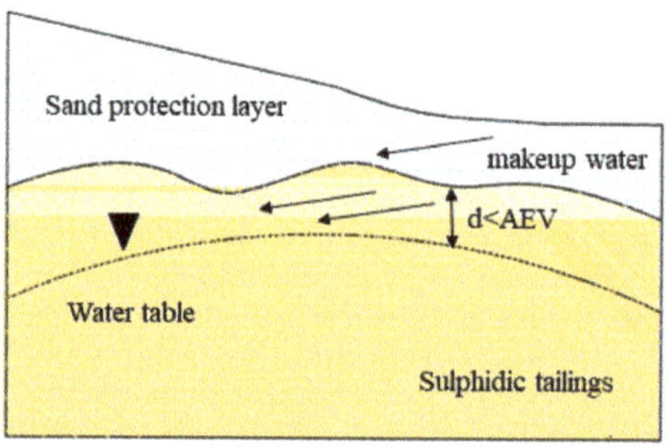

Figure 4. Schematic diagram of elevated water table coverage. (Note: the black triangle indicates the water level depth; the black dotted line indicates that the water level depth is less than the minimum value of AEV; and the black arrow indicates the flow direction of makeup water.

EWT technology is an effective auxiliary governance scheme, but its environmental and technical requirements are too high. The maintenance of high water level requires long-term funds investment, technical support, and personnel maintenance. The application of EWT technology can only be limited in humid areas with low evaporation, and artificial reconstruction will greatly increase the economic cost. This is a common problem with all wet cover techniques.

2.2. *Dry Covers*

Dry covers are low permeability preservative layers made of inorganic mineral materials. It can prevent oxygen and water from entering the tailings layer and reduce the oxidation rate of sulfide minerals [46]. The dry cover layer is composed of materials with different granulation characteristics, such as compacted clay, portland cement, fly ash, etc. [47].

2.2.1. Monolayer Cover

Monolayer cover is the simplest and most widely used method of tailing pond cover technology. It reduces the diffusion flux of oxygen and the accumulation of acid caused by oxidation by virtue of the low porosity and chemical properties of the surface covering material, such as limestone or phosphate minerals. In practical applications, monolayer cover usually refers to a mixture of two or more materials with different properties (Figure 5). In some studies, desulfurized tailings with non-acid producing activity were used as a single material covering layer, and the oxidation rate was reduced by 75~82% [21]. However, in the face of the huge storage of sulfide minerals, the production of AMD is still huge in sufficient oxidation time. Hence, the oxygen barrier performance of single layer covering is poor. Pabst et al. [48] put forward a similar view that a single material covering layer could not prevent the oxidation of sulfide minerals in the underlying tailings. TA tailings covered with CA produce a very acidic leachate (pH < 3) and, after 10 cycles, the pH was below 2.5. Meanwhile, the concentration of sulfate and iron reaches 40,000 $mg \cdot L^{-1}$ and 8000 $mg \cdot L^{-1}$, respectively. These condition inhibitions on AMD were limited. Therefore, more research tends to use a mixture of multiple materials. For example, the blending of tailings with waste rock [49], limestone [50], and neutralizing sludge [51] all showed better barrier performance as single-layer covering materials. Hakkou et al. [52] covered the surface of acid-producing tailings with 15% of alkaline phosphate waste mixed with coarse-size tailings and found that the acidity accumulation peaked at 3199 mg $CaCO_3 \cdot L^{-1}$ and stabilized at 280 mg $CaCO_3 \cdot L^{-1}$, and sulfate concentration caused by oxidation decreased significantly, from 4900 $mg \cdot L^{-1}$ to 480 $mg \cdot L^{-1}$, in a short time.

The availability of single-layer covering materials and the simplicity of technology implementation are favored by many researchers and mine managers. However, the performance of monolayer cover mainly depends on the thickness of the cover and the properties of mixed materials. Therefore, the increase in coverage thickness also means an increase in cost. In addition, the long-term stability of monolayer cover is also a concern. If the neutralization effect of surface alkalinity on the oxidation products of the lower layer is lost, the barrier effect will be greatly weakened. At the same time, the drying of the surface material can easily lead to the cracking of the covering layer. Dehydration cracks can provide preferential channels for air and water, making them lose their barrier function, providing a preferential channel for air and moisture [53]. Therefore, monolayer cover is more of a temporary protection measure.

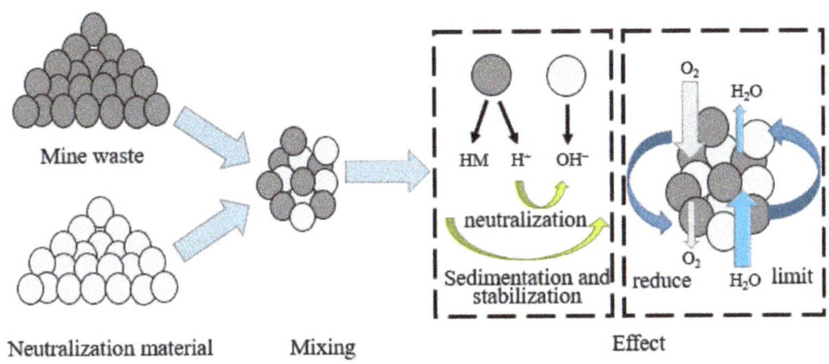

Figure 5. Mixing of mine waste and neutralizing materials [54]. This picture was adapted from [54] with permission from publisher Elsevier in March 2019.

2.2.2. Stabilization Cover

Stabilization treatment is mainly used to solidify sulfide minerals under the action of binders. The hard layer formed by solidifying can inhibit the reactivity of sulfide and the migration of harmful components, thereby eliminating the negative impacts on the receiving environment [55]. A commonly used binder is Portland cement. However, cement cannot be directly applied to high sulfur tailings because residual acid and sulfate salts inhibit its chemical stability [56]. In addition, cement is vulnerable to harmful effects caused by sulfate attack [57,58]. Therefore, as binders for stabilization cover processes, geopolymers have been recommended as a substitute [59,60]. Under the action of an alkali activator, solid silicoaluminate dissolves and releases Si and Al monomer and dimer.

Then, polymerization occurs to form a hardened material with three-dimensional structure (Figure 6). Silicoaluminate is the main component of ore, so the reuse of tailings and waste rock has become an advantage [61].

Ahn et al. [56] used the polymer prepared by lime, tailings, and sodium silicate as the covering layer of sulfur-containing tailings. The TCLP results indicate that the leached amounts of heavy metals Cd, Fe, Mn, and Zn of GP tailings were significantly reduced by 99.3%, 92.9, 98.9, and 99.7%, respectively, with 10% S/S materials compared to the original tailings. The stability of heavy metals was attributed to the carbonate-bound phases, and sulfide minerals were surrounded by calcium silicate generated from sodium silicate, inhibiting further reaction. Ash fly is rich in SiO_2 and Al_2O_3, which have the potential to prepare geopolymers. The research confirmed that the geopolymers prepared from fly ash have a great advantage in stabilizing $Cr(NO_3)_3$ [62]. In addition, the hard layer tends to be stable for a long time. Sarkkinen et al. [63] used the analytic hierarchy process, including life cycle assessment, to evaluate the three stable treatment schemes (Table 1). The results showed the advantages of stabilized cover technology in economy, technology, and ecology and highlighted the characteristics of sustainability.

Stabilization cover is a relatively new method, which can physically weaken the water penetration and oxygen diffusion while solidifying and sequestration pollutants in a three-dimensional structure, thus improving the long-term availability of the covering layer. However, due to the complexity of the site environment of mine wasteland, such as runoff drainage with strong acidity and high sulfate concentration, the stability of cover should be seriously concerning, especially sulfate erosion and drying cracking.

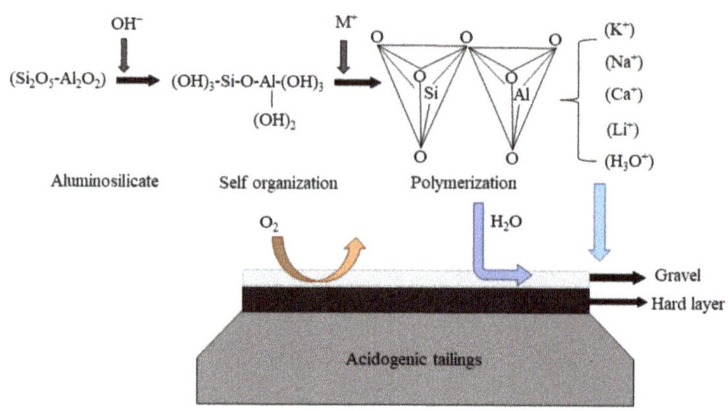

Figure 6. Schematic diagram of geopolymer reaction [64]. This picture was adapted from [64] with permission from publisher Elsevier in 2021.

Table 1. Multiple-criteria decision analysis of solidification/stabilization cover scheme [65]. This picture was adapted from [65] with permission from publisher Elsevier in 10 May 2017.

	Structure		Main Criteria Weight	Cost Accounting (€/m^2)	Normalized Cluster Value	Priority Order
	Above the Surface	Below Surface				
SPC [a]	Vegetation (100 mm)	PC stabilized tailings (1 m)	Economic, 33% Technical, 33% Social-Ecological, 33%	26.32	0.172	3
AAC [b]	Vegetation (100 mm)	AAC stabilized tailings (1 m)	Economic, 33% Technical, 33% Social-Ecological, 33%	26.78	0.238	2

Table 1. *Cont.*

	Structure		Main Criteria Weight	Cost Accounting (€/m^2)	Normalized Cluster Value	Priority Order
	Above the Surface	Below Surface				
AHCL [c]	Vegetation (100 mm) + Moraine (300 mm) + tailings (200 mm) + CaO + moraine (200 mm)	AAC stabilized tailings (300 mm)	Economic, 33% Technical, 33% Social-Ecological, 33%	22.09	0.262	1
Soil layer				500		
Multilayer			2000–3000			

[a] stabilization with cement; [b] MgO activated ground granulated blast furnace slag; [c] advanced hardpan cover liner.

2.2.3. Benign Material Cover

Two advantages make benign materials a competitive choice for cover layers. The first is the continuous alkaline release capacity, which reduces soluble metals and non-metals by consuming H^+ to produce precipitation. The second is low permeability, which reduces the oxidation rate by physically reducing the oxygen flux. In recent years, industrial by-products and residues (such as green liquor slag, fly ash, cement kiln dust, red mud bauxite, etc.) have become common benign materials for inhibiting the generation of AMD due to their high neutralization potential [66–68]. The addition of these alkaline materials occurs during neutralizing reactions. the formation of secondary minerals (sulfate, carbonate, and hydroxide) can fix the dissolved metals through adsorption and co-precipitation [69,70]. A column experiment was filled with the mixture of pyrite and fly ash. The kinetic study confirmed that the pyrite oxidation rate was zero under alkaline pH when Fe(III) coating formed on the mineral surface [70]. Olds et al. [71] mixed granite powder and cement kiln ash at a volume of 4:1 as the overlay, effectively limiting the diffusion of oxygen with a permeability of only 10^{-7}~10^{-6} m·s^{-1}. Cement kiln ash had an alkalinity of about 650 kg $CaCO_3$·t^{-1}, which was dissolved and permeated slowly by rainfall to inhibit acid accumulation of the lower tailings [72]. The mixture of waste rock with pulp and alkaline by-products from steel mills caused trace element concentrations in the leaching solution to be less than 100 μg·L^{-1} and the pH to be close to neutral (pH~2) [66].

However, the continuous consumption of alkalinity limits the long-term performance of these kinds of materials. Industrial by-products usually contain more pollution impurities, which easily cause the release of toxic substances and increase the environmental burden. The release of heavy metals and the generation of neutralized sludge are environmental threats that need to be paid more attention to. Abreu et al. [73] found that, although red mud as a covering material has sufficient alkalinity for AMD generation inhibition, heavy metals accumulated at the red mud–tailing interface, causing secondary pollution along with the permeation and migration of leachate. Generally, the applications of industrial by-products have achieved remarkable results in the short-term validation, but the secondary pollution problems are inevitable, especially for the areas that need to be reclaimed after remediation, and the migration and enrichment of heavy metals should be strongly considered.

2.3. *Organic Reactive Barriers (ORB)*

The use of ORB to prevent oxygen penetration into the underlying tailings pond is an advanced technology in covering schemes as of recently. On the one hand, the physical advantages of the material itself ensure the physical barrier of external water and oxygen [74]. On the other hand, the metabolic reactions by internal microorganisms consume diffusive oxygen [9]. These two reasons cause oxygen depletion in the contact gap between the underlying sulfide minerals (Figure 7). Compared with other cover materials, organic cover materials have the advantages of low permeability, high cation exchange capacity and high alkalinity, thus limiting sulfide mineral oxidation and AMD generation [9]. Organic materials such as sawdust [75], straw, paper pulp and municipal waste compost [76,77] were used as organic cover layers, playing significant roles as oxygen barrier and acid-base modulator. However, the degradation of organic matter is rapid, and the long-term effectiveness of the covering needs a frequent supply of organic carbon-rich materials. Nason et al. [78] comfirmed that 20% of biological sludge was consumed within two years. At present, although the ORB cover has prominent oxygen barrier effects than single-layer cover in theory, there are more constraints in practice, such as the influence of temperature and pH on biological activity [79], the reduction of degradable substances on the treatment efficiency, and the impact of mine environment on microbial proliferation and variation.

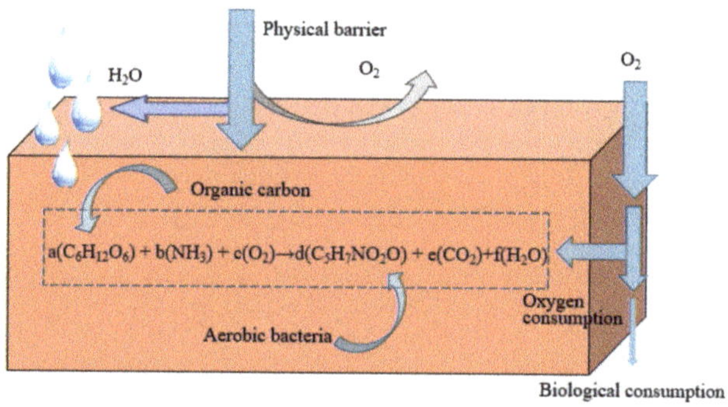

Figure 7. Schematic diagram of oxygen consumption coverage considering microbial growth.

Furthermore, organic reactive materials can be used as a benign soil environment regulator. The decomposition process of organic covering can provide necessary nutrients (N, P, K, Ca, Mg, etc.) for plant growth, increase soil organic matter, and provide fertility for the soil environment. It can also create a suitable environment in the covering area to promote the proliferation of microorganisms. However, the use of organic rich waste can lead to risks. For example, organic cover may induce the reductive dissolution of secondary minerals, such as iron hydroxide, resulting in the release of toxic elements (As, Cd, Cu, Pb and Se). It may also promote the activity and proliferation of pro-oxidation-bacteria, resulting in adverse effects opposite to the oxygen barrier, and more attention should be paid to the environmental risks, especially for the use of sewage sludge. For example, NH^{4+} in sludge will lead to the formation of nitrate [79], leading to the oxidation of pyrite and promoting the generation of AMD. In addition, pathogenic microorganisms also have the risk of migration along with the surface runoff.

3. Future Perspectives

With the progress of industrial development, increasing attention has been paid to the management of mine environment, and increasingly research has been performed with tailings covering technology. However, further development is needed from laboratory research to engineering application, and the main directions for further research are as follows:

(1) Most of the tailings covering schemes are limited to short-term research and lack of long-term problem monitoring. Therefore, it is necessary to conduct relatively long-term simulation experiments in the future research, and then it is necessary to predict the long-term treatment performance with appropriate geological simulation software so as to verify the lasting effectiveness of the scheme.

(2) The availability of water and oxygen is greatly reduced after physical covering treatment. However, the internal reactions caused by covering, such as side reactions, the

proliferation and activity changes of aerobic and anaerobic bacteria, and the protective effect of precipitation coating on the surface of particles, magnify the microscopic oxidation process of sulfide minerals. Thus, the micro-oxidation process after the implementation of the technology should be studied, and the periodic oxidation reactions of sulfide minerals under different conditions and different oxidation states are encouraged to be analyzed.

(3) The management of mine environment should focus on sustainability, in which ecological restoration is an important part. Therefore, the tailing covering system can be combined with vegetation restoration as synergistic technology for both pollution control and ecological restoration.

(4) The introduction of cheap materials into the treatment scheme has showed great economic advantages. The long-term stability, as well as the leaching and migration of harmful substances and harmful side reactions of new materials, especially industrial wastes, can be further investigated in depth. In addition, the application of new materials and new technologies can be comprehensively evaluated for their environmental impacts.

4. Conclusions

The oxidation of sulfide minerals will lead to the continuous production of AMD, and the pollution is serious and lasting. The most effective way to suppress this pollution is to remove one of the basic reactants (oxygen, water, iron, microorganisms, etc.) required for the oxidation reaction. From the technical consideration of source control, tailing cover technology is preferred. Wet cover, dry cover, and organic reactive barriers are the basic types of tailing cover schemes. Wet covers are suitable in areas with natural water lakes, dams, and low water evaporation because water is required as an oxygen barrier. The application of the dry covers scheme has strict requirements on materials, and the covering materials used require strong stability and little secondary pollution. Organic reactive barriers are an innovative method that combines physical barriers with microbial oxygen consumption metabolism. The double barrier of cover minimizes the oxidation efficiency of sulfide, but the degradation consumption of biosolids limits the sustainability of organic reactive barriers. Therefore, organic cover is suitable for a good environment in the covering area and promotes the proliferation of a large number of microorganisms. In addition, the organic covering material must have continuous alkaline penetration and microbial metabolism. In this review, we have aimed to provide an overview about different tailings covering schemes and suggest direction for future work. It can help in guiding future studies on the development of new covering materials and technologies to controlling pollution from AMD.

Author Contributions: Conceptualization, methodology, data curation, writing—original draft preparation, M.S.; software, formal analysis, Y.L.; validation, investigation, Y.C.; resources, visualization, C.L.; writing—review and editing, F.Z. and J.H.; supervision, S.T.; project administration, Q.Z.; funding acquisition, J.H. All authors have read and agreed to the published version of the manuscript.

Funding: The authors thank the financial support from National Natural Science Foundation of China (Grant No. 52060011) and the National Key Research and Development Program of China (Grant No. 2018YFC1802603).

Data Availability Statement: The data involved in this review are from published literature.

Acknowledgments: The authors thank all of the participants coordinators, and administrators for their support during the review.

Conflicts of Interest: The authors declare no conflict of interest.

References

1. Adiansyah, J.S.; Rosano, M.; Vink, S.; Keir, G. A framework for a sustainable approach to mine tailings management: Disposal strategies. *J. Clean. Prod.* **2015**, *108*, 1050–1062. [CrossRef]
2. Johnson, D.; Hallberg, K. Acid mine drainage remediation options: A review. *Sci. Total Environ.* **2005**, *338*, 3–14. [CrossRef] [PubMed]
3. Azapagic, A. Developing a framework for sustainable development indicators for the mining and minerals industry. *J. Clean. Prod.* **2004**, *12*, 639–662. [CrossRef]
4. Albanese, S.; Vivo, B.D.; Lima, A.; Frattasio, G.; Kříbek, B.; Nyambe, I.; Majer, V. Prioritising environmental risk at the regional scale by a GIS aided technique: The Zambian Copperbelt Province case study. *J. Geochem. Explor.* **2014**, *144*, 433–442. [CrossRef]
5. Anawar, H.M. Sustainable rehabilitation of mining waste and acid mine drainage using geochemistry, mine type, mineralogy, texture, ore extraction and climate knowledge. *J. Environ. Manag.* **2015**, *158*, 111–121. [CrossRef]
6. Shim, M.J.; Choi, B.Y.; Lee, G.; Hwang, Y.H.; Yang, J.S.; O'Loughlin, E.J.; Kwon, M.J. Water quality changes in acid mine drainage streams in Gangneung, Korea, 10 years after treatment with limestone. *J. Geochem. Explor.* **2015**, *159*, 234–242. [CrossRef]
7. Navarro, M.C.; Pérez-Sirvent, C.; Martínez-Sánchez, M.; Vidal, J.; Tovar, P.J.; Bech, J. Abandoned mine sites as a source of contamination by heavy metals: A case study in a semi-arid zone. *J. Geochem. Explor.* **2008**, *96*, 183–193. [CrossRef]
8. Niane, B.; Moritz, R.; Guédron, S.; Ngom, P.M.; Pfeifer, H.R.; Mall, I.; Poté, J. Effect of recent artisanal small-scale gold mining on the contamination of surface river sediment: Case of Gambia River, Kedougou region, southeastern Senegal. *J. Geochem. Explor.* **2014**, *144*, 517–527. [CrossRef]
9. Peppas, A.; Komnitsas, K.; Halikia, I. Use of organic covers for acid mine drainage control. *Miner. Eng.* **2000**, *13*, 563–574. [CrossRef]
10. Akcil, A.; Koldas, S. Acid Mine Drainage (AMD): Causes, treatment and case studies. *J. Clean. Prod.* **2006**, *14*, 1139–1145. [CrossRef]
11. Kaur, G.; Couperthwaite, S.J.; Millar, G.J. Alternative neutralisation materials for acid mine drainage treatment. *J. Water Process Eng.* **2018**, *22*, 46–58. [CrossRef]
12. Huang, W.; Wang, S. Research on treatment of mine wastewater using sulfide precipitation floatation. *Tech. Equip. Environ. Pollut. Control* **2004**, *5*, 60–62. (In Chinese)
13. Nariyan, E.; Wolkersdorfer, C.; Sillanpää, M. Sulfate removal from acid mine water from the deepest active European mine by precipitation and various electrocoagulation configurations. *J. Environ. Manag.* **2018**, *227*, 162–171. [CrossRef]
14. Pat-Espadas, A.M.; Portales, R.L.; Amabilis-Sosa, L.E.; Gomez, G.; Vidal, G. Review of Constructed Wetlands for Acid Mine Drainage Treatment. *Water* **2018**, *10*, 1685. [CrossRef]
15. Tan, L.C.; Papirio, S.; Luongo, V.; Nancharaiah, Y.V.; Cennamo, P.; Esposito, G.; Hullebusch, E.V.; Lens, P. Comparative performance of anaerobic attached biofilm and granular sludge reactors for the treatment of model mine drainage wastewater containing selenate, sulfate and nickel. *Chem. Eng. J.* **2018**, *345*, 545–555. [CrossRef]
16. Zeng, H.; Dong, Y.; Lin, H. Research progress of source control technologies of acid mine drainage. *Saf. Environ. Eng.* **2020**, *27*, 7. (In Chinese)
17. Swanson, D.A.; Barbour, S.L.; Wilson, G.W. Dry-site versus wet-site cover design. Proceeding of the 4th International Conference on Acid Rock Drainage, Vancouver, BC, Canada, 31 May–6 June 1997; Volume 4, pp. 1595–1610.
18. Garcilaso, I. *European Commission, 2004a. Reference Document on Best Available Techniques for Management of Tailings and Waste-Rock in Mining Activities, BREF Code MTWR, July*; European Commission: Brussels, Belgium, 2009.
19. Aachib, M.; Mbonimpa, M.; Aubertin, M. Measurement and Prediction of the Oxygen Diffusion Coefficient in Unsaturated Media, with Applications to Soil Covers. *Water Air Soil Pollut.* **2004**, *156*, 163–193. [CrossRef]
20. Gosselin, M.; Aubertin, M. Evaluation de l'effet du degre de saturation sur la diffusion et la consommation d'oxygene dans des residus miniers sulfureux. In Proceedings of the 8th Joint IAH-CNC and CGS Groundwater Specialty Conference and 90th Canadian Geotechnical Conference, Ottawa, ON, Canada, 21–24 October 2007.
21. Romano, C.G.; Mayer, K.U.; Jones, D.R.; Ellerbroek, D.A.; Blowes, D.W. Effectiveness of various cover scenarios on the rate of sulfide oxidation of mine tailings. *J. Hydrol.* **2003**, *271*, 171–187. [CrossRef]
22. Jackson, L.M.; Parbhakar-Fox, A. Mineralogical and geochemical characterization of the Old Tailings Dam, Australia: Evaluating the effectiveness of a water cover for long-term AMD control. *Appl. Geochem.* **2016**, *68*, 64–78. [CrossRef]
23. Yanful, E.K.; Verma, A. Oxidation of flooded mine tailings due to resuspension. *Can. Geotech. J.* **1999**, *36*, 826–845. [CrossRef]
24. Vigneault, B.; Campbell, P.; Tessier, A.; Vitre, R.D. Geochemical changes in sulfidic mine tailings stored under a shallow water cover. *Water Res.* **2001**, *35*, 1066–1076. [CrossRef] [PubMed]
25. Yanful, M.A.S.; Ernest, K. A design approach for selecting the optimum water cover depth for subaqueous disposal of sulfide mine tailings. *Rev. Can. Géotech.* **2005**, *42*, 207–228.
26. Lottermoser, B.G. *Mine Wastes: Characterization, Treatment and Environmental Impacts*; Springer: Berlin/Heidelberg, Germany, 2010.
27. Kalonji-Kabambi, A.; Bussière, B.; Demers, I. Hydrogeochemical Behavior of Reclaimed Highly Reactive Tailings, Part 2: Laboratory and Field Results of Covers Made with Mine Waste Materials. *Minerals* **2020**, *10*, 589. [CrossRef]
28. Aachib, M.; Aubertin, M. Essais en colonne sur des couvertures avec effets de barrière capillaire. In Proceedings of the 51st Canadian Geotechnical Conference, Edmonton, AB, Canada, 4–7 October 1998; Volume 2, pp. 837–844.

29. Bussière, B.; Aubertin, M.; Chapuis, R.P. The behavior of inclined covers used as oxygen barriers. *Can. Geotech. J.* **2003**, *40*, 512–535. [CrossRef]
30. Pabst, T.; Bussiere, B.; Aubertin, M.; Molson, J. Comparative performance of cover systems to prevent acid mine drainage from pre-oxidized tailings: A numerical hydro-geochemical assessment. *J. Contam. Hydrol.* **2018**, *214*, 39–53. [CrossRef]
31. Bussière, B.; Aubertin, M.; Mbonimpa, M.; Molson, J.W.; Chapuis, R.P. Field experimental cells to evaluate the hydrogeological behaviour of oxygen barriers made of silty materials. *Can. Geotech. J.* **2007**, *44*, 245–265. [CrossRef]
32. Kabambi, A.K.; Bussière, B.; Demers, I. Hydrogeological Behaviour of Covers with Capillary Barrier Effects Made of Mining Materials. *Geotech. Geol. Eng.* **2017**, *35*, 1199–1220. [CrossRef]
33. Molson, J.; Aubertin, M.; Bussière, B.; Benzaazoua, M. Geochemical transport modelling of drainage from experimental mine tailings cells covered by capillary barriers. *Appl. Geochem.* **2008**, *23*, 1–24. [CrossRef]
34. Dagenais, A.M.; Aubertin, M.; Bussiere, B.; Cyr, J. Performance of the Lorraine mine site cover to limit oxygen migration. In *Transactions—Society for Mining Metallurgy and Exploration Incorporated*; SME: Englewood, CO, USA, 2005; Volume 318.
35. Larochelle, C.G.; Bussiere, B.; Pabst, T. Acid-Generating Waste Rocks as Capillary Break Layers in Covers with Capillary Barrier Effects for Mine Site Reclamation. *Water Air Soil Pollut.* **2019**, *230*, 57. [CrossRef]
36. Dagenais, A.-M.; Aubertin, M.; Bussière, B.; Martin, V. Large scale applications of covers with capillary barrier effects to control the production of acid mine drainage. *Proc. Post-Min.* **2005**, 16–17.
37. Ouangrawa, M.; Aubertin, M.; Molson, J.W.; Bussière, B.; Zagury, G.J. Preventing Acid Mine Drainage with an Elevated Water Table: Long-Term Column Experiments and Parameter Analysis. *Water Air Soil Pollut.* **2010**, *213*, 437–458. [CrossRef]
38. Ouangrawa, M.; Molson, J.; Aubertin, M.; Bussière, B.; Zagury, G.J. Reactive transport modelling of mine tailings columns with capillarity-induced high water saturation for preventing sulfide oxidation. *Appl. Geochem.* **2009**, *24*, 1312–1323. [CrossRef]
39. Ouangrawa, M.; Molson, J.; Aubertin, M.; Zagury, G.; Bussière, B. The effect of water table elevation on acid mine drainage from reactive tailings: A laboratory and numerical modeling study. In Proceedings of the 7th International Conference on Acid Rock Drainage (ICARD), St. Louis, MO, USA, 26–30 March 2006.
40. Mbonimpa, M.; Aubertin, M.; Aachib, M.; Bussière, B. Diffusion and consumption of oxygen in unsaturated cover materials. *Can. Geotech. J.* **2003**, *40*, 916–932. [CrossRef]
41. Bussière, B.; Maqsoud, A.; Aubertin, M.; Martschuk, J.; Mcmullen, J.; Julien, M. Performance of the oxygen limiting cover at the LTA site, Malartic, Quebec. *CIM Bull.* **2006**, *1*, 1–11.
42. Dagenais, A.M.; Aubertin, M.; Bussière, B. Parametric study on the water content profiles and oxidation rates in nearly saturated tailings above the water table. In Proceedings of the 7th International Conference on Acid Rock Drainage, St. Louis, MO, USA, 26–30 March 2006.
43. Khire, M.V.; Benson, C.H.; Bosscher, P.J. Capillary Barriers: Design Variables and Water Balance. *J. Geotech. Geoenviron.* **2000**, *126*, 695–708. [CrossRef]
44. Dobchuk, B.; Nichol, C.; Wilson, G.W.; Aubertin, M. Evaluation of a single-layer desulphurized tailings cover. *Can. Geotech. J.* **2013**, *50*, 777–792. [CrossRef]
45. Dagenais, A.M. *Techniques de Controle du Drainage Minier Acide Basees Sur les Effets Capillaires*; Ecole Polytechnique: Montreal, QC, Canada, 2005.
46. Kleinmann, R. At-source control of acid mine drainage. *Mine Water Environ.* **1990**, *9*, 85–96. [CrossRef]
47. Holmes, P.R.; Crundwell, F.K. The kinetics of the oxidation of pyrite by ferric ions and dissolved oxygen: An electrochemical study. *Geochim. Cosmochim. Acta* **2000**, *64*, 263–274. [CrossRef]
48. Pabst, T.; Molson, J.; Aubertin, M.; Bussiere, B. Reactive transport modelling of the hydro-geochemical behaviour of partially oxidized acid-generating mine tailings with a monolayer cover. *Appl. Geochem.* **2017**, *78*, 219–233. [CrossRef]
49. Gorakhki, M.H.; Bareither, C.A.; Gorakhki, M.H.; Bareither, C.A. Sustainable reuse of mine tailings and waste rock as water-balance covers. *Minerals* **2017**, *7*, 128. [CrossRef]
50. Mylona, E.; Xenidis, A.; Paspaliaris, I. Inhibition of acid generation from sulphidic wastes by the addition of small amounts of limestone. *Miner. Eng.* **2000**, *13*, 1161–1175. [CrossRef]
51. Demers, I.; Bouda, M.; Mbonimpa, M.; Benzaazoua, M.; Bois, D.; Gagnon, M. Valorization of acid mine drainage treatment sludge as remediation component to control acid generation from mine wastes, part 2: Field experimentation. *Miner. Eng.* **2015**, *76*, 117–125. [CrossRef]
52. Hakkou, R.; Benzaazoua, M.; Bussiere, B. Laboratory Evaluation of the Use of Alkaline Phosphate Wastes for the Control of Acidic Mine Drainage. *Mine Water Environ.* **2009**, *28*, 206–218. [CrossRef]
53. Demers, I.; Benzaazoua, M.; Mbonimpa, M.; Bouda, M.; Bois, D.; Gagnon, M. Valorisation of acid mine drainage treatment sludge as remediation component to control acid generation from mine wastes, part 1: Material characterization and laboratory kinetic testing. *Miner. Eng.* **2015**, *76*, 109–116. [CrossRef]
54. Park, I.; Tabelin, C.B.; Jeon, S.; Li, X.; Seno, K.; Ito, M.; Hiroyoshi, N. A review of recent strategies for acid mine drainage prevention and mine tailings recycling. *Chemosphere* **2019**, *219*, 588–606. [CrossRef]
55. Benzaazoua, M.; Marion, P.; Picquet, I.; Bussière, B. The use of pastefill as a solidification and stabilization process for the control of acid mine drainage. *Miner. Eng.* **2004**, *17*, 233–243. [CrossRef]
56. Ahna, J.S.; Song, H.; Yim, G.J.; Sang, W.J.; Kim, J.G. An engineered cover system for mine tailings using a hardpan layer: A solidification/stabilization method for layer and field performance evaluation. *J. Hazard. Mater.* **2011**, *197*, 153–160. [CrossRef]

57. Benzaazoua, M.; Fall, M.; Belem, T. A contribution to understanding the hardening process of cemented pastefill. *Miner. Eng.* **2004**, *17*, 141–152. [CrossRef]
58. Benzaazoua, M.; Bussière, B.; Demers, I.; Aubertin, M.; Fried, é.; Blier, A. Integrated mine tailings management by combining environmental desulphurization and cemented paste backfill: Application to mine Doyon, Quebec, Canada. *Miner. Eng.* **2008**, *21*, 330–340. [CrossRef]
59. Komnitsas, K.; Zaharaki, D. Geopolymerisation: A review and prospects for the minerals industry. *Miner. Eng.* **2007**, *20*, 1261–1277. [CrossRef]
60. Pacheco-Torgal, F.; Castro-Gomes, J.; Jalali, S. Alkali-activated binders: A review Part 1. Historical background, terminology, reaction mechanisms and hydration products. *Constr. Build. Mater.* **2008**, *22*, 1305–1314. [CrossRef]
61. Tariq, A.; Yanful, E.K. A review of binders used in cemented paste tailings for underground and surface disposal practices. *Cheminform* **2013**, *131*, 138–149. [CrossRef] [PubMed]
62. Guo, X.L.; Zhang, L.Y.; Huang, J.B.; Shi, H.S. Detoxification and solidification of heavy metal of chromium using fly ash-based geopolymer with chemical agents. *Constr. Build. Mater.* **2017**, *151*, 394–404. [CrossRef]
63. Sarkkinen, M.; Kujala, K.; Gehör, S. Decision support framework for solid waste management based on sustainability criteria: A case study of tailings pond cover systems. *J. Clean. Prod.* **2019**, *263*, 117583. [CrossRef]
64. Raja, V.K.B.; Raj, S.K.; Sairam, M.D.; Kasyap, A.; Kumar, V.G.; Padmapriya, R.; Baalamurugan, J.; Sonawane, P.D. Geopolymer green technology. In Proceedings of the 28th International Conference on Processing and Fabrication of Advanced Materials (PFAM), Chennai, India, 7–9 December 2021; pp. 1003–1007.
65. Kefeni, K.K.; Msagati, T.A.M.; Mamba, B.E.B. Acid mine drainage: Prevention, treatment options, and resource recovery: A review. *J. Clean. Prod.* **2017**, *151*, 475–493. [CrossRef]
66. Alakangas, L.; Andersson, E.; Mueller, S. Neutralization/prevention of acid rock drainage using mixtures of alkaline by-products and sulfidic mine wastes. *Environ. Sci. Pollut. Res.* **2013**, *20*, 7907–7916. [CrossRef]
67. Rodriguez-Jorda, M.P.; Garrido, F.; Garcia-Gonzalez, M.T. Effect of the addition of industrial by-products on Cu, Zn, Pb and As leachability in a mine sediment. *J. Hazard. Mater.* **2012**, *213–214*, 46–54. [CrossRef]
68. Doye, I.; Duchesne, J. Neutralisation of acid mine drainage with alkaline industrial residues: Laboratory investigation using batch-leaching tests. *Appl. Geochem.* **2003**, *18*, 1197–1213. [CrossRef]
69. Bertocchi, A.F.; Ghiani, M.; Peretti, R.; Zucca, A. Red mud and fly ash for remediation of mine sites contaminated with As, Cd, Cu, Pb and Zn. *J. Hazard. Mater.* **2006**, *134*, 112–119. [CrossRef]
70. Perez-Lopez, R.; Nieto, J.M.; de Almodovar, G.R. Immobilization of toxic elements in mine residues derived from mining activities in the Iberian Pyrite Belt (SW Spain): Laboratory experiments. *Appl. Geochem.* **2007**, *22*, 1919–1935. [CrossRef]
71. Olds, W.; Weber, P.; Pizey, M. Alkalinity producing caps for minimisation of acid mine drainage generation in waste rock dumps. In Proceedings of the AusIMM New Zealand Branch Annual Conference, Nelson, New Zealand, 25–28 August 2013; pp. 253–262.
72. Smart, R.; Miller, S.D.; Stewart, W.S.; Rusdinar, Y.; Schumann, R.E.; Kawashima, N.; Li, J. In situ calcite formation in limestone-saturated water leaching of acid rock waste. *Sci. Total Environ.* **2010**, *408*, 3392–3402. [CrossRef]
73. De Abreu, A.T.; de Faria, E.M.; Guimaraes, J.A.C.; Leite, A.D.; de Lena, J.C. Laboratory evaluation of the use of alkaline covers to prevent acid mine drainage. *Rev. Bras. Cienc. Solo* **2012**, *36*, 787–801. [CrossRef]
74. Wang, B.; Dong, X. At-Source control methods for acid mine drainage from tailing impoundments. *China Min. Mag.* **2015**, *24*, 6. (In Chinese)
75. Germain, D.; Tassé, N.; Cyr, J. Rehabilitation of Mine Tailings by Simultaneous Prevention of Oxidation and Treatment of Acid Effluents Using a Wood-Waste Cover. In Proceedings of the 6th International Conference on Acid Rock Drainage, Cairns, Australian, 14–17 July 2003.
76. Cabral, A.; Racine, I.; Burnotte, F.; Lefebvre, G. Diffusion of oxygen through a pulp and paper residue barrier: Reply. *Can. Geotech. J.* **2001**, *38*, 661. [CrossRef]
77. Pierce, W.G.; Belzile, N.; Wiseman, M.E.; Winterhalder, K. Composted organic wastes as anaerobic reducing covers for long term abandonment of acid-generating tailing. In Proceedings of the 11th Annual Meeting of the American Society of Mining and Reclamation (ASMR), Pittsburgh, PA, USA, 24–29 April 1994; pp. 148–157. [CrossRef]
78. Nason, P.; Jia, Y.; Maurice, C.; Alakangas, L.; Öhlander, B. Biodegradation of Biosolids Under Aerobic Conditions: Implications for Cover Materials for Sulfide Mine Tailings Remediation. *Mine Water Environ.* **2016**, *35*, 273–282. [CrossRef]
79. Ribeta, I.; Ptacek, C.J.; Blowes, D.W.; Jambor, J.L. The potential for metal release by reductive dissolution of weathered mine tailings. *J. Contam. Hydrol.* **1995**, *17*, 239–273. [CrossRef]

toxics

MDPI

Article

Rapid and Effective Lead Elimination Using Cow Manure Derived Biochar: Balance between Inherent Phosphorus Release and Pollutants Immobilization

Huabin Wang [1,2], Yi Wen [1,2], Yu Ding [3], Zhiqiang Yue [4], Dan Xu [3], Ying Liu [1,2], Yong Zhang [1,2], Rui Xu [1,2,*] and Weiqing Zeng [1,4,*]

1 School of Energy and Environment Science, Yunnan Normal University, Kunming 650500, China
2 Yunnan Key Laboratory of Rural Energy Engineering, Kunming 650500, China
3 Baoshan City Longyang Rural Energy Workstation, Baoshan 678000, China
4 Yuxi Agricultural Environmental Protection and Rural Energy Workstation, Yuxi 653100, China
* Correspondence: ecowatch_xr@163.com (R.X.); zdfad707@163.com (W.Z.); Tel.: +86-27-87792151 (R.X.); Fax: +86-27-87792151 (R.X.)

Abstract: Cow manure derived biochar (CMBC) can serve as a promising functional material, and CMBC can be regarded as an ecofriendly approach compared to conventional ones. CM bioadsorbent can be employed for heavy metal immobilization (such as for lead) as well as an amendment to increase soil fertility (e.g., phosphorus). Few studies have examined the surface interactions between pollutants and bioadsorbents when inherent nutrient release is present. In this work, CMBC was prepared and applied for Pb(II) removal, and the vital roles of released phosphorus from CMBC were comprehensively disclosed. Furthermore, CMBC could immobilize part of the Pb(II) in soil and promote plant growth. CM400 was an effective adsorbent whose calculated Q_e reached 691.34 mg·g^{-1}, and it rapidly adsorbed 98.36 mg·g^{-1} of Pb(II) within 1 min. The adsorption mechanisms of Pb(II) by CMBC include ion exchange, physical adsorption, electrostatic attraction, chemical precipitation, surface complexation, and cation–π bond interaction. Based on the residual phosphorus content and adsorption effect, complexation rather than the chemical precipitation had a greater contribution toward adsorption. Besides, as the concentration of Pb(II) increased, the main adsorption mechanisms likely transformed from chemical precipitation to ion exchange and complexation. CMBC not only had a good effect on Pb(II) removal in the solution, but also immobilized the Pb(II) in soil to restrain plant uptake as well as promote plant growth. The main novelty of this work is providing more insights to the cow manure bio adsorbent on Pb immobilization and phosphorus release. This study is expected to serve as a basis and reference for analyzing the release effects of inherent nutrients and the interfacial behaviors with heavy metals when using CMBC and other nutrient–rich carbon–based fertilizers for pollution control.

Keywords: cow manure; biochar; lead pollution; phosphorus release; crop growth

check for updates

Citation: Wang, H.; Wen, Y.; Ding, Y.; Yue, Z.; Xu, D.; Liu, Y.; Zhang, Y.; Xu, R.; Zeng, W. Rapid and Effective Lead Elimination Using Cow Manure Derived Biochar: Balance between Inherent Phosphorus Release and Pollutants Immobilization. *Toxics* **2023**, *11*, 1. https://doi.org/10.3390/toxics11010001

Academic Editors: Junhao Qin, Peidong Su, Feng Zhu and Lin Ding

Received: 20 November 2022
Revised: 13 December 2022
Accepted: 16 December 2022
Published: 20 December 2022

1. Introduction

In recent decades, the risks posed by heavy metals have garnered widespread global attention owing to the persistence, toxicity, and bioaccumulation of these metals. Among them, lead (Pb) is persistent in the environment and can result in both decreased growth and reproduction in plants and animals, in addition to neurological effects in vertebrates, while the potential harm of these effects has attracted growing concerns [1]. In China, lead smelting operations are mostly centered in the south–central and southwest areas, such as at the Huize Pb–Zn deposit in Yunnan Province [2]. Pb(II) removal methods consist of membrane filtration, electrocoagulation, coagulation, photo-catalytic redox procedures, Fenton oxidation, and γ-ray irradiation [3,4]. In particular, it is widely used as an adsorbent to eliminate Pb(II), as the adsorption process can be easily handled at low-cost [5,6].

Therefore, it is essential to create rapid, efficient, and cost-effective adsorbents. Due to the abundance of functional groups, pore structure, and strong affinity for heavy metals in biochar, biochar has become widespread in Pb(II) removal [7].

Over recent years, the production of cow manure (CM) has increased with the developments of the large-scale and intensive cattle breeding industry in China [8]. CM is a valuable resource for the sustainable development of agriculture [9], and is mainly used as fertilizer, energy, and so on [10]. The contents of lignin and cellulose in CM are considerably higher than those in pig manure and chicken manure, which increases the fermentation time [11]. CM is rich in phosphorus, while the heavy metal content is lower than that of pig manure [12], so it is imperative to use CM with a high value. Researchers have found and confirmed that CM can promote the growth of crops and reduce phosphorus application [13]. However, the inherent phosphorus from CM is water extractable phosphorus, which can easily cause eutrophication [14]. Researchers have found that the slow release of phosphorus can be achieved by pyrolysis [15]. Furthermore, the cow manure derived biochar shows more efficient adsorption of heavy metals than wood-derived biochar, as the PO_4^{3-} and CO_3^{2-} contained and functional groups improve the adsorption via complexation and precipitation [16]. Thus, cow manure derived biochar is widely used to adsorb heavy metals [17,18]. Compared with compost, the cow manure derived biochar offers a lower cost, easier preparation, and better adsorption [18]. Modifications such as KH_2PO_4 or NaOH can significantly improve adsorption [19,20].

Even though the presence of phosphorus in cow manure derived biochar has been found [21], few studies are available regarding the phosphorus release during heavy metal removal or the effect of phosphorus fertilizers during adsorption. Few studies have analyzed the relationship between inherent phosphorus and Pb(II) adsorption by cow manure derived biochar. Therefore, this work focused on a phosphorus-contained biomass, CM, and used a two-step method to prepare biochar functional materials, named CMBC. In this work, the adsorption effects of CMBC on Pb(II) in solution and soil were investigated. Through the adsorption experiment in solution, the differences in adsorption capacity of three kinds of biochar prepared at different pyrolysis temperatures were compared, and the adsorption mechanism of Pb(II) adsorption by CMBC was analyzed by changing the reaction conditions. By simulating lead-contaminated soil and planting Chinese cabbage in lead-contaminated and pollution-free soil, the growth of Chinese cabbage with or without CMBC was compared. In order to compare and analyze the effect of CMBC on reducing the bioenrichment of Pb(II), the Pb(II) content in the roots, stems, and leaves of Chinese cabbage was detected after maturity. The ability of CMBC to promote crop growth was analyzed by comparing the phosphorus content in three parts of Chinese cabbage. Notably, this study aimed to explore the release mechanism of inherent phosphorus and the interactions between inherent phosphorus and Pb(II). Furthermore, CMBC was applied to lead-contaminated soil to verify whether CMBC can solidify heavy metals and promote crop growth.

2. Materials and Methods

2.1. Experimental Material

$Pb(NO_3)_2{\cdot}3H_2O$, HNO_3 (65–68%), and NaOH were used, and all chemicals were of analytical grade. Further, CM was collected from the Laohadu Demonstration Ranch in Mengzi City, Yunnan Province, China.

2.2. Preparation of CMBC

The preparation process for the CMBC is illustrated in Scheme 1. Fresh CM was dried at 85 °C and ground (100-mesh). Firstly, 4 g dried CM, 112 mL deionized water, and 38 mL of NaOH solution (1 mol·L^{-1}) were mixed. Secondly, the mixture was shaken with ultrasound for 30 min and stirred for another 2 h. Thirdly, the mixture was hydrothermally heated at 200 °C for 24 h and then rinsed with acetone. Finally, the mixture was dried at

85 °C. CMBC was obtained via pyrolysis under the protection of N_2 at 400 °C, 600 °C, and 800 °C for 1 h, denoted as CM400, CM600, and CM800, respectively.

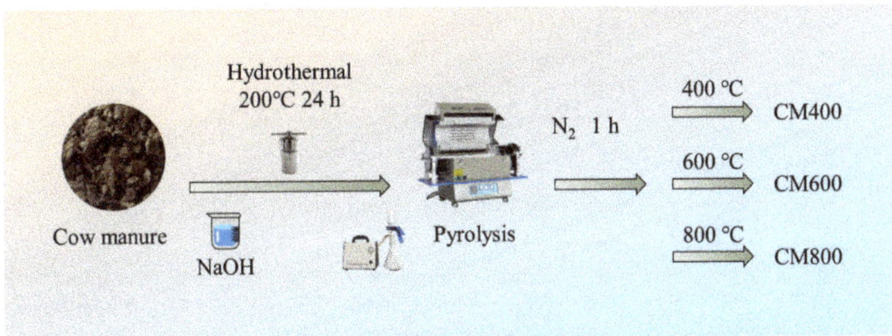

Scheme 1. The preparation process for CMBC.

2.3. Analysis Methods

The elemental composition of biochar was analyzed with elemental analyzer (Elementar Vario EL cube, Langenselbold, Germany). The morphology was analyzed with scanning electron microscopy (SEM) and SEM–EDX at 5 kV (Mira LMS, Tescan, Brno, Czech Republic). The specific surface area (SSA) of biochar was detected by N_2 adsorption isotherms at 77 K using a Micropore Analyzer (ASAP 2460, Micrometrics, Cumming, GA, USA). The electron binding energy and elemental valence were analyzed using X-ray photoelectron spectroscopy (XPS) (K-Alpha, Thermo Scientific, Waltham, MA, USA). The functional groups were qualitatively examined using a Fourier transform infrared spectrometer (FTIR) (Niolet iN10EA, Thermo Scientific, Waltham, MA, USA). The amount of the elements in the solutions was determined by inductively coupled plasma mass spectrometry (ICP) (Thermo Fisher–X series, Waltham, MA, USA). The contents of the elements in the soil and crop were determined via atomic absorption spectrophotometry (iCETM3500 AAS, Waltham, MA, USA).

2.4. Adsorption Experiment

Twenty mg of biochar was weighed, then 20 mL deionized water or Pb(II) solution was added. After oscillation (200 $r \cdot min^{-1}$, 25 °C), 5 mL solution was filtered (0.22 μm) at predetermined times (0.008, 0.083, 0.5, 1, 2, 6, 12, 18, and 24 h), and the phosphorus or Pb contents were determined via ICP. The adsorbent dosage was 1 $g \cdot L^{-1}$, then 20 mL Pb(II) solution was added (50, 100, 200, 300, 400, 500, 600, 800, and 1000 $mg \cdot L^{-1}$), and HNO_3 and NaOH solutions were used to adjust the pH of solution (pH = 7.0). After 24 h of oscillation (200 $r \cdot min^{-1}$), the sample solution was filtered (0.22 μm), and the Pb(II) content was determined via ICP. When the adsorbent dosage was 1 $g \cdot L^{-1}$ and the initial concentration of Pb(II) was 100 $mg \cdot L^{-1}$, the effects of pH (3.0, 5.0, 7.0, 9.0, and 11.0) and temperature (25 °C, 35 °C, and 45 °C) on the adsorption was explored.

The adsorption capacity of Pb(II) was calculated using Equation (S1) and the experimental results were fitted using the adsorption dynamic model (Equations (S2)–(S5)) and the isothermal adsorption model (Equations (S6) and (S7)). The equations and symbols in the equation are described in the Supplementary Materials (Text S1).

2.5. Pot Experiments

Chinese cabbage is a type of common model plant, and it has small, fast growth, growth ability, etc. [22]. Three experimental groups (T_1: 1 g CM400 application, T_2: 1 g CM600 application, T_3: 1 g CM800 application) and two control groups (CK_1: no pollution treatment, CK_2: no adsorbent application) were set up in the pot experiment. Firstly, 2 kg screened soil (5 mm) was placed in a plastic pot. Secondly, to simulate the actual conditions of Pb(II) pollution in Yunnan Province, China [23], the experimental groups and CK_2 were sprayed with $Pb(NO_3)_2$ solution (100 mL, 60 mg·L^{-1}), and CK_1 was sprayed with 100 mL water. We then ripened the loaded soil by pouring a permeable and dried it naturally for 40 days. Thirdly, Chinese cabbage seeds were directly sown into pots (10 seeds per pot), and 1 g CM400, CM600, or CM800 were applied to three experimental groups, respectively, and then covered with 1–2 cm of soil. Each pot was irrigated daily with 150 mL water, and the budding and growth of Chinese cabbage were observed and recorded. Once the Chinese cabbage was mature enough, the Pb(II) and P contents in roots, stems, and leaves were determined.

3. Results and Discussion

3.1. Characterization

The basic elements of CMBC were shown in Table 1. For CM400, the contents of carbon element (C%), oxygen element (O%), and hydrogen element (H%) were 28.11, 41.48, and 5.05% respectively, while they were 39.21, 37.89, and 1.63% for CM800. Obviously, O% and H% decreased with the increase of pyrolysis temperature, which mainly attributed to the pyrolysis of organic compounds [24]. During carbonization, both H/C and O/C ratios decreased due to the release of functional groups containing H and O [25], thus, the O/C ratios of CM400, CM600, and CM800 were 1.48, 0.07, and 0.04, respectively, and the H/C ratios for them were 0.18, 0.07, and 0.04. Notably, CM400 had the highest O/C and H/C ratios, which indicated that CM400 may have had the most prevalent functional groups and the best adsorption effect on hydrophobic pollutants [26], namely, CM400 may have had a stronger ability to adsorb Pb(II) via complexation [27]. Noteworthy, phosphorus element (P%) was 0.61, 0.34, and 0.27% for CM400, CM600, and CM800, and the results of elemental analysis proved the existence of inherent phosphorus in CMBC.

Table 1. Basic elemental composition, specific surface area, and pore structure of CMBC.

	CM400	CM600	CM800
C (%)	28.11	37.14	39.21
H (%)	5.05	2.53	1.63
O (%)	41.18	40.34	37.89
P (%)	0.61	0.34	0.27
O/C	1.48	1.09	0.97
H/C	0.18	0.07	0.04
BET (m^2·g^{-1})	2.51	11.78	12.93
Vtotal (m^3·g^{-1})	0.05	0.03	0.03
Average pore size (Å)	837.94	90.31	108.91

The SSA is presented in Table 1. The SSA were 2.51, 11.78, and 12.93 m^2·g^{-1} for CM400, CM600, and CM800, respectively, which showed that the SSA of CM800 was higher than that of CM400 and CM600. In the range of 400–800 °C, the SSA of CMBC increased with the preparation temperature, and in the range of 400–600 °C, SSA increased greatly, while in the range of 600–800 °C, SSA increased steadily but slightly. The SSA is associated with physical properties and significantly affects the physical adsorption capacity. Hence, CM800 had better physical adsorption capacity than the two other types of biochar [28]. Moreover, the average pore sizes of CM400, CM600, and CM800 were 837.94, 90.31, and 108.91 Å. The pore size of CM400 is larger than the others, so CM400 may have had a better adsorption effect via pore filling.

The SEM and SEM-EDS images are shown in Figure 1. The layered porous structure riddled with micropores is presented on CMBC, as CM contains cellulose [29]. The ledge structure of CMBC was related to pyrolysis temperature change; when the pyrolysis temperature $\leq$600 °C, CM400 and CM600 retained a part of the stereoscopic carbonaceous skeleton structure and bulk accumulation whereas the ledge structure of CM800 became thinner. More large pores were observed on CM800 than CM400 and CM600, thus, the CM800 may have had better physical adsorption capacity [30], which was consistent with the above analysis. Moreover, the CMBC contained Si, Ca, and P.

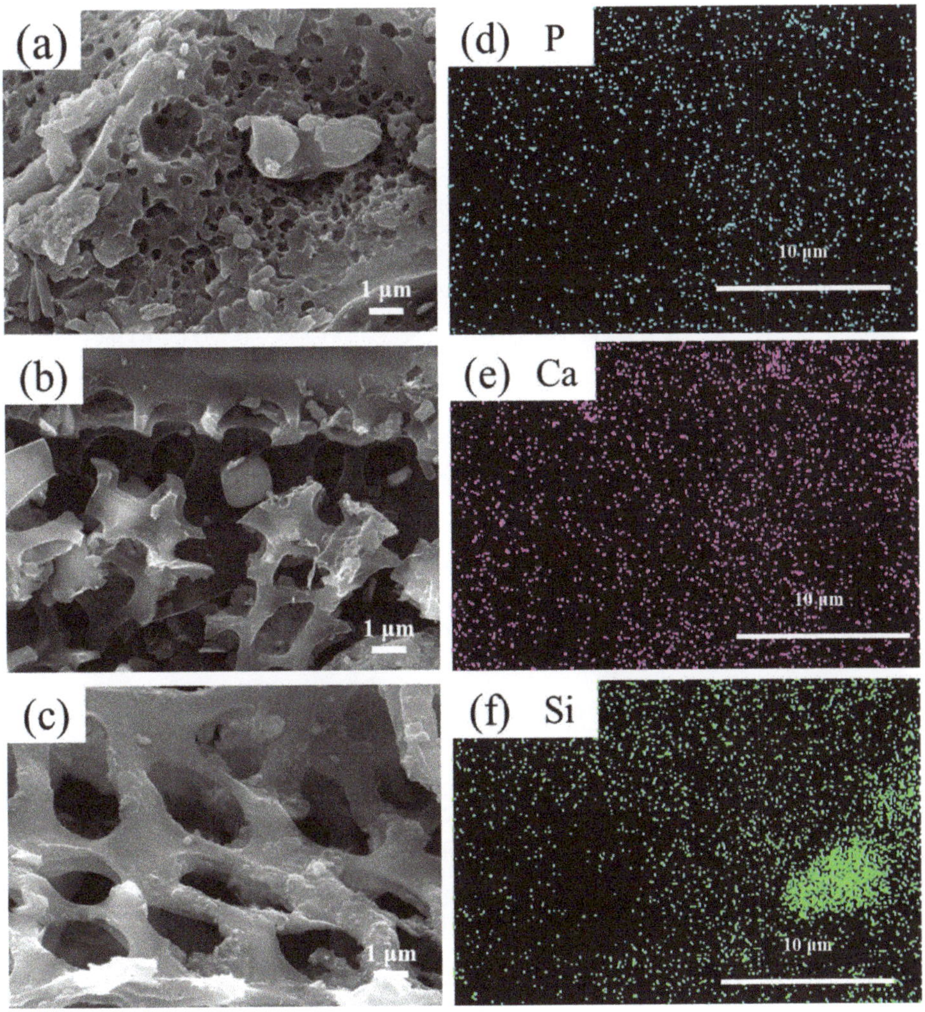

Figure 1. SEM and EDS images of CMBC (**a–c**): SEM images of CM400, CM600, and CM800; (**d–f**): EDS images for P, Ca, and Si of CM400.

The XRD spectrum is depicted in Figure 2. The three kinds of CMBC exhibited the typical irregular wide peaks of biochar at approximately 24.3°. In addition to SiO_2 (26.6°) (PDF#70–2517), the substances included a combination of embedded metal cations (Ca^{2+} and Mg^{2+}) and anions (PO_4^{3-} and CO_3^{2-}), such as $Ca_3(PO_4)_2$ (PDF#09–0169), $Mg_3(PO_4)_2$ (PDF#48–1167), and $CaCO_3$ (PDF#70–0095) [31]. The higher the pyrolysis temperature, the more evident were the peaks of $Ca_3(PO_4)_2$, $Mg_3(PO_4)_2$, and Mg_3PO_4OH (PDF#47–0957), as the high-temperature pyrolysis was conducive to the formation of cationic and anionic conjugates, namely, CM400 was more likely to release inherent substances into solution. The peak of $CaCO_3$ in CM800 decreased, which may be because high pyrolysis temperature (>600 °C) likely caused the decomposition of $CaCO_3$ [32].

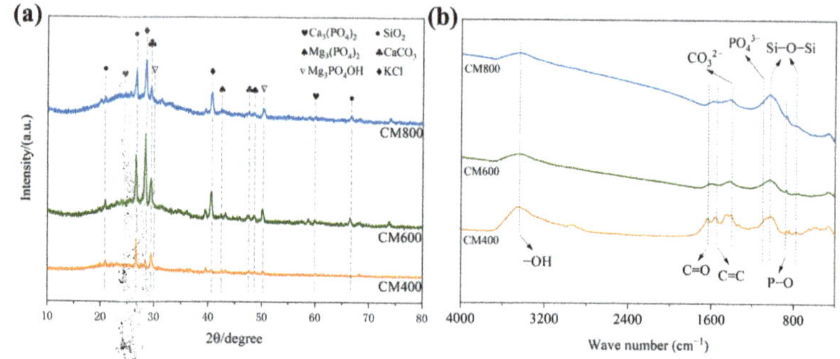

Figure 2. XRD and FTIR spectra of three kinds of CMBC. (**a**) XRD spectra and (**b**) FTIR spectra.

The FTIR spectrum was depicted in Figure 2. The –OH peak (3458 cm^{-1}), the C=C stretching peaks (1566 cm^{-1} and 1455 cm^{-1}), the C–O stretching vibration peak (1384 cm^{-1}), and the C–O–C peak (876 cm^{-1}) were observed [33]. Furthermore, the peak of $CO_3{}^{2-}$ and $PO_4{}^{3-}$ were located at 1415 cm^{-1} and 1096 cm^{-1} [34], as well as the peak of Si–O–Si at 1030 cm^{-1} and 780 cm^{-1} [35]. The results indicated that $CO_3{}^{2-}$, $PO_4{}^{3-}$, and SiO_2 were retained in CMBC. As shown in the FTIR spectra (Figure 2b), compared with CM400, the peak value of functional groups such as –OH, C=O, and C=C of CM800 was weakened. Hence, as the pyrolysis temperature rose from 400 °C to 800 °C, there was a decrease in the functional groups. To sum up, CM400 solidified less inherent phosphorus, so more inherent phosphorus could be released into the solution to participate; at the same time, its functional groups was rich, which made the complexation stronger.

3.2. *Pb(II) Adsorption*

3.2.1. Effect of pH

The effect of pH on adsorption is shown in Figure 3. When pH = 3, Q_e plunged (42.09, 28.95, and 8.77 $mg{\cdot}g^{-1}$ for CM400, CM600, and CM800, respectively). Although the strong acid condition promoted the release of $PO_4{}^{3-}$, which can form precipitation with Pb(II) [36], Q_e of CM400 was only 42.09 $mg{\cdot}g^{-1}$. Therefore, H^+ likely competed with Pb(II) for the adsorption sites, and the protonated surface functional groups also exhibited electrostatic repulsion [37]. Under alkaline conditions, there were more negative charges on the surface of CMBC and more free hydroxyl groups provided adsorption sites [38]. Besides, OH^- was involved in adsorption, such as $Pb_{10}(PO_4)_6(OH)_2$. The Q_e of CM400 was almost unchanged from pH = 5 to pH = 11 (97.01 vs. 98.47 $mg{\cdot}g^{-1}$). This not only showed that CM400 functions well in a wide range of pH, but also showed that although the adsorption sites increased steadily, Q_e did not vary because Pb(II) was adsorbed. In conclusion, the CMBC had the best adsorption effect for Pb(II) in an alkaline environment.

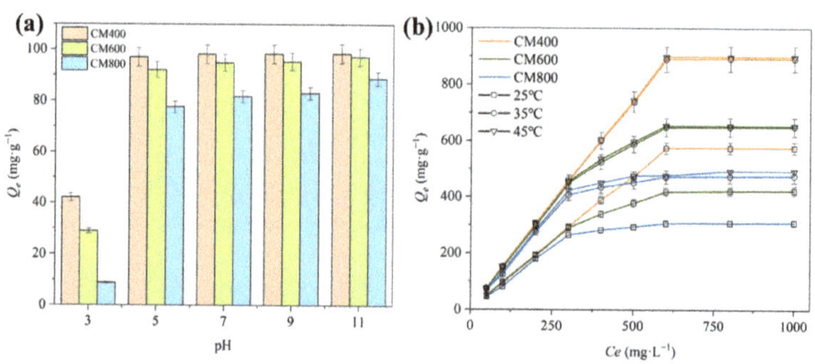

Figure 3. The effect of pH and temperature on the adsorption. (**a**) pH and (**b**) temperature. Experimental conditions: [initial concentration of Pb(II)] = 100 $mg{\cdot}L^{-1}$, [dosage] = 1 $g{\cdot}L^{-1}$.

3.2.2. Effect of Temperature

The temperature had an impact on adsorption (Figure 3). The effect of pH on adsorption is shown in Figure 4. When the original concentration of Pb(II) was ≤300 mg·L^{-1}, the removal rates were not affected by temperature (98.38% vs. 99.80% at 25 and 45 °C). Once the initial concentration was 500 mg·L^{-1} or 1000 mg·L^{-1}, the adsorption was enhanced with temperature. For example, when the initial concentration was 1000 mg·L^{-1}, Q_e of CM600 was 422.59 mg·g^{-1} and that of CM800 was 308.63 mg·g^{-1} at 25 °C, but Q_e of CM600 and CM800 were 425.1 and 320.1 mg·g^{-1} at 45 °C. Therefore, the adsorption by CMBC was endothermic, and CM600 and CM800 were more significantly affected by temperature.

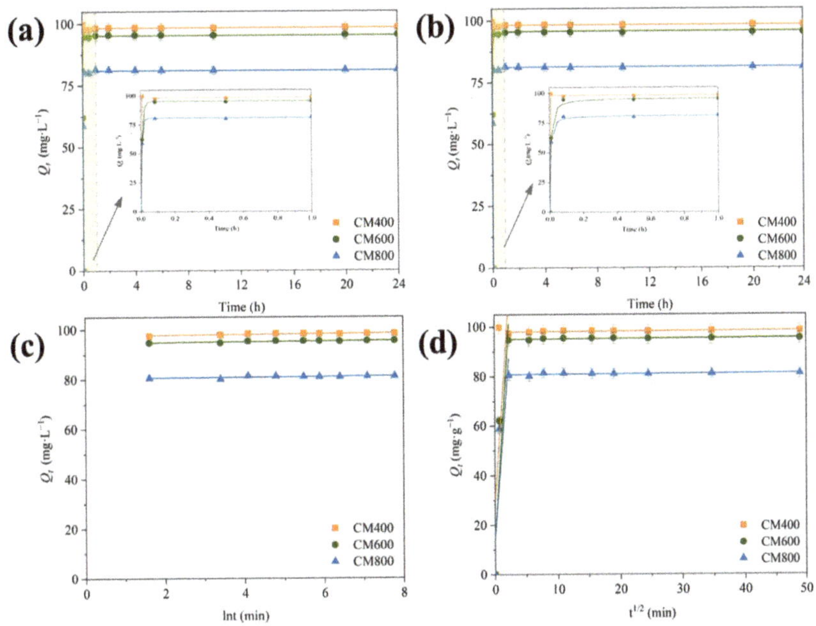

Figure 4. Adsorption kinetics of Pb(II) by CMBC. (**a**) Pseudo-first-order model, (**b**) pseudo-second-order model, (**c**) Elovich model, and (**d**) internal diffusion model. Experimental conditions: [initial concentration of Pb(II)] = 100 mg·L^{-1}, [dosage] = 1 g·L^{-1}, [temperature] = 25 °C.

3.2.3. Adsorption Kinetics

The fitting results of CMBC are shown in Table S1 and Figure 4. The Elovich model and internal diffusion model could not fit the adsorption of Pb(II) well, for the R^2 values of three kinds of CMBC were all less than 0.5. Thus, the surface adsorption energy was not uniformly distributed during adsorption, and the internal diffusion is not the sole rate-determining factor as $C \neq 0$. For CM400 and CM600, the pseudo-second-order kinetic model fitted the adsorptions better (0.9997 and 0.9998 vs. 0.9999 and 0.9999), by contrast, for CM800, the pseudo-first-order kinetic model described the adsorption slightly more accurately (0.9998 vs. 0.9988). Therefore, physisorption and chemisorption likely coexisted during adsorption by CM400 and CM600, while physisorption dominated Pb(II) adsorption by CM800 [39]. Physisorption was reversible and slow [1], which can be verified by the lower Q_e of CM800 than that of CM400 (98.21 vs. 81.28 mg·g^{-1}). Compared with other research data, three kinds of CMBC were all effective Pb(II) adsorbents; among them, CM400 showed the best effect, as its Q_e reached 691.34 mg·g^{-1} (Table 2).

3.2.4. Isothermal Adsorption

The isothermal adsorption model was presented in Figure 5. The R^2 values of the Langmuir model were 0.9981, 0.9966, and 0.9945 for CM400, CM600, and CM800, respectively. Those of the Freundlich model were 0.9934, 0.9916, and 0.9899 for CM400, CM600, and CM800, respectively. Clearly, Langmuir model fit the adsorption of Pb(II) better. Therefore, the adsorption of Pb(II) by CMBC was monolayer adsorption [40].

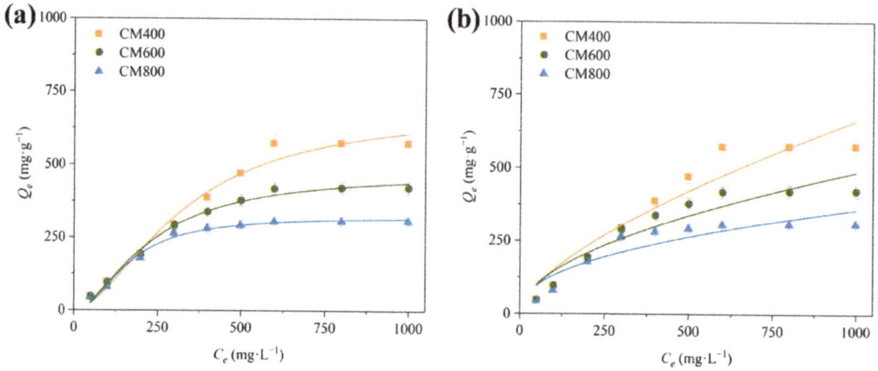

Figure 5. Isothermal adsorption model of Pb(II) by CMBC. (**a**) Langmuir model and (**b**) Freundlich model. Experimental conditions: [dosage] = 1 g·L^{-1}, [temperature] = 25 °C.

Table 2. Adsorption capacities of different adsorbents for Pb(II).

Adsorbent	Biomass	Pyrolyzation (°C)	Balance Time (min)	Q_{max} (mg·g^{-1})	Refs.
Nitrogen- and phosphorus-enriched biochar	*C. oleifera* shells	550	200	723.6	[41]
Biochar-supported phosphate-doped ferrihydrite	Corn stalks	400	360	247	[42]
Phosphorous-laden biochar	Sawdust	500	1440	568.22	[43]
Dairy manure-derived biochar	Cow manure	200	180	140.76	[44]
Biochar-supported nanoscale ferrous sulfide composite	Peanut shells	250	150	88.06	[45]
Sugarcane bagasse biochar	Sugarcane bagasse	600	360	2.5	[46]
Iron-sulfur co-doped biochar composite	rice straw	300	240	631.7	[3]
Cow manure biochar	Cow manure	400 600 800	0.5 5 5	691.34 473.36 323.83	This work

3.3. *Inherent Phosphorus Release*

Pyrolysis temperature was a decisive factor for morphology and crystal structure of phosphorus in biochar [47]. In water, *Qe*, the solution phosphorus content reached 0.620, 0.495, and 0.329 mg·g^{-1}. The results were summarized in Figure S1, and as shown in Table S2, the R^2 values of the pseudo-first-order model for phosphorus release by CM400, CM600, and CM800 were 0.9786, 0.9667, and 0.9296, respectively. Simultaneously, those of pseudo-second-order model were 0.9954, 0.9870, and 0.9786, respectively. Therefore, the phosphorus release by CMBC involved both physical and chemical reactions, which was consistent with the phosphorus release from reed straw biochar modified with phosphate [48].

During Pb(II) adsorption, the process of phosphorus release from CMBC was also affected by the pH and temperature (Figure S2). A higher reaction temperature promoted phosphorus release (0.166 mg·L^{-1} at 45 °C vs. 0.045 mg·L^{-1} at 25 °C), which inferred that phosphorus release was endothermic.

The higher the pH, the more of the phosphorus content remained (Figure S2). When pH = 3, the remaining phosphorus content was 0.047, 0.048, and 0.035 mg·L^{-1} for CM400, CM600, and CM800, respectively, while when pH = 11, they were 0.423, 0.274, and 0.197 mg·L^{-1} for CM400, CM600, and CM800, respectively. Under the initial alkaline reaction conditions, phosphorus mostly existed as $HPO_4{}^{2-}$ and $PO_4{}^{3-}$, whereas phosphorus mostly existed in the form of $HPO_4{}^{2-}$ and $H_2PO_4{}^{-}$ under neutral conditions [49]. Strong acidic conditions promoted the release of $PO_4{}^{3-}$ from CMBC, while in an alkaline environment, OH^- also precipitated with Pb(II) [50]. Thus, the amount of phosphorus consumed for precipitation was reduced, contributing to a higher phosphorus content at pH = 11.

Considering the adsorption capacity mentioned above, the adsorption capacity significantly exceeded that at pH = 3 and the residual phosphorus content at pH = 11 considerably exceeded that at pH = 3. Thus, under alkaline conditions, the chemical precipitation of phosphate likely contributed less toward adsorption, whereas complexation of functional groups and Pb(II) had a considerably greater contribution.

Excluding the pH and temperature, the Pb(II) concentration also affected the phosphorus release (Figure 6). For example, when the initial concentration of the Pb(II) solution was 100 mg·L^{-1}, Q_e of CM400 was more than that of CM800 (98.38 vs. 81.19 mg·g^{-1}), however, the residual phosphorus content of CM400 was less than that of CM800 (0.05 vs. 0.09 mg·L^{-1}). Thus, when Pb(II) was less than 100 mg·L^{-1}, the main adsorption mechanism likely was chemical precipitation, while the main mechanisms transformed into ion exchange and complexation. When Pb(II) solution was 100 mg·L^{-1}, the residual phosphorus content increased rapidly within 6 h and then decreased. Finally, the residual phosphorus content remained stable 12 h after the onset of adsorption. Because the solution became alkaline when CMBC was placed in, the inherent phosphorus was released slowly at first. Further, Pb(II) was adsorbed by electronegative CMBC via electrostatic attraction, and underwent a complexation reaction and ion exchange with the functional groups. During the adsorption, OH^- was consumed, and the reaction solution tended to be neutral. Thus, the phosphorus release accelerated, and the remaining content remained stable.

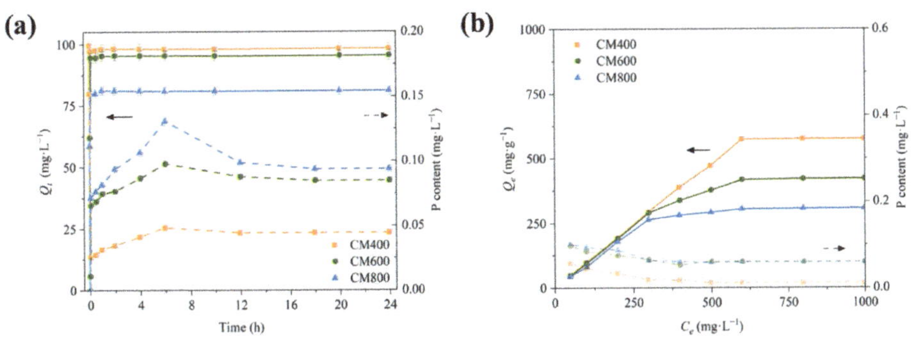

Figure 6. The relationship between the residual phosphorus content and Q_t. (**a**) the relationship between the residual phosphorus content and Q_t at different reaction times [initial concentration of Pb(II)] = 100 mg·L^{-1}, [dosage] = 1 g·L^{-1}, [temperature] = 25 °C, (**b**) The relationship between the residual phosphorus content and Q_t in reaction with different initial concentrations, [reaction times] = 24 h, [dosage] = 1 g·L^{-1}, [temperature] = 25 °C.

3.4. Adsorption Mechanism Analysis

The XRD of Pb(II)-loaded CMBC is shown in Figure 7. The characteristic peaks at 20.8°, 27.28°, and 42° represented $Pb_3(PO_4)_2$ (PDF#73–0834), which confirm that inherent phosphorus released and reacted [42]. The characteristic peak at 26.6° and 40° represented

$PbSiO_3$ (PDF#74–1101) and $PbCO_3$ (PDF#85–1088), indicating that CMBC adsorbed Pb(II) via precipitation and ion exchange [51]. The precipitation equilibrium constant (Ksp) of $PbCO_3$ was 3.3×10^{-14}, while those of $CaCO_3$ and $MgCO_3$ were 2.8×10^{-9} and 2.6×10^{-5}. Therefore, a few Pb ions were exchanged with the precipitates. Moreover, the peak at 27.4° represented $C_3H_6O_3Pb$, and the peak at 27.8° represented CH_2O_3Pb (PDF#14–0831), indicating an ion exchange [52]. There were $Pb_3(CO_3)_2(OH)_2$ (PDF#72–1144) and $Pb_{10}(PO_4)_6(OH)_2$ (PDF#87–2478), verifying that the hydroxyl group participated in adsorption [53].

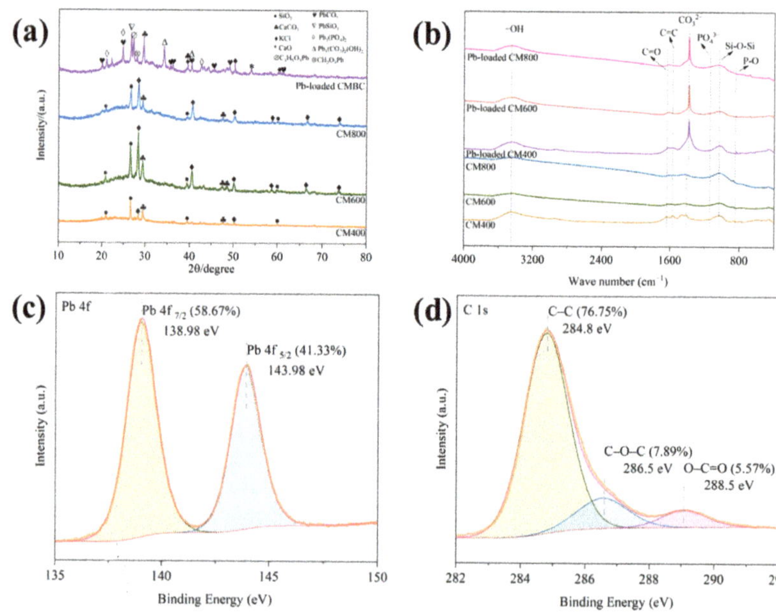

Figure 7. XRD, FTIR and XPS spectra of Pb(II)–loaded ((**a**): XRD spectrum, (**b**): FTIR spectrum, (**c**,**d**): XPS spectra).

The FTIR of Pb(II)-loaded CMBC is shown in Figure 7. The –OH group peak shifted, presumably because the hydroxyl group complexed with Pb(II). The C=O vibration peak of the carboxyl and carbonyl groups at 1640 cm^{-1} disappeared, possibly because the oxygen-containing functional group ligand complexed with Pb(II) [54]. The peak value of Si–O–Si was positively correlated with the adsorption, indicating that the cation–π bond interaction existed, because positron coordination improved the adsorption if the peak value positively correlated with the adsorption [55]. The C–O–C peak at 876 cm^{-1} shifted, indicating that H^+ provided by the carboxyl and hydroxyl groups exchanged with Pb(II) [56].

The XPS of Pb(II)-loaded CMBC were depicted in Figure 7. The peak of Pb 4f (138 eV) appeared after the adsorption and the valence of Pb(II) did not change, so there were no oxidation-reduction reactions that occurred in Pb(II). The characteristic peaks of C 1s at 284.80 eV, 286.35 eV, 288.60 eV, and 285.35 eV represented C–C, C–O, COOH, and C–OH groups, respectively [48], and the surface functional groups were crucial to adsorption.

CMBC adsorbed Pb(II) in the solution rapidly and efficiently. The adsorption mechanisms include physical adsorption, ion exchange, cation–π bond interaction, chemical precipitation, electrostatic attraction, and surface complexation (Figure 8). Pb(II) was adsorbed via electrostatic attraction because CMBC was electronegative, and the ion diameter of Pb(II) (0.12 nm) was adsorbed via pore filling. The variation in Si–O–Si peak after adsorption was positively correlated with the Pb(II) adsorption, thus, CMBC adsorbed Pb(II) via cation–π bond. Pb(II) was adsorbed onto the biochar surface via ion exchange, replacing H^+ from the basic functional groups. The inherent phosphorus reacted with Pb(II) to form precipitates, and the oxygen-containing functional group ligands on the surface of tCMBC were complexed with the Pb(II).

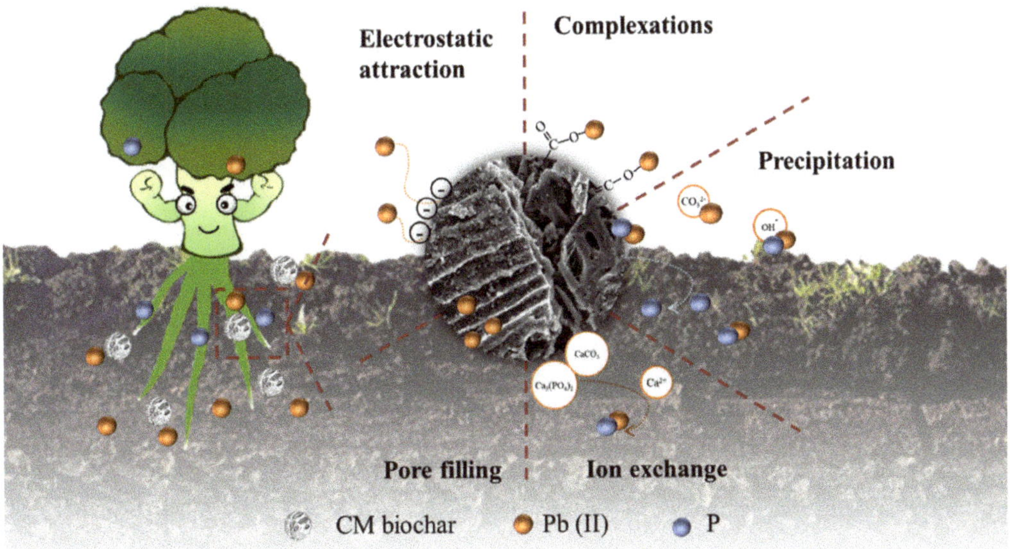

Figure 8. Surface interaction between Pb(II) and CM functional materials.

The adsorption mechanisms differed under diverse reaction conditions. Under alkaline conditions, the complexation had a greater contribution than precipitation, and the main adsorption mechanisms transformed from chemical precipitation to ion exchange and complexation.

The main adsorption mechanisms for CM400 were chemical precipitation, surface complexation, and ion exchange. By contrast, CM800 had a rich pore structure and a large pore size; hence, its physical adsorption ability was strong.

3.5. Analysis of Pot Experiment Results

The effects of CMBC on soil decontamination and fertilizer improvement were also investigated, and the contents are shown in Figure S3. On the third day after sowing, for T_1, T_2, T_3, and CK_1, three, three, two, and three buds were noted, however, none of the CK_2 groups sprouted at that time. Application of CMBC can reduce the germination inhibition of Chinese cabbage growing in lead-contaminated soil. It was speculated that CMBC can reduce the biotoxicity of Pb(II) to Chinese cabbage growth by solidifying soil heavy metals during the germination stage of Chinese cabbage. After the Chinese cabbages ripened, the roots, stems, and leaves were cut for digestion and detection. For CK_2, the Pb(II) content in leaves was 0.04 $mg{\cdot}L^{-1}$, while it was 0.03, 0.02, and 0.00 $mg{\cdot}L^{-1}$ for T_1, T_2, and T_3, a reduction of 26.82%, 40.50%, and 99.72%, respectively. The Pb(II) content in stems was 0.04 $mg{\cdot}L^{-1}$ for CK_2, while it was 0.03, 0.02, and 0.01 $mg{\cdot}L^{-1}$ for T_1, T_2, and T_3. Furthermore, the Pb(II) content in roots for CK_2 was 0.86 $mg{\cdot}L^{-1}$, while it was 0.06, 0.07, and 0.04 $mg{\cdot}L^{-1}$ for T_1, T_2, and T_3, a reduction of 92.57%, 91.48%, and 94.92%, respectively. After the Chinese cabbage planted in lead-contaminated soil ripened, by comparing the data of CK_2, T_1, T_2, and T_3, it was found that the Pb(II) content in three parts of the Chinese cabbage in the four groups was different. When biochar was applied, the Pb(II) content in the cabbage decreased. The cabbage treated with CM800 (T_3) had the least Pb(II) content. In other words, if biochar was applied to lead-contaminated soil and regardless of the type of biochar applied, both had an effect on the Pb(II) content in Chinese cabbage. The application of CMBC to lead-contaminated soil can achieve decontamination, and CM800 was most effective. Therefore, it is speculated that CMBC can reduce the bioenrichment of Pb(II) in Chinese cabbage by solidifying soil heavy metals during the growth stage of Chinese cabbage. In short, the biochar likely immobilized Pb(II) in the soil to reduce the amount of plant uptake [57].

The phosphorus content in leaves was 3.85, 1.96, 1.88, 6.65, and 1.41 mg·L^{-1} for T_1, T_2, T_3, CK_1, and CK_2, respectively. The phosphorus content in stems was 1.85, 1.88, 1.37, 2.27, and 0.41 mg·L^{-1}. The phosphorus content in roots was 1.83, 1.65, 1.14, 2.39, and 0.62 mg·L^{-1}, respectively. Compared with CK_2, application of CMBC increased phosphorus in plants. In summary, the application of CMBC adsorbed Pb(II) in soil, reduced the Pb(II) content in crops, and promoted plant growth effectively.

4. Conclusions

CM400 proved to be an effective adsorbent; its Q_e reached 691.34 mg·g^{-1}, and it rapidly adsorbed 98.36 mg·g^{-1} of Pb(II) within 30 s. The adsorption mechanisms of Pb(II) by CMBC include ion exchange, physical adsorption, electrostatic attraction, chemical precipitation, surface complexation, and cation–π bond interaction. CMBC released phosphorus and the phosphorus released from CM400 reached 0.620 mg·g^{-1}. The phosphorus release was endothermic, and the higher the pH, the more the phosphorus content remained. Based on the residual phosphorus content and adsorption effect, complexation rather than chemical precipitation had a greater contribution toward adsorption. Besides, as the concentration of Pb(II) increased, the main adsorption mechanisms likely transformed from chemical precipitation to ion exchange and complexation. CMBC not only had a good effect on Pb(II) removal in the solution, but also immobilized the Pb(II) in the soil to restrain the plant uptake and promoted plant growth.

Supplementary Materials: The following supporting information can be downloaded at: https://www.mdpi.com/article/10.3390/toxics11010001/s1, Figure S1: Phosphorus release kinetics in deionized water; Figure S2: The effect of pH (a) and temperature (b) on phosphorus release; Figure S3: Pb(II) and phosphorus contents in each part of the crops; T1: 1 g CM400 application, T2: 1 g CM600 application, T3: 1 g CM800 application, CK1: no pollution treatment, CK2: no adsorbent application. Table S1: Table of fitting parameters of kinetic model and isotherm model of Pb(II) absorbtion. Table S2: Table of fitting parameters of kinetic model, Elovich model, and Internal diffusion model of phosphorus release.

Author Contributions: H.W.: Conceptualization, Writing—Review and Editing; Y.W.: Formal Analysis, Writing—Original Draft; Y.D.: Writing—Review and Editing; Z.Y.: Data Curation, Investigation, Methodology; D.X.: Data Curation, Software, Methodology; Y.L.: Supervision; Y.Z.: Supervision; W.Z.: Conceptualization; R.X.: Funding Acquisition, Resources, Supervision. All authors have read and agreed to the published version of the manuscript.

Funding: This research was supported by the Yunnan Plateau Characteristics of Bio–fertilizer (202202AE090025), Applied Basic Research Foundation of Yunnan Province (202201AS070020, 202201AU070061), Yunnan Province Education Department Project (2022J0136), Yunnan Provincial Observation and Research Station (202105AM07003), and Yunnan Academy of Experts Workstation (YSZJGZZ–2021062).

Institutional Review Board Statement: Not applicable.

Data Availability Statement: Not applicable.

Conflicts of Interest: This paper is an original work of the authors, who have read and approved this version, and due care has been taken to ensure the integrity of the work. No part of this paper has been published or submitted elsewhere. No conflict of interest exists in the submission of this manuscript. A statement explaining why the manuscript is novel and significant is stated below.

References

1. Zhang, H.H.; Tian, S.C.; Zhu, Y.; Zhong, W.; Qiu, R.; Han, L. Insight into the adsorption isotherms and kinetics of Pb (II) on pellet biochar via in-situ non-destructive 3D visualization using micro-computed tomography. *Bioresour. Technol.* **2022**, *358*, 127406. [CrossRef]
2. Luo, X.; Wu, C.; Lin, Y.; Li, W.; Deng, M.; Tan, J.; Xue, S. Soil heavy metal pollution from Pb/Zn smelting regions in China and the remediation potential of biomineralization. *J. Environ. Sci.* **2023**, *125*, 662–677. [CrossRef] [PubMed]
3. Cheng, S.; Zhao, S.; Guo, H.; Xing, B.; Liu, Y.; Zhang, C.; Ma, M. High-efficiency removal of lead/cadmium from wastewater by MgO modified biochar derived from crofton weed. *Bioresour. Technol.* **2021**, *343*, 126081. [CrossRef]
4. Wang, H.; Wang, S.; Chen, Z.; Zhou, X.; Wang, J.; Chen, Z. Engineered biochar with anisotropic layered double hydroxide nanosheets to simultaneously and efficiently capture Pb^{2+} and CrO_4^{2-} from electroplating wastewater. *Bioresour. Technol.* **2020**, *306*, 123118. [CrossRef] [PubMed]
5. Hou, R.; Wang, L.; O'Connor, D.; Rinklebe, J.; Hou, D. Natural field freeze-thaw process leads to different performances of soil amendments towards Cd immobilization and enrichment. *Sci. Total Environ.* **2022**, *831*, 154880. [CrossRef] [PubMed]
6. Soudani, A.; Youcef, L.; Bulgariu, L.; Youcef, S.; Toumi, K.; Soudani, N. Characterizing and modeling of oak fruit shells biochar as an adsorbent for the removal of Cu, Cd, and Zn in single and in competitive systems. *Chem. Eng. Res. Des.* **2022**, *188*, 972–987. [CrossRef]
7. Xu, L.; Liu, Y.; Wang, J.; Tang, Y.; Zhang, Z. Selective adsorption of Pb^{2+} and Cu^{2+} on amino-modified attapulgite: Kinetic, thermal dynamic and DFT studies. *J. Hazard. Mater.* **2020**, *404*, 124140. [CrossRef]
8. Zhi, B.; Xiang, S.; Wang, Y.; Dai, Z.; Du, P.; Wang, R.; Li, X.; Yang, G.; Feng, Y.; Ren, G.; et al. Redeploy manure resources to enhance the agro-pastoral cycle. *Sci. Total Environ.* **2022**, *846*, 157439. [CrossRef]
9. Zhao, S.; Chen, Y.; Gu, X.; Zheng, M.; Fan, Z.; Luo, D.; Luo, K.; Liu, B. Spatiotemporal variation characteristics of livestock manure nutrient in the soil environment of the Yangtze River Delta from 1980 to 2018. *Sci. Rep.* **2022**, *12*, 7353. [CrossRef]
10. Behjat, M.; Svanström, M.; Peters, G. A meta-analysis of LCAs for environmental assessment of a conceptual system: Phosphorus recovery from dairy wastewater. *J. Clean. Prod.* **2022**, *369*, 133307. [CrossRef]
11. Fajobi, M.O.; Lasode, O.A.; Adeleke, A.A.; Ikubanni, P.P.; Balogun, A.O. Investigation of physicochemical characteristics of selected lignocellulose biomass. *Sci. Rep.* **2022**, *12*, 2918. [CrossRef] [PubMed]
12. Fu, H.; Wang, B.; Wang, H.; Liu, H.; Xie, H.; Han, L.; Wang, N.; Sun, X.; Feng, Y.; Xue, L. Assessment of livestock manure-derived hydrochar as cleaner products: Insights into basic properties, nutrient composition, and heavy metal content. *J. Clean. Prod.* **2021**, *330*, 129820. [CrossRef]
13. Nguyen, T.T.; Sasaki, Y.; Katahira, M.; Singh, D. Cow Manure Application Cuts Chemical Phosphorus Fertilizer Need in Silage Rice in Japan. *Agronomy* **2021**, *11*, 1483. [CrossRef]
14. Chen, G.; Wang, J.; Yu, F.; Wang, X.; Xiao, H.; Yan, B.; Cui, X. A review on the production of P-enriched hydro/bio-char from solid waste: Transformation of P and applications of hydro/bio-char. *Chemosphere* **2022**, *301*, 134646. [CrossRef] [PubMed]
15. Holatko, J.; Hammerschmiedt, T.; Mustafa, A.; Kintl, A.; Radziemska, M.; Baltazar, T.; Jaskulska, I.; Malicek, O.; Latal, O.; Brtnicky, M. Carbon-enriched organic amendments differently affect the soil chemical, biological properties and plant biomass in a cultivation time-dependent manner. *Chem. Biol. Technol. Agric.* **2022**, *9*, 52. [CrossRef]
16. Liao, W.; Zhang, X.; Ke, S.; Shao, J.; Yang, H.; Zhang, S.; Chen, H. Effect of different biomass species and pyrolysis temperatures on heavy metal adsorption, stability and economy of biochar. *Ind. Crop. Prod.* **2022**, *186*, 115238. [CrossRef]
17. Zhang, P.; Zhang, X.; Li, Y.; Han, L. Influence of pyrolysis temperature on chemical speciation, leaching ability, and environmental risk of heavy metals in biochar derived from cow manure. *Bioresour. Technol.* **2020**, *302*, 122850. [CrossRef]
18. Mulyani, O.; Joy, B.; Kurnia, D. The Various Forms of Cow Manure Waste as Adsorbents of Heavy Metals. *Appl. Sci.* **2022**, *12*, 5763. [CrossRef]
19. Sarmah, M.; Borgohain, A.; Gogoi, B.B.; Yeasin, M.; Paul, R.K.; Malakar, H.; Handique, J.G.; Saikia, J.; Deka, D.; Khare, P.; et al. Insights into the effects of tea pruning litter biochar on major micronutrients (Cu, Mn, and Zn) pathway from soil to tea plant: An environmental armour. *J. Hazard. Mater.* **2023**, *442*, 129970. [CrossRef]
20. Naeem, M.A.; Imran, M.; Amjad, M.; Abbas, G.; Tahir, M.; Murtaza, B.; Zakir, A.; Shahid, M.; Bulgariu, L.; Ahmad, I. Batch and Column Scale Removal of Cadmium from Water Using Raw and Acid Activated Wheat Straw Biochar. *Water* **2019**, *11*, 1438. [CrossRef]
21. Goldfarb, J.L.; Hubble, A.H.; Ma, Q.; Volpe, M.; Severini, G.; Andreottola, G.; Fiori, L. Valorization of cow manure via hydrothermal carbonization for phosphorus recovery and adsorbents for water treatment. *J. Environ. Manag.* **2022**, *308*, 114561. [CrossRef] [PubMed]
22. Wei, X.; Chen, H.; Lin, D.; Xu, H.; Wang, J.; Zhang, J.; Hu, Z.; Deng, J.; Gao, J.; Li, H.; et al. A field study of nano-FeS loaded lignin hydrogel application for Cd reduction, nutrient enhancement, and microbiological shift in a polluted paddy soil. *Chem. Eng. J.* **2023**, *451*, 138647. [CrossRef]
23. Zhang, Y.; Chang, F.; Duan, L.; Wu, H.; Zhang, H. The Industrial Pollution History Inferred by Stable Pb Isotope in the Chronological Sedimentary Record of a Plateau Lake, Yunnan Province, Southwestern of China. *J. Coast. Res.* **2020**, *115*, 641–647. [CrossRef]

24. He, J.; Yang, Z.; Guo, M.; Gu, L.; Zhang, L.; Yan, Y.; Ran, J. Experimental study on the key factors affecting the gasification performance between different biomass: Compare citrus peel with pine sawdust. *Int. J. Hydrogen Energy* **2022**, *47*, 30428–30439. [CrossRef]
25. Yan, C.; Jin, J.; Wang, J.; Zhang, F.; Tian, Y.; Liu, C.; Zhang, F.; Cao, L.; Zhou, Y.; Han, Q. Metal-organic frameworks (MOFs) for the efficient removal of contaminants from water: Underlying mechanisms, recent advances, challenges, and future prospects. *Coord. Chem. Rev.* **2022**, *468*, 214595. [CrossRef]
26. Ahmad, S.; Gao, F.; Lyu, H.; Ma, J.; Zhao, B.; Xu, S.; Ri, C.; Tang, J. Temperature-dependent carbothermally reduced iron and nitrogen doped biochar composites for removal of hexavalent chromium and nitrobenzene. *Chem. Eng. J.* **2022**, *450*, 138006. [CrossRef]
27. Zhang, Z.; Li, Y.; Zong, Y.; Yu, J.; Ding, H.; Kong, Y.; Ma, J.; Ding, L. Efficient removal of cadmium by salts modified-biochar: Performance assessment, theoretical calculation, and quantitative mechanism analysis. *Bioresour. Technol.* **2022**, *361*, 127717. [CrossRef]
28. Shakya, A.; Vithanage, M.; Agarwal, T. Influence of pyrolysis temperature on biochar properties and Cr(VI) adsorption from water with groundnut shell biochars: Mechanistic approach. *Environ. Res.* **2022**, *215*, 114243. [CrossRef]
29. Anacleto TMOliveira, H.R.; Diniz, V.L.; de Oliveira, V.P.; Abreu, F.; Enrich-Prast, A. Boosting manure biogas production with the application of pretreatments: A meta-analysis. *J. Clean. Prod.* **2022**, *362*, 132292. [CrossRef]
30. Adhikari, S.; Timms, W.; Mahmud, M.P. Optimising water holding capacity and hydrophobicity of biochar for soil amendment—A review. *Sci. Total. Environ.* **2022**, *851*, 158043. [CrossRef]
31. Wu, J.; Li, Z.; Huang, D.; Liu, X.; Tang, C.; Parikh, S.J.; Xu, J. A novel calcium-based magnetic biochar is effective in stabilization of arsenic and cadmium co-contamination in aerobic soils. *J. Hazard. Mater.* **2020**, *387*, 122010. [CrossRef] [PubMed]
32. Wu, J.; Huang, D.; Liu, X.; Meng, J.; Tang, C.; Xu, J. Remediation of As(III) and Cd(II) co-contamination and its mechanism in aqueous systems by a novel calcium-based magnetic biochar. *J. Hazard. Mater.* **2018**, *348*, 10–19. [CrossRef] [PubMed]
33. Wang, F.; Li, J.; Su, Y.; Li, Q.; Gao, B.; Yue, Q.; Zhou, W. Adsorption and recycling of Cd(II) from wastewater using straw cellulose hydrogel beads. *J. Ind. Eng. Chem.* **2019**, *80*, 361–369. [CrossRef]
34. Sun, Z.; Wang, X.; Xia, S.; Zhao, J. Treatment of Pb(II) pollution in livestock wastewater by $MgFe_2O_4$ modified manure-biochar derived from livestock itself: Special role of endogenous dissolved organic matter and P species. *Chem. Eng. J.* **2022**, *446*, 137068. [CrossRef]
35. Kumaraswamy, R.V.; Saharan, V.; Kumari, S.; Choudhary, R.C.; Pal, A.; Sharma, S.S.; Rakshit, S.; Raliya, R.; Biswas, P. Chitosan-silicon nanofertilizer to enhance plant growth and yield in maize (*Zea mays* L.). *Plant Physiol. Biochem.* **2021**, *159*, 53–66. [CrossRef]
36. Ul Ain, Q.; Zhang, H.; Yaseen, M.; Rasheed, U.; Liu, K.; Subhan, S.; Tong, Z. Facile fabrication of hydroxyapatite-magnetite-bentonite composite for efficient adsorption of Pb(II), Cd(II), and crystal violet from aqueous solution. *J. Clean. Prod.* **2020**, *247*, 119088. [CrossRef]
37. Zhao, Z.; Wang, B.; Zhang, X.; Xu, H.; Cheng, N.; Feng, Q.; Zhao, R.; Gao, Y.; Wei, M. Release characteristics of phosphate from ball-milled biochar and its potential effects on plant growth. *Sci. Total. Environ.* **2022**, *821*, 153256. [CrossRef]
38. Zhang, Y.; Wang, F.; Cao, B.; Yin, H.; Al-Tabbaa, A. Simultaneous removal of Pb and MTBE by mixed zeolites in fixed-bed column tests. *J. Environ. Sci.* **2022**, *122*, 41–49. [CrossRef]
39. Chowdhury, I.R.; Chowdhury, S.; Mazumder, M.A.J.; Al-Ahmed, A. Removal of lead ions (Pb^{2+}) from water and wastewater: A review on the low-cost adsorbents. *Appl. Water Sci.* **2022**, *12*, 185. [CrossRef]
40. Xu, J.; Hu, C.; Wang, M.; Zhao, Z.; Zhao, X.; Cao, L.; Lu, Y.; Cai, X. Changeable effects of coexisting heavy metals on transfer of cadmium from soils to wheat grains. *J. Hazard. Mater.* **2021**, *423*, 127182. [CrossRef]
41. Fan, Y.; Wang, H.; Deng, L.; Wang, Y.; Kang, D.; Li, C.; Chen, H. Enhanced adsorption of Pb(II) by nitrogen and phosphorus co-doped biochar derived from Camellia oleifera shells. *Environ. Res.* **2020**, *191*, 110030. [CrossRef] [PubMed]
42. Li, H.; Jiang, Q.; Zhang, J.; Wang, Y.; Zhang, Y. Synchronization adsorption of Pb(II) and Ce(III) by biochar supported phosphate-doped ferrihydrite in aqueous solution: Adsorption efficiency and mechanisms. *Colloids Surf. A Physicochem. Eng. Asp.* **2022**, *648*, 129230. [CrossRef]
43. Pei, L.; Yang, F.; Xu, X.; Nan, H.; Gui, X.; Zhao, L.; Cao, X. Further reuse of phosphorus-laden biochar for lead sorption from aqueous solution: Isotherm, kinetics, and mechanism. *Sci. Total Environ.* **2021**, *792*, 148550. [CrossRef] [PubMed]
44. Yuan, Z.; Liu, M.; Chen, Z. Dairy manure biochar modified with sodium hydroxide and its effect on lead removal in aqueous solution. In Proceedings of the 4th International Conference on Sensors, Measurement and Intelligent Materials (ICSMIM), Shenzhen, China, 27–28 December 2015.
45. Chen, C.G.; Qiu, M.Q. High efficiency removal of Pb(II) in aqueous solution by a biochar-supported nanoscale ferrous sulfide composite. *RSC Adv.* **2021**, *11*, 953–959. [CrossRef] [PubMed]
46. Inyang, M.; Gao, B.; Zimmerman, A.; Zhou, Y.; Cao, X. Sorption and cosorption of lead and sulfapyridine on carbon nanotube-modified biochars. *Environ. Sci. Pollut. Res.* **2014**, *22*, 1868–1876. [CrossRef]
47. Zhu, Y.; Zhao, Q.; Li, D.; Li, J.; Guo, W. Performance comparison of phosphorus recovery from different sludges in sewage treatment plants through pyrolysis. *J. Clean. Prod.* **2022**, *372*, 133728. [CrossRef]
48. Cui, H.; Dong, T.; Hu, L.; Xia, R.; Zhou, J.; Zhou, J. Adsorption and immobilization of soil lead by two phosphate-based biochars and phosphorus release risk assessment. *Sci. Total Environ.* **2022**, *824*, 153957. [CrossRef]

49. Bruun, S.; Harmer, S.L.; Bekiaris, G.; Christel, W.; Zuin, L.; Hu, Y.; Jensen, L.S.; Lombi, E. The effect of different pyrolysis temperatures on the speciation and availability in soil of P in biochar produced from the solid fraction of manure. *Chemosphere* **2017**, *169*, 377–386. [CrossRef]
50. Liu, Z.; Ou, T.; Su, M.; Peng, H.; Song, G.; Kong, L.; Chen, D. U(VI) sequestration by Al-rich minerals: Mechanism on phase dependence and the influence of natural organic matter. *Chem. Eng. J.* **2021**, *415*, 128858. [CrossRef]
51. Liu, M.; Hu, B.; Zhang, C.; Wang, Q.; Sun, Z.; He, P.; Chen, Y.; Chen, D.; Zhu, J. Effect of sodium silicate on the flotation separation of chalcopyrite and galena using sodium sulfite and sulfonated lignin as depressant. *Miner. Eng.* **2022**, *182*, 107563. [CrossRef]
52. Li, J.; Zheng, L.; Wang, S.-L.; Wu, Z.; Wu, W.; Niazi, N.K.; Shaheen, S.M.; Rinklebe, J.; Bolan, N.; Ok, Y.S.; et al. Sorption mechanisms of lead on silicon-rich biochar in aqueous solution: Spectroscopic investigation. *Sci. Total. Environ.* **2019**, *672*, 572–582. [CrossRef] [PubMed]
53. Qu, J.; Wei, S.; Liu, Y.; Zhang, X.; Jiang, Z.; Tao, Y.; Zhang, G.; Zhang, B.; Wang, L.; Zhang, Y. Effective lead passivation in soil by bone char/CMC-stabilized FeS composite loading with phosphate-solubilizing bacteria. *J. Hazard. Mater.* **2021**, *423*, 127043. [CrossRef] [PubMed]
54. Wang, H.; Cai, J.; Liao, Z.; Jawad, A.; Ifthikar, J.; Chen, Z.; Chen, Z. Black liquor as biomass feedstock to prepare zero-valent iron embedded biochar with red mud for Cr(VI) removal: Mechanisms insights and engineering practicality. *Bioresour. Technol.* **2020**, *311*, 123553. [CrossRef] [PubMed]
55. Feng, C.; Huang, M.; Huang, C.-P. Specific chemical adsorption of selected divalent heavy metal ions onto hydrous γ-Fe_2O_3-biochar from dilute aqueous solutions with pH as a master variable. *Chem. Eng. J.* **2023**, *451*, 138921. [CrossRef]
56. Wang, H.; Liu, Y.; Ifthikar, J.; Shi, L.; Khan, A.; Chen, Z.; Chen, Z. Towards a better understanding on mercury adsorption by magnetic bio-adsorbents with gamma-Fe_2O_3 from pinewood sawdust derived hydrochar: Influence of atmosphere in heat treatment. *Bioresour. Technol.* **2018**, *256*, 269–276. [CrossRef]
57. Ali, A.; Shaheen, S.M.; Guo, D.; Li, Y.; Xiao, R.; Wahid, F.; Azeem, M.; Sohail, K.; Zhang, T.; Rinklebe, J.; et al. Apricot shell- and apple tree-derived biochar affect the fractionation and bioavailability of Zn and Cd as well as the microbial activity in smelter contaminated soil. *Environ. Pollut.* **2020**, *264*, 114773. [CrossRef]

MDPI
St. Alban-Anlage 66
4052 Basel
Switzerland
www.mdpi.com

Toxics Editorial Office
E-mail: toxics@mdpi.com
www.mdpi.com/journal/toxics

www.ingramcontent.com/pod-product-compliance
Lightning Source LLC
Chambersburg PA
CBHW061809210926
51850CB00006B/656
9783725813988